1971
Rep't.
20th

AN INTRODUCTION TO PLANE GEOMETRY

AN INTRODUCTION TO PLANE GEOMETRY

WITH MANY EXAMPLES

BY

H. F. BAKER, Sc.D., LL.D., F.R.S.

*Emeritus Professor of Astronomy and Geometry
and Fellow of St John's College in the
University of Cambridge*

CHELSEA PUBLISHING COMPANY
BRONX, NEW YORK

THE PRESENT WORK IS A REPRINT, WITH CORRECTIONS, OF
A WORK ORIGINALLY PUBLISHED IN 1943. IT IS REPRINTED
WITH THE PERMISSION OF CAMBRIDGE UNIVERSITY PRESS.

LIBRARY OF CONGRESS CATALOG CARD NUMBER 70-141879

INTERNATIONAL STANDARD BOOK NUMBER 0-8284-0247-7

LIBRARY OF CONGRESS CLASSIFICATION NUMBER QA455

DEWEY DECIMAL CLASSIFICATION NUMBER 516'.22

PRINTED, 1971, ON SPECIAL LONG-LIFE ALKALINE PAPER

PRINTED IN THE UNITED STATES OF AMERICA

PREFACE

It is hoped that scholarship candidates in our schools, and first year students in our universities, will be able to read this volume. They will find at least a large number of informative examples, which they can solve by any method they have been taught. Only a general acquaintance with the theorems obtained in Euclid's theory is assumed, which, in practice, most students obtain by experiment. Much trouble, with some repetition, has been given to make the logical basis of the theory clear; it is hoped this may be sufficient.

From conservative readers some objection to the plan followed may be expected, at first. After long consideration, the writer believes that such a plan is not only logically sound, but pardonable, even indispensable, if the two aims are to be reached, of doing without the notion of distance (or congruence) as an axiom, and of obtaining a theory in which the so-called imaginary elements have equal standing with the real elements. Moreover, a considerable shortening of the time usually spent in reaching a competent knowledge of elementary geometry should be possible in this way.

In one respect the volume is recognised to be deficient, namely in the consideration of the topology of real plane figures; this deficiency belongs also to all the elementary treatises, where the attempt is made to supply the need *tacitly*, by drawing diagrams. Here, for reasons of space under present conditions, diagrams are given only sparingly; but the verbal descriptions are, it is hoped, clear enough for the reader to make his own.

The writer is under very great obligations to W. L. Edge, Sc.D., of the University of Edinburgh, for his assistance in reading the proof sheets. He would add, also, that the manuscript of the volume was prepared in the years previous to the autumn of 1939; and further, that, while the proofs were being corrected, the writer had the opportunity of many talks with Mr J. H. Grace; evidence of this will be found in the numerous interesting Examples to which his name has been attached. Finally he would express his gratitude to the Staff of the University Press for their continued helpfulness.

H. F. B.

June, 1942

CONTENTS

Contents

CHAPTER I

EUCLID'S THEORY OF PARALLEL LINES

1. In Euclid's theory of plane geometry, carried through on a metrical basis, distinction must necessarily be made, in dealing with two lines of the plane, between the case when the lines have a common point and the case when the lines have no such common point, that is, are *parallel*. But it was recognised long ago, at the beginning of the seventeenth century, by Kepler and Desargues, that it is possible to use a phraseology, when the lengths of lines are not considered, which covers both these cases. This is based directly on Euclid's axiom that, through a given point in the plane, only one line can be drawn, in the plane, to be parallel to a given line. It is the object of this chapter to shew how naturally this phraseology is suggested; and, what is important, to make it clear that nothing of definiteness is lost by adopting this phraseology, except in propositions which state the equality of length (or *congruence*) of two lines. In this volume we regard geometry as concerned with the coincidence of points (or lines); we regard length as a derived notion, introduced by definition. It will be seen that all Euclid's propositions in regard to length can be obtained, unaltered in form; but with a more general content.

It is supposed that the reader, from his experience of rigid bodies, has a general familiarity with Euclid's results. Thus the explanatory remarks in this chapter are not put forward as formal demonstrations. On a similar ground, reference is not limited to figures which lie in a plane. The present chapter serves as an introduction to Chapter II; but that chapter is formulated independently, and the consideration of this chapter may be postponed, if desired.

2. In Euclid's system a line is generally limited by two points; and, when the line is to be produced, it is so stated. For us, a line is essentially unlimited in length; and, as in Euclid's system, it contains an infinite number of points. It is assumed that the line is determined by any two, arbitrarily chosen, of its points, say A and B; if, on the line so determined, C be another point, then the line is equally determined by A and C, and B is said to lie on the line so determined.

In a similar way, a *plane* is supposed here to be extended indefinitely in all directions, and contains an infinite number of points. And it is assumed that the plane is determined by any three of its points, say A, B and C, provided these do not all lie in one line; if, on the plane so determined, D be another point, not lying on the line determined by A and B, then the plane is equally determined by A, B and D, and C lies on the plane. It is also assumed that all points of the line which is determined by any two points of the plane equally lie on the plane. Thus a plane is determined by any line and any point which does not lie in the line; or, also, a plane is determined by any two lines which have a common point.

3.   We have considered the possibility of two lines which lie in a plane having (or not having) a common point. We consider also the possibility of two planes (not coinciding with one another) having, or not having, common points. We make the assumption: *Two planes either have no point in common, or, if they have one point in common, then they have an infinite number of points in common, all lying on a line which contains this point.* In saying this, we are considering the ordinary space of our experience, in which both the planes are supposed to lie. By No. 2, two planes cannot have, besides the points of a line, any other common point, or they would be the same plane.

4.   Two lines of a plane which have no point in common are said to be *parallel*, and either is said to be parallel to the other. Conversely, when we say that two lines are parallel, we imply, not only that they have no common point, but also that they lie in a plane. If a be a line, and B be a point not belonging to a, there is, by No. 2, a plane containing B and a; we may denote this plane by (B, a). Euclid assumes that, in this plane, there is a definite line, say b, containing B, which is parallel to a. As the line b lies in the plane (B, a), any point, say B', of the line b, lies in this plane. Thus the plane (B, a) is determined by the line a and the point B'. Similarly, this plane is determined by the line b and any point of the line a. Also, the lines a, b, being parallel, and, therefore, by definition, in the same plane, if a line c be determined as containing any point A of a, and any point B of b, then the plane of the two lines a, b is that determined by the lines a, c, which have a point in common, and is the same plane as that determined by the lines c, b, which have a point in common.

By definition, if a line a be parallel to a line b, then b is parallel to a. Now let c be a line, other than a, in the plane of the lines a, b, which is parallel to b. Then c is also parallel to a. For, if c, a had a common point, there would be *two* lines, a and c, both parallel to b, lying in the plane determined by this point and the line b, both passing through this point; and this would be contrary to Euclid's Axiom. Thus if, of three lines a, b, c, lying in the same plane, the line a is parallel to b, and b is parallel to c, then a is parallel to c.

5.   It is also true that if three lines a, b, c, not all lying in the same plane, be such that a is parallel to b, and b is parallel to c, then a is parallel to c. By the definition, the lines a, b lie in a plane, and the lines b, c lie in a plane; but, in the case now to be considered, these planes are not the same.

To prove this, let H be a point not lying in either of the lines a, b; when H is in the plane of a, b, it follows from No. 4 that there is a definite line containing H which is parallel to each of a and b; we first prove that this is also true when H does not lie in the plane of the parallel lines a, b. Consider the two planes (H, a) and (H, b); by the assumption made in No. 3, these planes, having the point H in common, have in common all the points of a line, say h, which contains H. This line h cannot have a point common with a. For, if there were such a point, say P, this, lying in the line h, which lies in the plane (H, b), would itself lie in the plane (H, b); then, as h contains H, the point H would lie in the plane (P, b); this plane, however, as P lies on a, is the plane containing the parallel lines a, b (No. 4); and we have supposed that H does not lie in this plane. Thus P does not exist, wherefore h, which lies in the plane (H, a), is parallel to a. Similarly, h is parallel to b.

Returning then to the hypothesis, of a line c, parallel to b, but not in the plane of the lines a, b, take H on this line c. It is then clear that c is the same as the line h, and is, hence, parallel to a; as we wished to prove.

6.   As a consequence of No. 5, if a, b be two lines which have a common point (or, as we often say, which *meet*, or *intersect*), then any third line c which lies in the plane determined by a and b must meet one at least of these lines. For else, both a and b would be parallel to c, and would therefore be parallel to one another, and not have a common point.

7.   Let a plane, in which there is a line a, be denoted by ϖ; let H be a point which does not lie in the plane ϖ. By Euclid's Axiom a line h can be drawn through H to be parallel to a; we prove that h contains no point which lies in the plane ϖ. For the plane of the parallel lines $h\,a$, having the line a common with the plane ϖ, has no points common with ϖ other than those on the line a; and all the points of the line h lie in this plane of the parallel lines h, a. Thus, any point common to the line h and the plane ϖ must lie in the line a; and there is no such point, because h is parallel to a.

8.   If a line h has no point in common with the plane ϖ, either is said to be *parallel to the other*. Let H be a point of this line h, and A be a point of the plane ϖ. The plane (A, h), having the point A in common with the plane ϖ, has also, in common with ϖ, a line through A, say the line a (No. 3). This line a, lying in the plane (A, h), and in the plane ϖ, has no point in common with the line h, because h has no point in common with ϖ. Thus the line a is parallel to h. Wherefore, in a given plane, a line can be drawn through an arbitrary point of the plane, so as to be parallel to an arbitrary line, outside the plane, which is parallel to the plane.

9.   Let a, b be two lines which lie in a plane ϖ, and have a point in common. Let H be a point not lying in the plane ϖ. Let a line be drawn through H parallel to the line a; say the line h; also, let another line be drawn through H parallel to the line b; say the line k. These lines h, k, having the point H in common, determine another plane; say the plane ψ. We prove that the plane ψ has no point in common with the plane ϖ.

For, if possible, let O be a common point of the planes ψ, ϖ. Then (No. 3) these planes have in common the points of a line, which contains O; denote this line by p. Then the line p, in the plane ψ, must (No. 6) have a point in common either with the line h, or the line k; say the point P. This point P, lying on the line p of the plane ϖ, must itself be in ϖ. Thus one at least of the two lines h, k contains a point which lies in the plane ϖ. This, however, is not so, by No. 7. Wherefore O does not exist.

10.   Two planes which have no point in common are said to be parallel. Let ψ and ϖ be two such planes. Let H be a point of the plane ψ, and A a point of the plane ϖ. Let a be a line, in the plane ϖ, which contains the point A. Then the plane (H, a), having the point H in common with the plane ψ, meets this plane in a line containing H (No. 3); say the line h. This line h is parallel to the line a; because any common point of the lines h, a would be a common point of the parallel planes ψ, ϖ, and the lines h, a are in the same plane (H, a). Now, keeping H the same, and drawing various lines a in the plane ϖ, through the point A, there are lines through H, parallel respectively to these various lines a, each constructed in one of the planes (H, a); and every line through H parallel to a line a arises in this way. These lines all lie in the plane ψ. Thus the plane ψ contains all the lines through H which are parallel to lines in the plane ϖ drawn through the point A. This is true when H, A are any two points respectively in the planes ψ, ϖ.

11.   Now let a, h be any two lines not lying in the same plane. Thus they have no point in common, and are not parallel. We prove, (i), that a plane exists passing through either line and parallel to the other line; (ii), that only one such plane exists through either line; and (iii), that the two planes, so described, one through each line, are parallel to one another.

Take three points H, K, L on the line h. Let a line be drawn through H, in the plane (H, a), to be parallel to a, say the line p. Similarly, let q, r be two other lines, parallel to a, drawn respectively through K and L. By No. 5, any two of the lines p, q, r are parallel to one another, and q and r are parallel to one another. Further, the plane containing q and r, is, by No. 4, the same as the plane containing q and h, because h meets q and r; and, then, for a similar reason, it is the same as the plane containing q and p. Thus the three lines p, q, r lie in one plane, say the plane ψ, this plane containing the line h. This plane ψ, by the same reasoning, contains all lines drawn through points of h, to be parallel to the line a. Moreover, any point P, of this plane ψ, is on one of these parallel lines, namely on the line drawn through P, in the plane ψ, to be parallel to the line p. It follows that the plane ψ is parallel to the line a; for any point, P, common to the plane ψ and the line a, would be common to a and a line, in the plane ψ, parallel to a.

Conversely, any plane ψ', drawn through h to be parallel to a, must contain the line p drawn, parallel to a, through any point H of h; for this line p is that common to ψ' and the plane (H, a), by reasoning already given. Thus ψ' coincides with ψ.

Similarly, a unique plane ϖ can be drawn through the line a, to be parallel to the line h. This is constructed by lines through the points of the line a, all parallel to the line h.

And the planes ψ, ϖ have no point in common. For, any such point, say O, as lying in the plane ψ, would lie on a line, say r, in the plane ψ, meeting the line h in a point, this line r being parallel to a; and, like-wise, this point O, as lying in the plane ϖ, would lie on a line, say n, in the plane ϖ, meeting the line a in a point, this line n being parallel to h. The plane containing O and the line h would thus contain the line r; and, as n is parallel to h, would contain the line n. Therefore, the plane containing the lines r, n, which have a point (O) in common, would contain the line h. By parity of reasoning, this plane would contain the line a. By hypothesis, however, the lines h, a are not in the same plane.

The argument, assuming that the lines r, n determine a plane, as-sumes that these lines are not the same. And they cannot be, because the lines h, a are not parallel.

12.   We come now to a phraseology by which we may succinctly summarise and reproduce the properties of parallelism from which we have chosen the preceding results.

Compare the fact that two lines in the same plane, if not parallel, have a point in common, taken with Euclid's Axiom (No. 4 above), with the fact that, through a point B not lying in a line a, a definite line can be drawn to contain an assigned point of this line; the com-parison suggests that we extend the meaning of the word *point*, and say that two parallel lines *have* a point in common. So long as it ap-pears necessary, we may say that this common point of two parallel lines is *at infinity*. A consequence is that a line has a single point at infinity; this we may also express by saying that a line has a single *inaccessible* point. Euclid's Axiom is, in fact, that if, through a point B not in the line a, all possible lines be drawn in the plane (B, a), every such line will meet a except one such line.

It may be remarked that it is possible to construct a self-consistent geometry of the plane with the assumption that there is an infinite number of lines through B, in the plane (B, a), which do not meet the line a. In such a theory there is, on the line a, an infinite number of inaccessible points. This possibility seems to have been present to the minds of some of the Greek geometers; but it was not until approxi-mately the beginning of the nineteenth century that a systematic scheme was constructed to embody this. The success of the scheme so constructed shews that the decision between the two points of view is not a matter of ascertainable fact, as was long thought, to be settled by logic, or by measurement (especially with instruments made in accordance with Euclid's theory). But a further consequence is that the relations between the lengths of lines, which are of different forms in the two schemes, cannot be an essential part of the analysis of the geometrical figures which are possible.

13.   Consider now the application of the phraseology suggested in No. 12. We have proved (Nos. 4, 5) that all lines which are parallel to one line (and not necessarily in the same plane) are parallel to one another. This may be summarised by saying that all lines parallel to one line have a common point; or, that the inaccessible points of these lines, one on each line, coincide with one another. The argument of No. 5 suggests that if a, b be two given lines, and H a point not lying on either, there can be drawn through H a line having a common point with both a and b, this being the line common to the two planes (H, a) and (H, b). When a, b have a common point, this is obviously true, the line in question being that from H to this common point. If a, b are parallel, and we adopt the phraseology of No. 12, this statement is that proved in No. 5. When a, b are not parallel, the line common to the planes (H, a), (H, b) meets a because it is in the plane (H, a), though the point of meeting may be the inaccessible point of a; and for a similar reason this line meets b. Conversely, any line through H to meet a and b must lie in the plane (H, a), even when this point of meeting is the inaccessible point of a, and this line is parallel to a; and it must similarly lie in the plane (H, b). In all cases, therefore, when H does not lie on a, or b, there is one line through H to meet a and b. A more general form of No. 6 is then that, if a, b be two lines in the same plane, whether parallel or not, then any line lying in this plane meets both a and b. And No. 7 becomes the fact that, if a line a lies in a plane ϖ, and H be a point not in this plane, then the line joining H to the inaccessible point of the line a meets the plane ϖ only in this point. Further, No. 8 considers the relation of a plane ϖ, and a line h not lying in this plane, when it is possible to draw, in the plane ϖ, through an arbitrary point A lying in this plane, a line containing the inaccessible point of the line h. We are led to say that a line h, parallel to a plane ϖ, is a line whose inaccessible point lies in this plane; and thence, to say that any line, not lying in a plane, contains a single point which lies in this plane. The Nos. 9, 10 then state that, if in a plane ϖ there be a system of lines all passing through a point, and, from a point H not lying in this plane, the lines h, k, ... be drawn, each to the inaccessible point of one of the system of lines in the plane ϖ, then these lines h, k, ... lie in a plane; and this new plane has, common with ϖ, only these inaccessible points. As the inaccessible points of two parallel lines in the plane ϖ are the same point, we thus reach the conclusion (using No. 3 in an extended sense) that the inaccessible points of all the lines of a plane ϖ lie upon another plane, and hence lie on a line; this line contains a single point of any line of the plane, namely the inaccessible point of this line. Accepting this, we may call the line in any plane which contains the inaccessible points of all the lines of the plane, *the inaccessible line of the plane* (or, *the line at infinity* of the plane). Thence we naturally say that two parallel planes have, for common points, all the points of the inaccessible lines of both planes, these inaccessible lines being the same line. Wherefore, as two planes either have a point in common, or are parallel, we can say that any two planes meet in a line; that is, have in common all the points of a line. No. 11 is then briefly stated by saying that, if two lines a, h do not lie in the same plane, there exists a plane containing either of these lines and also the inaccessible point of the other line; the two planes so obtainable, which both contain the inaccessible points of both the given lines a, h, meet one another in the line containing these inaccessible points; and this common line of the two planes is the inaccessible line of either.

14.   We can go further, and regard the inaccessible points of all the possible lines of space, and also, all the inaccessible lines of all the possible planes of space, as lying in a single plane. This we may call the *inaccessible plane of space* (or, *the plane at infinity*). In justification of this phraseology, however, we content ourselves with the remark that this hypothesis of a plane of inaccessible points is clearly in accord with the existence of a single inaccessible point on every line, and of a single inaccessible line in every plane. For, the statement that a line either meets a plane or is parallel to it, has become, in the phraseology adopted, the statement that a line not lying in a plane has one point common with it; while, similarly, a plane meets any other plane in a line.

15.   It is a common method in scientific theory to introduce a new phraseology in order to describe at once two sets of facts previously regarded as separate, when it has been shewn that such a unity of description is possible without loss of essential precision. On the basis of what has been shewn here we shall therefore in future say that *any* two lines in a plane have a point in common. Thereby we neglect the distinction between the inaccessible point of a line, and the other points of the line. The developments of the theory will shew amply that nothing of geometrical importance is lost by this. We shall also, in dealing incidentally with figures not lying in a plane, suppose that any line, not lying in a plane, has one point which lies in this plane; and also that any two planes have in common the points of a line. We see that the point of view is a direct consequence of Euclid's assumption that a line has only one inaccessible point, and might have been adopted by him save for his conception that a line has a definite length.

CHAPTER II

THE FUNDAMENTAL ELEMENTS. THE PROPOSITIONS OF INCIDENCE. THE PRINCIPLE OF DUALITY

In the space with which we deal there are certain fundamental elements: *Points, Lines* and *Planes.*

A line contains points, in infinite number, and is assumed to be determined by any two of these. If A, B be any two points of the line, and C be any other point of the line, the line determined by the points A, C is the same line, and contains the point B.

A *plane* contains points and lines, in infinite number; and is assumed to be determined by any three of its points which do not lie in a line. If A, B, C be any three points of the plane, not lying in a line, and D be any other point of the plane, not lying in the line determined by A and B, the plane determined by the points A, B, D is the same plane, and contains the point C. If A, B be any two points of a plane, all the points of the line determined by A and B are assumed to lie in the plane. Thus, a plane is determined by a line, say a, and a point C which does not lie in this line. This is the plane which is determined by the point C and any two points which lie in the line a. Likewise, a plane is determined by two lines which have a common point; for, if C be this common point, and A, B be two other points, one in each of the two lines, the plane is determined by the points A, B, C, and contains both the lines.

It is assumed that any two lines which lie in a plane have a point in common. Also, that any line which does not lie in a plane contains a point which is also a point of the plane. The consistence of these assumptions with Euclid's theory is explained in Chapter I. It follows that any two planes have in common the points of a line, which thus lies in both planes. For, taking a point O in one of the planes, not lying in the other plane, draw two lines in the first plane with O as common point; each of these lines will contain a point, other than O, which lies in the second plane. The line which contains both these points will lie in both planes. As the two planes are different, they cannot have a point in common besides the points of this common line. Further, *three* planes, unless they have in common the points of a line, have a single point in common. For, the line common to two of the planes, which contains all the points common to these two planes, unless this line lies wholly in the third plane, has a point lying in the third plane. Three planes, however, which have *two* points in common, have in common all the points of the line determined by these two points. We remark also that, when three planes have a single point in common, the three lines common to the respective pairs of these planes all contain the common point of the three planes.

Conversely, if three lines be such that every two of them have a point in common, so that every two of them lie in a plane, then either the three lines have one point in common, or the three lines all lie in one plane. For, if the three lines do not have one point in common, there will be three distinct points, each common to two of the lines; of these three points, there will be two on each of the three lines; wherefore the plane determined by these three points will contain every one of the lines. Likewise, if the three lines do not all lie in one plane, there will be three distinct planes, each determined by two of the lines; of these planes there will be two passing through each of the three lines; the point common to these three planes will be common to all the three lines.

The theorems for points, lines and planes, such as those just stated, constitute what are called the *Propositions of Incidence*. A point and a line are called *incident* when the point lies in the line; likewise a point and a plane are called incident when the point lies in the plane.

Duality of elements and figures. A consideration of the Propositions of Incidence reveals a certain duality of relations, both in a plane, and, in a different way, in space. In the plane, this depends upon the fact that, just as two points determine a line, in which these points lie, so two lines determine a point, which lies on both the lines. Thus, if we have any plane figure, consisting of points and lines, there exists also a figure consisting of lines and points, such that, to any point of the first figure, lying on a line of that figure, there corresponds a line of the second figure passing through a point of this figure which corresponds to the line of the first figure; while, to any line of the first figure, containing, say, two points A, B, there corresponds a point of the second figure common to the lines, say a, b, of the second figure, which correspond to the points A, B of the first. The correspondence is *descriptive*, being that between the incidence of points on lines in the one figure, and the passage of corresponding lines through points in the other figure. One figure does not determine the positions of the elements in the other figure, but only the existence of the incidences referred to. For example, to three points, A, B, C, of one figure, will correspond three lines, a, b, c, of the other figure; then, to the line containing B, C in one figure will correspond the point of intersection (b, c) in the other figure; to a line, say p, containing A, which meets the line BC in a point D, in the one figure, will correspond, in the other figure, a point, P, lying on the line a, and this point P, with the point (b, c), determines a line d, which corresponds to the point D of the other figure. Again, to three lines p, q, r of the first figure, containing respectively the points A, B, C, and meeting in a point, L, will correspond, in the second figure, three points P, Q, R, lying respectively on the lines a, b, c, and these points P, Q, R will lie on a line, say l. More generally, just as the plane of the first figure, regarded as an element of space, is determined by three points A, B, C lying therein, which do not lie in a line, so the plane may be determined by three lines a, b, c, lying therein, which do not meet in a point.

Two such figures as we have described are said to be *duals* of one another. They may be the same figure, which is then dual to itself; for instance, the figure constituted by the angular points and sides of a triangle is dual to the figure constituted by the sides and angular points of the same triangle. The conception of dual figures in planes is simple; but it requires the assumption that any two lines in a plane have a point in common. This may probably be the reason why the explicit recognition of the principle was so long delayed.

Though this volume is concerned with geometry in a plane, it is not desirable to avoid reference to figures in space. We consider therefore the principle of duality which likewise exists in space.

In space, a line is determined by two of its points. But it is also determined by two planes which both contain the line. Likewise, a plane is determined by three of its points which do not lie in a line. But a point is also determined by three planes which contain this point and have not a line in common. Again, two lines which have a point in common lie in a plane; and, similarly, two lines which lie in a plane have a point in common. Such facts may suggest the existence of a duality in space, in which a *point is dual to a plane*, and a *line is dual to a line*. This suggestion is fully borne out by more detailed examination. To any figure consisting of points, lines, and planes, in space, corresponds a figure consisting of planes, lines, and points, in which, to any point, A, of the one figure, corresponds a plane, α, of the other figure; to any point A lying on a line l, of one figure, corresponds a plane α passing through a corresponding line, l', of the other figure; to any point A, lying on a plane β, of one figure, corresponds a plane α, passing through a point, B, of the other figure; to any line l, of one figure, containing two points A, B, corresponds a line l', of the other figure, lying in two corresponding planes α, β; and so on. As simple examples, if three points lie in a line, then a plane exists containing this line which contains any further point not lying in the line; dually, if three planes have a line in common, a point exists lying on this line which lies on any further plane not containing the line. Or again, two lines in space in general have no point in common; dually, two lines in space in general do not lie in a plane. Or, a geometrical statement may be self-dual; as, for instance, that already made, that, if three lines in space be such that every two have a point in common, so that every two of the lines lie in a plane, then either the three lines have a point in common, or they all lie in the same plane.

We give now some detailed examples, exhibiting the duality in a plane, or in space:

(1) Let A, B, C, D be four points in a plane, of which no three lie in a line; these determine six lines, each containing two of the points. These six lines form three pairs, of which each pair contains all the

four points A, B, C, D; for instance, the lines BC, AD form such a pair. We shall speak of such a pair as forming a *complementary* pair of joins of the four points A, B, C, D; and of these four points as forming a *quadrangle*. The three points, P or (BC, AD); Q or (CA, BD); and R or (AB, CD), of which each is the common point of the two lines of a complementary pair of joins, will be called the *vertices* of the quadrangle. A particular theorem, which will be proved below, is that the three points of intersection (BC, QR), (CA, RP), (AB, PQ), say respectively L, M, N, lie in a line.

Dually, let a, b, c, d be four lines in a plane, of which no three meet in a point; the intersections of pairs of these lines form six points. These six points consist of three pairs, of which each pair uses the four lines for its definition; for instance, the two intersections (b, c), (a, d) form such a pair. We shall speak of such a pair as being a *complementary* pair of intersections of the four given lines; and of these four lines as forming a *quadrilateral*. The three lines, p, q, r, of which each contains a pair of complementary intersections, will be called the *diagonals* of the quadrilateral; for instance, p joins the two points (b, c), (a, d). A particular result, which will be proved later, is that if l be the line joining the points (b, c), (q, r), likewise m the line which joins the points (c, a), (r, p), and n the line which joins the points (a, b), (p, q), then the lines l, m, n have a common point.

It is of the essence of the principle of duality, that a proof of the fact that the points L, M, N lie in a line suffices to shew that the lines l, m, n meet in a point. Every step of the former proof has a dual counterpart in the latter proof. This remark applies to any two dual figures.

(2) Let a, b, c, d, e be any five lines in a plane, of which no three meet in a point. We can determine a point, say E, on one of these lines, e, by the following construction: consider the points (e, a) and (e, d), common to e and a, d, respectively; join the former point, (e, a), to the point (c, d), by a line; and the latter point (e, d), to the point (a, b), by another line. Let these joining lines meet in P. The line which joins the point (b, c) to P meets e in the point E.

Dually, let A, B, C, D, E be five points in a plane, of which no three lie in a line. We can determine a line, say e, passing through one of these points, E, by the following construction: take the point common to the lines EA, CD; also the point common to the lines ED, AB; let p be the line joining these points. Take now the point common to the line BC, and the line p. The line e, which is to be constructed, is the joining line of E to this common point.

It will be found that both these constructions have importance.

(3) Turning now to the duality in space, consider four points which do not lie in a plane, but lie in threes in four planes, no three of the four points being in a line. There are then six lines, each joining two of the points, each of these lines being common also to two of the four planes. Dually, consider four planes, which have no point common to all, but meet in threes in four points, no three of the planes having a

line in common. There are then, again, six lines, each common to two of the planes, each of these lines being also the joining line of two of the four points. Each figure is in fact self-dual. We generally speak of it as forming a tetrahedron, having four angular points, four faces, and six edges.

(4) Given, in space, two lines l, m, which do not lie in a plane; and also a plane ϖ which does not contain either of the lines l, m; there is a line, lying in the plane ϖ, which has a point common with both l and m. For a line, in the plane ϖ, which meets the line l, must necessarily pass through the single point common to the plane ϖ and the line l, say the point L; likewise, a line in the plane ϖ, which meets the line m, must necessarily pass through the single point, say M, which is common to the plane ϖ and the line m. The line in the plane ϖ which meets l and m is thus the line LM. The points L, M do not coincide because l, m have no point in common.

Dually, given two lines l, m, in space, which have no common point, and also a point, P, which does not lie in either of the lines l, m; there is a single line, through the point P, which meets both l and m. For a line, through the point P, which meets the line l, must necessarily lie in the plane (P, l), which contains P and the line l, say the plane α. Likewise, a line, through the point P, which meets the line m, must necessarily lie in the plane (P, m), which contains P and the line m, say the plane β. Conversely, a line in the plane α has a point on the line l, and a line in the plane β has a point on the line m. Thus the line common to the planes α, β meets both l and m; and it passes through the point P which lies in both these planes. The two planes α, β must have a line in common, that is, these planes do not coincide, because, otherwise, the lines l, m, lying respectively in these planes, would have a point in common.

CHAPTER III

THE SYMBOLIC REPRESENTATION OF THE PROPOSITIONS OF INCIDENCE. PAPPUS' THEOREM

THE propositions we have explained in Chapter II are capable of a symbolical representation which is so simple and suggestive that it seems proper to give a preliminary account of this representation at once. We shall finally adopt this symbolism, and shall make the assumption that it is precisely representative of the geometry; for the present, however, we are mainly concerned only to illustrate the correspondence of the symbolism with the fundamental assumptions we have made. We do not restrict ourselves to figures lying in one plane.

With a point, say A, of a geometrical figure under consideration, we associate a symbol; this symbol may conveniently be denoted by (A). This is not a number; and the symbols of two points cannot be multiplied, or divided one by the other. But we attach a meaning to a linear aggregate of such symbols, written as a sum; in such an aggregate each symbol may occur multiplied by a number; and we suppose that the commutative and associative laws hold for the component terms of such an aggregate. The numbers spoken of are such (real or complex) as occur in ordinary algebraical analysis. A symbol $p\,(A)$, in which p is such a number, other than zero, corresponds to the same point as does the symbol (A); correspondingly, in a linear aggregate, say $p\,(A)+q\,(B)+r\,(C)+\ldots$, which also represents a point, all the numbers $p, q, r, \ldots$ may be multiplied, or divided, by the same number, without effect upon the point represented by this aggregate.

We then make the convention that the symbols, (A), (B), (C), corresponding to three different points which lie in the same line, are connected by an equation of the form $p\,(A)+q\,(B)+r\,(C)=0$, in which p, q, r are numbers, of which no one is zero. Thus the symbol (C), of a point which lies in the line determined by the two points A, B, is expressible, in terms of (A) and (B), by a linear equation, say $(C)=m\,(A)+n\,(B)$, in which m, n are numbers. This obviously corresponds to the geometrical property of a line, that it is determined by any two points of itself. But we also make a converse assumption, namely, that any point in a figure which has a symbol of the form $m\,(A)+n\,(B)$, in which m and n are numbers, is a point of the line joining the points A and B.

We have said that a plane is determined by any three points of it which are not in line. Correspondingly, we suppose that, if (A), (B), (C), (D) be the symbols of four points of a plane, of which the points

A, B, C are not in a line, then (*D*) is expressible in terms of the other symbols, in a form (*D*)=p (*A*)+q (*B*)+r (*C*), wherein p, q, r are numbers. And we make also the converse assumption, that any point *D*, of which the symbol is expressible in this form, is a point of the plane which is determined by the points *A, B, C* (or a point of the line *AB*, if *A, B, C* be in line).

Finally, we suppose, when we are dealing with figures in space, that the symbol of any point is linearly expressible in terms of the symbols of any four points of the space, arbitrarily chosen, which do not lie in a plane (or, therefore, any three in a line).

It is intended that two symbols do not represent the same point, unless one of these is obtainable from the other by attaching a numerical multiplier; and also that two different points have not the same symbol, or symbols differing from one another by a numerical multiplier. Thus, in the line joining two points *A, B*, two symbols m (*A*)+n (*B*), and m' (*A*)+n' (*B*), in which m, n, m', n' are numbers, do not represent the same point unless

$$m' \, (A) + n' \, (B) = k \, [m \, (A) + n \, (B)],$$

in which k is a number, namely

$$m' \, (A) + n' \, (B) = \; km \, (A) + kn \, (B);$$

or $\qquad\qquad (m' - km) \, (A) = (kn - n') \, (B);$

and, if (*A*), (*B*) represent different points, this can only be true if $m' = km$, and $n' = kn$. Conversely, however, the symbols m (*A*)+n (*B*) and km (*A*)+kn (*B*) do represent the same point, whatever number k may be (other than zero). Similarly, in the plane determined by three points *A, B, C* which are not in line, two symbols p (*A*)+q (*B*)+r (*C*) and p' (*A*)+q' (*B*)+r' (*C*), in which p, q, r, p', q', r' are numbers, do not represent the same point unless there be a number k (other than zero) such that $p' = kp$, $q' = kq$, $r' = kr$; but, when this is so, the two symbols represent the same point. Likewise, in space, if the points *A, B, C, D* be not in a plane, two symbols p (*A*)+q (*B*)+r (*C*)+s (*D*) and p' (*A*)+q' (*B*)+r' (*C*)+s' (*D*) represent the same point if, but only if, $p'/p = q'/q = r'/r = s'/s$.

The meaning of the assumptions we have described will be made clear most easily by applying them to the Propositions of Incidence, and to some simple geometrical theorems:

(1) If *D, E* be any two points in the plane which is determined by the points *A, B, C*, which do not lie in a line, then both the symbols (*D*), (*E*) are expressible as linear functions of the symbols (*A*), (*B*), (*C*). Thus, any linear function of (*D*) and (*E*) is also expressible as a linear function of (*A*), (*B*), (*C*). This corresponds to the geometrical fact that any point in the line *DE* lies in the plane *ABC*. (As in other cases, it is understood that the coefficients in the linear functions are numbers.)

(2) The theorem that two lines in a plane, determined respectively by the points *A, B* and the points *C, D*, have a point in common,

follows at once from the assumption that the symbols of any four points in a plane are connected by a linear equation. For, if this linear equation, for the points A, B, C, D, be $p\,(A)+q\,(B)=r\,(C)+s\,(D)$, where p, q, r, s are numbers, then, in the first member of this equation, we have the symbol of a point which, by the assumptions made, lies on the line AB, and, in the second member of this equation, we have the symbol of a point lying on the line CD. As these symbols are equal, it follows by our assumptions that these two points coincide.

(3) Similarly, the theorem in space, that a plane has a point in common with a line which does not lie in the plane, follows from the assumption that the symbols of any five points in space are connected by a linear equation. For, if the plane be determined by three points A, B, C, which do not lie in a line, and the line spoken of be determined by the points D, E, the necessary equation connecting the symbols of the five points, say $p\,(A)+q\,(B)+r\,(C)=s\,(D)+t\,(E)$, where p, q, r, s, t are numbers, has, in the first member, the symbol of a point of the plane, and, in the second member, the symbol of a point of the line. As these symbols are equal, the two points are the same point, common to the plane and line. If the line were entirely in the plane, both of (D), (E) would be linear functions of (A), (B), (C).

(4) The theorem that two planes in space have a line in common follows with equal simplicity. For, if the first plane be determined by the points A, B, C, not lying in a line, and the second plane be determined by the points D, E, F, equally not lying in a line, then the symbols of the five points of space, A, B, C, D, E, are connected by a linear equation, as are those of the five points A, B, C, D, F. These equations give rise, as in (3), to points lying in the plane ABC and lying respectively in the lines DE, DF. Both these points lie, not only in the plane ABC, but also in the plane DEF. We can clearly suppose A, B, C to be taken so as not to lie in the plane DEF, and D, E, F to be taken so as not to lie in the plane ABC. Then the line joining the two points where the lines DE, DF meet the plane ABC is the line common to both planes.

(5) That three planes in space, which have not a line in common, have a point in common is similarly obtained, the argument being only a translation of the simple geometrical proof which was given earlier. If O be this common point of the three planes, and A, B, C be points, other than O, respectively on the three lines through O which are common to the pairs of the planes, the separate planes are determined respectively by the points B, O, C; C, O, A; A, O, B; and the symbol of any point of the space in which the three planes lie can be expressed in terms of the symbols of O, A, B, C.

(6) Another simple illustration of the use of the symbols is the formulation of the proof that, through an arbitrary point, a line can be drawn to meet both of two arbitrary lines in space which have no point in common (and, therefore, do not lie in a plane). Let one of these given lines be determined by two of its points, A and B, and the other

line by two of its points, C, D. As the lines do not lie in a plane, there is no linear equation with numerical coefficients connecting the symbols, (A), (B), (C), (D), of the four points taken. Now let (O) be the symbol of an arbitrary point not lying on either of the two given lines. Then, by our assumptions, the symbols of the five points in space, A, B, C, D, O, will be connected by a linear equation, which we may write in the form $(O)=(U)+(V)$ wherein (U) stands for an aggregate $p\,(A)+q\,(B)$, and (V) stands for an aggregate $r\,(C)+s\,(D)$, in which p, q, r, s are numbers. In this form, however, the equation expresses that the point O is on the line joining two points U, V, lying respectively on the given lines; and as there is no ambiguity as to the values of the ratios of the numbers p, q, r, s, the points U, V are quite definite. This is the proposition to be proved.

(7) Let A, B, C, D be four points in space, of which no three are in line. Suppose, for the present, that D is not in the plane ABC. Let P, Q, R be arbitrary points lying respectively in the lines DA, DB, DC. We usually speak of P, Q, R as being *in perspective*, from the *centre D*, respectively with A, B, C. The line QR, in the plane DBC, has a point on the line BC, say L. So the line RP, in the plane DCA, meets CA, say in M; and the line PQ, in the plane DAB, meets AB, say in N. These points L, M, N are in line. For, as we are supposing that D does not lie in the plane ABC, the plane PQR does not coincide with the plane ABC; these planes have therefore a common line, and no common points save in this line. This line contains L, because this, lying in QR and BC, lies in both planes; likewise this line contains M and N. Thus the points L, M, N lie in a line, as stated. This line is called *the axis of perspective* of the two perspective triads A, B, C and P, Q, R.

We now formulate the proof of this result in terms of the symbols: As D is on each of the lines AP, BQ, CR, the symbol of D must, in accordance with the assumptions we have made, be capable of expression in each of the three forms

$$(D)=a\,(A)+p\,(P), \quad (D)=b\,(B)+q\,(Q), \quad (D)=c\,(C)+r\,(R),$$

where a, b, c, p, q, r are numbers. These equations, however, lead to $b\,(B)-c\,(C)=-q\,(Q)+r\,(R)$, which equates the symbol of a point of the line BC with the symbol of a point of the line QR. These lines therefore meet in a point, say L, whose symbol may be given by $(L)=b\,(B)-c\,(C)$. Likewise, the symbol of the point, M, common to the lines CA, RP, is given by $(M)=c\,(C)-a\,(A)$; and the symbol of the point, N, common to the lines AB, PQ, is given by $(N)=a(A)-b(B)$. These three, however, are such that $(L)+(M)+(N)=0$. Wherefore, by the assumptions we have made for the symbols, the points L, M, N lie in a line.

But now it is to be remarked that the proof of this result by means of the symbols does not require the condition that D is not in the plane ABC, as did the geometrical proof. The theorem is in fact true, and the assumptions made for the symbols are so far vindicated, when D does lie in the plane ABC. Of this a geometrical proof, using only the Propositions of Incidence which we have given for the elements, *point, line, plane*, is given below, under the heading *Desargues' theorem.*

Here we touch a point of logic which is of importance, though the reader may prefer to leave the full consideration of it until later. The theorem that the points L, M, N are in line, even when D is a point of the plane ABC, is a theorem of plane geometry. It can be deduced, geometrically, as will be seen, by using the Propositions of Incidence for points, lines *and* planes; it is a fact that it cannot be deduced geometrically from the Propositions of Incidence for points and lines in a plane only, without some further axiom of plane geometry. That the theorem can be obtained so simply, by the use of the symbols, shews that the assumptions we have made for the symbols involve more than the Propositions of Incidence in the plane. This conclusion will presently (in No. 8) be enforced by consideration of another theorem of plane geometry which can be proved by the use of the symbols, but not from the Propositions of Incidence even when the properties of the elements in space are assumed.

(8) We pass now to consider a theorem of plane geometry which we call *Pappus'* theorem. This is found in his *Collection of Geometrical Theorems* (? A.D. 300); it can be deduced, as we shall indicate, from results proved by Euclid. But it cannot be deduced from the geometrical axioms we have so far given, by geometrical reasoning. On the other hand, it has the importance that if it be assumed to be true, and added as an Axiom to the geometrical axioms so far given, all our geometrical results can be deduced geometrically, without further assumptions.

The theorem in question may be stated thus: *Let l, l' be any two lines in a plane, and R, Q be any two points in this plane, not lying on either of the lines l, l'. Let A be any point of the line l, and A' be any point of the line l', neither of these lying at the common point of l and l'. Let the common point of the line AR and the line l' be called B', and the common point of the line $A'R$ with the line l be called B. Likewise, let AQ, $A'Q$ meet the lines l', l respectively in C', C. We thus have three points A, B, C on the line l, and three points A', B', C' on the line l'. The theorem is that the common point, say P, of the lines BC' and $B'C$, lies on the line QR.* We may, however, take A, B, C arbitrarily on the line l, and take A', B', C' arbitrarily on the line l', and define P, Q, R, respectively, as the intersections $(BC', B'C)$, $(CA', C'A)$, $(AB', A'B)$; the theorem then is that P, Q, R lie in a line.

It is suggested to the reader that he should make an approximate test of the theorem by drawing diagrams. Save, however, for the particular case when the three lines AA', BB', CC' meet in a point, a theoretical proof of geometrical kind does not follow from the geometrical theorems of incidence we have given, even with help of projections from space of three dimensions.

We prove the theorem now with the use of the symbols: As the points B, B' lie respectively on the lines RA' and RA, the symbols are connected by equations of the forms

$$(B) = m\,(R) + p'\,(A'), \quad (B') = m'\,(R) + p\,(A), \qquad \ldots\ldots(1)$$

where m, p', m', p are numbers. Likewise, because C, C' lie respectively on the lines AB, $A'B'$, that is l, l', we have equations

$$(C) = q\,(B) + a\,(A), \quad (C') = q'\,(B') + a'\,(A'), \qquad \ldots\ldots(2)$$

where q, a, q', a' are numbers. From these two sets of equations we can deduce, eliminating (A), (A') and (R), that

$$a'k\,(B) + m'p'a\,(C') = ak'\,(B') + mpa'\,(C), \qquad \ldots\ldots(3)$$

where k, k' are, respectively, $k = mpq - m'a$, and $k' = m'p'q' - ma'$. A direct method of carrying out the algebra is, from the equations (1) and (2), to express both $a'k\,(B) - ak'\,(B')$ and $mpa'\,(C) - m'p'a\,(C')$ in terms of (A), (A') and (R). In the equation (3), the two members are respectively symbols of a point on the line BC', and of a point on the line $B'C$. Thus either of these is the symbol, (P), of the point P. We also find, from equations (1), (2), that

$$m'q'\,(C) - k'q\,(A') = mq\,(C') - kq'\,(A), \qquad \ldots\ldots(4)$$

and either of the two members of this equation is thus the symbol, (Q), of the point Q. The symbol, (R), of the point R, is given in terms of (B), (A'), and in terms of (B'), (A), by equation (1).

Thence, by carrying out the elementary algebra, we find

$$(P) = pp'\,(Q) - kk'\,(R).$$

From this equation, on the basis of the assumptions made for the symbols, we infer that the points P, Q, R lie in a line; as we wished to prove.

(9) The dual of Pappus' theorem should also be recognised: *Let a, b, c be lines in a plane which meet in a point, O, and a', b', c' be three other lines in this plane which meet in another point, O'. Denote the points of intersection (b, c') and (b', c), respectively, by L and L'; and the points of intersection (c, a') and (c', a), respectively, by M and M'; and also the points (a, b'), (a', b), respectively, by N and N'. Then the lines LL', MM', NN' meet in a point.* These lines are the respective duals of the points P, Q, R in Pappus' theorem.

(10) In order to increase the reader's familiarity with Pappus' theorem, we now make, for a moment, a diversion from the orderly development of the logical theory of this volume. We shew how Pappus' theorem may be deduced from familiar results of Euclid's theory. And, in No. 11, we shew how the dual of Pappus' theorem would follow from the usual theory of the equilibrium of forces acting on a rigid body in one plane.

Take any two lines in one plane, on which there lie respectively the points A, B, C and A', B', C'. Suppose that the lines AB', $A'B$ are parallel, and that the lines AC', $A'C$ are parallel. We prove that the lines BC', $B'C$ are parallel. This is not a statement of Pappus' theorem in its general form; but the reader of Chapter I will recognise that it is at least a particular case of the general theorem.

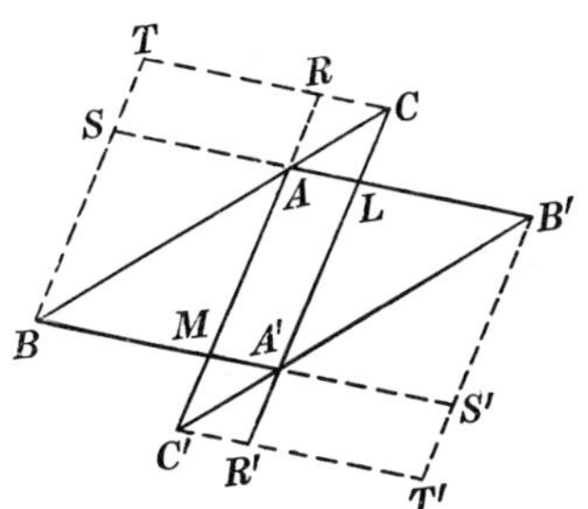

Draw through B, B' lines both parallel to the parallel lines AC', $A'C$, meeting the parallel lines AB', $A'B$, respectively, in S and S'. Also, draw through C, C' lines both parallel to the parallel lines AB', $A'B$, meeting the parallel lines AC', $A'C$, respectively, in R and R'. Let CR, BS meet in T, and $C'R'$, $B'S'$ meet in T'. Also, let AB', $A'C$ meet in L, and AC', $A'B$ meet in M.

Then $AC'T'B'$ is a parallelogram, of which $C'B'$ is a diagonal. Thus, by one of Euclid's results, the parallelograms $A'S'T'R'$ and $A'MAL$ are equal.

Again, $TBA'C$ is a parallelogram, of which BC is a diagonal. Thus the parallelograms $A'MAL$ and $ASTR$ are equal.

Wherefore the parallelograms $A'S'T'R'$ and $ASTR$ are equal; and the sides of one of these are parallel to the sides of the other.

From this, by one of Euclid's later theorems, the ratio of the lengths of the lines AR, $A'S'$, belonging to these parallelograms respectively, is equal to the ratio of the lengths of the lines $A'R'$, AS. As opposite sides of a parallelogram are equal, it follows that CL, LB' are in the same ratio as $C'M$, MB, of which CL, $C'M$ are parallel, and LB', MB are parallel. We see then, again by Euclid's theory, that the lines CB', $C'B$ are parallel; as we wished to prove.

The reader may be interested to prove that the three lines BB', CC', LM meet in a point.

(11) Now let C, M, C', L be the angular points of a parallelogram, taken in order; also A, B' be points in the sides LC, MC', respectively, such that AB' is parallel to the other pair of sides; and A', B be points in the sides $C'L$, CM, respectively, such that $A'B$ is parallel to the other pair of sides. Let $A'B$, AB' meet in O.

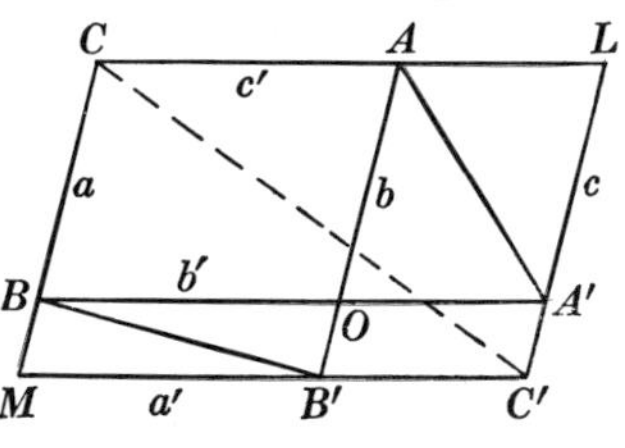

The figure can be constructed when the segments AA', BB' are arbitrarily given, provided BA', $B'A$ are not parallel.

If we assume the result known as the parallelogram of forces, it can be shewn that forces represented in position and sense by the segments AA' and BB' are equivalent to the force represented in all respects by CC'. Thus CC' passes through the common point of AA' and BB'.

This is that case of the dual of Pappus' theorem in which the lines a, b, c of that dual are the parallel lines MC, $B'A$, $C'L$, and the lines a', b', c' are the parallel lines MC', BA', CL. The fact that AA', BB', CC' meet in a point can be proved, of course, by Euclid's results, so that in effect we have another proof of (the dual of) Pappus' theorem.

(12) As an example of the application of the dual of Pappus' theorem, the reader may prove the following result: In space, let a closed hexagon be taken, whose sides, in order, are the lines a, b', c, a', b, c', these not lying in a plane. Let P be a point lying upon a line which meets all of the three lines, a, b, c, and also upon a line which meets all three of the lines a', b', c'. The hexagon has three diagonals, namely the line joining the pair of opposite vertices (b, c'), (b', c); the line joining (c, a') and (c', a); and the line joining (a, b') and (a', b). It can then be shewn that P also lies on a line which meets all these three diagonals. This can be proved, from the dual of Pappus' theorem, by projecting the figure from P upon an arbitrary plane.

(13) Let A, B, C be three points, of a line l, lying in a plane ϖ; and A', B', C' be three points, of another line l', lying in the same plane. From A, B, C, respectively, draw arbitrary lines a, b, c not lying in the plane ϖ; from A', B', C', respectively, draw lines a', b', c', of which a' meets b and c, while b' meets c and a, and c' meets a and b. Let the plane containing the intersecting lines b, c' be called α, and the plane containing the intersecting lines b', c be called α'; and let p be the line common to the planes α, α'. Similarly, let q be the common line of the planes β, or (c, a'), and β', or (c', a), and let r be the common line of the planes γ, or (a, b'), and γ', or (a', b). Deduce from Pappus' theorem that the lines p, q, r meet the plane ϖ in three points which lie in a line.

(14) We have emphasised that Pappus' theorem cannot be deduced from the Propositions of Incidence we have given, in the general case. It can, however, be so deduced when the lines AA', BB', CC', in the enunciation of Pappus' theorem (No. 8), meet in a point. This will appear below when we consider the theory of harmonic ranges and pencils. Meanwhile, the reader may be interested to examine this case with the help of the symbols.

CHAPTER IV

THEOREMS PROVED FROM THE PROPOSITIONS OF INCIDENCE

As the Propositions of Incidence have evidently a very fundamental character, it is of interest to see that many geometrical results, which are of importance, can be proved from these alone. Though, as has been said, we shall also need, in future, to add the use of such further axiom as would be provided by assuming Pappus' theorem; or, alternatively, shall need the use of the assumptions involved in the use of the symbols.

Besides the geometrical deductions to which we now pass, we shall introduce illustrations expressed in terms of the symbols. This will familiarise the reader with the use of the symbols; but, in several cases, will also suggest geometrical results which might not have appeared.

We use the Propositions of Incidence for elements in space; and it does not appear desirable to exclude examples of construction in space. The reader may prefer to postpone the consideration of some of these.

(1) **Desargues' theorem.** This consists of two parts, which are converses of one another: (*a*) *If two triangles ABC, $A'B'C'$, not necessarily lying in the same plane, be such that the joins of corresponding angular points, namely the lines AA', BB', CC', meet in a point, say O, then the three points P, Q, R, of which P is the point of intersection of the sides BC, $B'C'$, and Q is the point $(CA, C'A')$, and R is the point $(AB, A'B')$, lie in a line.* The point O is called the centre of perspective of the two triangles, and the line PQR is called the axis of perspective. That, in the case in which the triangles ABC, $A'B'C'$ do not lie in the same plane, the sides BC, $B'C'$ have a point, P, in common, follows because BC, $B'C'$ lie in the plane determined by the two lines OBB' and OCC'. So for the existence of the points Q and R. The theorem is first proved for the case in which ABC, $A'B'C'$ do not lie in a plane, and the other case deduced from this. The converse statement is, (*b*) *If two triangles ABC, $A'B'C'$ be such that the sides BC, $B'C'$ have a point P in common, also the sides CA, $C'A'$ have a point Q in common, and the sides AB, $A'B'$ have a point R in common, and it be also true that the three points P, Q, R lie in a line, then the three lines AA', BB', CC', joining corresponding angular points of the two triangles, meet in a point,* say O. This is true whether the two triangles be in the same plane, or not.

The proof of (*a*), for the case when the triangles ABC, $A'B'C'$ do not lie in the same plane, is very simple. The lines BC, $B'C'$, both lying in the plane $OBB'CC'$, have a point in common, say P. This point lies in both the planes ABC, $A'B'C'$. Likewise, the point Q, common to the lines CA, $C'A'$, lies in both these planes, as does the point R, common to AB and $A'B'$. The common points of the planes ABC, $A'B'C'$, however, lie in a line. Thus P, Q, R lie in a line. It is assumed here that no one of the six sides of the two triangles lies in the line common to the planes of the two triangles.

The proof of (a) for the case when the triangles ABC, $A'B'C'$, and therefore the point O common to the lines AA', BB', CC', lie in one plane, may be given as follows: Draw through O an arbitrary line, h, not lying in this plane; on this line take two points H, H' arbitrarily. As h contains the point O of the line AA', the lines h, AA' lie in a plane; thus the two lines HA, $H'A'$ have a common point, say A''.

In a similar way, the lines HB, $H'B'$ have a common point, say B'', and the lines HC, $H'C'$ have a common point, say C''. Then, as the points B'', C'', respectively on the lines HB, HC, lie in the plane HBC, it follows that the line $B''C''$ has a point in common with the line BC. Similarly, the line $B''C''$ has a point in common with the line $B'C'$. While the lines BC, $B'C'$, in the common plane of the original triangles, have the point, P, in common. The three lines BC, $B'C'$, $B''C''$ are therefore such that every two of them have a point in common. Wherefore, by a theorem already remarked (Chapter II), these three lines either lie in the same plane, or have a point in common. But, as the planes HBC, $H'B'C'$ are different from the plane ABC, it is clear that the line $B''C''$ does not lie in the plane containing the lines BC, $B'C'$. Thus the lines BC, $B'C'$, $B''C''$ meet in a point, namely, in P. In a similar way $C''A''$ contains the point Q common to the lines CA, $C'A'$; and $A''B''$ contains the point R common to the lines AB, $A'B'$. Whence, the points P, Q, R lie in a line, namely the line common to the plane $A''B''C''$, and the original plane of ABC and $A'B'C'$. This is what we desired to prove. It is seen that an essential step of the proof lies in using the fact that two planes in space meet in a line.

Coming next to the part (b) of the theorem, we have two triangles ABC, $A'B'C'$, supposed at first to be in different planes, such that the sides BC, $B'C'$ have a common point, P, the sides CA, $C'A'$ have a common point, Q, and the lines AB, $A'B'$ have a common point, R. As all of P, Q, R lie in both the planes ABC, $A'B'C'$, the points P, Q, R are in the line common to these planes. Consider the three lines AA', BB', CC': By hypothesis, these do not lie in a plane; but every two of them lie in a plane; for instance BB' and CC' both lie in the plane $PBCB'C'$; so that BB' and CC' meet in a point. The three lines AA', BB', CC', not lying in a plane, are then such that every two of them meet in a point. Whence, by the theorem already quoted, these three lines meet in a point. This is the conclusion we wished to draw.

Last, if ABC, $A'B'C'$ be two triangles in the same plane, with the property that the three points common to the pairs of corresponding sides, namely the points P, or $(BC, B'C')$, Q, or $(CA, C'A')$, and R, or $(AB, A'B')$, lie in a line, then we prove that the three lines joining corresponding angular points of the two triangles, namely the lines

AA', *BB'*, *CC'*, meet in a point. This is the dual, in the plane, of the result proved in the second part of (*a*), the sides of the triangles being dual each to its opposite angular point. Thus no further proof is really necessary, since every step in the former proof can be replaced by its dual. But we may neglect this, and proceed as follows: Draw an arbitrary plane through the line *PQR*, not coinciding with the plane of *ABC* and *A'B'C'*. In this new plane draw, respectively through the points *P*, *Q*, *R*, three arbitrary lines forming a triangle *A"B"C"*, so that *PB"C"*, *QC"A"*, *RA"B"* are these lines. By the preceding first case of (*b*), the lines *AA"*, *BB"*, *CC"* have a point in common, say *H*; and the lines *A'A"*, *B'B"*, *C'C"* have a point in common, say *H'*. Neither *H*, nor *H'*, is in the plane of *ABC*, *A'B'C'*. The lines *HH'*, *AA'* have a point in common, because *HA*, *H'A'* meet in *A"* and so lie in a plane. Thus *AA'* passes through the point where the line *HH'* meets the plane of *ABC*, *A'B'C'*. The lines *BB'*, *CC'*, also, similarly pass through this point. Thus *AA'*, *BB'*, *CC'* meet in a point, as we desired to prove.

The whole of Desargues' theorem is thus established.

(2) We now give some examples of the application of Desargues' theorem. But, before doing this, as an interesting exercise in constructions in space, we consider what is the *dual, in space*, of the first part of (*a*) in No. 1. In this we have two triangles *ABC*, *A'B'C'*, not lying in the same plane, which are in perspective from a centre, which we now call *T*; and we deduce the existence of an axis of perspective, say *t*, containing the three points of intersection, *P*, or (*BC*, *B'C'*); *Q*, or (*CA*, *C'A'*); and *R*, or (*AB*, *A'B'*). In taking the dual figure, we should consider, instead of the three points *A*, *B*, *C* and the plane determined by these, three planes, say α, β, γ, and the common point of these planes, say *H*. Likewise, instead of the points *A'*, *B'*, *C'*, and the plane of these, consider three planes α', β', γ', and the common point of these planes, say *H'*. We suppose *H'* not to coincide with *H*. The hypothesis then, of the dual figure, is that the three lines of intersection of the pairs of planes (α, α'), (β, β'), (γ, γ') lie in a plane. The deduction to be made is that the two lines of intersection (β, γ) and (β', γ') lie in a plane, say ϑ; the lines of intersection (γ, α) and (γ', α') lie in a plane, say ϕ; and the lines (α, β), (α', β') lie in a plane, say ψ; and that the three planes ϑ, ϕ, ψ meet in a line. This result we can prove as follows: let the lines (α, α'), (β, β'), (γ, γ'), which lie in a plane, form a triangle whose angular points are *U*, *V*, *W*, the side *VW* of this triangle being the line (α, α'), etc. Then, by the hypothesis, we have two tetrahedra, whose angular points are respectively *U*, *V*, *W*, *H* and *U*, *V*, *W*, *H'*. The line (β, γ) is the line *HU*, and the line (β', γ') is the line *H'U*. Whence the lines (β, γ), (β', γ') lie in the plane *HH'U*, which is ϑ. So the planes ϕ, ψ are respectively the planes *HH'V* and *HH'W*. Thus, as desired, the planes ϑ, ϕ, ψ have a line in common; namely, the line *HH'*.

Now consider the section of the dual figure we have just dealt with, by an arbitrary plane, say ϖ. In this plane, the planes α, β, γ give three lines, (ϖ, α), (ϖ, β), (ϖ, γ), or, say, respectively, a, b and c. So the planes α', β', γ' give three lines (ϖ, α'), (ϖ, β'), (ϖ, γ'), or say, respectively, a', b' and c'. The lines a, a' meet in a point, say P, where the line (α, α'), or VW, meets the plane ϖ. Similarly, in this plane ϖ, there is a point Q, common to the lines b, b', which lies on the line WU; and there is a point, R, common to the lines, c, c', which lies on the line UV. The three points P, Q, R thus lie in a line, namely, that in which the plane ϖ is met by the plane UVW. Let the triangle, in the plane ϖ, formed by the lines a, b, c, have angular points A, B, C, of which A is the common point (b, c), etc. So, let the triangle formed by the lines a', b', c' be A', B', C'. Of these two triangles, we have seen, the common points of corresponding sides, namely P of a and a', Q of b and b', and R of c and c', lie in a line. Further, by the construction, the line HA is the line HU, so that H, A, U are in a line. So H, B, V are in a line, and H, C, W are in a line. In the same way there are three lines containing, respectively, H', A', U; H', B', V; and H', C', W. Thus the plane UHH' meets the plane ϖ in the line AA'; likewise, BB', CC' are the lines in the plane ϖ which lie respectively in the planes VHH' and WHH'. Hence the lines AA', BB', CC', joining corresponding angular points of the two triangles ABC, $A'B'C'$, meet in a point (where the line HH' meets the plane ϖ).

We have thus deduced the figure which occurred above in the second part of (b) as a plane section of the figure which is the dual of the figure occurring in the first part of (a).

Ex. 1.   Two triangles PQR, $P'Q'R'$, in the same plane, are said to have a centre of perspective when the lines joining P, Q, R each to one of P', Q', R' meet in a point. The points P', Q', R' may be taken in one of six different orders. The existence of the perspectivity for one of these orders does not exclude the possibility of its occurrence for another of these orders.

For example, in the figure for the dual of *Pappus'* theorem occurring in chapter III (in (9)), we have two sets of three lines a, b, c and a', b', c'. Denote the six intersections (b, c'), (c, a'), (a, b'), (b', c), (c', a), (a', b), respectively, by P, Q, R, P', Q', R'. Shew that the dual of Pappus' theorem may be stated by saying that if P, Q, R are respectively in perspective with Q', R', P', and are also respectively in perspective with R', P', Q', then they are also respectively in perspective with P', Q', R'.

In dual way, two triangles in a plane, whose sides are determined respectively by the lines p, q, r and p', q', r', may have an axis of perspective containing the three intersections (p, q'), (q, r'), (r, p'), or, say, respectively, C', A', B', and also a second axis of perspective, the three points (p, r'), (q, p'), (r, q'), or, say, respectively, B, C, A, being in line. Shew that Pappus' theorem may be stated by saying that when both these are true, then also the three points (p, p'), (q, q'), (r, r') are in line.

Ex. 2.   In a plane, let a be a line determined by two points A, B', and b be a line which is determined by two points B, A'. Let R be an arbitrary point of the plane. We shew how to construct a line passing through R which contains the common point, O, of the two given lines a, b, without using the point O itself for the construction.

For this we assume (1) that A, B, A', B' have such positions that the common point, X, of the lines AB, $A'B'$ can be constructed, (2) that a line can be drawn through X to meet the lines RB', RA', respectively, in such points, L, M, that the common point, H, of the lines AL, BM can be constructed.

Then, the two triangles ALB', BMA' have X as centre of perspective. Thus these have an axis of perspective, containing the three intersections of corresponding sides, namely (LB', MA'), (AB', BA'), (AL, BM). Whence, the line from R, which contains H, passes through O. And, to find H, we have not used the point O.

This is the solution which follows from Desargues' theorem, and the proof of this, we have seen, requires only the Propositions of Incidence for points, lines and planes.

But it is of interest to remark that, if we assume Pappus' theorem, we can, still without using O, construct another point besides H in the line RO. For, let RA' meet AB in C, and RA meet $A'B'$ in C'. Also let BC' meet $B'C$ in K. Then, on the two lines XAB, $XA'B'$, respectively, we have the

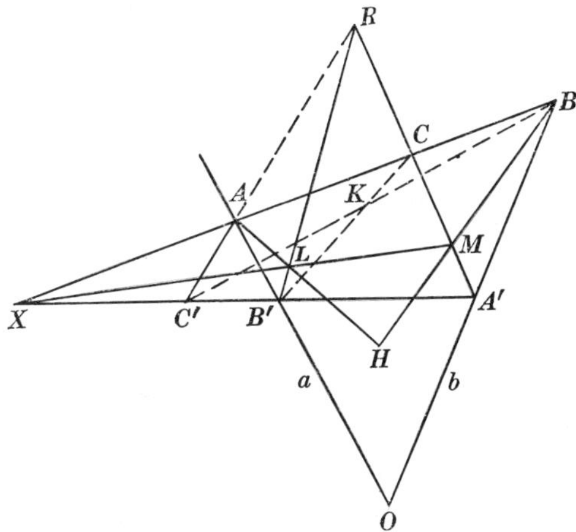

two triads of points A, B, C and A', B', C'. Whence, by Pappus' theorem, the three intersections $(BC', B'C)$, $(CA', C'A)$, $(AB', A'B)$ are in a line. These are, respectively, the points K, R, O. Thus the line RK passes through O. Cf. Ex. 3, following.

We have, however, emphasised that Pappus' theorem does not follow from Desargues' theorem; nor from the Propositions of Incidence.

Ex. 3. Let T, Y, Q be three points lying on a line, and O, M be two other points, and X any point of the line OM. Let TM, OY meet in N, and XT, XY meet QO, QM, respectively, in T' and Y'. By Pappus' theorem, applied to the two sets T, Y, Q and O, M, X, it follows that N, T', Y' are in line. Also, if any line $OM'X'$ be drawn meeting $Y'M$ and $Y'X$ in M' and X', respectively, then, by applying Pappus' theorem to the two sets O, M', X' and Y', T', N, it follows that the lines NM', $T'X'$ meet on TYQ.

Now let ABC and $A'B'C'$ be two triangles such that AA', BB', CC' meet in O. Let CA, $C'A'$ meet in Y, and AB, $A'B'$ meet in Z. Let OBB' meet ZY in Q. Next let QA meet OCC' in M and meet $C'A'Y$ in L, and let LB' meet AB in N; let MN meet ZYQ in T. Then, as above, applying Pappus' theorem to the two sets O, A, A' and L, B', N, we see that ON contains Y. Also, taking the point X of the first figure to be at C, and considering the two sets T, Y, Q and O, M, C, it follows, by Pappus' theorem, that CB contains T; while, taking X at C', and considering the sets T, Y, Q and O, M, C', we see that $C'B'$ also passes through T, lying on YZ. Or (Hessenberg, *Math.*

Annal. LXI (1904), p. 161), if two triangles in a plane have a centre of perspective, we can deduce from Pappus' theorem that they have an axis of perspective; that two triangles in a plane with an axis of perspective also have a centre of perspective can easily be deduced from this.

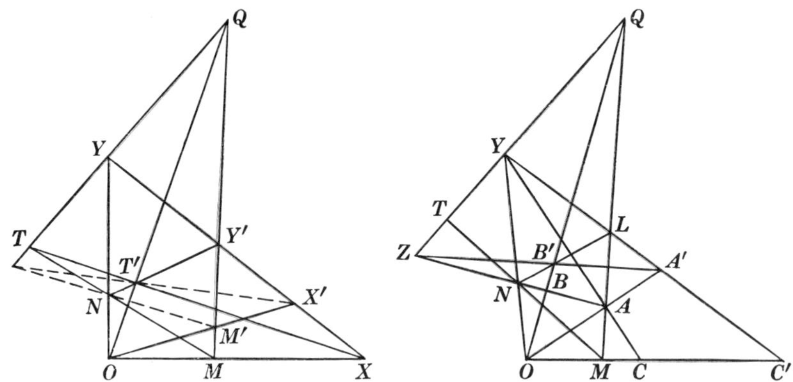

Ex. 4. We now consider a very simple example of a figure in space. Let A, B, C, D and A', B', C', D' be two sets of four points in space, so situate that the four lines AA', BB', CC', DD' all pass through a point O. We express this by saying that the two tetrahedra $ABCD$ and $A'B'C'D'$ have a *centre of perspective*, O. Prove that the edge BC of the former tetrahedron has a point in common with the edge $B'C'$ of the latter, and, in general, that every edge of the former tetrahedron meets the corresponding edge of the latter. Prove further, that the six intersections, $(BC, B'C')$, $(CA, C'A')$, $(AB, A'B')$, $(DA, D'A')$, $(DB, D'B')$, $(DC, D'C')$, lie in a plane, and are the intersections, in pairs, of four lines lying in this plane. This plane is called *the plane of perspective* of the two tetrahedra.

Ex. 5. The figure in Ex. 4 furnishes also a simple illustration of the use of the symbols. Since the points A', A, O are in a line, the symbols of these points are connected by a linear equation; this we may write in the form $a'(A') = a(A) + (O)$, where a, a' are numbers. In accordance with our convention that a numerical multiple of a symbol represents the same point as does the original symbol, we may alter the significance of (A) and (A'), and write this equation in the form $(A') = (A) + (O)$. This is not necessary; but it adds to the simplicity of the formulae. There are three other similar equations, which we write in the forms

$$(B') = (B) + (O); \ (C') = (C) + (O); \ (D') = (D) + (O).$$

From these there follow the six equations such as the two

$$(B') - (C') = (B) - (C); \ (A') - (D') = (A) - (D).$$

The former of these equations expresses that the edges $B'C'$ and BC have a point in common; and the six points of intersection of corresponding edges of the two tetrahedra are thus the points with symbols such as (P), (U), where $(P) = (B) - (C)$ and $(U) = (A) - (D)$. The three points whose symbols are (U), (V), (W) determine a plane. The joins of these points are three lines containing, respectively, the points whose symbols are $(V) - (W)$; $(W) - (U)$; $(U) - (V)$. These, however, are, respectively, $(B) - (C)$, or (P), together with (Q) and (R). As $(P) + (Q) + (R) = 0$, these points are in line. This completes the proof of the statements made, as given by the symbols.

It is significant that there is no need to use the fact that the symbol (O) is a linear function of (A), (B), (C), (D).

Ex. 6.   We state the dual of the result given above in Ex. 2. In a plane, let a point U be defined as the common point of two given lines a, a'; and a point V be defined as the common point of two given lines b, b'. Then, we can construct points lying on the line UV, without using U or V, in the following way: Let A be the common point of the lines a, b; let A' be the common point of the lines a', b'; let S be any point on the line AA'. Draw two lines through the point S, one of these meeting the lines a, a', respectively, in the points P, P', the other meeting the lines b, b', respectively, in the points Q, Q'. Then, considering the two triangles $PAQ, P'A'Q'$, it follows by Desargues' theorem, that PQ and $P'Q'$ meet in a point lying on the line UV.

We can also employ Pappus' theorem: Let a line through the point A meet the lines a', b', respectively, in the points B, C; and a line through A' meet a, b, respectively, in B', C'. Then, considering the two triads of points A, B, C and A', B', C', each lying on a line, the intersection $(BC', B'C)$ lies on the line joining the points $(AB', A'B)$ and $(AC', A'C)$; and this is the line UV.

(3) **The harmonically conjugate point of any point of a line, with respect to two other points of the line.** Let A, B, C be any three points in a line. Draw any plane containing this line. From each of

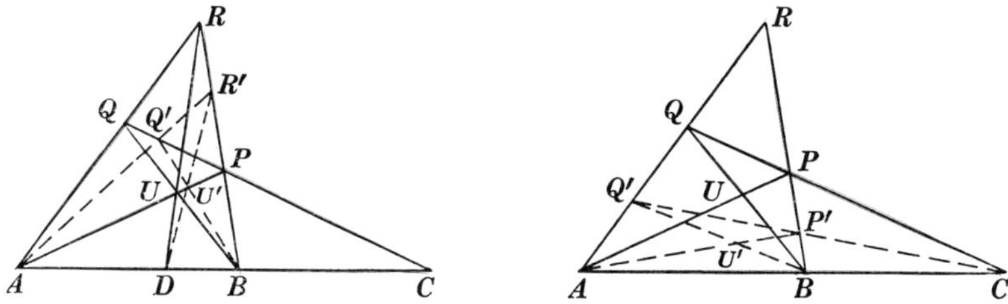

A, B, C draw, in this plane, a line. Let P be the common point of these lines drawn from B, C; Q the common point of these lines drawn from C, A; and R the common point of these lines drawn from A, B. For distinctness, regard the three points A, B, C as consisting of the couple (A, B), with a third point C. Let U be the common point of the lines AP, BQ; let D be the point of the original line which lies on RU. We denote D by $(A, B)/C$.

The point D is independent of the plane through the line ABC in which the construction is made; it is also independent of what lines AQR, BRP, CPQ, through the given points A, B, C, are used to make the construction in the plane chosen.

The fundamental direct proof of this statement is by drawing another plane through the given line ABC, making therein an exactly similar construction; and then shewing, by the coincidence of the points A, B, C in the two planes, that the figure in the second plane is in perspective, from a point of space, with the figure in the first plane. This is a quite easy application of the Propositions of Incidence in space.

But, possibly, it may be preferred to regard the theorem as belonging to geometry in the original plane only, and to deduce it from

Desargues' theorem in that plane. For this, we shew that the point D is unaltered by varying any one of the lines AQR, BRP, CPQ; it will then be unaltered if all these are varied. First, keep the lines already drawn, through B and C both unaltered, and draw a different line through A, to meet CP, BP, respectively, in Q' and R'; let the common point of AP and BQ' be U'. Then, the triangles QRU, $Q'R'U'$ have a centre of perspective at P; thus they have an axis of perspective, this containing the point A where QR meets $Q'R'$, and also the point B where QU meets $Q'U'$. Thus $R'U'$ meets AB in the point D, in which AB is met by RU. A precisely similar argument applies to the case in which we vary the line drawn through B. Now, keep the lines already drawn through A and B both unaltered, but draw a different line through C to meet RB in P' and RA in Q'; let the common point of AP' and BQ' be U'. Then, the triangles PQU, $P'Q'U'$ have the line CBA as axis of perspective. Thus they have a centre of perspective at the common point of the lines PP' and QQ'; this shews that RUU' are in a line, which passes through D.

This completes the proof that the point D, $=(A, B)/C$, is the same whatever be the lines AQR, BRP, CPQ drawn in the plane we have taken. As has been said, it can be shewn that D is the same whatever plane be taken through the given line ABC.

The point D is called *the harmonic conjugate of C in regard to A and B*. It will occur very frequently in the sequel.

There are many fundamental properties of the relation; these follow at once from the construction:

(a) It is clear from the figure that the interchange of A, B, the point C remaining the same, leads to the same point D; so that

$$(B, A)/C = (A, B)/C.$$

(b) The point C is the harmonic conjugate of D in regard to A and B; so that $C = (A, B)/D$. To see this, we have only to rename the points P, Q, R, U, C, D, calling them, respectively, R, U, P, Q, D, C. The passage from D to C is then descriptively identical with that, followed before, from C to D. Further, as A, B are interchangeable, so are C and D.

(c) Not only are C and D harmonic conjugates in regard to A and B, but A, B are harmonic conjugates in regard to C and D. For, in the original figure, let QD meet BP in P', and CP' meet AR in Q'. Then it follows, by Desargues' theorem, that P, D, Q' are in line. To see this, consider the triangles QPU and $P'Q'R$; these have the line ABC as axis of perspective; whence PQ', QP' meet in the point D in which RU meets QP'. Wherefore, applying the

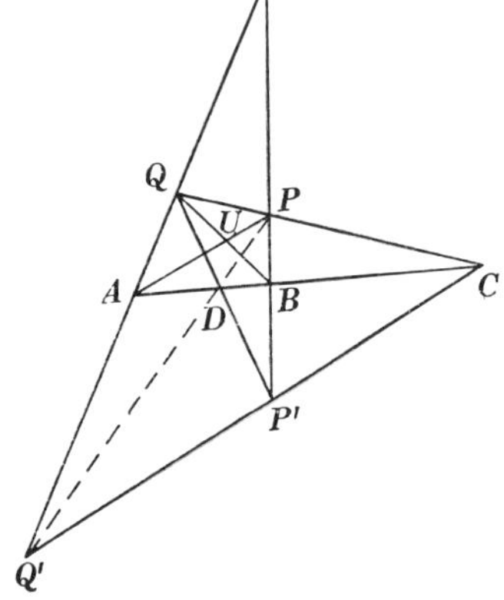

original construction to find $(C, D)/B$, we may draw, respectively through C and D, the lines CP, and DP', which meet in Q; then, through B, the line meeting CP and DP' in P and P'; then, taking the point Q' common to CP' and DP, we are to find the point in which the line QQ' meets CD. This is the point A; so that $A = (C, D)/B$.

We may sum up the relations (a), (b), (c) by writing $A, B \sim C, D$, in which A, B are interchangeable and C, D are interchangeable, and an equivalent statement is $C, D \sim A, B$. But, when this form is used, it must be remembered that $A, B \sim C, D$ and $A', B' \sim C, D$, together, do *not* involve $A, B \sim A', B'$.

(d) We have already spoken of two figures as being *in perspective from a point*, or as having *a centre of perspective*, when there corresponds to every point of one figure a definite point of the other, lying with the former point upon a line through the centre of perspective. We may likewise speak of two figures, which lie in the same plane, as having *an axis of perspective* when there corresponds to every line of one figure a definite line of the other, meeting the former line in a point lying on the line which is the axis of perspective. Also, we often speak of a set of points which lie on a line as constituting a *range* on that line; and of a set of lines, lying in one plane, which pass all through a point, as constituting a *pencil* of lines, through that point, in that plane.

With these definitions, we can prove that, if A, B, C, D and A', B', C', D' be two ranges, on two lines in the same plane, which are in perspective; and, if the former be in harmonic relation, or $D = (A, B)/C$, then also the latter are in harmonic relation, or $D' = (A', B')/C'$.

By hypothesis, the lines AA', BB', CC', DD' meet, in the centre of perspective. Suppose first that C' coincides with C, no one of A', B', D' coinciding with A, B, D. Recalling then the construction we have used to find D or D', we can take, for the point R of that construction, the centre of perspective common to the lines AA' and BB'; thence, clearly, since DD' passes through the centre of perspective, it follows that, if $D = (A, B)/C$, then also $D' = (A', B')/C'$. In a quite similar way the result follows when A' coincides with A. In general, let the lines joining the centre of perspective of the two ranges to D and B meet the line $A'C$, respectively, in D_1 and B_1. Then, as the two ranges A, D, B, C and A', D_1, B_1, C are in perspective and have C in common, we have $D_1 = (A', B_1)/C$. And, as the two ranges A', D_1, B_1, C and A', D', B', C' are in perspective and have A' in common, we have, from $D_1 = (A', B_1)/C$, the result $D' = (A', B')/C'$.

(e) The converse of (d) is true in the form: If two ranges A, B, C, D and A', B', C', D', in the same plane, have $D = (A, B)/C$ and $D' = (A', B')/C'$, and, also, the points C, C' coincide, then the lines AA', BB', DD' meet in a point. More generally, if any two corresponding points of the two ranges coincide with one another, the ranges are in perspective. This follows at once from (d).

From this it is possible to deduce that, if, on two lines, not necessarily lying in the same plane, we have two ranges A, B, C, D and A', B', C', D', both in harmonic relation among themselves, then there exists an intermediate range, of four points on a line, also in harmonic relation among themselves, with which both the original ranges are in perspective, from different centres. But we postpone the proof of this.

Another remark should be made: It is clear from the figure by which we deduce the point $D=(A, B)/C$ that, when C coincides with A or B, then D coincides with C. But, it has tacitly been assumed that D does not coincide with C for every position of C. We shall in fact, in a subsequent chapter, and for a particular purpose, assume that, if A, B be two *real* points on a line, then the other real points of the line fall into two classes which have no point in common; or, in other words, that the two points A, B define two *segments* upon the line; and then we shall add the further assumption that, if C be a real point of one of these segments, then D, $=(A, B)/C$, is a point of the other segment. This can be expressed by saying that C is *separated* from D by the points A, B; and A is separated from B by the points C, D. Such assumptions, suggested by the diagrams we can draw, belong to the topological properties of a figure of real points; they do not enter into the enunciation of the Propositions of Incidence, but are additional thereto.

(4) It is interesting, and important for the sequel, to put the geometrical theory of harmonic points into connexion with the treatment of the same matter with the help of the symbols.

Let the symbol of C, which is any point on the line AB, be expressed in terms of the symbols of A and B by the equation $(C)=\lambda\,(A)-\mu\,(B)$, where λ, μ are numbers. Recurring to the figure used in the geometrical theory, the symbol of the point P is a linear function of the symbols of R and B; if we modify properly the symbol of P, that is, write (P) for a certain $p(P)$, where p is a number, we can suppose that the symbol of P is given by $(P)=\mu(B)-\nu(R)$, where ν is a number, and μ is the same number as before. From the two equations we have

$$(C)+(P)=\lambda(A)-\nu(R),$$

where the two members are symbols of points on the lines CP and AR, respectively. We may thus suppose that the symbol of the common point, Q, of these two lines, is given by $(Q)=\nu(R)-\lambda(A)$. To find the symbol of the point U, it is necessary to find a linear function of (A) and (P), that is, of (A) and $\mu(B)-\nu(R)$, which is also a linear function of (B) and (Q), that is, of (B) and $\nu(R)-\lambda(A)$. This is evidently given by $\lambda(A)+\mu(B)-\nu(R)$, which is thus the symbol of U. We infer that the symbol of the point D, common to the lines RU and AB, is

$$(D)=\lambda(A)+\mu(B).$$

This we compare with the symbol, $\lambda(A)-\mu(B)$, of the point (C); namely, *when two points C, D are harmonic conjugates in regard to two points A, B, or C, $D \sim A$, B, the symbols of D, C, in terms of those of*

A, B, are of the forms $\lambda(A) \pm \mu(B)$. The various geometrical results proved above can thence be expressed at once; in particular, $2\lambda(A) = (D) + (C)$, $2\mu(B) = (D) - (C)$ express that

$$B = (D,\ C)/A, = (C,\ D)/A.$$

Ex. 1.   We consider the dual of the figure we used for constructing $(A,\ B)/C$. In that figure, denote the line ABC by x, the line CPQ by l, and the line RU by k. In a dual figure, suppose we are given three lines a, b, c, meeting in a point X. Take an arbitrary line r, and an arbitrary point L on the line c; then take the line p which joins the point L to the intersection $(b,\ r)$, and the line q which joins the point L to the intersection $(a,\ r)$. These determine the intersections $(a,\ p)$ and $(b,\ q)$. Let the line u, joining these two points, meet the line r in the point K. The line d, joining the points X, K, is, by the construction, the dual of the point D of the original figure. We may denote it by $(a,\ b)/c$.

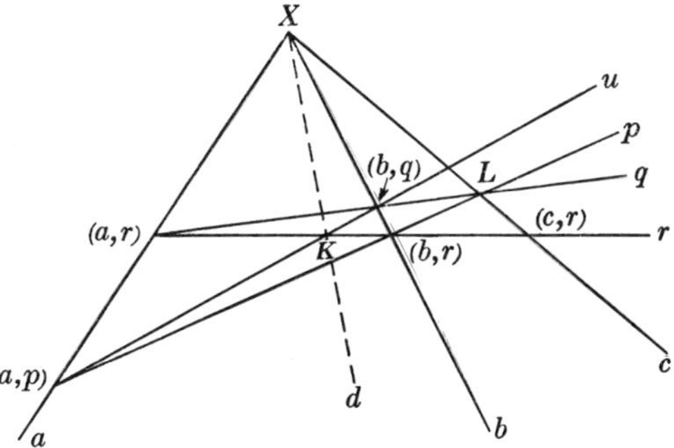

The dual figure shews that the point K, or $(d,\ r)$, is the harmonic conjugate of the point $(c,\ r)$ in regard to the two points $(a,\ r)$ and $(b,\ r)$. For the point $(a,\ p)$ functions as did the point R in the original construction, and the line c as did the line CPQ.

Four lines, a, b, c, d, meeting in a point, for which $d = (a,\ b)/c$, *are said to form a harmonic pencil of lines. They are duals, we have proved, of four points lying on a line which are in harmonic relation. They meet an arbitrary line* (r) *in four points which are in harmonic relation.* Dually, it follows that if four points on a line, which are in harmonic relation, be joined to an arbitrary point, the joining lines form a harmonic pencil.

Ex. 2.   Let x, x' be two lines in a plane. From an arbitrary point O, in this plane, let three lines, OAA', OBB', OCC', be drawn, meeting the line x respectively in the points A, B, C, and the line x' in the respective points A', B', C'. Prove that the three intersections $(BC',\ B'C)$, $(CA',\ C'A)$, $(AB',\ A'B)$ lie in a line, which passes through the common point of x and x'. This line may be called *the harmonic line of the point O in regard to the lines x, x'.*

This is the case of Pappus' theorem, already mentioned, which can be deduced from the Propositions of Incidence only.

Ex. 3.   If any line meet the sides BC, CA, AB of a triangle respectively in the points P, Q, R, prove that the symbols of these points, in terms of the symbols of A, B, C, may be taken to be

$$(P) = q(B) - r(C),\quad (Q) = r(C) - p(A),\quad (R) = p(A) - q(B),$$

where p, q, r are numbers. If then the points P', Q', R' be taken, on the respective sides, which are given by $P' = (B,\ C)/P$, $Q' = (C,\ A)/Q$,

$R' = (A, B)/R$, find the symbols of P', Q', R'; and shew that the lines AP', BQ', CR' meet in a point. Conversely, if this point be given, construct P, Q, R; and shew that these lie in a line.

It is sometimes convenient to speak of the point in which AP', BQ', CR' meet as *the pole of the line PQR in regard to the triangle ABC*; and of the line *PQR* as the polar of this point in regard to the triangle.

Ex. 4.   Let C, D be two arbitrary points in the line AB, neither of these coinciding with B. We can then suppose, appropriately absorbing multiplying numbers in the symbols of C and D, as has been explained, that the symbols of C and D, in terms of those of A and B, are $(C) = (A) + m(B)$, $(D) = (A) + n(B)$, where m, n are numbers. Then take two points X, Y, of the line, with symbols $(X) = (A) + k(B)$, $(Y) = (A) - k(B)$, in which k denotes one of the two numbers $m^{\frac12} . n^{\frac12}$, or say $(mn)^{\frac12}$. We can then verify at once that

$$(n^{\frac12} + m^{\frac12})\,(X) = n^{\frac12}\,(C) + m^{\frac12}\,(D), \quad (n^{\frac12} - m^{\frac12})\,(Y) = n^{\frac12}\,(C) - m^{\frac12}\,(D).$$

In accordance with what has been proved, these equations lead to the conclusion that *there exist two points X, Y, of the line AB, which are harmonic conjugates in regard to A and B, and are, at the same time, harmonic conjugates in regard to the other arbitrary points C, D of this line*; say $X, Y \sim A, B$ and also $X, Y \sim C, D$.

This proof, of the geometrical fact, depends on the assumption of the validity of the conventions made for the use of the symbols. The fact is important, and a geometrical proof will be given below from the theory of conics; but it is the case that no proof of the existence of X and Y can be made by linear constructions from the Propositions of Incidence. We shall not scruple however to assume, from the proof given, that the points X, Y do exist, and are expressible as here. Provided A, B, C, D are in general positions on the line, we can shew, in the same way, that there is only one such pair, X, Y.

Ex. 5.   We enunciate now a simple harmonic property of a tetrahedron. Let A, B, C, D be four arbitrary points in space, of which no three are in a line. Let the six joining lines, BC, CA, AB, DA, DB, DC, meet an arbitrary plane respectively in the points U, V, W, P, Q, R. Then take the harmonic conjugates

$$(B, C)/U, \ (C, A)/V, \ (A, B)/W, \ (D, A)/P, \ (D, B)/Q, \ (D, C)/R,$$

denoting these respectively by U', V', W', P', Q', R'; and, let the three joining lines $U'P', V'Q', W'R'$ meet the arbitrary plane respectively in the points X, Y, Z. Then prove that the three harmonic conjugates $(U', P')/X$, $(V', Q')/Y$, $(W', R')/Z$ coincide with one another.

It will be clear later that this result is equivalent with the theorem of metrical geometry that, if the mid point of each edge of a tetrahedron be joined to the mid point of the opposite edge, the three joining segments have a common mid point.

Ex. 6.   Let the symbols of three points D, E, F, lying respectively in the sides BC, CA, AB of a triangle, be given, in terms of the symbols of A, B, C, respectively, by

$$(D) = (B) + x(C), \ (E) = (C) + y(A), \ (F) = (A) + z(B),$$

where x, y, z are numbers. Prove that the necessary and sufficient condition, for the lines AD, BE, CF to meet in a point, is $xyz = 1$.

This may be obtained by expressing that there are numbers λ, μ, ν, such that the symbols $\lambda(A) + (D)$, $\mu(B) + (E)$, $\nu(C) + (F)$ represent the same point.

Shew that $xyz = 1$ is also the condition that the points L, M, N, respectively on the sides BC, CA, AB, with symbols

$$(L) = (B) - x(C), \ (M) = (C) - y(A), \ (N) = (A) - z(B),$$

should lie in a line. In effect, this example is the same as Ex. 3.

Ex. 7.   In a given plane, let O be a given point, and l a given line. Let P be any point of the plane. Let the line OP meet the line l in H, and on OH take the point P' such that $P'=(O, H)/P$. The points P, P' are called *harmonic images* of one another, in regard to the point O and the line l.

Now, let ABC be any triangle in the plane, and let the lines BC, CA, AB be called a, b, c, respectively. Of any point P of the plane, take the harmonic image Q, in regard to the point B and the line b; also take the harmonic image, R, of the same point P, in regard to the point C and the line c. It can then be proved that the line QR passes through A, and that Q is the harmonic image of R in regard to the point A and the line a.

For, let AP meet BC in D, and also let AQ, AR meet BC in Y, Z, respectively. Recall the theorem, proved above (in (3)), that if two ranges, on two lines, be in perspective from a point, and one of them be a harmonic range, so likewise is the other. Thence we prove that $Y=(C, B)/D$, and $Z=(B, C)/D$, so that Y, Z coincide; and then, taking B, or C, as centre of perspective, that $Y=(Q, R)/A$.

If the symbol of P, in terms of the symbols of A, B, C, be

$$(P)=x(A)+y(B)+z(C),$$

where x, y, z are numbers, prove that the symbols of Q, R are

$$(Q)=x(A)-y(B)+z(C) \text{ and } (R)=x(A)+y(B)-z(C).$$

(5) Construction of a sixth point on a line, given two couples of points, and a fifth point, upon the line. Couples in involution.

For the construction of the fourth harmonic point, we were given a

single couple A, B of points upon a line, and a third arbitrary point C upon the line; and we determined then a fourth point D corresponding to C. In the construction now to be considered, we are given two couples, A, B and O, U, of points on a line, and a fifth arbitrary point E; we determine a sixth point P, of the line, corresponding to E.

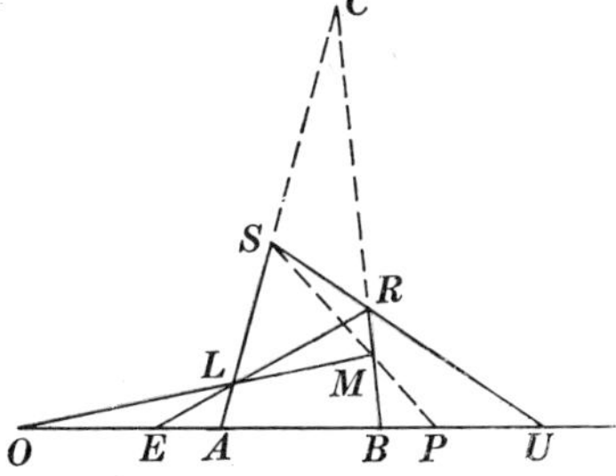

Let an arbitrary plane be put through the line. In this plane draw through A and B two arbitrary lines; on the second of these lines, let R be an arbitrary point; and let RE meet the first of these lines in L; then let OL meet the second of these lines in M, and UR meet the first of these lines in S. Let SM meet the original line in P. We are to prove that, A, B, O, U, E being given, the point P is independent of what plane is drawn through the original line, of what lines are drawn through A, B in this plane, and of what point R is taken on the second of these lines.

To prove this in a fundamental way, we take a second plane through the original line, then determine therein, as in the former plane, a quadrangle $R'S'L'M'$, and prove that the point P', in which $S'M'$ meets the original line, coincides with P.

For, the triangles $LMR, L'M'R'$ have an axis of perspective, B, E, O, the pairs of corresponding sides of these two triangles meeting in these points, respectively. Whence, by Desargues' theorem, the lines LL',

MM', RR' meet in a point. So, the triangles LSR, $L'S'R'$ are such that pairs of corresponding sides meet, respectively, in the points U, E, A; thus the lines LL', SS', RR' meet in a point. The two lines MM', SS' thus have a common point (where LL', RR' meet). It follows hence that the lines SM, $S'M'$, lying in the common plane of MM' and SS', have a common point. These two lines, lying, respectively, in the two planes drawn through the original line, can only meet on the line of points common to these two planes. This shews that $S'M'$ contains the point P in which SM meets the original line; as we wished to prove.

(6) We now consider the figure, in No. 5 preceding, with the help of the symbols. In (4) above (Ex. 4), the use of the symbols has led to two points X, Y, whose existence we now assume, which are such that both X, $Y \sim A$, B and X, $Y \sim O$, U. We prove now that also X, $Y \sim E$, P; so that P is given as $(X, Y)/E$, when X, Y have been found from A, B and O, U.

The proof of this is not as brief as could be wished; but serves as an exercise in the use of the symbols. In the figure employed, let C denote the common point of the lines AS, BR. As the points B, R, C are in a line, as also the points A, L, C, there are relations among the symbols of the points which may be written $\rho(R) = (C) + r(B)$, $\lambda(L) = (C) + l(A)$, in which ρ, r, λ, l are numbers. These lead to $\lambda(L) - \rho(R) = l(A) - r(B)$; and from this we can write the symbol of the point E, where the lines LR, AB meet, in the form $(E) = l(A) - r(B)$. Similarly, because B, M, C are in a line, we have an equation $\mu(M) = (C) + m(B)$, wherein μ, m are numbers; taken with $\lambda(L) = (C) + l(A)$, this leads to

$$\lambda(L) - \mu(M) = l(A) - m(B).$$

Thus, for the symbol of the point O, where the lines LM, AB meet, we have $(O) = l(A) - m(B)$. Again, because A, S, C are in line, we have an equation $\sigma(S) = (C) + s(A)$, in which σ, s are numbers; taken with $\rho(R) = (C) + r(B)$, this leads to the symbol of U, namely $(U) = s(A) - r(B)$. Finally, from the equations $\sigma(S) = (C) + s(A)$, $\mu(M) = (C) + m(B)$, we can put for the symbol of P the value $(P) = s(A) - m(B)$.

Now, let the equations we have found for (E), (O), (U), (P) be written in the forms

$$l^{-1}(O) = (A) + x(B), \qquad l^{-1}(E) = (A) + y(B),$$

$$s^{-1}(U) = (A) + x_1(B), \qquad s^{-1}(P) = (A) + y_1(B),$$

where $x = -ml^{-1}$, $x_1 = -rs^{-1}$, $y = -rl^{-1}$, $y_1 = -ms^{-1}$. Three of these equations determine the points O, U, E from A and B by means of the numbers x, x_1, y, respectively, the multipliers l^{-1}, s^{-1} being immaterial to the points represented. The point P is then determined from A, B by the fourth equation by means of the number y_1, which is not independent of x, x_1, y but such that $yy_1 = xx_1$.

This result includes the statement made in regard to the points X, Y which are such that X, $Y \sim A$, B and X, $Y \sim O$, U. For these points have symbols $(A) \pm (xx_1)^{\frac{1}{2}} (B)$; while the points which are harmonic conjugates both in regard to A, B and in regard to E, P have symbols $(A) \pm (yy_1)^{\frac{1}{2}} (B)$. (Cf. (4), Ex. 4 above.)

We may thus say: *If three complementary pairs of joins of four points in a plane meet an arbitrary line in the pairs of points F, F′; G, G′; and H, H′, then, there exists a pair of points X, Y, on the line, in regard to which the points of all the pairs F, F′; G, G′; H, H′ are harmonic conjugates.*

A geometrical interpretation for the points X, Y when A, B and O, U are given is as follows: Let ABC be a triangle, and O, R, S be arbitrary points respectively on the sides AB, BC, CA of this. There exist then two other triangles XML and $YM'L'$ with angular points lying respectively on the same sides AB,
BC, CA (X and Y on AB, M and M' on BC and L, L' on CA), which are such that the sides of these, LM, LX, XM, pass respectively through the given points O, R, S, and also $L'M'$, $L'Y$, YM' pass respectively through these points O, R, S. The point T, where CX meets the side $L'M'$, lies with the point T', where CY meets the side LM, on the line SR. If this figure be assumed, it is clear, because $M'S, RL'$ meet in Y, that CT meets SM' in the point $(S, M')/Y$, so that $X, Y \sim A, B$; and also, if SR meet AB in U, by projecting the range on $M'S$, from T, on to

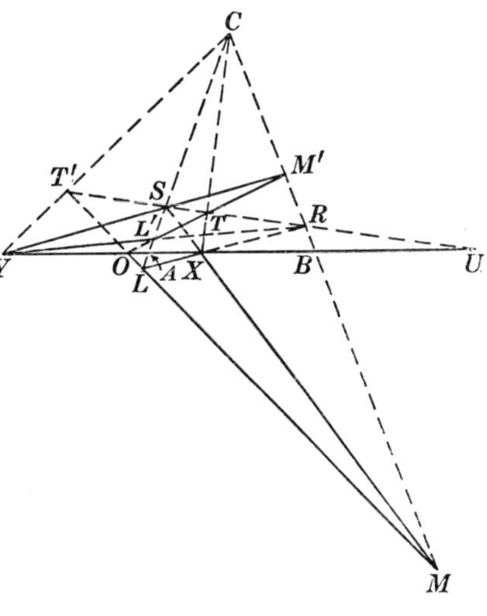

AB, that $X, Y \sim O, U$. The figure thus constructs the points X, Y which are in this double relation with the arbitrary points A, B and O, U of the fundamental line. When X, Y are thus found, the position of P, corresponding to an arbitrary E, would be given by $P = (X, Y)/E$; in particular P would coincide with E when this is at X or Y. But, the only proof of this interpretation open to us at this stage is with the help of the symbols.

It is important to notice that the symbols give information which is not at once apparent from the geometrical construction we have followed to obtain the point P from the points A, B, O, U, E. For, the symbols shew that the same point P is obtained from the given point E when, keeping A, B the same, the points O, U are interchanged; or, more generally, that A, B may be interchanged, with or without the interchange of O and U. The symbols also shew that, with the same points A, B, O, U, if E be taken where P is found to be, then P falls where E was; that is, that the points E, P may be interchanged.

These results follow geometrically however from the construction given if, in addition to the Propositions of Incidence, we use Pappus' theorem, as follows:

(*a*) When O, U are interchanged, the construction given leads to new points S', M', on the lines AS, BR, respectively. The fact that SM, $S'M'$ meet AB in the same point, P, follows by considering the

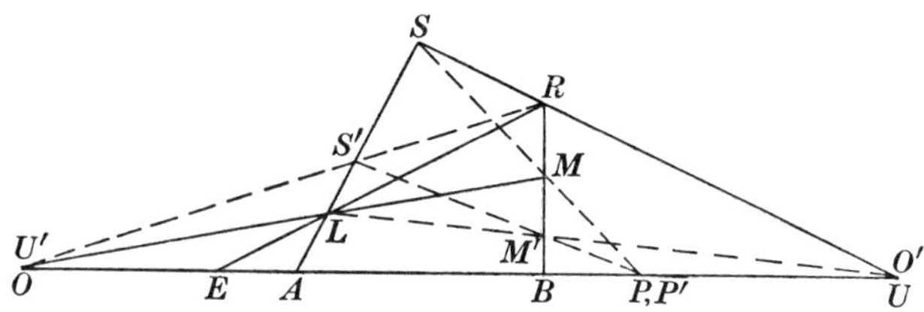

two triads of points L, S', S and R, M, M', each on a line, of which the joins LM, RS' meet in O, the joins LM', RS meet in U, and, hence, by Pappus' theorem, the joins SM, $S'M'$ meet on AB.

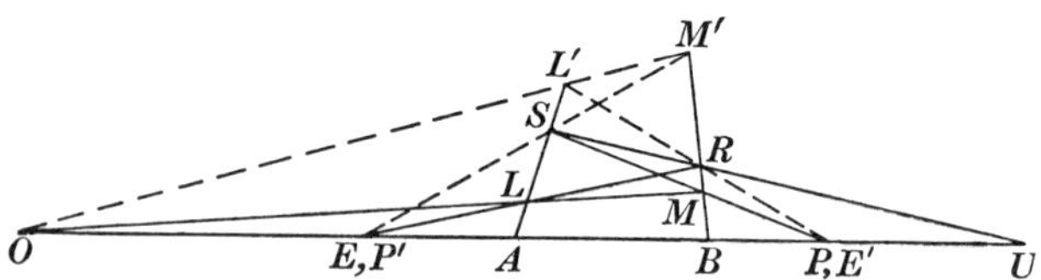

(*b*) Similarly, if E be taken at P, by considering the two triads S, L', L and R, M, M', since SM, RL' meet in P, and LM, $L'M'$ meet in O, it follows that SM', RL meet on AB, in E.

Ex. 1. Let four planes in space have a common line, t; and let these planes be met by two other lines a, b, which do not meet the axis t, respectively, in the points A, A', X, X' and B, B', Y, Y'. If then $A, A' \sim X, X'$, it is true that also B, $B' \sim Y$, Y'.

For, take two arbitrary points P, Q, on the axis t, and let the planes (P, a) and (Q, b) meet in the line c; also, let the line c meet the four planes, respectively, in C, C', Z, Z'. Then the two ranges (A, A', X, X'), (C, C', Z, Z') are in perspective from P, and the two ranges (B, B', Y, Y'), (C, C', Z, Z') are in perspective from Q. Hence the result follows from (3), section (*d*), preceding.

Ex. 2. Let n, n', x be three lines in space, of which no two have a common point. It has been noticed already (Chap. II (4)) that, from any point X, on the line x, a line can be drawn to meet both the lines n, n', say in A, A', respectively. On this transversal line, let the point X' be taken, given by $X' = (A, A')/X$. Then, as X varies on the line x, the locus of X' is a fourth line.

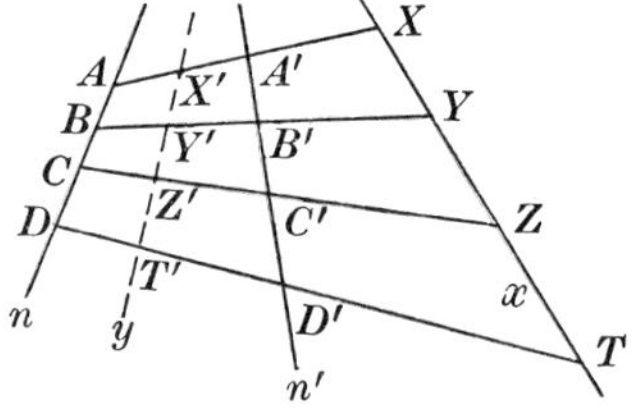

For, draw, from a second point Y, on the line x, the transversal Y, B, B', meeting the lines n, n', respectively, in B and B'; and, from a third point Z, on the line x, draw the transversal meeting n, n' in C, C', respectively. Then, from X', draw the transversal line to meet the lines BB', CC', respectively, in Y', Z'. Denote this last transversal by y. Now consider the four planes, all containing the line $CZ'C'Z$, which are defined as containing, respectively, also the lines n, n', x and y. These planes are met by the lines AA', BB' respectively in the points A, A', X, X' and in B, B', Y, Y'. Thus, from A, $A' \sim X$, X', we infer, by the preceding example, that B, $B' \sim Y$, Y'. Conversely then, if Y' be defined on the line BB' by $Y' = (B, B')/Y$, it follows that the line $X'Y'$ meets any transversal such as C, C', Z. And hence, that it meets this in a point Z' given by $Z' = (C, C')/Z$. For, we have only to repeat the preceding argument with four planes, containing the line AA', and, respectively, also the lines n, n', x, y. This proves the result which has been stated.

This theorem forms an easy illustration of the use of the symbols. As above, draw from the points X, Y, Z of the line x, the transversals XAA', YBB', ZCC' of the lines n, n'. On the first two of these take the points $X' = (A, A')/X$ and $Y' = (B, B')/Y$, respectively. As the lines n, n' do not lie in a plane, the symbols of the points A, A', B, B' are linearly independent. The symbols of X, Y being, respectively,

$$(X) = h(A) + h'(A'), \quad (Y) = k(B) + k'(B'),$$

where h, h', k, k' are numbers, the symbols of X', Y', by a result proved above, will be

$$(X') = h(A) - h'(A'), \quad (Y') = k(B) - k'(B').$$

The symbol $(Z) = p(X) + q(Y)$, of any third point Z on the line x, wherein p and q are numbers, is then

$$(Z) = [ph(A) + qk(B)] + [ph'(A') + qk'(B')].$$

In this, $ph(A) + qk(B)$ is the symbol of a point on the line n, say C, and $ph'(A') + qk'(B')$ is the symbol of a point on the line n', say C'; thus $(Z) = (C) + (C')$. There is, however, a point on the line $X'Y'$ with symbol $p(X') + q(Y')$; if we substitute in this the forms found above for (X') and (Y'), we see that this point has the symbol $(C) - (C')$. This is then the point on the transversal ZCC' which is given by $(C, C')/Z$; and it lies on the line $X'Y'$. This again proves the theorem.

It would only be a change of notation for the symbols of A, A', B, B' to suppose the numbers h, h', k, k' all equal to 1; and this would simplify the formulae.

Ex. 3. Let the sides BC, CA, AB of a triangle be met by a line respectively in the points L, M, N; let D be a further point, not necessarily in the plane of the triangle. On the lines DL, DM, DN, respectively, let A', B', C' be arbitrary points. The lines $B'C'$, $C'A'$, $A'B'$ lie in the plane $DLMN$; let them meet the line LMN respectively in the points L', M', N'. We can then prove that the lines AL', BM', CN' meet in a point, whatever be the points A', B', C' taken on DL, DM, DN.

In terms of the symbols of A, B, C we can suppose, without loss of generality (Ex. 3, (4) above), that the symbols of L, M, N are, respectively, $(L) = (B) - (C)$, $(M) = (C) - (A)$, $(N) = (A) - (B)$; thence the symbols of A', B', C' are, respectively, by absorbing suitable numerical multipliers in these,

$$(A') = l(D) + (B) - (C), \quad (B') = m(D) + (C) - (A), \quad (C') = n(D) + (A) - (B),$$

where l, m, n are numbers. The symbol of L', wherein $B'C'$ meets MN, is then such a linear function of (B') and (C') as is also a linear function of (M) and (N); this is thus given by

$$(L') = n(B') - m(C') \quad \text{or} \quad n\,[(C) - (A)] - m\,[(A) - (B)],$$

which is $$(L') = l(A) + m(B) + n(C) - (l + m + n)\,(A).$$

Similar forms hold for (M') and (N'). These shew that the lines AL', BM', CN' meet in the point, say D', whose symbol is $l(A) + m(B) + n(C)$.

We thus obtain a result stated by Moebius (*Werke*, i, p. 441), namely: *If A, B, C, D be a tetrahedron in space, it is possible to construct another tetrahedron of which the angular points A', B', C', D' lie respectively on the faces BCD, CAD, ABD, ABC of the former, while A, B, C, D lie respectively on the faces B'C'D', C'A'D', A'B'D', A'B'C' of the latter.*

Without loss of generality, as the reader may shew, it is possible to suppose the symbols of the points A', B', C', D' expressed in terms of those of A, B, C, D by equations of the forms

$$
\begin{aligned}
(A') &=   r(B) - q(C) + p(D), \\
(B') &= -\, r(A) + p(C) + q(D), \\
(C') &= q(A) - p(B) + r(D), \\
(D') &= -\, p(A) - q(B) - r(C),
\end{aligned}
$$

wherein the numerical coefficients p, q, r are such as to form a skew symmetrical matrix; and any such four equations determine a suitable tetrahedron $A'B'C'D'$, the two ratios of p, q, r being arbitrary.

When the point D is in the plane of A, B, C we have the proposition of plane geometry: *Let A, B, C and A', B', C' be two triangles in a plane; let k be a line in the plane such that the lines joining A', B', C', respectively, to the points where the line k meets the sides BC, CA, AB meet in a point; then the lines joining A, B, C, respectively, to the points where the line k meets the sides B'C', C'A', A'B' also meet in a point.*

In particular, if the lines through A', B', C', respectively parallel to BC, CA, AB, meet in a point, so do the lines through A, B, C, respectively parallel to $B'C'$, $C'A'$, $A'B'$.

The general result is in fact contained in the theorem of (5) above. For we have a quadrangle, A', B', C', D, whose pairs of complementary joins meet the line in the pairs of points L, L', M, M', N, N'; let D' be defined as the common point of $M'B$ and $N'C$; then the consideration of the pairs of complementary joins of the four points A, B, C, D' shews that $D'A$ passes through L'. As is there indicated, the geometrical proof, independent of the use of the symbols, cannot be completed without use of Pappus' theorem.

Ex. 4. We have emphasized the fact that we cannot prove all the results of plane geometry in a geometrical manner unless we add to the Propositions of Incidence some geometrical theorem of incidence which is independent of these. We have said that the theorem of Pappus would be such a sufficient geometrical assumption. There is a very simple result in the geometry of space, equally incapable of proof by the Propositions of Incidence alone, which could be assumed, and would be sufficient. The theorem of Pappus is deducible from this; and, conversely, this result can be deduced geometrically from the theorem of Pappus.

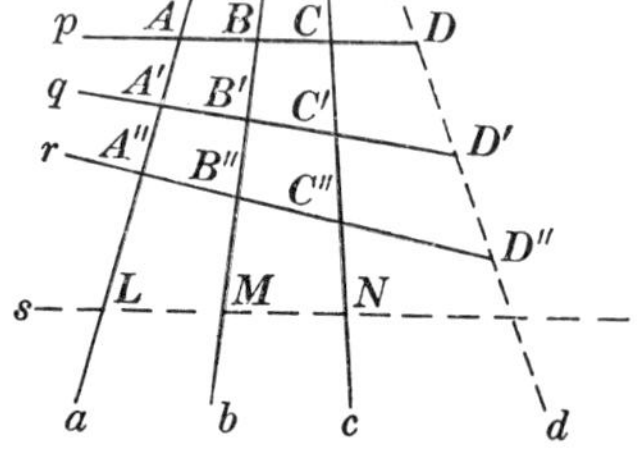

We give here a statement of this geometrical result, and a proof of it based on the use of the symbols. For the equivalence of this result with Pappus' theorem the reader may refer to *Principles of Geometry*, Vol. i, p. 49.

Let a, b, c be three lines in space, of which no two have a common point. From any point of either of these lines it is possible to draw a line to meet the other two lines. There exists therefore an infinite number of lines each meeting all of the lines a, b, c. Let p, q, r be three such transversals of a, b, c; because no two of a, b, c have a common point, it follows that no two of p, q, r have a common point. Let the lines p, q, r meet the line a, respectively, in A, A', A''; meet the line b, respectively, in B, B', B''; and meet the line c in C, C', C''.

The theorem to be considered is: *If s be any line, besides p, q, r, which meets all of a, b, c; and d be any line, besides a, b, c, which meets all of p, q, r, then the lines s, d have a common point.* Let the line s meet the lines a, b, c, respectively, in the points L, M, N; and the line d meet the lines p, q, r, respectively, in the points D, D', D''. As the points A, B, A', B' are not in one plane, the symbol of C'' can be expressed in terms of those of A, B, A', B'; and we can suppose the numerical multipliers of A, B, A', B' to be so chosen that the expression is $(C'') = (A) + (B) + (A') + (B')$. This equation shews that the point C'' is on the line joining the two points whose symbols are $(A) + (B)$ and $(A') + (B')$; from the forms of these symbols, these points are, respectively, on the lines AB and $A'B'$, or p and q; and only one transversal can be drawn to p and q from the point C''. This proves that these two points are the points C and C', whose symbols are therefore $(C) = (A) + (B)$ and $(C') = (A') + (B')$. By like reasoning we infer that the symbols of A'' and B'' are $(A'') = (A) + (A')$, and $(B'') = (B) + (B')$.

Now consider the point D''. It is on the line $A''B''$, and its symbol is therefore of the form $(D'') = \lambda(A'') + \mu(B'')$, where λ, μ are numbers. This, however, is the same as $(D'') = \lambda(A) + \mu(B) + \lambda(A') + \mu(B')$, wherein $\lambda(A) + \mu(B)$ and $\lambda(A') + \mu(B')$ are symbols of points lying, respectively, on the lines AB and $A'B'$, or p and q. Thus, as before, the symbols of D and D', where the transversal from D'' meets p and q, are given by $(D) = \lambda(A) + \mu(B)$ and $(D') = \lambda(A') + \mu(B')$; while $(D'') = (D) + (D')$.

Next consider the point N. Being on the line CC', this point has a symbol of the form $(N) = n(C) + n'(C')$, where n, n' are numbers; and, by what has been shewn, this is $(N) = n(A) + n'(A') + n(B) + n'(B')$. As before, therefore, we infer that the symbols of L, M are $(L) = n(A) + n'(A')$, $(M) = n(B) + n'(B')$, while $(N) = (L) + (M)$.

From the results now obtained we have

$$n(D) + n'(D') = n\,[\lambda(A) + \mu(B)] + n'\,[\lambda(A') + \mu(B')],$$

which is the same as $\quad \lambda\,[n(A) + n'(A')] + \mu\,[n(B) + n'(B')]$,

or $\qquad\qquad\qquad\qquad\qquad \lambda(L) + \mu(M)$.

But $n(D) + n'(D')$ and $\lambda(L) + \mu(M)$ are symbols of points lying, respectively, on the lines d and s. Hence these two lines have a point in common; as we wished to shew.

Ex. 5. We may prove here the following result also: If four lines be arbitrarily taken in space, of which no two have a common point, there exist in general two lines which meet all four. It may happen that these two coincide with one another. When the four given lines are met by three other lines, there is an infinite number of lines which meet the four, as we have just shewn for a, b, c, d.

To simplify the statement of the proof, we may suppose three of the four given lines to be the lines a, b, c of the preceding example (4), and the fourth line to be that joining two general points P and P_1. We may suppose the symbols of P and P_1 to be expressed in terms of the symbols of the four independent points A, B, A', B' in the respective forms

$$(P) = a(A) + a'(A') + b(B) + b'(B'), \quad (P_1) = a_1(A) + a_1'(A') + b_1(B) + b_1'(B'),$$

where a, a', b, b' and a_1, a_1', b_1, b_1' are numbers. By what we have seen in the preceding example, a general line meeting the lines a, b, c is that which joins two points having the symbols $n(A) + n'(A')$ and $n(B) + n'(B')$. For this line to contain a point of the fourth line, say the point of symbol $(P) + \xi(P_1)$, where ξ is a number, or

$$(a + \xi a_1)(A) + (a' + \xi a_1')(A') + (b + \xi b_1)(B) + (b' + \xi b_1')(B'),$$

it is clearly necessary and sufficient that the multipliers of (A) and (A') in this should have the ratio of the multipliers of (A) and (A') in $n(A)+n'(A')$, and the multipliers of (B) and (B') should have the ratio of the multipliers of (B) and (B') in $n(B)+n'(B')$. This requires

$$(a'+\xi a_1')/(a+\xi a_1)=n'/n=(b'+\xi b_1')/(b+\xi b_1).$$

These two equations give two possible values for the number ξ; and to each of these belongs a value of n'/n. The point where the transversal meets the line PP_1 has then a symbol $n(A)+n'(A')+x\,[n(B)+n'(B')]$, where $x=(b+\xi b_1)/(a+\xi a_1)$. This proves the result enunciated.

CHAPTER V

THE FUNDAMENTAL HYPOTHESIS
MADE IN THIS VOLUME

THE reader will by now have become sufficiently familiarised with the use, and the power, of the symbols which we have used to represent points. It seems proper to make a more formal statement, in general terms, of the place we assign to these symbols in the present volume.

We suppose a system of objects, each represented by a symbol, such as (A), with the understanding that, if p be any number, real or complex, other than zero, subject to the distributive, associative, and commutative laws of ordinary algebraical analysis, the symbol $p\,(A)$ shall equally represent the object. We suppose that an expression of the form $p\,(A)+q\,(B)+\dots$, of a limited number of terms, in which p, q, ... are numbers, equally represents one of the objects considered, with the understanding that two objects, among those considered, which are both represented by expressions of this form, with the same number of terms, are the same object if the ratios of the numbers p, q, ..., in one of these expressions, are the same as the ratios of the corresponding numbers in the other expression. It is also understood that the associative, distributive and commutative laws apply to the terms, in (A), (B), ..., of such an expression; so that, if k be a number, the expression $k\,[p\,(A)+q\,(B)+\dots]$ is the same as $kp\,(A)+kq\,(B)+\dots$.

In particular, we say, at first, that the objects represented by all expressions of the form $x\,(A)+y\,(B)$, in which (A), (B) remain the same, but x, y are different numbers, constitute a *linear aggregate*. The symbol of any particular object C of this aggregate, other than A or B, is then of the form $(C)=x_0\,(A)+y_0\,(B)$, in which x_0, y_0 are particular numbers, neither of which is zero. Thence we can express (B) linearly in terms of (A) and (C), with definite numerical coefficients, and any object of this linear aggregate has a symbol representable in terms of (A) and (C). Wherefore, the linear aggregate consists of an infinite number of objects, for which the symbol of any one can be expressed by the symbols of any two of the objects of the aggregate. But, in the expression, $(P)=x\,(A)+y\,(B)$, for any object of the aggregate in terms of (A) and (B), there is an indefiniteness, arising from the fact that, if p, q be any numbers, the symbols $p\,(A)$ and $q\,(B)$ represent the same objects as do (A) and (B) respectively. This indefiniteness we avoid by supposing the values of x and y to be assigned for a third object of the linear aggregate, which, like A and B, is a definite given object of the aggregate. When the values of the numbers x, y for this third object are assigned, the ratio of the unspecified multipliers p, q, arising for (A) and (B), is fixed; and hence, in the expression $x(A)+y(B)$ for any object (P), the ratio of x to y is definite. As the object represented by $k(P)$, where k is a number, is the same as that represented by (P), the object represented by $x(A)+y(B)$, when the ratio of x to y is definite, is a definite object of the aggregate.

Similarly, we define what, for the moment, we call a *planar aggregate*, depending on *three* objects. The symbol for any object of this planar aggregate is of the form $x(A)+y(B)+z(C)$, where (A), (B), (C) represent three objects of the planar aggregate which do not belong to a linear aggregate, and x, y, z are numbers. Then, if (D) be the symbol of any object of the planar aggregate which does not belong to the linear aggregate defined by (A) and (B), we shall have an equation of the form $(D)=x_0(A)+y_0(B)+z_0(C)$, in which x_0, y_0, z_0 are numbers, of which z_0 is not zero. From this, (C) can be expressed linearly in terms of (A), (B) and (D), and, hence, the symbol of *any* object of the aggregate can be expressed in terms of (A), (B) and (D) instead of in terms of (A), (B) and (C). Thus the planar aggregate depends on any three of its members which do not belong to a linear aggregate. And, further, it is clear, that all members of the linear aggregate which is determined by any two members of the planar aggregate, belong to the planar aggregate. It is supposed, in what has been said, that the symbols (A), (B), (C) are a definite choice from among the respective symbols $p(A)$, $q(B)$, $r(C)$, where p, q, r are numbers, all of which represent the same objects, respectively, as (A), (B), (C). This choice is fixed if we assign the values of x, y, z, in the expression $x(A)+y(B)+z(C)$, for a fourth given definite object of the aggregate. Then, in the expression of all other objects of the aggregate, in terms of (A), (B), (C), the ratios of x, y, z are definite.

We can then proceed further, defining what, for the moment, we may call a *spatial* aggregate, depending on four objects which do not lie in a planar aggregate. With the assumption that two objects which have the same symbol, or symbols differing only by a numerical multiplier, are the same object, we can investigate theorems as to objects which are common to aggregates contained in the spatial aggregate. As, for instance, we can prove that a linear aggregate, constructed from two objects of the spatial aggregate, has a member which also belongs to a planar aggregate, constructed from three members of the spatial aggregate (unless all the members of the linear aggregate belong to the planar aggregate). The proof depends on the assumption that the symbols of any five members of the spatial aggregate are connected by a linear equation with numerical coefficients (or, by several such equations).

We can obviously use the words *point*, *line*, *plane* and *space* for the words *object*, *linear aggregate*, *planar aggregate* and *spatial aggregate*, respectively, provided that the laws assigned for the symbols lead to the same rules of operation and combination, for these points, lines, etc., as arise from our geometrical conceptions of the properties of points, lines, etc., as usually named. These geometrical conceptions, founded on our experience of the world, are primarily for what we call the *real* points, lines, etc. For these real elements, we assume here that all the geometrical relations can be deduced from certain Propositions of Incidence, taken with one other proposition or geometrical theorem, which is conveniently taken to be that which we have called Pappus' theorem. The method of deduction of the geometrical relations is by the operation of *joining* (two points by a line, three points by a plane, etc.), and the use of *section* (of two lines in a point, of two planes in a line, etc.). We seek to establish an exact correspondence between the realm of *real* points, and the realm of objects defined by symbols, as above, in such a way that every geometrical operation is represented by a statement for the symbols of the objects, and conversely. It may be sufficiently obvious from what has been already given in the preceding chapters, including the proof given for Pappus' theorem by use of the symbols, that any geometrical theorem has a counterpart in the relations of the symbols of the objects. It is not so clear, perhaps, that the converse is true; it may be thought that there can exist relations for the symbols whose geometrical interpretation would lead to geometrical statements which are not true; for example, to take an extreme case, it may be feared that two points of a geometrical figure which are different may, by applying the rules of operation for the symbols, come to be represented by objects whose symbols are the same (or differ only by a numerical multiplier). This does not seem the place to enter upon a formal proof that such failures of correspondence, between the two realms, do not in fact arise; such a proof, to be convincing, would need to be given in great detail. Moreover, an *a posteriori* proof is furnished by the success of the following developments. Instead, we introduce what, if desired, may be regarded only as a working hypothesis, to be tested by experience. *The fundamental hypothesis we make in this volume is, that the correspondence, between the realm of real points, and their geometrical relations, on the one hand, and the realm of objects, as represented by their symbols, and the relations of these, on the other hand, is exact in both senses.*

But, there is very strong reason for the use of the symbols which is independent of this. The remarkable work of Poncelet, and his successors, has abundantly shewn that it is necessary, even for a complete development of the geometry of real elements, to regard these real elements as part of a wider realm of so-called *imaginary* or *ideal* elements. In this volume, these ideal elements, and their relations, are regarded as being *defined*, so as to be in exact correspondence with the symbols which represent them, and the relations of these symbols. The fundamental laws for the symbols are so enunciated as to apply equally to ideal and to real elements. For instance, there is no statement in these fundamental laws, as so far given, which implies that the points of a line can be arranged *in order*; this property, which can be assigned to the *real* points of the line, by taking account of certain topological properties of the plane, does *not* hold for the ideal points. A great logical advantage of the treatment followed in this volume, by means of the symbols, appears to lie in the fact that the ideal elements, of which the real elements form only part, are, from the first, given the importance which belongs to them, instead of being introduced only as suggestions arising from the theory of the real elements.

Remark. The history of the early efforts of mathematicians to deal with the fact that the roots of an algebraical equation, with real coefficients, may not be real; and the reluctance, at a much later date, to accept a proof of a property of real numbers, which is conducted with the use of imaginary numbers, forms an interesting study. Reference to many of the early papers, beginning with Wallis (1685), will be found in an article by Cayley, with the title "Multiple Algebra" (*Quart. Journ. of Math.* June 1887, p. 270; *Papers*, xii, pp. 459–489). It may be that only after Cauchy's use of imaginary numbers, early in the nineteenth century, to establish physical results, was tolerance of such imaginary numbers seen to be inescapable, though Gauss' work on complex integers must have had its influence. It was later that it was recognised that the numbers necessary in Analysis are primarily such imaginary (or complex) numbers, the real numbers being a particular case of these.

After Descartes' suggestion (1637), of computing the properties of geometrical curves from their algebraic equations*, it was inevitable that, when complex numbers had obtained recognition in algebra, the question of imaginary points should arise in geometry. Poncelet's work (1822) placed the utility of taking account of such points beyond any doubt; apart from his discovery of the "imaginary points at infinity", for circles and spheres, he has constantly in mind, in his great treatise (*Propriétés projectives*, 1822; 2nd Ed. 1865), the "ideal" intersections of lines and curves. For instance, having proved a theorem for two circles which have two real common points, he at once enquires whether the same result remains true for two circles which meet in "ideal" points. He is led to assume a "principle of continuity" which, reliable in his hands, may seem to a reader somewhat indefinite, and introduced only to save the invention of a proof requiring the use of only real points.

Confessedly, imaginary points are more abstract than real points— just as negative numbers in arithmetic are more abstract than positive numbers. And the location of imaginary points in a plane of real points cannot be shewn in a diagram. Their definition must, in the first place, be by shewing how to use them precisely. In this volume, this precision is obtained by the use of the symbols.

---

* With Descartes (1596–1650) it is right to name Fermat (1601–1665); and also Desargues (1593–1662), and Pascal (1623–1662).

CHAPTER VI

THE SYMBOLS OF THE REAL POINTS OF A LINE

(1) The real points of a line have a property not belonging to the complex (or imaginary) points, in that there is associated with our conception of the real points an idea of "betweenness", and of the order in which these points are arranged on the line. It is to be expected then that, if we take account of this, the symbols which represent the real points will have a corresponding property.

Let two definite real points of the line have the symbols (O), (U). These points are equally represented by $p(O)$, $q(U)$, in which p, q are any numbers. Let the indefiniteness so arising be removed by assigning a third definite real point E of the line, and choosing the ratio p/q so that the symbol for this third point is $(E) = (O) + (U)$—or any numerical multiple of this. Then, by the laws we have imposed for the symbols, any symbol $(O) + \lambda(U)$, where λ is a number, represents a definite point of the line; and, conversely, λ is definite when this point is given. Conventionally, $\lambda = \infty$ gives the point U. The points so arising for real values of λ we may call real points of the line. But we introduce real points also by a different definition, obtained by real constructions from other real points; we must shew that points so arising correspond to real values of λ. For this, we utilise properties of *order* among the real points; these are derived ultimately from our observations, especially of diagrams; such properties we regard here as axioms. The result we arrive at is that, *all the real points of the line, in their natural order, correspond, each to each, to the real values of λ, taken in order, from $\lambda = -\infty$ to $\lambda = +\infty$, the symbol for any point being $(O) + \lambda(U)$.*

We make very little use of this result in the succeeding chapters, though it is interesting in itself, and becomes especially so in the chapter below (Chap. xv), in which *Distance* is defined; and, the argument by which we reach the result is, in some steps, only sketched. The reader, having appreciated the general trend, may prefer to pass over the details for the present.

(2) We assume that two points, P, Q, of a line, regarded as real points, determine *two segments* of the line, having no real points in common, save the points P, Q. If R be a further real point of the line, belonging to either segment, we say that the three points P, R, Q, considered in the order of the letters, determine an *order* on the line. There is then meaning in the statement that four real points of the line, points P, R, Q, S, are *in order*. Namely, this means that the point S, and the point R, are in different segments PQ, or, that S is not in that segment PQ which contains R. In this case we also say that R, S are *separated* by P and Q, and that P, Q are separated by R and S. If, as in the dis-

cussion of Euclid's theory of parallel lines, in Chapter I, we recognise the *inaccessible* point of a line, the two segments determined by two points P, Q of the line (neither of which is at the inaccessible point), may be distinguished by the fact that one of these segments contains the inaccessible point.

(3) For what follows we shall make use of two assumptions, or axioms, suggested by the diagrams we draw:

(*a*) If two real points R, S of a line be harmonic conjugates of one another in regard to two other real points P, Q of the line, or $S = (P, Q)/R$, then we assume that R and S are separated by P and Q (so that P and Q are separated by R and S).

(*b*) If the points P, S, Q, R remain in the harmonic relation, and P, Q are fixed, but R, and therefore S, move on the line, then, to put the statement in an informal but readily intelligible way, *the points R, S move in opposite directions on the line.*

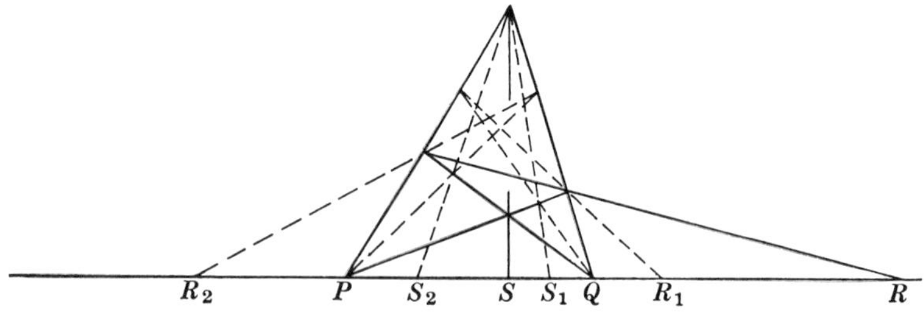

More precisely, if R_1 is in that segment QR which does not contain P, then the point S_1, given by $(P, Q)/R_1$, is in that segment SQ which does not contain P. And, if R_2 be in the segment PR which does not contain Q, then S_2, given by $S_2 = (P, Q)/R_2$, is in that segment PS which does not contain Q. When R_1 is at Q, so is S_1; and when R_2 is at P, so is S_2.

(4) Now suppose three definite real points, O, U and E, are given on the line, wherein, as has been explained, the symbols are so chosen that $(E) = (O) + (U)$. Further, for simplicity, let us denote the symbols (O), (U), (E) simply by O, U, E. The position of the point E is quite arbitrary; but, if we wish, we may call that segment OU, in which E lies, the *interior* segment OU; and, in drawing diagrams, we may suppose that E is separated by O and U from what, in Chapter I, was called the inaccessible point of the line; these conventions are not essential, but may serve to make the following argument clearer.

For the symbol of any point of the line we may take $(O) + \lambda (U)$, or, as we propose to write, $O + \lambda U$, or any numerical multiple of this; where λ is a number. This number is zero when the point is at O, it is unity when the point is at E, and, conventionally, it is ∞, or infinite, when the point is at U. For any two real points of the line which are harmonic conjugates of one another in regard to O and U, the values of λ are equal but of opposite sign (Chap. IV, (4)). By assumption (*a*), these two conjugate points are in different segments OU. In particular O, U are harmonic conjugates of one another in regard to E, and the point given by $E_1 = (O, U)/E$; and, as the symbol of E is $O + U$, the symbol of E_1 is $O - U$.

We desire to prove, first, that for all real points in that segment OU which contains E, the number λ is real and positive; then λ will be real and negative for all real points in that segment OU which contains E_1. But, further, considering the two points which have symbols $O + \lambda U$ and $\lambda O + U$, and noticing that

$$(\lambda O + U) + (O + \lambda U) = (\lambda + 1)\, E, \quad (\lambda O + U) - (O + \lambda U) = (\lambda - 1)\, E_1,$$

we see that the points represented by $\lambda O + U$ and $O + \lambda U$ are harmonic conjugates in regard to the points E and E_1. Hence, as O, U are harmonic conjugates in regard to E and E_1, it follows, by assumption

$\overline{\quad\underset{O}{\;}\quad\;\underset{E}{\overset{\lambda O + U}{\;}}\quad\;\underset{U}{\overset{O + \lambda U}{\;}}\quad\;\underset{E_1}{\;}\quad}$

(b), that if $O + \lambda U$ be the symbol of a real point in that segment EU which does not contain O, then $\lambda O + U$ is the symbol of a real point in that segment OE which does not contain U. Remarking, then, that the point of symbol $\lambda O + U$ is equally represented by the symbol $O + \lambda^{-1} U$, we infer that, if we can shew that all real points in that segment EU which does not contain O have symbols $O + \lambda U$, in which λ is real and $1 \leqslant \lambda \leqslant \infty$, then it will follow that all real points, in that segment OE which does not contain U, have symbols $O + \mu U$, wherein μ is real and $0 \leqslant \mu \leqslant 1$. We consider then, particularly, that segment EU which does not contain O.

(5) In regard to this, remark that if Q, P be real points in this segment, for which E, Q, P, U are in order, and we take the real point Q' given by $Q' = (P, U)/Q$, then, by assumption (a), the points E, Q, P, Q', U are in order. If the symbols of Q, P, Q' be $O + qU$, $O + pU$,

$\overline{\underset{O}{\;} \;-\;-\;-\;-\; \underset{E}{\;}\;\;\underset{}{\overset{Q}{\;}}\quad\;\underset{P}{\;}\quad\;\underset{}{\overset{Q'}{\;}}\quad\;\underset{U}{\;}\;}$

$O + q'U$, respectively, then we can prove that $q' = 2p - q$. For, from $O + qU = O + pU - (p - q)\, U$, we infer $O + q'U = O + pU + (p - q)\, U$. Thus, if it were proved that p, q are positive real numbers with $p > q$, it would follow that q' is a positive real number, with $q' > p$.

To apply this, take the point F_1 given by $F_1 = (E, U)/O$, which lies in that segment EU which does not contain O; since $O = E - U$, the

$\overline{\underset{O}{\;} \;-\;-\;-\;-\;-\; \underset{E}{\;}\;\;\underset{}{\overset{F_1,\, F_2}{\;}}\qquad\qquad\;\underset{U}{\;}\;}$

symbol of F_1 is $E + U$, or $O + 2U$. Then take the point F_2, given by $F_2 = (F_1, U)/E$, which lies in that segment $F_1 U$ which does not contain E; since $E = F_1 - U$, the symbol of F_2 is $F_1 + U$, or $O + 3U$. And, in general, take the point F_h, given by $F_h = (F_{h-1}, U)/F_{h-2}$, which lies in that segment $F_{h-1} U$ which does not contain F_{h-2}. The process shews that $F_{h-2} = F_{h-1} - U$, and hence $F_h = F_{h-1} + U$, and that the symbol of F_h is $O+(h+1)\,U$. We thus have a series of real points, in order upon the line,

$$E,\ F_1,\ F_2,\ \ldots,\ F_{h-1},\ F_h,\ \ldots,\ U,$$

with symbols $O+\lambda U$, in which the values of λ are a series of ascending positive integers.

From this it follows, after what has been said, that in that segment OE which does not contain U, we have a series of real points, also in order upon the line,

$$E,\ G_1,\ G_2,\ \ldots,\ G_{k-1},\ G_k,\ \ldots,\ O,$$

with symbols $E=O+U$, $G_1=2O+U$, ..., $G_k=(k+1)\,O+U$, ..., in which k is a positive integer.

(6) Now, denote a point of symbol $mO+nU$, in which m and n are positive integers, by $P_{m,n}$. Taking m arbitrarily we see at once, as above, that

$$P_{m,n+1}=(P_{m,n},\ U)/P_{m,n-1},$$

and that

$$O,\ P_{m,1},\ P_{m,2},\ \ldots,\ P_{m,n-1},\ P_{m,n},\ P_{m,n+1},\ \ldots,\ U.$$

are in order upon the line. In particular $P_{m,m}$, or $m\,(O+U)$, is the point E; and the points

$$E,\ P_{m,m+1},\ P_{m,m+2},\ \ldots,\ P_{m,n},\ P_{m,n+1},\ \ldots,\ U,$$

where $n>m$, are in order. Therefore, if p, q be positive integers, with $q/p>n/m$, the points E, $P_{m,n}$, $P_{p,q}$, U are in order; for we can suppose the symbols of $P_{m,n}$ and $P_{p,q}$ to be, respectively,

$$pmO+pnU,\quad mpO+mqU,$$

where the numerical multipliers of O are the same, and $mq>np$. Wherefore, for any succession of positive numerical rational fractions λ, with $1<\lambda<\infty$, there is a series of points $O+\lambda U$; and therein, when $\lambda<\lambda_1$, the points of symbols E, $O+\lambda U$, $O+\lambda_1 U$, U are in order, lying in that segment EU of the line which does not contain O.

(7) It can now be further proved that such points $O+\lambda U$ occur everywhere in this segment; namely, that there is no segment FG, lying in this segment EU, with E, F, G, U in order, within which there is *no* point of symbol $O+\lambda U$, with λ a positive numerical fraction which is >1. Or, as we may say, there is no such *empty* segment FG. For the proof of this we refer to another place (*Principles of Geometry*, Vol. I (1929), pp. 130–136). But, in rough, the proof depends on three steps: (*a*) Let us call a point of symbol $O+\lambda U$, in which λ is a numerical fraction, a *section point*. Then, not both of the end points F, G, of an assumed empty segment FG, can be section points. For, then, as appears sufficiently from the preceding argument, we could, by choosing a suitable section point outside the segment FG, and taking the harmonic conjugate of this in regard to F and G, arrive at a section point lying within the segment FG. (*b*) If G be not a section point, the segment FG can be extended, to become FG', with E, F, G, G' in order, the segment FG' being also an empty segment. The point G' may be a section point, or may be the limiting point of a series of points T, all of them section points, in which E, F, G', T are in order.

Similarly, in case F is not a section point, a similar extension is possible, to a point F', with F', F, G, U in order, F'' being a point of the same character as G'. Thereby the empty segment FG is replaced by an empty segment $F'G'$; and, as E, U are section points, E, F', G', U will be in order. (*c*) But this conclusion is impossible. For whether F', G' be section points, or one or both be limiting points of such section points, it is possible, taking a suitable section point outside the segment $F'G'$, to find a section point, within the segment $F'G'$, which is the harmonic conjugate of the outside point, in regard to section points in the two series of points of which F', G' are limiting points (one of these possibly coinciding with F', or G').

(8) Whence, recalling what was said, as to the relation of the segment OE with the segment EU, and of the relation of that segment UO which does not contain E with the segment OEU, we reach the conclusions: (*a*) With μ a numerical fraction, there are points of symbol $O+\mu U$ in every segment of the line; (*b*) Three points of the line which are in the geometrical order OEU, and have such symbols $O+\mu U$, have their appropriate values of μ in ascending arithmetical order. And that, conversely, the arithmetic order determines the geometric order.

We have then two aggregates: (*a*) of points of symbols $O+\mu U$, in which μ is a numerical fraction, which can be constructed from the given points O, E, U by continued application of the process of finding the harmonic conjugate of a given point in regard to two given points, all the points involved belonging to the aggregate in question. Such points occur in every segment of the line; and (*b*) the aggregate of values of μ. The orders of corresponding elements of these two aggregates are the same. Now, we may express the fact that points with symbols $O+\mu U$, in which μ is a real numerical fraction, are found everywhere in the line, by saying that any real point of the line can be approximated to, with arbitrary accuracy, by means of such points $O+\mu U$. On the other hand, it is a theorem of arithmetic that every real number can be approximated to, with arbitrary accuracy, by means of real numerical fractions. We finally conclude therefore that *there is an exact correspondence between the points expressed by symbols* $O+xU$, *in which x is any real number, and the totality of real points of the line, the ascending order of magnitude of the numbers x corresponding to a definite order of the real points of the line.*

(9) *Ex.* 1.   Let p, q, r be real numbers in ascending order of magnitude. Take the points of symbols $P=O+pU$, $Q=O+qU$, $R=O+rU$, and the point $P'=(Q, R)/P$. If the symbol of this be $P'=O+p'U$, shew that p, q, p', r are in ascending order of magnitude.

Ex. 2.   If x, y, z, t be real numbers, and $f(\lambda)$ denote the function

$$(a\lambda+b)/(c\lambda+d),$$

in which a, b, c, d are general real given numbers, and if

$$x'=f(x),\ y'=f(y),\ z'=f(z),\ t'=f(t),$$

prove that, if x, z be arithmetically separated by y and t, then also x', z' are arithmetically separated by y' and t'.

CHAPTER VII

INVOLUTIONS. HARMONIC RANGES.
ALGEBRAICAL RELATIONS

(1) By the fundamental hypothesis we have made, any point on a line OU has a symbol expressible in terms of the symbols of O and U in the form $(O)+\lambda(U)$, where λ is a number. For brevity we shall denote the symbols (O), (U) simply by O and U.

Consider any two points, A, B, of the line, with respective symbols $O+\alpha U$, $O+\beta U$; let the quadratic equation whose roots are the two numbers α, β be $ax^2+2hx+b=0$, in which a, h, b are numbers. Let A', B' be two other points of the line, with respective symbols $O+\alpha'U$, $O+\beta'U$, the numbers α', β' being the roots of the quadratic equation

$$a'x^2+2h'x+b'=0.$$

Then, *the condition that A', B' shall be harmonic conjugates of one another in regard to A, B is*

$$ab'+a'b-2hh'=0.$$

For, the symbols of A', B', in terms of those of A, B, must be capable of the forms $O+\alpha U\pm\lambda(O+\beta U)$ in which λ is a number. With a proper value of λ, and proper values for the numbers m, n, there must, therefore, exist identities

$$m(O+\alpha'U)=O+\alpha U+\lambda(O+\beta U),\quad n(O+\beta'U)=O+\alpha U-\lambda(O+\beta U).$$

These give $m=1+\lambda$, $n=1-\lambda$, and

$$\alpha'=(\alpha+\lambda\beta)/(1+\lambda),\qquad \beta'=(\alpha-\lambda\beta)/(1-\lambda).$$

From these we deduce the equivalent equations

$$\frac{\alpha'-\alpha}{\alpha'-\beta}\Big/\frac{\beta'-\alpha}{\beta'-\beta}=-1,\quad \alpha'\beta'-\tfrac{1}{2}(\alpha'+\beta')(\alpha+\beta)+\alpha\beta=0,$$

of which the latter is the same as that we have stated in terms of the coefficients in the quadratic equations. This equation, for the geometrical relation $B'=(A,B)/A'$, must evidently persist if the points O, U of the line, and consequently the two quadratic equations, are taken differently.

(2) Suppose we have two pairs of points on the line, A, B and A_1, B_1, with respective symbols $O+\alpha U$, $O+\beta U$, $O+\alpha_1 U$, $O+\beta_1 U$, and corresponding quadratic equations satisfied, respectively, by α, β and by α_1, β_1, say $ax^2+2hx+b=0$ and $a_1x^2+2h_1x+b_1=0$. Let A', B' be another pair of points of the line, with symbols $O+\alpha'U$, $O+\beta'U$, where α', β' satisfy the equation $a'x^2+2h'x+b'=0$. Then the conditions that A', B' should be harmonic conjugates both in regard to A, B and in regard to A_1, B_1, or say, the conditions that A', $B'\sim A$, B and also A', $B'\sim A_1$, B_1, are

$$a'b-2h'h+b'a=0,\qquad a'b_1-2h'h_1+b'a_1=0.$$

Thus the quadratic equation satisfied by α', β' is

$$\begin{vmatrix} 1, & -x, & x^2 \\ a, & h, & b \\ a_1, & h_1, & b_1 \end{vmatrix} = 0.$$

(3) A particular case of the preceding result, already remarked in a former chapter (Chap. IV (5)), arises when A_1, B_1 are the points O, U, respectively, so that $\alpha_1 = 0$, $\beta_1 = \infty$, and $a_1 = 0$, $b_1 = 0$. Then, the symbols of A, B being $O + \alpha U$, $O + \beta U$, there is a pair of points A', B' such that both A', $B' \sim O$, U and A', $B' \sim A$, B, the symbols of A', B' being $O \pm (\alpha\beta)^{\frac{1}{2}} U$.

For the sake of later developments it is important to remark on the *reality* of these points A', B'. When α, β are real, the number $(\alpha\beta)^{\frac{1}{2}}$ is real only when α, β are of the same sign. Regarding O and U as real points, let us speak of a point whose symbol is $O + \lambda U$ as real when λ is real. Then, from the results obtained in the preceding chapter, *the common harmonic conjugates for the two pairs of real points, O, U and A, B, are real only if both the points A, B are in the same segment OU, determined by the points O and U. They are imaginary when the pair A, B are separated by the pair O, U.*

The two harmonic conjugates for the pairs O, U and A, B will also be real if, when O, U are real, the symbols of A, B are of the respective forms $O + (m + in) U$ and $O + (m - in) U$, where m, n are real numbers. In this case we may speak of A, B as being conjugate imaginaries in regard to O and U. This case is of great importance in what follows.

More generally, from the quadratic equation above, it can be directly verified, that, if A, B, of respective symbols $O + (m + in) U$ and $O + (m - in) U$, where m, n are real numbers, and A_1, B_1, of respective symbols $O + (m_1 + in_1) U$ and $O + (m_1 - in_1) U$, be two pairs of points, both of which are pairs of so-called conjugate imaginaries in regard to O and U, then *the pair A', B', such that A', $B' \sim A$, B and A', $B' \sim A_1$, B_1, are real points* (O and U being regarded as real). For, by computation, we find that the condition that the quadratic equation, which gives A', B', should have real roots, is that the product

$$(\alpha - \alpha_1)\,(\alpha - \beta_1)\,(\beta - \alpha_1)\,(\beta - \beta_1),$$

or, say, with α_1 real, the product

$$(a_1\alpha^2 + 2h_1\alpha + b_1)\,(a_1\beta^2 + 2h_1\beta + b_1),$$

should be positive. This includes all the cases. Evidently, however, the geometrical formulation of the results depends on the separation theorems assumed in the preceding chapter. As a linear construction with real lines, each defined as joining two real points, takes no cognisance of the relative positions of the points, it is clear that there can be no such construction for the common harmonic conjugates of two pairs of real points. The construction, with the help of conics, which arises below, involves such assumptions as that a given conic and a given line determine the common points of these.

Ex. If three pairs of points on a line be such that each pair consists of points which are harmonic conjugates for *both* the other pairs, then the symbols of the pairs of points can be taken to be of the respective forms O, U; $O+U, O-U$ and $O+iU, O-iU$.

(4) Involutions of pairs of points on a line.

The aggregate of all the pairs of points on a line which are harmonic conjugates of one another, in regard to a certain pair of points upon the line, is said to form an *involution of pairs of points*. Of the points of such a pair one can be taken arbitrarily, the other being determined thereby if the involution is given. The two points in regard to which all the pairs are harmonic conjugates are called *the foci of the involution*. At each of the two foci two points of a pair of the involution coincide with one another.

After what precedes, an involution is determined by any two of its pairs of points. If A, A' and B, B' be two pairs of points upon the line, there are two points, X, Y, which are harmonic conjugates of one another both in regard to A, A', and in regard to B, B'; or say $X, Y \sim A, A'$ and $X, Y \sim B, B'$. Any pair of points C, C' which are harmonic conjugates of one another in regard to X and Y, or, say, such that $C, C' \sim X, Y$, is a pair of the involution determined by the pairs A, A' and B, B'.

As we have said, when A, A' and B, B' are given, there exists no linear construction for finding X and Y. But if A, A' and B, B' be given, and also C, there is a linear construction for C'. We have in fact shewn, in a previous chapter (Chap. IV (5)), that, if a quadrangle of four points be given in a plane, and an arbitrary line of the plane be met by one pair of complementary joins of the four points in the points A and A', and the same line be met by another pair of the complementary joins of the four points in B and B', and be met by the remaining pair of complementary joins of the four points in C and C', then there exists a pair of points X, Y on the line such that $A, A' \sim X, Y$; $B, B' \sim X, Y$; and $C, C' \sim X, Y$. The proof given was with the help of the symbols. But, alternatively, given two pairs of points, A, A' and B, B', upon a line, we can construct pairs, C, C', of the involution on the line determined by A, A' and B, B', by constructing, in an arbitrary plane which contains the line, a quadrangle, of which two pairs of complementary joins contain, respectively, A, A' and B, B', and one other join of the four points contains C. We have shewn, in the place referred to, that then the remaining join of the four points is determined; and this constructs the point C' to be paired with C. We have thus another definition of an involution upon a line, regarded as determined by two given pairs of the involution. But, as indicated above, the question of the interchangeability of two points of a pair requires examination; and it appears more serviceable to define an involution, with the help of the symbols, in the manner adopted here.

There are certain algebraic results in regard to involutions which it is useful to bear in mind:

If, in terms of the symbols of two points O, U of the line, the symbols of the foci, X, Y, of the involution be $O+\xi U$ and $O+\eta U$; and the symbols of any pair, A, A', of the involution be $O+\alpha U$ and $O+\alpha' U$, then ((1), above), as A, $A' \sim X$, Y, we have

$$\alpha\alpha' - \tfrac{1}{2}(\alpha+\alpha')(\xi+\eta)+\xi\eta=0.$$

Thus the symbols, $O+\alpha U$, $O+\alpha' U$, of a pair of the involution, are such that α, α' satisfy an equation

$$l\alpha\alpha' + m(\alpha+\alpha')+n=0,$$

in which l, m, n are numbers, these being the same for every pair of the involution. The ratios of l, m, n can evidently be chosen so that two arbitrary pairs of points, A, A' and B, B', are pairs of the involution. Further, the condition that three pairs of points $O+\theta U$, $O+\theta' U$, with θ, θ' respectively equal to α, α'; to β, β'; and to γ, γ', should belong to an involution, is

$$\begin{vmatrix} \alpha\alpha', & \alpha+\alpha', & 1 \\ \beta\beta', & \beta+\beta', & 1 \\ \gamma\gamma', & \gamma+\gamma', & 1 \end{vmatrix} = 0.$$

If α, α', and β, β', and γ, γ' be respectively the roots of the quadratic equations $a_0x^2+2a_1x+a_2=0$, $b_0x^2+2b_1x+b_2=0$, $c_0x^2+2c_1x+c_2=0$ (where a_0, a_1, etc. are numbers), this is equivalent to saying that there must exist numbers p, q, r such that an equation

$$p(a_0x^2+2a_1x+a_2)+q(b_0x^2+2b_1x+b_2)+r(c_0x^2+2c_1x+c_2)=0$$

holds identically for all values of x.

There are two particular cases which are of frequent occurrence: *(a) When the points O, U of the line, in terms of which any pair A, A' of an involution are expressed, are themselves a pair of the involution, then the symbols $O+\alpha U$, $O+\alpha' U$, of any pair of the involution, are such that the product $\alpha\alpha'$ is the same for every pair; say $\alpha\alpha'=\lambda^2$. The foci of the involution are then the two points whose symbols are $O\pm\lambda U$. (b) When the points O, U are the foci of the involution, the relation for a pair of points $O+\alpha U$, $O+\alpha' U$, belonging to the involution, is $\alpha+\alpha'=0$. In this case, the general pair of the involution have symbols $O\pm\alpha U$.*

It is true also that *two involutions of pairs of points upon a line have one common pair.* This is clear because $\alpha\alpha'$ and $\alpha+\alpha'$ can be determined to satisfy two equations

$$l\alpha\alpha' + m(\alpha+\alpha')+n=0 \text{ and } l_1\alpha\alpha' + m_1(\alpha+\alpha')+n_1=0,$$

in which the ratios $l:m:n$ and $l_1:m_1:n_1$ are given. It is also clear because, if X, Y and X_1, Y_1 be the foci, respectively, of the two involutions, there exists a pair P, P' satisfying both the relations P, $P' \sim X$, Y and P, $P' \sim X_1$, Y_1. The points P, P' are the common pair of the two involutions.

It will be found that the consideration of ranges in involution, as of harmonic ranges, is very frequently of use. The following examples may serve to make the ideas familiar.

Ex. 1.   It has already been proved (Chap. iv (3)) that if two ranges, each of four points, lying on two lines in a plane, be in perspective, from a point of the plane, and the points of one of the ranges be in harmonic relation, then the points of the other range are equally in harmonic relation, taken in the same order. We have defined four lines, which join a point to four points of a line which are in harmonic relation, as a *pencil* of lines in harmonic relation; and have remarked that such a pencil is the dual of four points of a line which form a harmonic range.

It follows that if three pairs of points on a line, which are in involution, be in perspective, from a point, with three pairs of points on another line, then these latter pairs are also in involution. We speak of the lines, which join a point to the pairs of points of a line which are in involution, as forming a *pencil in involution*. This is then dual to a range of pairs of points on a line which are in involution. The rays from the centre of perspective to the foci of the involution on the line may be called *the focal rays* of the pencil.

Ex. 2.   It has been seen that the three pairs of complementary joins of four points in a plane meet an arbitrary line of the plane, respectively, in three pairs of points belonging to an involution. The dual result will then also be true, namely, the three pairs of lines joining an arbitrary point of a plane to the three pairs of complementary intersections of four given lines of the plane form a pencil in involution.

The reader may welcome a direct verification of this. Let A, B be one pair of complementary intersections of the four given lines, and O be an arbitrary point of the plane. The symbols of another pair, U, V, of complementary intersections of the four given lines, may be expressed in terms of the symbols of A, B, O, it being supposed that O does not lie on the line AB. The representation may be taken in the form

$$U = O + mA + nB, \quad V = O + pA + qB,$$

where m, n, p, q are numbers. Another intersection, M, of two of the four lines, say of the lines AU and BV, has then a symbol of the form $M = O + pA + nB$; for this is at once $U + (p-m)A$ and $V + (n-q)B$. Likewise, the complementary intersection, L, namely of the lines AV and BU, has a symbol $L = O + mA + qB$; for this is at once $V + (m-p)A$ and $U + (q-n)B$. The pairs of lines joining O to the complementary pairs of intersections A, B; U, V and L, M, thus meet the line AB in the points of respective symbols A, B; $mA + nB$, $pA + qB$; and $mA + qB$, $pA + nB$. Since the product of the numbers n/m and q/p is the same as the product of the numbers q/m and n/p, it follows, by what is remarked above, that these three pairs of points in the line AB are in involution. Thus the lines joining O to the three pairs of points A, B; U, V and L, M meet any line whatever, in the plane, in a range of three pairs of points belonging to an involution; and themselves form a pencil of three pairs of lines in involution.

Ex. 3.   Important converses of the dual results given in Ex. 2 are: *(a) Let lines l, m, n be drawn in the plane of a triangle, respectively through the angular points A, B, C. Let a line, in the plane of the triangle, not passing through any of the angular points, meet the sides BC, CA, AB, respectively, in D, E, F and meet the lines l, m, n, respectively, in L, M, N. Then a sufficient condition for the lines l, m, n to meet in a point is that the pairs of points D, L; E, M; F, N, on the line, should belong to an involution.* To prove this, we may consider the intersection with the line in question of the line joining A to the common point of the lines m, n. *(b)* The correlative condition, *that points L, M, N, lying respectively on the sides BC, CA, AB of a triangle, should be in a line, is the existence of a point O, in the plane of the triangle, not lying on any of the sides, such that the three pairs of lines OA, OL; OB, OM; OC, ON should belong to a pencil in involution.*

Ex. 4.   If A, V, B, U be four points of a plane, we have already emphasised the importance of the three points which are the intersections of the pairs of complementary joins of the four points, say L, common to AV and BU; M, common to AU and BV, and N, common to AB and UV. We have called L, M, N the *vertices* of the quadrangle $AVBU$. It is sometimes useful to recognise that, in terms of the symbols of L, M, N, the symbols of B, U, V, A are respectively of the forms

$$B = lL + mM + nN, \quad U = -lL + mM + nN, \quad V = lL - mM + nN,$$

$$A = lL + mM - nN,$$

where l, m, n are numbers. By altering the notation for the symbols L, M, N we may suppose all of l, m, n to have the value 1. The proof of this is a direct example of the theory of harmonic ranges.

Ex. 5.   Two points, P, P', are said to be *harmonic images* of one another, in regard to a given point O, and a given line n, when O, P, P' are in line and the line PP' meets n in the point $(P, P')/O$. If the point P move on a line, p, then the corresponding point P' also moves on a line, p', and the lines p, p' meet in a point of the given line n; also the lines p, p', of which either may be taken arbitrarily, are said to be harmonic images of one another in regard to the given elements (O, n). It is easy, further, to see that, if q, q' be two other lines which are harmonic images of one another in regard to (O, n), then the common point of p and q is the harmonic image of the common point of p' and q'.

If we take the harmonic image of an arbitrary point, lying in the plane of a triangle ABC, first in regard to the point A and the line BC, then in regard to the point B and the line CA, and then in regard to the point C and the line AB, we obtain, altogether, four points which are interchanged among themselves by any combination of the three ways of taking the harmonic image which we have referred to. If the symbol of the original point, in terms of the symbols of A, B, C, be $\xi A + \eta B + \zeta C$, where ξ, η, ζ are numbers, the other three points have respective symbols $-\xi A + \eta B + \zeta C$, $\xi A - \eta B + \zeta C$, $\xi A + \eta B - \zeta C$. Cf. Ex. 4 preceding. The theorem has a dual, when we start with an arbitrary line instead of an arbitrary point.

Ex. 6.   As the four points of a quadrangle give rise to a triangle formed by the three vertices of the quadrangle, so the four lines of a quadrilateral give rise to a triangle formed by the three diagonals of the quadrilateral, as has been explained (Chap. II (1)). Let the four lines of the quadrilateral be b, c and b', c', and the angular points of the diagonal triangle be P, Q, R, the notation being such that the two intersections (b, c) and (b', c') lie on the diagonal QR. Let any line through P meet the lines b, c, respectively, in E and F. Prove that the two intersections (QF, RE) and (QE, RF) lie on the lines b' and c' (or c' and b'), and are in line with the point P.

Shew how to construct a quadrilateral whose diagonals are along the sides of the vertex triangle of a given quadrangle.

Ex. 7.   We have defined the harmonic line of a point, lying in the plane of two given lines, in respect to these lines (Chap. IV (4), Ex. 2). Now consider a quadrangle of four points A, V, B, U in a plane; let O be any point of the plane. Take the harmonic line of O in regard to the complementary joins AV, BU, which meet, say, in the vertex L of the quadrangle. Also take the harmonic line of O in regard to the complementary joins AU, BV, which meet, say, in the vertex M of the quadrangle. Let O' be the common point of these harmonic lines. Then, the harmonic line of O in regard to the third pair of complementary joins AB, UV, which meet, say, in the vertex N, also passes through O'.

For, the points O, O' are the foci of the involution in which the line OO' is met by the three pairs of complementary joins of the points A, V, B, U.

In the particular case in which O is at one of the vertices, say L, of the quadrangle, the point O' is an undetermined point of the join, MN, of the other two vertices.

If, as in Ex. 4, the symbols of B, U, V, A, in terms of those of L, M, N, be respectively $L+M+N$, $-L+M+N$, $L-M+N$, $L+M-N$, and the symbols of O, O' be respectively $xL+yM+zN$ and $x'L+y'M+z'N$, where x, y, z, x', y', z' are numbers, then we have $xx'=yy'=zz'$. To prove this, let the line LO meet the line BA in a point whose symbol, in terms of the symbols of B and A, is $qB+pA$, where q, p are numbers; since then, if ξ be an appropriate number, there exists an equation $(\xi-x)L+O=qB+pA$, which is in fact the same as

$$\xi L+yM+zN=q\,(L+M+N)+p\,(L+M-N),$$

we infer that $q+p=y$ and $q-p=z$. Thus the intersection (LO, BA) has the symbol $(y+z)\,B+(y-z)\,A$. Similarly, the intersection (LO', BA) has the symbol $(y'+z')\,B+(y'-z')\,A$. As O' is on the harmonic line of O in regard to the lines LB, LA, we thus have

$$(y+z)/(y-z)= -(y'+z')/(y'-z').$$

This gives $yy'=zz'$. Considering the vertex M instead of L, we similarly find $xx'=zz'$.

If we replace the symbols L, M, N by lL, mM, nN, respectively, where l, m, n are numbers, and write the symbols of O, O' in the respective forms $xlL+ymM+znN$, $x'lL+y'mM+z'nN$, we still have $xx'=yy'=zz'$.

Two such points as O, O' are said to be *conjugate* to one another in regard to the quadrangle A, V, B, U.

Ex. 8. There is an exact dual of the results in Ex. 7, when we start with four lines forming a quadrilateral. If X, Y be given points in a plane, and any line meet the line XY in the point O, the *harmonic point* of this line, in regard to the points X, Y, is the point O' of the line XY given by $O'=(X, Y)/O$. The theorem, dual to that of Ex. 7, is that the three harmonic points of an arbitrary line of the plane, taken in turn in regard to the three pairs of complementary intersections of the four lines of the quadrilateral, are three points which lie in another line. This line, and the original line, are the focal rays of the pencil in involution which is formed by joining the common point of the two lines to the three pairs of complementary intersections of the quadrilateral. The two lines are said to be *conjugate to one another* in regard to the quadrilateral.

Ex. 9. Let A, A'; B, B' and C, C' be three pairs of points on a line which are in involution. Suppose further that it happens that A, A' are harmonic conjugates of one another in regard to B, B'; or A, $A' \sim B$, B'. Prove that, then, the points $(A, A')/C$ and $(B, B')/C'$ are the same point; and the points $(A, A')/C'$ and $(B, B')/C$ are the same point.

Ex. 10. Let A, B, C, D be four arbitrary points on a line. An involution on the line exists of which B, C and A, D are pairs; another involution is determined by the pairs C, A and B, D; and a third involution by the pairs A, B and C, D. Let the pairs of foci of these three involutions be respectively X, X', and Y, Y', and Z, Z'. Prove that Y, $Y' \sim Z$, Z'; also Z, $Z' \sim X$, X'; and X, $X' \sim Y$, Y'.

Ex. 11. Let A, A'; B, B' and C, C' be three pairs of points on a line which are in involution. Suppose, also, that B, $C \sim A$, A'. Prove that, then, B', $C' \sim A$, A'.

If A, A' and B, B' be two arbitrary pairs of points on a line, and the points C, C' be defined by $C=(A, A')/B$ and $C'=(A, A')/B'$, prove that the three pairs A, A'; B, B'; C, C' are in involution.

Ex. 12. If A, A'; B, B'; C, C' and D, D' be four pairs of points of an involution on a line, and A_1, B_1, C_1 be the three other points of the line given by $A_1=(D, D')/A'$, $B_1=(D, D')/B'$ and $C_1=(D, D')/C'$, then the four pairs of points A, A_1; B, B_1; C, C_1 and D, D' belong to an involution.

Ex. 13.   If A, A'; B, B' and C, C' be three pairs of an involution of points on a line, and P, Q, R be the points given by $P = (B', C')/A$; $Q = (C', A')/B$ and $R = (A', B')/C$, then the three pairs A, P; B, Q and C, R belong to an involution.

Ex. 14.   Let A, B, C, I, J be five arbitrary points of a line. There are three involutions, determined respectively by the three pairs of points B, C and I, J; C, A and I, J; and A, B and I, J. Let X, X'; Y, Y'; Z, Z' be the pairs of foci of these involutions, respectively. Prove, with proper choice of notation, that the three pairs A, X; B, Y and C, Z are in involution. By interchange of points in the couples X, X'; Y, Y'; Z, Z', three other sets of three pairs of points in involution are obtainable.

Ex. 15.   Let a, b, c, a', b', c' be numbers, generally different. Prove that the quadratic polynomial in x

$$(x - b)(c - a) . (x - c')(a' - b') - (x - c)(a - b) . (x - b')(c' - a')$$

is unaltered by replacing a, b, c, respectively, by b, c, a and, at the same time, a', b', c', respectively, by b', c', a'. Hence, or otherwise, prove that if A, B', C, A', B, C' be six arbitrary points of a line, then the three involutions determined, respectively, by the three pairs of points: B, C' and B', C; C, A' and C', A; A, B' and A', B, have one pair of points common to these three involutions.

Ex. 16.   Let A, B, C, D and A', B', C', D' be eight arbitrary points of a line. The two involutions determined, respectively, by the two pairs of points B, C and B', C', and by A, D and A', D', have a common pair of points, say P_1, Q_1. Likewise the two involutions determined, respectively, by the two pairs $(C, A; C', A')$ and $(B, D; B', D')$ have a common pair, say P_2, Q_2; and the two involutions determined, respectively, by the two pairs $(A, B; A', B')$ and $(C, D; C', D')$ have a common pair, say P_3, Q_3. Prove that the three pairs P_1, Q_1; P_2, Q_2; P_3, Q_3 belong to an involution.

Of this result, found by Steiner by the consideration of a cubic curve, a proof is given below (Misc. Ex. 125) from the theory of conics; and likewise an account of Steiner's proof (Note 5).

All the results given here for involutions are capable of direct algebraic proof, using the symbols of the points; but many of them find their best interpretation in connexion with the theory of conics; in particular, the result in Ex. 15 is equivalent with the famous theorem associated with the name of Pascal.

Ex. 17.   Let A, A'; B, B' and C, C' be three pairs of points in involution on a line. If the two points M, N given by $M = (A, A')/C$ and $N = (B, B')/C$ be such that $M, N \sim C, C'$, then either $A, B \sim C, C'$ and $A', B' \sim C, C'$ or else $A, B' \sim C, C'$ and $A', B \sim C, C'$. Conversely, if A, A'; B, B'; C, C' be pairs of an involution and $A, B \sim C, C'$, then the points M, N given by $M = (A, A')/C$ and $N = (B, B')/C$ are such that $M, N \sim C, C'$.

Ex. 18.   Let the nine points of a line X, X', Y, Y', Z, Z', L, M, N be such that the three involutions determined, respectively, by the pairs $(M, N; X, X')$; $(N, L; Y, Y')$; $(L, M; Z, Z')$ have a common pair of points, while the three pairs L, X; M, Y; N, Z belong to an involution. Prove that the three pairs L, X'; M, Y'; N, Z' also belong to an involution.

Ex. 19.   Let A, A'; B, B' and C, C' be the pairs of complementary intersections of four lines in a plane; let O be an arbitrary point of the plane. Take the two points of intersection (OB, AB') and (OB', AB); and let the line joining these two points meet the line CC' in U. Prove that U lies also on the join of the two points $(OB, A'B')$ and $(OB', A'B)$, which are obtained from the two former points by replacing A by A'. Further, let the line joining the two points (OA, BA') and (OA', BA) meet the line CC' in V. Prove that V lies also on the join of the two points $(OA, B'A')$ and $(OA', B'A)$. Then shew that $U, V \sim C, C'$. Verify also, without using the symbols for the points, that the results may be obtained by applying the dual of Pappus' theorem.

Ex. 20. Let ABC and $A'B'C'$ be two triangles in a plane. Let $B'C'$ meet CA, AB respectively in B_1 and C_1; let $C'A'$ meet AB, BC, respectively, in C_2 and A_2; and let $A'B'$ meet BC, CA, respectively, in A_3 and B_3. Then A_2, A_3 lie on BC, also B_3, B_1 lie on CA, and C_1, C_2 lie on AB. Now, let D_1, D_2 be the points on BC such that D_1, $D_2 \sim B$, C and D_1, $D_2 \sim A_2$, A_3. Likewise, on CA, let E_1, E_2 be such that E_1, $E_2 \sim C$, A and also E_1, $E_2 \sim B_3$, B_1. Also, on AB, let F_1, F_2 be such that F_1, $F_2 \sim A$, B and F_1, $F_2 \sim C_1$, C_2. Thus, on each side of the triangle ABC we have a pair of points which are the foci of a certain involution of pairs of points lying on this side. Let the two foci on a side be, for the moment, called *associated*.

Prove that, if the two triangles ABC and $A'B'C'$ be in perspective from a point, the six foci lie in threes upon four lines, no two of the foci upon any one of these four lines being associated. Prove also that the diagonal triangle of the quadrilateral formed by the four lines is the triangle ABC.

Further, considering the six lines such as AD_1, AD_2, joining the angular points A, B, C each to the pair of foci on the opposite side of the triangle, prove that these six lines meet in threes in four points, each of these points arising from three foci which are not associated. Prove also that the vertex triangle of the quadrangle formed by these four points is the triangle ABC.

Conversely, if three of the six foci, of which no two are associated, lie in a line; or, if the three joins of A, B, C, to three not associated opposite foci, meet in a point; then the triangles ABC, $A'B'C'$ are in perspective.

We may easily formulate the dual theorem for two triangles ABC, $A'B'C'$, in which we consider the six lines AB', AC'; BC', BA' and CA', CB'.

Ex. 21. Let O be any point in the plane of the triangle ABC; let AO, BO, CO meet BC, CA, AB, respectively, in D, M and N. Let the lines NM, MD, DN meet the sides BC, AB, CA, respectively, in L, E, F. Thus EF and MN meet BC in L, which is $(B, C)/D$. Now join an arbitrary P of the line BC to A, M and N, by lines meeting the sides EF, FD, DE, of the triangle DEF, respectively, in X, Y, Z. Prove that X, Y, Z are in line.

The proof can be given as an example of the important principle enunciated in Ex. 3, namely, by shewing that the three pairs of lines PX, PD; PY, PE and PZ, PF belong to a pencil in involution. For, let PE and PN meet AC respectively in T and U; we see at once that F, $M \sim A$, C and T, $U \sim A$, C. Whence, by Ex. 11, the pairs A, C; M, T and F, U are in involution.

Ex. 22. Let A, B, C, D be four arbitrary points in a plane. Let a line meet AB, CD, BC, AD respectively in E, F, H, K. Let O, P, Q be, respectively, the intersections (AC, BD), (BF, CE), (CK, DH). Prove that OP, OQ are harmonic conjugates in regard to OA and OB.

CHAPTER VIII

RELATED RANGES AND PENCILS

It conduces to brevity in the statement of many geometrical results to use the idea of Related Ranges (sometimes called Projective Ranges, or Projectively Related Ranges). The theory of such ranges can be established very easily by employing the symbols for points.

(1) Let O, U be two given points on a given line; the symbols of these points may be denoted, also, by O and U. It is understood, as in all similar cases, that the ratio of the numerical multipliers inherent in these symbols O, U has been fixed; as by assigning a third point of the line and taking $O+U$ (or a numerical multiple of this) for the symbol of this third point, as has been explained (Chap. v). But, and this has importance, the theory is developed so as to be independent of the value of this ratio of the multipliers.

The symbol of any point of the line may then be taken to be $O+xU$, where x is a number, the point U being conventionally given by $x=\infty$. The points obtained by different values of x are said to form a *Range* on the line.

Suppose that on another line, which may coincide with the former, we likewise have a range of points, with symbols of the form $O'+x'U'$, in terms of the symbols O', U' of two given points of this other line, where x' is a number.

Then, when the two numbers x, x' are connected by a general relation of the form
$$\alpha xx' + \beta x + \gamma x' + \delta = 0,$$
in which α, β, γ, δ are numbers whose ratios are fixed, so that either of x and x' determines the other, the ranges described by the points of respective symbols $O+xU$, $O'+x'U'$ are said to be *Related*, either point being said to *correspond* to the other. The ranges may also be said to be in (1, 1) algebraic correspondence. Either of x and x' is then a fractional linear function of the other, say
$$x' = (ax+b)/(cx+d), \quad x = (-dx'+b)/(cx'-a),$$
in which a, b, c, d are numbers whose ratios are fixed.

The nature of the relation between the two ranges is independent of the origins O, U and O', U' by which the points are defined. If, for instance, in the second range, we take, instead of O', U', two fixed points O_1 and U_1, with such symbols that
$$O' = pO_1+qU_1, \quad U' = mO_1+nU_1,$$
where p, q, m, n are numbers for which the ratios q/p and n/m are fixed, and not the same, then a point of symbol $O'+x'U'$ is expressed by
$$(p+mx')\,O_1+(q+nx')\,U_1,$$

or, essentially, by $O_1 + x_1 U_1$, where

$$x_1 = (q + nx')/(p + mx');$$

if then x is a fractional linear function of x', with fixed coefficients, it is a fractional linear function of x_1, also with fixed coefficients; and conversely.

More generally, in a similar way, any two ranges which are related to the same range are related to one another.

The definition of related ranges which we have given is equivalent to saying that, if A, B, C be any three given points of one range, a related range can be found on (the same or) another line, such that three arbitrarily assigned points A', B', C' of the new range shall correspond to A, B, C, respectively. For the three equations (for $i = 1, 2, 3$),

$$x_i' = (ax_i + b)/(cx_i + d),$$

in which x_1, x_2, x_3, x_1', x_2', x_3' are given numbers, enable us to determine appropriate values for the three ratios of a, b, c, d.

This may be stated differently, by saying that, if x_1, x_2, x_3, x be the parameters associated, respectively, with four points A, B, C, P of one range, and x_1', x_2', x_3', x' the parameters associated, respectively, with the four corresponding points A', B', C', P' of the other range, then the fractions

$$\frac{x' - x_1'}{x' - x_2'} \Big/ \frac{x_3' - x_1'}{x_3' - x_2'}, \quad \frac{x - x_1}{x - x_2} \Big/ \frac{x_3 - x_1}{x_3 - x_2}$$

are equal to one another.

And, it is noticeable, and important, that this statement remains true whatever be the origins—O, U, O', U'—taken on the two lines. For, it is easy to verify that, if each of x_1, x_2, x_3, x be replaced by the *same* linear fractional function of itself, say x_i by $\xi_i = (px_i + q)/(rx_i + s)$, where p, q, r, s are the same for each of x_1, x_2, x_3, x, then the fraction

$$(x - x_1)(x - x_2)^{-1}/(x_3 - x_1)(x_3 - x_2)^{-1}$$

remains unaltered. A particular case of this statement, which is significant, is that if the symbols O, U be altered, being replaced by numerical multiples, hO, kU, of these, the fraction considered is unaltered. For this change is equivalent only to multiplying all of x_1, x_2, x_3, x by the same number.

Wherefore, *the fraction*

$$\frac{x - x_1}{x - x_2} \Big/ \frac{x_3 - x_1}{x_3 - x_2},$$

arising from the four points A, B, C, P, has a geometrical meaning depending only on these four points; and its value is the same as for the corresponding points A', B', C', P' in any other range which is related to the former. And this fact defines the relation in which the ranges stand. We shall denote this fraction by $(P, C/A, B)$, and call it the cross ratio of the points P, C in regard to the points A, B. Thus, the necessary and sufficient condition for two ranges to be related is that the cross ratio of any four points of one of these ranges shall be equal to the cross ratio of the four corresponding points of the other range, the order in which the points are taken being the same in the two cases.

A particular case, of frequent use, which shews the simplicity of the ideas, is that in which we take the cross ratio, in regard to two points O, U, of two points whose symbols are $O + xU$ and $O + x_3U$. For this we have to take, in the general form, $x_1 = 0$ and $x_2 = \infty$. The result is, denoting the points of symbols $O + xU$, $O + x_3U$, respectively, by P, C, that
$$(P, C/O, U) = x/x_3';$$

and, as we have said, this is true independently of the ratio of the numerical multipliers inherent in the symbols O, U. Still more particularly, if P, C are harmonic conjugates in regard to O and U, so that $x_3 = -x$, we have $(P, C/O, U) = -1$.

Two corollaries from what has been said may perhaps be repeated: (*a*) Two ranges, of which each is related to the same third range, are related to one another; (*b*) Two ranges, which are related, are wholly coincident if three points of one coincide with the three corresponding respective points of the other.

Ex. 1.   If we interchange the points P, C, in the range A, B, C, P; or, if we interchange the points A, B, the cross ratio $(P, C/A, B)$ is changed into its inverse. Namely,
$$(C, P/A, B).(P, C/A, B) = 1, \quad (P, C/B\ A).(P, C/A, B) = 1.$$
Thus, if *both* P, C be interchanged, and also A and B, the cross ratio is unaltered in value.

Ex. 2.   There are 24 ways in which the four points A, B, C, P may be arranged; each of these gives a cross ratio. There are, however, only six different values of the cross ratio, in general, each of which arises from four different orders of the four points. And these six different values are expressible in terms of any one of them, say ϵ, the expressions being
$$\epsilon,\ \epsilon^{-1},\ 1 - \epsilon,\ (1 - \epsilon)^{-1},\ 1 - \epsilon^{-1},\ \epsilon\,(\epsilon - 1)^{-1}.$$
These six values are obtainable, from any one of them, by continued application of the two processes of replacing x by x^{-1}, or by $1 - x$, where x denotes any one of the six values. For definiteness the complete list of the 24 cross ratios may be given; but the effective result can be summarised as in Ex. 3 following:

$$
\begin{aligned}
(1) \quad & \epsilon = (P, C/A, B) = (A, B/P, C) = (C, P/B, A) = (B, A/C, P), \\
(2) \quad & \epsilon^{-1} = (C, P/A, B) = (A, B/C, P) = (P, C/B, A) = (B, A/P, C), \\
(3) \quad & 1 - \epsilon = (P, A/C, B) = (C, B/P, A) = (A, P/B, C) = (B, C/A, P), \\
(4) \quad & (1 - \epsilon)^{-1} = (P, A/B, C) = (B, C/P, A) = (A, P/C, B) = (C, B/A, P), \\
(5) \quad & 1 - \epsilon^{-1} = (P, B/C, A) = (C, A/P, B) = (B, P/A, C) = (A, C/B, P), \\
(6) \quad & \epsilon\,(\epsilon - 1)^{-1} = (P, B/A, C) = (A, C/P, B) = (B, P/C, A) = (C, A/B, P).
\end{aligned}
$$

Ex. 3.   In each of these 24 cross ratios, the four letters A, B, C, P occur in a certain order, the two written first in this order being separated, by a solidus, from those written third and fourth. The four letters can be so divided into two couples in three ways. The explicit table shews that if, without paying regard to the solidus, the two letters of any couple be interchanged, and, at the same time, the two letters of the complementary couple be interchanged, the value of the cross ratio is unaltered. This gives three new arrangements of the letters which lead to cross ratios equal to the original one.

Further, as remarked in Ex. 1, the interchange of the two letters which precede the solidus, or of the two letters which follow the solidus, changes a cross ratio ϵ into a cross ratio ϵ^{-1}. While, as the table shews, the interchange of the letter preceding the solidus with the letter following the solidus changes the cross ratio ϵ into the cross ratio whose value is $1-\epsilon$.

These three remarks summarise the table given.

Ex. 4.   If A, B, C, P be any four points in a line, prove that

$$(P,\ A/B,\ C).(P,\ B/C,\ A).(P,\ C/A,\ B) = -1.$$

(2) The simplest example of two related ranges is that of two ranges which are in perspective from a point. From a point, H, in the plane of two lines which meet in a point O, let a line, HUU', be drawn, meeting the lines, respectively, in the points U and U'; and also another line, HPP', meeting these lines, respectively, in P and P'. Let the symbol of P, in terms of the symbols of O and U, be $P = O + xU$, where x is a number. We can suppose the symbol of U' to be $U' = U + a^{-1}H$, where a is a number. Then, the point whose symbol is $O + xU'$ is a point of the line OU'; but, because

$$ax^{-1}\left(O+xU'\right) = ax^{-1}O + aU' = ax^{-1}O + aU + H = ax^{-1}\left(O+xU\right) + H$$

is the same as $ax^{-1}P + H$, this point of symbol $O + xU'$ is on the line HP; it is thus the point P'. To the range of points $P = O + xU$, on one line, corresponds therefore the range of points $P' = O + xU'$, on the other line; these two ranges are then related. By choosing the common point O of the two lines as an origin for both ranges, and the two other origins, U, U', to be in perspective from H, the corresponding points P, P' are characterised by the same parameter x.

Conversely, suppose we have two related ranges on lines which have a common point. Suppose also that the common point of the two lines, regarded as belonging to one of the lines, corresponds to the same point regarded as belonging to the other line. We can then prove that the two ranges are in perspective from a point. For, let P, Q be two points of one range, and P', Q' the respectively corresponding points of the other range; and let the lines PP' and QQ' meet in the point H. When we project from H, on to the first line, the range which lies on the second line—that is, draw lines from H to the points of the second range and take the intersection of each with the first line—we shall obtain, on the first line, by what we have just proved, a range which is related to the given range on the second line; this is therefore related to the given range on the first line. With this range, however, the projected range has three corresponding points in common, namely the common point O of the two lines, and also the points P, Q. Wherefore, by what we have said, the projected range coincides with the first range. Thus the two given ranges are in perspective from H.

(3) More generally, let a range, (R), lying on one line, l, be projected from a point, H_1, on to another line, l_1, giving thereon a related range, (R_1). Also, let the same range (R) be projected, from another point, H_2, on to another line, l_2, giving thereon a related range, (R_2). It is assumed that the lines l, l_1 lie in a plane, and that H_1 lies in this plane;

and, also, that the lines l, l_2 lie in a plane, and that H_2 lies in this plane; so that the line l meets both the lines l_1 and l_2. By what we have proved, the ranges (R_1), (R_2) are related to one another.

Conversely, suppose we have two related ranges (R_1), (R_2), on two lines l_1, l_2, respectively; these lines may, or may not, meet one another. Then it is true that, in an infinite number of ways, a centre H_1, and a centre H_2, can be found, and also a line l meeting both l_1 and l_2, such that the ranges (R_1) and (R_2) project, respectively from H_1 and H_2, into the same range on the line l. To prove this, let A_1, C_1, B_1 be three points of the range (R_1), and A_2, C_2, B_2 be the respectively corresponding points of the range (R_2). As we have said, the assignment of this triplet of pairs of corresponding points determines the relation of the two ranges (R_1) and (R_2). Let the line A_2B_1 be l, and C be any point on this line. As the lines l_1, l are in a plane (since these lines meet in B_1), the lines A_1A_2 and C_1C have a common point, say H_1; as the lines l, l_2 are in a plane (since these meet in A_2), the lines B_1B_2 and CC_2 have a common point, say H_2. The range (R_1), on l_1, given by A_1, C_1, B_1, is thus projected from H_1 into the range $(A_2, C, B_1, ...)$ on l; and the range (R_2), on l_2, determined by A_2, C_2, B_2, is projected from H_2 into the same range $(A_2, C, B_1, ...)$ on l. This range (R) is thus in perspective with both the ranges (R_1), (R_2), respectively, from H_1 and H_2.

(4) The preceding argument holds whether the lines l_1, l_2 be in the same plane or not. When these lines meet, say in T, there is reason for taking a certain definite line l, upon which to project the two given ranges. To the point T, on l_1, regarded as belonging to the range (R_1), there will correspond, in the related range (R_2) on l_2, a definite point O_2; to the point T, on l_2, regarded as belonging to the range (R_2), there will correspond, in the related range (R_1) on l_1, a definite point O_1. We can omit the case, considered above, when O_1 and O_2 coincide. Take then the line O_1O_2 for the line l. Let the points of the range (R_2) which correspond, respectively, to points A_1, P_1 of the range (R_1) be called A_2, P_2. Also, let A_1A_2 meet the line l in H, and A_1P_2, A_2P_1 meet the line l, respectively, in P and P'. The points P_1, A_1, O_1, T, of the range (R_1), then correspond, respectively, to the points P_2, A_2, T, O_2 of the range (R_2). From A_2, however, the points P_1, A_1, O_1, T are, respectively, in perspective with the points P', H, O_1, O_2 of the line l; and, from A_1, the points P_2, A_2, T, O_2 are, respectively, in perspective with the points P, H, O_1, O_2. As the ranges (P', H, O_1, O_2), (P, H, O_1, O_2) are thus related, and have the three corresponding points H, O_1, O_2 coincident, we infer that P and P' coincide.

Thus, *the line l has the property of containing the common point of the cross-joins, A_1P_2, A_2P_1, of any two pairs of points, A_1, P_1 of the range (R_1), and A_2, P_2 of the range (R_2), which correspond to one another. And, in particular, the two ranges on l_1, l_2 are the projections of the same range on l, with centres of projection at any two corresponding points A_2, A_1, on l_2 and l_1, respectively.*

This result evidently includes Pappus' theorem, formerly referred to, that, if A_1, B_1, C_1 be three arbitrary points of a line l_1, and A_2, B_2, C_2 be three arbitrary points of another line l_2, which meets l_1, then the three points of intersection (B_1C_2, B_2C_1), (C_1A_2, C_2A_1) and (A_1B_2, A_2B_1) lie in a line. And, if we determine two related ranges, on the lines l_1 and l_2, by the condition that the points A_1, B_1, C_1 of the first of these correspond, respectively, to the points A_2, B_2, C_2 of the second, then the line given by Pappus' theorem is the line containing the points O_1, O_2, of l_1 and l_2, respectively, which correspond to the common point T of l_1 and l_2, regarded as belonging, respectively, to l_2 and l_1. In the particular case in which both O_1 and O_2 coincide with T, the lines A_1A_2, B_1B_2, C_1C_2, joining, respectively, corresponding pairs of points of the two ranges, meet in a point, say H, as has been proved above. The line given by Pappus' theorem is then the harmonic line of H in regard to l_1 and l_2, and passes through T.

(5) Conversely, let Pappus' theorem be assumed to be true. Then we can define two related ranges, on two lines which have a common point, in which the relation is determined by the correspondence of three arbitrary points of one range to three arbitrary points of the other, without using the symbols of points.

For, in the enunciation of Pappus' theorem, repeated above, let P denote the intersection (B_1C_2, B_2C_1), which lies on the line l given by the Pappus' theorem. Let D_1 be any point on the line l_1; let the line B_2D_1 meet the line l in V, and B_1V meet l_2 in the point D_2; and let C_1D_2 and C_2D_1 meet in W. Then, by Pappus' theorem, applied to the two triads of points B_1, C_1, D_1 and B_2, C_2, D_2, the points P, V, W lie in a line. This is then the line l. We thus deduce D_2, on the line l_2, to correspond, in the second range, to the arbitrarily taken D_1 of the range on l_1. And we can use C_1, C_2, or A_1, A_2, instead of B_1, B_2, for this purpose.

(6) In a plane, the dual of a range of points, on a line, is *a pencil of lines* all passing through the same point of the plane. By what we have proved, any two transversal lines, in the plane, not passing through the central point of the pencil, meet the rays of the pencil in two ranges which are related to one another. Thus, *two pencils of lines in a plane, with different centres, may be spoken of as related when the ranges which they determine on two transversal lines (or on the same line) are themselves related ranges*. This may be extended to two pencils of lines lying in two different planes. And we may speak of the cross ratio of a pencil of four lines, meaning that of the range which these four lines determine on any transversal. If the pencil consist of the lines joining a point O, respectively, to points P, C, A, B of the plane, we may denote this cross ratio by $O(P, C/A, B)$.

> *Ex.* If I, J be any two points in the plane of a triangle ABC, prove that
> $$A(B, C/I, J) . B(C, A/I, J) . C(A, B/I, J) = 1.$$

(7) In space, the dual of a range of points, lying on a line, is a series of planes, all containing a certain line. This we may call a *pencil of*

planes. We can prove that two transversal lines, not meeting the axis line of the pencil of planes, meet the planes of the pencil in two ranges of points which are related to one another. For, let these transversal lines be l_1 and l_2; let a particular plane, ϖ, of the pencil meet l_1, l_2, respectively, in P_1 and P_2. Take two arbitrary fixed points A, B, lying on the axis line of the pencil. The lines AP_1 and BP_2, both lying in the plane ϖ, meet in a point, say P. These lines AP_1 and BP_2 lie also, respectively, in the planes $[A, l_1]$ and $[B, l_2]$, which join A to l_1, and B to l_2, respectively. Let the line common to these planes $[A, l_1]$ and $[B, l_2]$ be called l. Then, the point, P, common to the lines AP_1 and BP_2, must be among the points common to these two planes, namely, P is on l. Wherefore, the range (R_1) determined on l_1 by the planes of the pencil, and the range (R_2) determined on l_2 by the planes of the pencil, are both in perspective with a range (R), of points P on the line l, the centres of these two perspectivities being, respectively, the points A and B. It follows that the ranges (R_1), (R_2) are related, both being related to the range (R) of points P in which the line l is met by the planes of the pencil.

We may therefore speak *of two pencils of planes in space, with different axes, or the same axis, as being related, when they are met by two transversal lines in two ranges of points which are related. Or also, if the pencils of planes have different axes, when they meet the same line in coincident ranges.*

Ex. For an interesting corollary, let a, b, c be three lines in space, no two having a common point. Take points P_1, P_2, ... on the line a; from these let the transversals $P_1Q_1R_1$, $P_2Q_2R_2$, ... be drawn, to meet the line b in the range Q_1, Q_2, ..., and the line c in the range R_1, R_2, ... Then the three ranges $(P_1, P_2, ...)$, $(Q_1, Q_2, ...)$, $(R_1, R_2, ...)$, respectively on the lines a, b, c, are related. For, let the transversals $P_1Q_1R_1$, $P_2Q_2R_2$, ... be called l_1, l_2, ...; then the ranges $(Q_1, Q_2, ...)$, $(R_1, R_2, ...)$, on b and c, are sections, by b and c, respectively, of the pencil of planes, having a as axis, $[a, l_1]$, $[a, l_2]$, ... Thus these two ranges are related. Similarly, any other two of the three ranges are related to one another.

(8) Two related ranges of points may lie on the same line; there are then properties arising from the coincidence on this line of corresponding points of the two ranges. Let O, U be the symbols of two fixed points on the line, and the symbols of two other points of the line be, respectively, $O + xU$, $O + yU$, where x, y are numbers. If the point $O + xU$ vary on the line, and the point $O + yU$ describe a range on the line related to that described by the former point, then x, y are connected by an equation of the form

$$ayx + by + cx + d = 0,$$

where a, b, c, d are numbers whose ratios are constant.

There will, therefore, in general, be two points, X, Y, of the line, at which the corresponding points $O + xU$, $O + yU$ coincide with one another, these being the points of symbols $X = O + \xi U$ and $Y = O + \eta U$, wherein ξ, η are the roots of the quadratic equation in t given by $at^2 + (b + c)\,t + d = 0$.

First, suppose these points X, Y are different. Then, we may refer the points of the two ranges on the line to these points X, Y, denoting the symbols of corresponding points by $X+\lambda Y$ and $X+\mu Y$, where λ, μ are numbers. The fixed linear relation connecting λ and μ must then be such that $\mu=0$ when $\lambda=0$, and $\mu=\infty$ when $\lambda=\infty$; thus the linear relation takes a form $\mu=m\lambda$, where m is a fixed number. The symbols of corresponding points of the two ranges are then of the forms $X+\lambda Y$ and $X+m\lambda Y$, where λ varies.

It follows from this (No. (1) above), if corresponding points of the two related ranges be denoted by P and Q, respectively, that the cross ratio $(Q, P/X, Y)$ is equal to the constant number m. Conversely, this equation defines what is meant by saying that two ranges of corresponding points P, Q, on the line, are related, in case there are two distinct points of coincidence of corresponding points. By direct algebra we may verify that, if x, y be variable numbers, which are connected by the equation $ayx+by+cx+d=0$, in which a, b, c, d are numbers, of fixed ratios, and if ξ, η be the two values arising from this equation when $x=y$, supposed to be distinct, then the equation is equivalent to

$$\frac{y-\xi}{y-\eta}\Big/\frac{x-\xi}{x-\eta}=m,$$

where $m=(c-b+D)/(c-b-D)$, the number D being such that $D^2=(c-b)^2+4\,(cb-ad)$.

When the roots of the quadratic equation which determines ξ and η are equal, for which the condition is $D=0$, it can be proved that the relation connecting x and y can be put into the form

$$(y-\xi)^{-1}=(x-\xi)^{-1}+h,$$

where ξ is the repeated root of the quadratic equation, and h is a constant number, equal to $2a/(b-c)$. For this case, the points, X, Y, of coincidence, of corresponding points of the two ranges, coalesce with one another. If we choose this point of coalescence, X, for one of the two origins on the line, the other origin being denoted by O, and so write the symbols of corresponding points of the two related ranges in the form $O+x_1X$, $O+y_1X$, then the quadratic equation which gives the coincidences of corresponding points must have both its roots equal to ∞. Thus we see that the relation connecting x_1 and y_1 is of the form $y_1=x_1+k$, where k is a constant number, the numbers a, b, c, d used in general being such that $a=0$, $b+c=0$. Corresponding points of the two related ranges then have symbols $P=O+\lambda X$ and $Q=O+(\lambda+k)\,X$, where λ is a variable number. The point O may be taken arbitrarily on the line. Both in this case, and in the preceding case in which the points of coincidence are distinct, we see that the correspondence is determined by the assignment of the points (or point) of coincidence, of corresponding points of the two related ranges, together with the assignment of a particular pair of corresponding points. For these data enable us to determine the fixed numbers, m, or k, which enter into the formulae for the general pair of corresponding points.

(9) We may illustrate the two cases, considered in (8), geometrically. Let us suppose first, as is suggested by what we have proved above, that the two related ranges of points $(P, \ldots)$, $(Q, \ldots)$, on a line l, arise by projection of a range of points $(A, \ldots)$, on a line m, respectively from the centres of projection H, K; both l and m, and the points H, K, being in a plane. Let X be the point common to the lines l, m, and the line HK meet the line l in the point Y. Then, the corresponding points P, Q coincide in X, when the point A, of the line m, of which they are projections, is at X, on the line m; they also coincide, in Y, when A is at the point Z of the line m which lies on HK. And these points of coincidence, X and Y, coalesce when the line HK passes through X. In the general case, the cross ratio $(Q, P/X, Y)$ is equal to $(K, H/Z, Y)$, as we see by projecting from A; and this is the same for all points A of the line m. Conversely, this construction shews how two related ranges, on the same line, can be obtained by projection of a single range on another line, from two different centres.

(10) Taking the case when the points X, Y, of coincidence of corresponding points, of two related ranges on a line, do not coalesce, there is a particular case which is of great interest. If P_0, Q_0 be corresponding points, respectively, of the two ranges, not coincident with one another, it may happen that when the varying point P of the first range is at Q_0, then the corresponding point Q of the second range is at P_0.

If this is so for one particular pair of corresponding points P_0, Q_0, it is so for every pair. For, if the equations

$$ay_0x_0 + by_0 + cx_0 + d = 0, \quad ax_0y_0 + bx_0 + cy_0 + d = 0,$$

which express that it is so, are both satisfied, then, by subtraction of these, $(b - c)(x_0 - y_0) = 0$. We have supposed that $x_0 - y_0$ is not zero. Hence $b = c$. The relation connecting any two corresponding points of the two ranges, whose symbols are $O + xU$, $O + yU$, is then the symmetrical relation $ayx + b(y + x) + d = 0$.

In this case, the pair P, Q, of any two corresponding points of the two ranges, constitute a pair of an involution of pairs of points on the line. The two coincidences X, Y, of corresponding points of the two ranges, are then the foci of the involution. And, if we apply the formulae found above, for the symbols of two corresponding points, referred to the points of coincidence X, Y, namely $P = X + \lambda Y$, and $Q = X + m\lambda Y$, the condition for the particular case considered is that when $P = X + m\lambda Y$, then $Q = X + \lambda Y$. This requires $m^2 = 1$. The ranges would be coincident if $m = 1$. Thus $m = -1$, and the two corresponding points have symbols $P = X + \lambda Y$ and $Q = X - \lambda Y$; that is, the corresponding points are harmonic conjugates in regard to X and Y.

This also leads to a form of the condition that three pairs of points on a line, say A, A'; B, B' and C, C', should be in involution, which it is often convenient to use. For the condition that the two ranges P_1, P_2, P_3, P_4 and Q_1, Q_2, Q_3, Q_4 should be related is, we have seen, that $(P_1, P_2/P_3, P_4) = (Q_1, Q_2/Q_3, Q_4)$. Thus, for the case when corresponding pairs of the two related ranges are a pair of an involution, we have $(A, B/C, C') = (A', B'/C', C)$. And this is sufficient for A, A'; B, B'; C, C' to be pairs of an involution. We may obtain this result also directly with use of the symbols. For, if the symbols of A, B, A', B', in terms of the symbols of C, C', be written

$$A = C + \alpha C', \quad B = C + \beta C', \quad A' = C + \alpha' C', \quad B' = C + \beta' C',$$

where α, β, α', β' are numbers, the cross ratios $(A, B/C, C')$ and $(A', B'/C', C)$ are, respectively, α/β and β'/α'. The equality of these requires $\alpha\alpha' = \beta\beta'$; and this expresses that A, A'; B, B' and C, C' are pairs of an involution (Chap. VII (4)).

The deduction of an involution from two symmetrically related ranges fails, if the points of coincidence of corresponding points coalesce in one point. But, we can consider pairs of points, each consisting of a certain fixed point taken with a varying point of the line, as forming a degenerate involution.

Ex. As an example of the method of regarding an involution which has just been explained, we may consider the proof of a result stated in the preceding chapter (VII (4), Ex. 15). If A, B', C, A', B, C' be six points in a line, three involutions may be defined: (a) One determined by the two pairs of points B, C' and B', C; (b) Another determined by the two pairs of points C, A' and C', A; and (c) A third determined by the two pairs A, B' and A', B. Two involutions on the same line have a common pair of points; let X, X' be the pair common to the two latter involutions, (b) and (c). Then, because C, A'; C', A; X, X' are three pairs in involution, we have

$$(C, C'/X, X') = (A', A/X', X);$$

also, because A, B'; A', B; X, X' are three pairs in involution, we have $(B, B'/X, X') = (A', A/X', X)$. Whence we have

$$(C, C'/X, X') = (B, B'/X, X'),$$

of which the latter is equal to $(B', B/X', X)$. Then, from

$$(C, C'/X, X') = (B', B/X', X),$$

we infer that the three pairs B, C'; B', C; X, X' are in involution. Thus the three involutions (a), (b), (c) have a common pair, namely X, X'.

Also, from $(A, A'/X, X') = (B, B'/X, X') = (C, C'/X, X')$, it follows by (8), that if we determine two related ranges on the line by the condition that the three points A, B, C, of one of these ranges, correspond respectively to the points A', B', C' of the other range, then the points X, X' are the points of coincidence of corresponding points of these two ranges. Conversely, if X, X' be defined by this condition, the fact that they are a common pair of the involutions (a), (b), (c) is clear.

(11) We now give various examples of the preceding theory, partly for the sake of the results obtained, partly to make the ideas familiar.

Ex. 1. The pair of coincidences of two related ranges on the same line, determined by the assigned correspondence of points A, B, C of one range to points A', B', C' of the other, just considered (10), has importance when we consider two ranges on a line which are related by the fact that points A, B, C of one range correspond, respectively, to the same points of the line, taken in another order, considered as points of the other range. Suppose, for instance, that the points A, B, C of one range correspond, respectively, to the points B, C, A of the second range. Then, if H be one of the points of coincidence of corresponding points of the two ranges, the cross ratios $(H, A/B, C)$, $(H, B/C, A)$ are equal. Let the symbols of A, B, C, H, referred to two arbitrary points, O, U, of the line, be, respectively, $A = O + aU$, $B = O + bU$, $C = O + cU$, $H = O + hU$, where a, b, c, h are numbers. Then the condition of equality of the two cross ratios is

$$\frac{h-b}{h-c} \Big/ \frac{a-b}{a-c} = \frac{h-c}{h-a} \Big/ \frac{b-c}{b-a}.$$

Now, for brevity, denote $(h-a)(b-c)$, $(h-b)(c-a)$, $(h-c)(a-b)$, respectively, by p, q, r; so that $p + q + r = 0$; then the condition put down is $pq = r^2$, which is $p^2 + pq + q^2 = 0$; thus $p/q = \omega$, or $= \omega^2$, where ω is an imaginary cube root of unity. The two equal cross ratios, which are $-q/r$ and $-r/p$, are therefore both equal to $-\omega$, or to $-\omega^2$; and the cross ratio $(H, C/A, B)$ has the same value. Thus we may distinguish H from the other point of coincidence, K, for the two ranges $(A, B, C, \ldots)$, $(B, C, A, \ldots)$, by putting

$$(H, A/B, C) = (H, B/C, A) = (H, C/A, B) = -\omega,$$

$$(K, A/B, C) = (K, B/C, A) = (K, C/A, B) = -\omega^2.$$

Ex. 2.   Let X, X' be the points of coincidence of corresponding points of two related ranges $(A, B, C, \ldots)$, $(A', B', C', \ldots)$, on the same line. If we apply to every one of the parametric numbers which give the positions of $A, B, C, A', B', C', X, X'$ on the line, the same fractional linear substitution, with numerical coefficients, we shall obtain two other ranges $(A_1, B_1, C_1, \ldots)$, $(A_1', B_1', C_1', \ldots)$, on the line, also related to one another; and the coincident corresponding points X, X' will be changed into the coincident corresponding points X_1, X_1' of the new ranges. This is clear from what has been said, since a cross ratio is unaltered by applying the same fractional linear substitution to each of the four numbers by which the cross ratio is defined. Such a substitution, as we have seen, may be defined by assigning the points A_1, B_1, C_1 which correspond respectively to A, B, C.

Ex. 3.   Returning to the particular relation considered in Ex. 1, there will, after Ex. 2, be no essential loss of generality in supposing that the points A, B, C, there considered, are the points whose symbols, in regard to two fixed origins, O, U, of the line are, respectively, $O + U, O + \omega U, O + \omega^2 U$. It can then be shewn that the coincident corresponding points H, K, there considered, are the points O and U. If, for brevity, we denote the points by the parametric numbers which, respectively, define them, we find in fact for the cross ratios

$$(0, 1/\omega, \omega^2) = (0, \omega/\omega^2, 1) = (0, \omega^2/1, \omega) = -\omega,$$

$$(\infty, 1/\omega, \omega^2) = (\infty, \omega/\omega^2, 1) = (\infty, \omega^2/1, \omega) = -\omega^2.$$

The coincident corresponding points H, K, of the two related ranges $(A, B, C, \ldots)$, $(B, C, A, \ldots)$, are the points which are also called *the Hessian points* of the triad A, B, C. In terms of the symbols of these points H, K, as we have now proved, the points of the triad have the symbols $H + K$, $H + \omega K$, $H + \omega^2 K$. But, if m be an arbitrary number, the triad of points whose symbols are $H + mK$, $H + m\omega K$, $H + m\omega^2 K$, has also the points H, K for Hessian points. Denoting $H + mK$, etc., by A, B, C, the reader may easily verify that the three harmonic points A', B', C', given by

$$A' = (B, C)/A, \quad B' = (C, A)/B, \quad C' = (A, B)/C,$$

are, respectively, the same as the three harmonic points $(H, K)/A$, $(H, K)/B$,

$(H, K)/C$. The algebraic relations will later receive geometrical interpretations which simplify them.

Ex. 4. Let A, B, C be any three different points of a line. Then, by suitable choice of the point P on the line, the cross ratio $(P, C/A, B)$ can be made to take any assigned value. In particular, the cross ratio takes the values $0, 1, \infty$ when P is respectively at A, or C, or B. These values $0, 1, \infty$ also arise, respectively, when, for any general point P, the points B, C coincide, or the points A, B coincide, or the points C, A coincide.

Ex. 5. If A, B, C, P be points of a line, the equation

$$(P, C/A, B) = (C, P/A, B),$$

when P does not coincide with C, leads to $(P, C/A, B) = -1$, and $P = (A, B)/C$. For four points in this special relation, the six values generally possible for the cross ratios of four points (as in the table above, in (1)) then reduce to only three values, which are $-1, 2$ and $\frac{1}{2}$.

And, if A, B, C, P be four points of a line such that $(P, C/A, B) = -\omega$, where ω is one of the imaginary cube roots of unity, the six values in question reduce to only two values, which are $-\omega$ and $-\omega^2$.

Ex. 6. It is of some interest to form the sextic equation satisfied by the six values of the cross ratios of four general points of a line. By what was remarked (in (1)), this equation, if satisfied by a root ϵ, will also be satisfied by $\epsilon^{-1}, 1-\epsilon, (1-\epsilon)^{-1}, 1-\epsilon^{-1}, \epsilon(\epsilon-1)^{-1}$. The equation, if the four points be of respective symbols $O + aU, O + bU, O + cU$ and $O + hU$, is in fact

$$I_1^3/J_1^2 = 8\,\phi^3/\psi^2,$$

where ϕ denotes $\epsilon^2 - \epsilon + 1$ and ψ denotes $(\epsilon + 1)(\epsilon - 2)(2\epsilon - 1)$, while, in terms of p, q, r, which denote, respectively, $(h - a)(b - c), (h - b)(c - a), (h - c)(a - b)$, the functions I_1, J_1 are given by

$$I_1 = p^2 + q^2 + r^2, \quad J_1 = (q - r)(r - p)(p - q).$$

This result may be obtained by direct computation; or, having proved that the equation is independent of the origins O, U, by taking these origins at A, B, and supposing the symbols of C, P to be $A + B$ and $A + \epsilon B$. Further, if the quartic equation in t whose roots are a, b, c, h be written

$$c_0 t^4 + 4c_1 t^3 + 6c_2 t^2 + 4c_3 t + c_4 = 0,$$

then, with $\lambda = c_0^2/24$ and $\mu = c_0^3/432$, the functions λI_1 and μJ_1 are, respectively,

$$c_0 c_4 - 4c_1 c_3 + 3c_2^2 \quad \text{and} \quad \begin{vmatrix} c_0, & c_1, & c_2 \\ c_1, & c_2, & c_3 \\ c_2, & c_3, & c_4 \end{vmatrix}.$$

It is true too, as follows from a comparison of Exx. 1 and 6, that the Hessian points H, K of a triad of points A, B, C are such that the function I_1 vanishes for each of the two sets of points A, B, C, H and A, B, C, K; while J_1 vanishes when two of the four points A, B, C, H are harmonic conjugates in regard to the other two. These facts enable us to forecast the form of the sextic equation for ϵ.

Ex. 7. Let the sides BC, CA, AB of a triangle be met by three lines, in points, respectively, L_1, M_1, N_1 on the first line, points L_2, M_2, N_2 on the second line, and points L_3, M_3, N_3 on the third line. We can then determine three related ranges, one on each of these sides, by the conditions that L_1, L_2, L_3 on BC correspond, respectively, to M_1, M_2, M_3 on CA and to N_1, N_2, N_3 on AB. Conversely, *suppose we have three related ranges, one on each of these sides. There are then three lines each meeting the respective sides in corresponding points of these ranges.* For, (see Chap. IV (4), Ex. 6), the condition that the points, lying respectively on BC, CA, AB, of symbols $B - xC, C - yA, A - zB$, should lie in a line is $xyz = 1$. If x vary, but y and z are given fractional linear functions of x, the equation $xyz = 1$ is a cubic equation in x, with given numerical coefficients; to each root corresponds a definite value for each of y and z.

Dually, if we have three related pencils of lines in a plane, with centres at the points D, E, F, there are three cases in which three corresponding rays, of the three pencils, meet in a point. Conversely, suppose D, E, F and I, J, K be given points in general positions in a plane. Then, we may determine three related pencils of lines, with centres, respectively, at D, E, F, by the condition that the rays EI, EJ, EK correspond, respectively, to the rays DI, DJ, DK, as do the rays FI, FJ, FK. To any arbitrary line, p, through D, there will then correspond definite lines, q and r, respectively, through E and F. Denoting the respective intersections (q, r), (r, p), (p, q) by A, B, C, we thus have an infinite number of triangles, ABC, whose sides pass, respectively, through D, E, F. If one of these triangles be given, with the points D, E, F lying, respectively, on the sides BC, CA, AB of this, and the two points I, J be also given, the third point K can be determined.

Ex. 8.   We have seen that two involutions, of pairs of points, upon the same line, have a single pair in common. Such an involution arises from two ranges upon the line which are related to one another in a symmetrical manner.

Now consider two ranges upon the line, which are related to one another, but not in this symmetrical manner. Let us think of a general point, P', of the second of these ranges, as arising from the point, P, of the first range, to which it corresponds, by a definite operation; the numerical parameter, x', which defines P', is in fact a definite fractional linear function of the parameter x, by which P is defined. We may express this by writing $P' = \vartheta\,(P)$, and then $P = \vartheta^{-1}\,(P')$. Suppose, further, that upon this same line, we have another pair of related ranges, respectively of points Q and Q', denoted by $Q' = \phi(Q)$.

It happens then twice that when P and Q coincide, also P' and Q' coincide. For, taking the points P', Q', arising in these two cases, from the point of coincidence of P and Q, which will be denoted by $P' = \vartheta(P)$, and $Q' = \phi(P)$, we can, from the fractional linear equation denoted by $P = \vartheta^{-1}(P')$, obtain the fractional linear equation connecting the parameters of Q' and P', and so write $Q' = \psi(P')$, where ψ denotes $\phi\vartheta^{-1}$. Thus the points P', Q', as $P(=Q)$ varies, describe related ranges on the line; and two related ranges have two points of coincidence of corresponding points.

Thus we may say, adopting a definite order of consideration for two related ranges of points on a line, that, *if we have two pairs of related ranges on the line, there are two pairs of points, of which each pair consists of a point and the point corresponding to it in both modes of relation.* When both pairs of related ranges consist of pairs of points of an involution (not identical involutions), the two pairs of corresponding points are the same.

Ex. 9.   Consider two related ranges on a line, the corresponding points P, Q of these ranges being given respectively by the parameters x and y. Suppose, first, that the points of coincidence of corresponding points are different, being given by the parameters ξ and η. Then, as we have seen, (8), the numbers x, y are connected by an equation

$$\frac{y-\xi}{y-\eta} \Big/ \frac{x-\xi}{x-\eta} = m,$$

where m is a fixed number. Introducing a third parameter z, and an arbitrary fixed number c, this equation is obtainable as a consequence of the two relations

$$\frac{y-\xi}{y-\eta}\cdot\frac{z-\xi}{z-\eta}=c, \qquad \frac{z-\xi}{z-\eta}\cdot\frac{x-\xi}{x-\eta}=c/m,$$

of which the first expresses that the point Q, and a point R given by the parameter z, are together a pair of a definite involution on the line, and the second expresses that the point R and the point P are together a pair of another definite involution on the line. Wherefore, *the correspondence of the points P, Q, in the two related ranges, may, in an infinite number of ways, be regarded as arising from two involutions, in which P, R are a pair of one of these, and R, Q a pair of the other.*

If the coincident corresponding points, of the two given related ranges, coalesce with one another, the equation expressing that the ranges are related, which is now of the form $(y-\xi)^{-1}=(x-\xi)^{-1}+h$, as we have seen, may be similarly regarded as arising from the two equations

$$(y-\xi)^{-1}+(z-\xi)^{-1}=c, \quad (z-\xi)^{-1}+(x-\xi)^{-1}=c-h,$$

in which c is an arbitrarily chosen constant number, and z is a parameter. Both these are equations expressing an involution.

We may also deduce the result otherwise, with the symbols. First, when the coincident corresponding points of the related ranges are different, if X, Y be the symbols of these, the symbols of a general pair of corresponding points, in the two ranges, may be taken as $P=X+\lambda Y$, and $Q=X+m\lambda Y$, where m is a fixed number, and λ varies. Take then a point R, corresponding to P, of symbol $R=X+c\lambda^{-1}Y$, where c is an arbitrary fixed number. Then, as λ varies, the pair P, R, with respective symbols $X+\lambda Y$, $X+\lambda'Y$, in which $\lambda\lambda'$ is equal to the constant c, belong to an involution (above, Chap. VII (4)), of which X, Y form a pair; and R, Q, with respective symbols $X+c\lambda^{-1}Y$, $X+m\lambda Y$, also belong to (another) involution, of which X, Y form a pair. Next, when the coincident corresponding points, of the given related ranges, coalesce, let X be this point of coalescence, and O another arbitrary point of the line. Then the symbols of corresponding points P, Q of the two ranges may be taken to be (above, in (8)) $P=O+\lambda X$, and $Q=O+(\lambda+k)X$, where k is a fixed number, and λ varies. Take then a point R of symbol $O+(c-\lambda)X$, where c is an arbitrary fixed number. Then, as λ varies, the pair P, R, with symbols $O+\lambda X$, $O+\lambda_1 X$, in which $\lambda+\lambda_1$ is equal to the constant c, belong to an involution; while R and Q, of symbols $O+(c-\lambda)X$, $O+(\lambda+k)X$, likewise are a pair of (another) involution. We may write $P=O+\tfrac{1}{2}c\,X+(\lambda-\tfrac{1}{2}c)\,X$, and $R=O+\tfrac{1}{2}c\,X-(\lambda-\tfrac{1}{2}c)\,X$, so that P and R are harmonic conjugates in regard to the point X and the point whose symbol is $O+\tfrac{1}{2}c\,X$. We may also write

$$R=O+\tfrac{1}{2}\,(c+k)\,X+\tfrac{1}{2}\,(c-k-2\lambda)\,X$$

and

$$Q=O+\tfrac{1}{2}\,(c+k)\,X-\tfrac{1}{2}\,(c-k-2\lambda)\,X,$$

so that R and Q are harmonic conjugates in regard to X and the point whose symbol is $O+\tfrac{1}{2}\,(c+k)\,X$.

Ex. 10.   Let A, B, C, D be any four points of one plane, of which no three are in a line, and A', B', C', D' be any four respectively corresponding points of another (or the same) plane, of which, also, no three are in a line. Shew then that, to an arbitrary point P of the former plane, there corresponds a point P' of the latter plane, such that the pencil of four lines, joining any one of the five points A, B, C, D, P to the other four, is related to the pencil of four lines joining the corresponding point of the second plane to the respectively corresponding points of this second plane.

We may suppose the symbols of A, B, C, D and of A', B', C', D' to be so chosen that $D=A+B+C$, and $D'=A'+B'+C'$. Then, if the symbol of the general point P be written $xA+yB+zC$, where x,y,z are numbers, the symbol of the corresponding point P' will be capable of the form $xA'+yB'+zC'$.

CHAPTER IX

INTRODUCTION TO THE THEORY OF CONICS. FUNDAMENTAL GENERAL THEOREMS

So far, the only loci in a plane we have dealt with have been lines. A line is met by any general line, in its plane, in *one* point. We consider now certain curves, in a plane, of which any one is met by a general line, of its own plane, in *two* points. These curves are those which were recognised by the Greeks as sections of a cone, made by a plane. They are then, properly, *Conic Sections*; we shall speak of them as *Conics*. The notion of related ranges, and related pencils, dealt with in the preceding chapter, furnishes a very simple approach to the theory of these curves.

Besides the introductory properties of these curves, we prove also, in the examples at the end of the present chapter, some theorems, less immediately obvious, which we use in later chapters.

(1) **Definition of a Conic.** We have explained what is meant by saying that a range of points on a line is related to a range of points on another (or the same) line; and also what is meant by related pencils of lines in a plane. Two such pencils are related when the range, which one pencil determines on any line of the plane, is related to the range which the other pencil determines on another (or the same) line of the plane.

Let A, B be the respective centres of two such related pencils of lines, lying in the same plane. To any ray of the first pencil corresponds then a definite ray of the second pencil, and conversely. Let P be the point common to two such corresponding lines. The locus of the point P, for all pairs of corresponding lines through the centres A, B, is a curve; this is the curve which we call here a *Conic*. The locus passes through the point A; for, to the line BA of the second pencil, drawn through B, corresponds a certain line through A; and the common point of these two lines is the point A. Similarly the locus passes through B.

Thus, if P be any point of the conic, this curve meets the line AP in two points, namely at A and P; likewise the curve meets the line BP in two points. But, more generally, the curve meets *any* line in the plane in two points (though these may coalesce). For, the related pencils, with centres at A and B, meet this line, respectively, in two ranges, which are related; and, we have seen, for two such ranges there are two points where corresponding points coincide with one another. The line, therefore, contains two of the points P where corresponding rays of the two pencils meet one another.

It may happen, however, that the line chosen is such that the two coincidences of corresponding points, which lie upon it, coalesce with one another. Then the conic has in common with the line only this one point of coalescence. Such a line is called a *tangent line* of the conic, or, simply, a *tangent*. It will appear that, through every point P of the conic, there passes a single line having this property, the point of coalescence on this line being the point P itself. Thus we say that, at every point P of the conic, the conic has a tangent; and that the conic and the line *touch one another* at this point, the point P itself being called the *point of contact*. For the points A and B, in particular, which will be found to be quite general points of the conic, it is easy to specify the tangents. For, we have remarked that the line AP, joining the centre A to any point P of the conic, meets the conic in the two points A, P; here AP is the ray of the pencil (A), of centre A, which corresponds to the ray BP of the other pencil (B). If we take the ray BA of this pencil (B), the point P coalesces with A. Thus *the tangent of the conic at the point A is the line of the pencil (A) which corresponds to the line BA of the pencil (B)*. Similarly, the tangent of the conic at B is the line of the pencil (B) which corresponds to the line AB of the pencil (A). We shall shew how to construct the tangent at any other point of the curve.

(2) **The case when the conic degenerates into two lines.** There is however a case in which the reasoning just given clearly fails, namely, when the pencils (A), (B) are so related that the ray AB of the pencil (A) corresponds to the ray BA of the pencil (B), as may happen. Then the conic breaks up into two lines, the line AB itself being one of these. For, let P be the common point of one pair of corresponding lines of the pencils (A), (B), and Q another such point; and let the line PQ meet the line AB in the point R. Then, on the line PQ, the related pencils determine related ranges in which P and Q are coincidences of corresponding points; and R is a third such co-incidence on this line, since, by the hypothesis now being made, the line AB is a self-corresponding ray of the two pencils. Thus, by the theory of such ranges, the two ranges on the line PQ entirely coincide, and every point of the line PQ is a common point of corresponding rays of the two pencils. Thus, the conic, as defined, contains the line PQ as part of itself; and it is clear, conversely, that the conic contains no points which do not lie on this line, except points of the line AB.

It remains true that an arbitrary line of the plane meets the de-generate conic in two points. Also, it follows in a similar way that a conic containing three points which lie in a line breaks up into two lines, of which this line is one. In what follows, when we speak of a conic, we suppose that it is not degenerate, unless the fact is mentioned.

(3) **A conic is determined by any five of its points.** If A, B, A', B', C' be any five points of a plane, of which no three are in line, there exists a conic passing through these five points. For, we have only to take pencils of lines, with centres at two of these points, say A and B, which are related to one another by the condition that the rays joining

A to the points A', B', C' shall correspond respectively to the rays joining *B* to these same three points. The locus of the common point of two corresponding rays of these pencils is then a conic to which all the five points belong. The points *A*, *B* may, for a moment, be called the *centres*, and A', B', C' the *determining points*, in this process of defining the conic.

But, in fact, there is only one conic containing the five given points; if *any* two of the five points be chosen as centres, and the remaining three as determining points, the same conic is found.

To prove this, we shew first that the points A', B', which are any two of the determining points, may be taken as centres instead of *A*

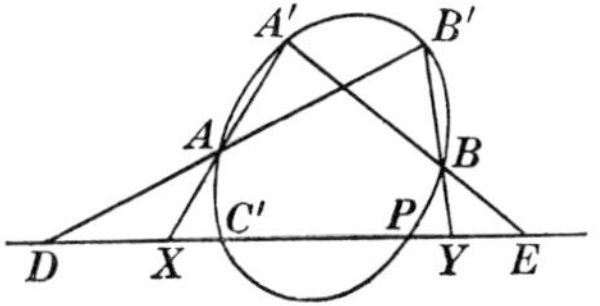

and *B*; then *A*, *B*, C' will be the determining points in the new process. Let *P* be any point of the conic which arises when *A*, *B* are the centres; so that the pencil of four rays joining *A* to the points A', B', C', *P*, say the pencil $A (A', B', C', P)$, is related to the pencil of four rays which join *B* to the same points, say the pencil $B (A', B', C', P)$. We shew that *P* belongs also to the conic which is determined by taking A', B' as the centres. For this, it is sufficient to shew that the pencil of four lines joining A' to the points (A, B, C', P) is related to the pencil $B' (A, B, C', P)$. Consider the line PC'; let this be met by the lines $A'A$ and $B'B$, respectively, in *X* and *Y*, and be met by the lines $A'B$ and $B'A$, respectively, in the points *E* and *D*. Then, because the pencils $A (A', B', C', P)$ and $B (A', B', C', P)$ are related, the ranges (X, D, C', P) and (E, Y, C', P) are related; of these (Chap. VIII (1), Ex. 3), the latter is related to the range (Y, E, P, C'). Whence the range (X, D, C', P) is related to the range (Y, E, P, C'); and from this it follows (Chap. VIII (10)), that the three pairs of points *X*, *Y*; *D*, *E*; *P*, C', on the line PC', belong to an involution. This, however, shews that the ranges (X, E, C', P) and (Y, D, P, C') are related; of which, again, the latter is related to the range (D, Y, C', P). The two ranges (X, E, C', P) and (D, Y, C', P) are, however, respectively on the pencils $A' (A, B, C', P)$ and $B' (A, B, C', P)$. These two pencils are therefore related; which is what we desired to prove.

The two points A', B' may then be taken as centres instead of *A*, *B*. In the same way, instead of A', B', we may take any two of the three points *A*, *B*, C' as centres, and find the same conic. And so on. The five points *A*, *B*, A', B', C' thus play symmetrical parts in the determination of the conic.

But, further, if *P* be any point of the conic besides these five points, the conic may be determined equally well from any five of the six points *A*, *B*, A', B', C', *P*. To prove this, it is sufficient to shew that the conic determined by the five points *A*, *B*, A', B', *P* contains the point C'; for, by what we have just shewn, the conic determined by any five points is unique. Now, we have proved that the conic

(A, B, A', B', C') meets the line $C'P$ in a point P such that the three pairs of points X, Y; D, E; P, C' are in an involution, whereof the points X, Y, D, E, on this line, are determined by the points A, B, A', B' alone. Thus, the conic (A, B, A', B', P) meets this line in the point C'. As we can thus replace the point C' by any point P of the conic, as originally defined, we can similarly replace any point of the five A, B, A', B', P by any other point Q of the conic.

By such process we infer that, *if any five points be taken on the conic, the curve can be determined from these, with any two of them taken as centres;* and, *if any six points be taken in a plane, of which no three are in a line, the necessary and sufficient condition that these six points lie on a conic is that, if A, B be any two of the six points, the pencil of lines joining A to the other four points shall be related to the pencil of lines joining B to the same four points, taken in the same order.*

(4) **Desargues' theorem for conics having four points in common.** We have proved (Chap. VII (4)) that the three complementary pairs of joins, of four points A, B, A', B', lying in a plane, with no three in a line, meet an arbitrary line of the plane in three pairs of points which belong to an involution. This involution is determined by the two pairs of points arising from the pair of joins AA', BB' and the pair of joins AB', $A'B$. Let C be any point of this arbitrary line; then a conic can be drawn through the five points A, B, A', B', C, as we have just proved. This conic will meet the line in a further point, P; and, by what has just been shewn, the pair of points C, P will belong to the involution on the line determined by the pairs of lines (AA', BB') and $(AB', A'B)$. Thus, the *conics containing four given points A, B, A', B' of a plane, of which no three lie in a line, meet an arbitrary line of the plane, each in a pair of points, such that all these pairs belong to an involution on the line; and among such conics are the degenerate conics each formed by a complementary pair of joins of the four given points.*

For the three degenerate conics, the theorem was essentially known to the Greeks (Pappus, *Collection*, A.D. 300). The general theorem was given by Desargues, essentially, but with a metrical proof.

(5) **Pascal's theorem for six points of a conic.** It is easy now to obtain the celebrated theorem, ascribed to Pascal (1640), that, *if six points on a conic be arranged in any order to form a hexagon, and the three points of intersection of the three pairs of opposite sides of this hexagon be taken, then these three points are in a line.* In other words, if the six points, arranged in this order, be named $UV'WU'VW'$, then the three points of intersection $(VW', V'W)$, $(WU', W'U)$, $(UV', U'V)$ lie in a line.

Consider the conic defined as before from five given points, of which no three are in a line; these points we denote now by P, Q, R, A, B; and we suppose the conic given as the locus of the point of intersection of corresponding rays of two pencils, whose centres are A and B, which are related to one another by the fact that the rays AP, AQ, AR

correspond, respectively, to the rays BP, BQ, BR. Let the given lines AR, PQ meet in R', and BQ, PR meet in Q'.

The pencil A $(P, Q, R, ...)$ determines, on the line PQ, a range $(P, Q, R', ...)$; the pencil B $(P, Q, R, ...)$ determines, on the line PR, a range $(P, Q', R, ...)$; these two ranges are re-lated, and the point P of one range corre-sponds to itself regarded as belonging to the other range. Wherefore (Chap. VIII (2)), the joins of corresponding points of these two ranges all pass through a point. This point, O, may be defined as the intersection of two of these joins of corresponding points, namely, QQ' and RR', which are the lines BQ and AR.

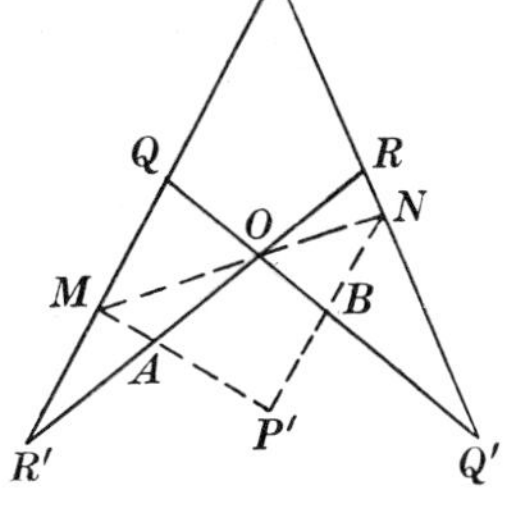

It follows, then, that if any arbitrary line, drawn through O, meets the lines PQ, PR, respectively, in M and N, then, as this line varies, the ranges deter-mined on the lines PQ, PR by the points M, N are those determined by corresponding rays of the two related pencils whose centres are A and B. And, thus, the point, say P', common to MA and NB, is a point of the conic. Wherefore:

(*a*) *When five points P, Q, R, A, B in a plane are given, of which no three are in a line, we have a construction, by intersections of lines only, for determining any number of points P', of the conic determined by these five points. The point P' is the third angular point of a variable triangle, MNP', whose other angular points M, N lie on the given lines PQ, PR, the sides MN, MP', NP' of this triangle, passing respectively through the given points O, A, B, of which O is the common point of the lines AR, BQ. From these given lines PQ and PR, and these given points O, A, B, the points Q, R are determined, respectively, as the intersections (OB, PQ) and (OA, PR).* This formulation is associated with the name of Maclaurin (1719).

(*b*) *If A, P', B, Q, P, R be six points of a conic, regarded as forming a hexagon when taken in this order, the three intersections of opposite sides of this hexagon, namely (AP', QP), $(P'B, PR)$, (BQ, RA), or, respectively, M, N and O, lie upon a line.* This is Pascal's theorem. The converse is also true, and follows from the figure, that the six angular points of a hexagon, whose pairs of opposite sides meet in three points which lie upon a line, lie upon a conic. It is supposed that no three of the six angular points lie on a line.

Six given points form essentially sixty—that is $\frac{1}{12}$ (6!)—hexagons. For, any of the six points can be taken as the first point, and we can describe a hexagon, from the first point, in either direction. There will therefore be 60 lines such as that obtained to contain the three inter-sections of opposite sides of the hexagon. These are generally called the *Pascal lines* arising from the given six points. Some account of the many theorems which hold for the intersections of these Pascal lines is given subsequently, in **Note 3**.

(6) **Pappus' theorem.** In our statement of Pascal's theorem, we have supposed that no three angular points of the hexagon lie on a line. But other cases may arise. In particular, three alternate angular points of the hexagon may lie on one line, and the other three (also alternate) angular points lie on another line. Then Pascal's theorem becomes that which we have called Pappus' theorem (Chap. III (8)). And this theorem can be proved if we assume, as here, that the correspondence of two related ranges of points, each on a line, is determined by the assignment of three points of one range, to correspond to three points of the other range.

(7) **The construction of the tangent of a conic at any one of five points which determine the conic.** If, in the figure by which we construct the point P' of the conic, by means of the variable triangle MNP', whose sides pass through the fixed points O, A, B, we suppose the point N taken at the point, N', common to the lines AB, PR, then the point P' comes to coincidence with the point A. And, if M' be the point common to the lines $N'O, PQ$, then AM' is the ray of the pencil (A) which corresponds to the ray BA (or BN') of the pencil (B). Thus, AM' is the line we have called the *tangent of the conic at the point A.*

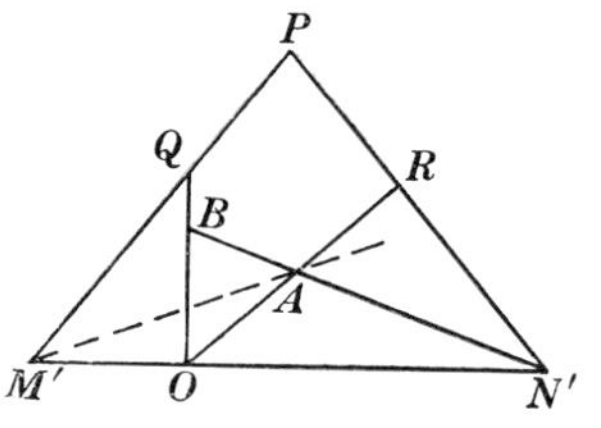

We see that this is suggested by Pascal's theorem when applied to a hexagon A, A, B, Q, P, R, of which two angular points coincide at A. Regarding the line AA as the line of the pencil (A) which corresponds to the line BA of the pencil (B), the three points

$$M', \text{ or } (AA, QP); \quad N', \text{ or } (AB, PR), \quad \text{and} \quad O, \text{ or } (BQ, RA)$$

are in a line; and these are intersections of opposite sides of the particular hexagon chosen.

(8) **The conic as a rational locus.** In our development, the possibility of related ranges on a line is made to depend on the fact that, when three points of the line are given, it is possible to characterise every point of the line by the value of a parametric number. The same possibility holds for the points of a conic.

For, on the conic, take an arbitrary point A; take also an arbitrary fixed line, l, in the plane of the conic. There corresponds then, to any point P of the conic, a definite point, P', of the line l, this being the point where the line AP meets l; in particular, to the point A of the conic corresponds the point where the tangent of the conic, at A, meets the line l. Conversely, any point P' of the line l gives rise to a point P of the conic, this being the point, other than A, where the line AP' meets the conic. In this way, a range of points P', on the line l, corresponds to a pencil of lines passing through A. If B be any other fixed point of the conic, the conic has been defined so that the pencil of lines AP, arising as P varies on the conic, is related to the pencil of lines BP, arising as P varies on the conic. It is clear then that we may speak of a range of points, P, on the conic as being related to another range of points, Q, on the conic. In the first place, this means that the pencil of lines AP, as P varies on the conic, A being an arbitrary fixed point of the conic, is related to the pencil of lines AQ, as Q varies; but we may take, instead of the pencil $A\,(Q)$, the pencil $B\,(Q)$, where B is any other fixed point of the conic. *A conic then, like a line, may be regarded as the seat of related ranges of points; and, also, a varying point of a conic may be characterised, like a varying point of a line, by the value of a number.*

(9) **Involutions on a conic.** One of the primary applications of the theory of related ranges, of points on a line, is to the theory of involutions. If we make the analogous application for related ranges on a conic, this will furnish an illustration of what is said in (8); but it leads also to an interesting geometrical fact.

On a line, we saw that an involution might be regarded as the aggregate of the pairs of points P, Q, which correspond to one another in two ranges which are related to one another in a particular *symmetrical* way; namely, so that the position of Q, when P arrives at the point where Q now is, is the present position of P (Chap. VIII (10)). From this there follows the necessary and sufficient condition for three pairs of points P_1, Q_1; P_2, Q_2 and P_3, Q_3, to belong to an involution, which is expressed, in one of its forms, by the equality of the cross ratios $(P_1, P_2/P_3, Q_3)$, $(Q_1, Q_2/Q_3, P_3)$.

This description of an involution on a line can be transferred in detail to a conic, since we have defined the meaning of related ranges on a conic. We prove now the geometrical fact that, *if P_1, Q_1; P_2, Q_2 and P_3, Q_3 be three pairs of an involution on a conic, as so defined, then the three lines P_1Q_1; P_2Q_2 and P_3Q_3 meet in a point.*

Let the lines P_1Q_1 and P_2Q_2 meet the line P_3Q_3, respectively, in L_1 and L_2; and the lines P_1Q_2 and P_2Q_1 meet the same line P_3Q_3, respectively, in M_1 and M_2. By hypothesis, the pairs of lines which join any point, O, of the conic, to P_1, Q_1; to P_2, Q_2 and to P_3, Q_3, are a pencil in involution. Thus the range on any line determined by the four lines $O\,(P_1, P_2, P_3, Q_3)$ is related to the range determined on this line by the severally corresponding four lines $O\,(Q_1, Q_2, Q_3, P_3)$; and, by the definition of the conic, this latter range may equally be determined by the four lines $O'\,(Q_1, Q_2, Q_3, P_3)$, where O' is any other point of the conic.

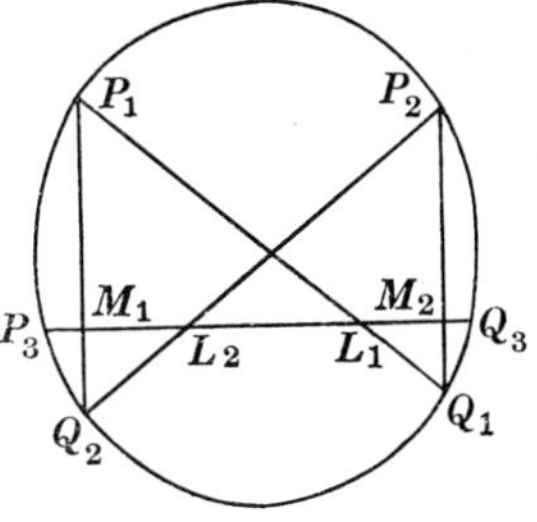

Take, for O, the point Q_2, and, for O', the point P_1. Thus the ranges determined on the line P_3Q_3 by the two pencils $Q_2\,(P_1, P_2, P_3, Q_3)$ and $P_1\,(Q_1, Q_2, Q_3, P_3)$ are related. Whence, the ranges (M_1, L_2, P_3, Q_3) and (L_1, M_1, Q_3, P_3) are related, whereof (Chap. VIII) the latter range is related to (M_1, L_1, P_3, Q_3). Comparing then (M_1, L_2, P_3, Q_3) and

(M_1, L_1, P_3, Q_3), wherein there are three corresponding points which are common to both, we infer that L_2 and L_1 coincide. Thus the line P_3Q_3 contains the common point of the two lines P_1Q_1 and P_2Q_2; which is what we wished to shew.

A very obvious consequence of this is that two given pairs of points of an involution on a conic determine the involution. This is true for an involution on a line, but there is then no such simple construction for the other pairs; the construction we gave was by a quadrangle.

(10) **The foci of an involution on a conic.** In an involution on a line, and, therefore, also on a conic, there are two points at each of which the two points of a pair of the involution coincide with one another. These we have called the *foci* of the involution. In the case of an involution on a line, any pair consists of two points which are harmonic conjugates of one another in regard to these foci.

It follows, then, from what precedes, if T be the common point of the lines P_1Q_1 and P_2Q_2, joining two pairs of points on a conic, say, the intersection of the two *chords* P_1Q_1, P_2Q_2 of the conic, that there are two lines passing through T each of which meets the conic in two points which coincide with one another. If these points be X and Y, the lines TX, TY will be what we have called the tangents of the conic, respectively at X and Y; and further, if O be any point of the conic, the rays OP_1, OQ_1 will be harmonic conjugates in regard to the rays OX, OY. Thus we may also say that, through any point T of the plane of the conic, there pass two (and only two) tangents of the conic; and, if the points of contact of these tangents be X and Y, and TPQ be any line through T which meets the conic in the points P and Q, then, regarded as points of a range on the conic, the points P, Q are harmonic conjugates in regard to X and Y. As this last is a symmetrical relation between the pairs X, Y and P, Q, we can make the inference that the tangents of the conic at P and Q meet on the line XY.

(11) **Alternative view of an involution on a conic.** We can arrange the ideas somewhat differently, and obtain the notion of an involution of pairs of points on a conic in another way.

Let the tangents at two points X, Y, of a conic, meet in a point T; and let a line be drawn through T to meet the conic in two points P, Q; let this line meet the line XY in the point R. The pencils of lines which join the four points X, Y, P, Q, on the conic, to any two points of the conic, are related to one another. Let the centres of these pencils be taken at X and Y; and recall that the ray of the pencil (X), which corresponds to the ray YX of the related pencil (Y), is the tangent of the conic at X, with a similar statement when X, Y are interchanged. Consider the ranges which these related pencils (X), (Y) determine on the line $TPRQ$; they give, respectively, the points T, R, P, Q and R, T, P, Q, arising, respectively, from $X(X, Y, P, Q)$ and $Y(X, Y, P, Q)$; so that we have $(T, R/P, Q) = (R, T/P, Q)$. As T, R do not coincide, we infer from this (Chap. VIII (11), Ex. 5) that the points T, R are harmonic conjugates in regard to the points P, Q, or, say, $R = (P,Q)/T$; and hence, if O be any point of the conic, that the pencil $O\,(X,\ Y,\ P,\ Q)$ is harmonic, the rays OP, OQ being harmonic conjugates in regard to the rays OX, OY. Thus, *regarded as points of a range on the conic, the points P, Q are harmonic conjugates in regard to the points X, Y.* Here, P, Q are the points of the conic which lie on any line drawn through the point T which is common to the tangents of the conic at X and Y. We may then define the pairs of points P, Q, so obtainable, as constituting an involution on the conic, with X, Y as foci.

(12) **Pascal's theorem derived from involutions on a conic.** The notion of an involution on a conic enables us to give a proof of Pascal's theorem, which we have already proved in a different way. It is clear, from what has been said, that, on a conic, an involution of pairs of points is determined by the assignment of two pairs, as was the case for an involution on a line. Let then A, B', C, A', B, C' be six arbitrary points of a conic. We can define three involutions on the conic, one by the two pairs B, C' and B', C; another by the two pairs C, A' and C', A; and the third by the two pairs A, B' and A', B. For six points on a line, it has been remarked (Chap. VII (4), Ex. 15) that these three involutions have a common pair of points. This is then also true for the three involutions on the conic. Let this pair consist of the points X, Y on the conic. Then, by (11), the line XY contains the common point of the lines BC' and $B'C$; and likewise it contains the common point of the lines CA' and $C'A$, and also the common point of the lines AB', $A'B$. That these three common points lie in a line is Pascal's theorem.

(13) **Pascal's theorem derived from related ranges on a conic.** Another proof of Pascal's theorem may be given, which illustrates the fact that we can transfer the theory of related ranges of points on a line to related ranges on a conic. We have shewn that the relation of two corresponding ranges of points on a line is determined by assigning points A', B', C', of one range, to correspond, respectively, to given points A, B, C, of the other range; and we have shewn that there are two points of the line at which corresponding points of the two ranges coincide with one another; these two points may coalesce. All this is then true for points of a conic. Suppose then that A, B, C, A', B', C' are six points on a conic. Determine two related ranges of points on the conic by the condition that the points, A', B', C', of one range shall correspond, respectively, to the points, A, B, C, of the other. Let X, Y be the two self-corresponding points of the two ranges. Let the intersections $(BC',\ B'C)$ and $(AB',\ A'B)$ be denoted, respectively, by L and N, and the intersection $(BB',\ LN)$ by H. There are then two related pencils of lines, one with centre B', joining this to the points $(A,\ B,\ C,\ \ldots)$ of one range on the conic, the other with centre B, joining this to the corresponding points $(A',\ B',\ C',\ \ldots)$ of the related range on the conic. These related pencils give rise to related ranges on the line LN; the pencil (B') gives on this line the range

$(N, H, L, ...)$; the pencil (B) gives the range $(N, H, L, ...)$. The two related ranges on LN thus have three corresponding points which coincide. Thus these two ranges coincide entirely. Wherefore, the line LN contains the two points X, Y of the conic where the related ranges thereon have two coincident common points. Thus the line XY contains the two intersections $(BC', B'C)$ and $(AB', A'B)$; by like reasoning it also contains $(CA', C'A)$. That these three intersections should lie in a line is Pascal's theorem. If the two points X, Y coalesce, the line XY is still definite, being the tangent of the conic at X; this follows from our original definition of the tangent, in (1).

(14) **Symmetrically related ranges of points on the conic.** Consider in particular, in (13), the case when the relation of the ranges $(A, B, C, ...)$, $(A', B', C', ...)$ on the conic is symmetrical, so that the pairs A, A'; B, B'; C, C'; ... belong to an involution as defined in (9). As, then, the points B, B' are interchangeable, the relation of two ranges may be defined by the fact that the points A, B', C, of one of them, correspond, respectively, to the points A', B, C' of the other. Whence, by (13), the two intersections $(BC, B'C')$, $(AB, A'B')$ lie on the line, XY, joining the self-corresponding points of the two ranges. These points X, Y are the foci of the involution A, A'; B, B'; .... By a similar argument, the line XY also contains the intersection $(CA, C'A')$. Wherefore, the two triangles $ABC, A'B'C'$ are in perspective; for they have the line XY as axis of perspective. These two triangles, by Desargues' theorem, have then a centre of perspective; namely, the lines AA', BB', CC' meet in a point. We thus obtain, again, the property of pairs of points in involution on the conic.

(15) **The polar line of a point in regard to a conic.** Let an involution of pairs of points on the conic be determined by the two pairs A, A' and B, B'. Denote the intersections (AA', BB'), $(AB', A'B)$, $(AB, A'B')$, respectively, by T, N and U. By the properties of the quadrangle A, A', B, B', the points R, S, in which the line NU meets the lines AA', BB', are respectively $R=(A, A')/T$ and $S=(B, B')/T$; similarly, the line NT meets the lines $AB, A'B'$, respectively, in $H=(A, B)/U$ and $K=(A', B')/U$.

Wherefore, by what is shewn above, the line NU meets the conic in the two points X, Y, which are harmonic conjugates, in the range of points of the conic, both in regard to the two A, A', and in regard to the two B, B'. And, if any line drawn through T meets the conic in the points P, Q, and meets the line NU in T', then $T'=(P, Q)/T$. We define *the polar of any point T, in regard to the conic, as the locus of the point T' so determined on any line through T which meets the conic in P and Q*; and it appears that this polar is a line; this may be defined, either as the join of the two points X, Y of the conic at which the tangents pass through T; or, as the join of the two intersections

$$N=(AB', A'B) \quad \text{and} \quad U=(AB, A'B'),$$

if TAA', TBB' be any two lines through T which meet the conic, respectively, in A, A' and in B, B'. In the same way, the polar of the point U is the line TN. But, also, as appears from the relation $T' = (P, Q)/T$, or $T, T' \sim P, Q$, it is true that, *if the polar of a point T contains the point T', then the polar of T' contains the point T.* As a consequence of this, because N lies on the polar of T, and also lies on the polar of U, it follows that the line TU is the polar of N.

A triangle, such as this triangle TNU, of which each angular point has the opposite side for polar, in regard to the conic, is called *a self-polar triangle.* We see also that, if four arbitrary points be taken on a conic, the vertex triangle of the quadrangle formed by these is a self-polar triangle in regard to the conic.

There is only one point of which a given line is the polar line; namely, the common point, T, of the tangents of the conic, at the points X, Y in which the line meets the conic. This point is called the *pole of the line*, in regard to the conic. This point can, by what is said, be constructed without use of the common points X, Y of the given line and the conic; namely, as the common point of the polar lines of any two points of the given line. Also, as we have seen, the polar line of a given point can always be constructed, by drawing two suitable chords of the conic through the given point.

Two points in the plane of a conic, of which either lies on the polar line of the other, are said to be *conjugate in regard to the conic.* Likewise, *two lines* are said to *be conjugate* when each contains the pole of the other. These definitions are often applied.

(16) **Interpretation of theorems for involutions by means of a conic.** We have regarded the properties of related ranges, and of involutions, on a conic, as deduced from the properties of ranges on a line. But the converse procedure is often convenient, and useful; especially when the properties of conics have been further developed. In particular, theorems for involutions on a line have often interesting geometrical interpretations on a conic. We illustrate this by considering some of the theorems for involutions given in the examples at the end of Chapter VII.

Ex. 1.   Let NTU be any self-polar triangle, in regard to a conic; and P be any point of the conic. Let the lines TP, UP meet the conic again, respectively, in Q and R. Then (15), the points P, Q are what we have (Chap. IV (4), Ex. 7) called harmonic images of one another in regard to the point T and the line NU; likewise, the points P, R are harmonic images of one another in regard to the point U and the line NT. Whence (*ibid*), the points Q, R are harmonic images of one another in regard to the point N and the line TU, and the chord QR passes through N. *It is possible, therefore, to inscribe in the conic an infinite series of triangles PQR, with sides passing through the angular points of a self-polar triangle, any one of the points P, Q, R being an arbitrary point of the conic.* This, with a different notation, is the result of Ex. 9, Chap. VII, when interpreted on a conic.

It should be remarked also that, if we take the pairs of intersections, with the conic, of the respective sides NU, NT, TU of the self-polar triangle, then every one of these pairs consists of points, harmonically conjugate, in the range of points on the conic, with respect to both the other pairs. This follows from what has been proved above. It is the interpretation, on the conic, of the result of Ex. 10, Chap. VII. The points of a line denoted by A, B, C, D in that example correspond to the points of the conic which, in the theory given above, were denoted by A, A' and B, B'.

Ex. 2. With another notation, the theorem of the preceding example may be stated by saying that, if TDD' be a fixed line drawn through a point T of the plane, which meets the conic in D and D', and the tangents of the conic at D and D' meet in the point H; and a variable point A', of the conic, be joined to T and H, by lines which meet the conic again respectively in A and A_1, then all the lines AA_1 pass through the same point of the line DD' (namely, the pole of the line TH). This gives the result of Ex. 12, Chap. VII.

Ex. 3.   Deduce, from the result of Ex. 13, Chap. VII (4), that, if a triangle, $A'B'C'$, inscribed in a conic, be in perspective with another inscribed triangle ABC, then the tangents of the conic at the points A', B', C', respectively, meet the sides BC, CA, AB in three points which lie in a line. Shew that Ex. 14, there given, leads to a converse of this result; while Ex. 15, as we have seen above, leads to a geometrical result, capable of easy proof, for the conic, which establishes Pascal's theorem.

Ex. 4.   Noticing that, from what has been proved here, the polars, in regard to a conic, of the points of a line, all pass through the pole of this line, shew that Ex. 17, Chap. VII, leads to the following result: Let a chord MN, and the tangent at a point C, of a conic, meet in R; let a line be drawn through R which meets the chords CM, CN respectively in P and Q; let the polar of P meet the conic in A, A'; and the polar of Q meet the conic in B, B'. Then the point R is either the common point of the lines $AB, A'B'$, or the common point of the lines $AB', A'B$.

Ex. 5.   We can prove that the correspondence of two related ranges of points on a conic may be built up by the succession of two involutions on the conic—a result which follows from a previous theorem for ranges on a line (Chap. VIII)—in the following way: Let P, P' be fixed corresponding points of two related ranges on a conic, of which the common corresponding points are X and Y. When X and Y coincide, the chord XY becomes the tangent at X. Take an arbitrary fixed point O on the chord XY, and let the line PP' meet this chord in Z_0. Let the line PO meet the conic again in K, and $P'K$ meet XY in O'; so that O' is also a fixed point. Now let Q be any point on the conic; let the line QO meet the conic again in L; and the line LO' meet the conic again in Q'. By applying Pascal's theorem to the hexagon $PKP'QLQ'$, we see that the lines $PQ', P'Q$ meet in a point of the line OO', or XY, say in Z.

Then, as Q, and therefore also Q', varies on the conic, consider the pencils $P'(Q)$ and $P(Q')$, with centres at P', P, formed by rays $P'Q$ and PQ'. Because corresponding rays of these pencils meet in a point Z of the line XY, these pencils are related. So then are also the ranges $(Q), (Q')$, of corresponding points on the conic. In particular, if Q be taken at P, then Z becomes Z_0, and Q' becomes P'; hence P, P' are corresponding points of these two related ranges. When Q is at X, then L is at Y, and Q' is also at X; thus X is a common corresponding point of the two ranges $(Q), (Q')$. So, likewise, is Y. When Y coincides with X, this is a point of coalescence of the common corresponding points of the ranges $(Q), (Q')$. It appears then that the two originally given related ranges, from which P and P' were taken, are the ranges $(Q), (Q')$.

This correspondence, of the original ranges, is then built up by two involutions. By the first involution, the point Q is paired with L, through the fixed point O, of the line XY; by the second involution, the point L is paired with Q', through the fixed point O', of the line XY. Of these O can be taken arbitrarily.

In this demonstration, there is the incidental result that two related ranges on a conic are in perspective with the same range on a line, from two centres of perspective (for PQ' meets $P'Q$ on XY). This has been remarked above (13), and used to prove Pascal's theorem. The same result for two related ranges on different lines (which together form a degenerate conic) has also been proved (6).

Ex. 6.   The poles, in regard to a conic, of the lines of a pencil, whose centre is T, lie on a line, the polar of T. On this line, these poles form a range which is related to the pencil of lines (and, therefore, related to the range of points in which these lines meet the polar of T). Dually, the polars, in regard to a conic, of the points of a line, constitute a pencil of lines, whose centre is the pole of the given line; and this pencil is related to the range of points.

Ex. 7.   Let the sides BC, CA, AB of a triangle inscribed in a given conic be met by a line respectively in the points L, M, N. Let lines LXX', MYY', NZZ' be drawn through L, M, N, meeting the conic, respectively, in X, X', in Y, Y', and in Z, Z'. Suppose that these lines are such that the lines AX, BY, CZ meet in a point. Prove that the lines AX', BY', CZ' also meet in a point. (Cf. Ex. 18, Chap. VII.)

Ex. 8.   Let B_1, C_1, C_2, A_2, A_3, B_3 be six points in order upon a conic. Shew that Pascal's theorem is the same as that the triangle whose sides are the three alternate lines A_2A_3, B_3B_1, C_1C_2 is in perspective with that whose sides are the three alternate lines B_1C_1, C_2A_2, A_3B_3. Hence (Chap. VII (4), Ex. 20) there follows a necessary and sufficient condition that three pairs of points, lying respectively on the sides of a triangle, should lie upon a conic.

Ex. 9.   As an application of the construction given for the tangent at any point of a conic, the following result may be proved: Let A, A'; B, B' and C, C' be the pairs of complementary intersections of four lines in a plane; and O be an arbitrary point of the plane. Then the tangents, at O, of the two conics described, respectively, through the two sets of five points, O, B, B', C, C' and O, A', A, C, C', meet the line CC' in two points which are harmonic conjugates in regard to C and C'.

The application of Pascal's theorem to the hexagon O, O, B, C, C', B' shews that the tangent at O of the first conic meets CC' in a point lying on the join of the two points $(OB, C'B')$, (OB', CB); likewise, the hexagon O, O, A, C, C', A' shews that the tangent at O of the second conic meets CC' in a point lying on the join of the two points $(OA, C'A')$, (OA', CA). (Cf. Ex. 17, Chap. VII.)

(17) **The expression of the points of a conic by a parametric number.** It is part of the fundamental hypothesis adopted in this volume, that the symbol of any point of a plane is expressible linearly by the symbols of three fixed chosen points in the plane. Thus, if X, Y, Z be the symbols of the points of reference, and the ratios of the algebraic multipliers which may be attached to X, Y, Z, without changing the points which they represent, be so chosen that a definite, arbitrarily chosen, point of the plane has the symbol $X + Y + Z$, then the symbol of any point P of the plane has the symbol

$$P = xX + yY + zZ,$$

where x, y, z are numbers whose ratios are definite. These numbers (or their ratios) are called the *coordinates* of the point P, with reference to X, Y, Z. In many investigations it is immaterial what point is taken for the *unit* point, whose symbol is $X + Y + Z$; the variation of this point only leads to using px, qy, rz, instead of x, y, z, for the coordinates of a point P, where p, q, r are numbers *which are the same for all points P.*

In particular, the points of a conic are expressible in the manner explained; and, as the conic can be regarded as being in correspondence with a line, the expression must involve a single number which varies from point to point of the conic.

A conic, we have seen, is determined when five general points of it are given; it is, therefore, also determined when three points of it are given, and also the tangents of the conic at two of these points. Let the three given points be X, Z and E, the last being the point to which we attach the symbol $X + Y + Z$, where Y is the point of the plane which is the common point of the tangents of the conic at X and Z. The points X, Z, E and the tangents YX, YZ determine the conic. Let P be any point of the conic; let the lines XP, ZP meet the sides YZ, YX of the triangle YXZ which are opposite to X and Z, respectively, in M and L. Let the symbol of P be $xX + yY + zZ$. The points of the conic are projected from X and Z by two related pencils, giving rise to two related ranges on the tangents YZ and YX. In particular, the three points X, Z, P of the conic give rise, in these ranges respectively, to the points Y, Z, M and X, Y, L. Herein, the symbols of M and L will be $P - xX$ and $P - zZ$, namely $yY + zZ$ and $xX + yY$, respectively. As, when P varies on the conic, these points M, L describe related ranges on the lines YZ and YX, in which Y, Z, of the first line, correspond to X, Y of the second line, it follows that the ratio y/z is a constant multiple of the ratio x/y; by supposing P to be at E, we see that these ratios are in fact equal to one another. Putting then $x/y = \theta$, we have $y/z = \theta$, and the coordinates x, y, z, of the point P of the conic, have the ratios $\theta^2 : \theta : 1$. Here θ is a parametric number, varying as P varies on the conic. The symbol of any point of the conic may then be written $\theta^2 X + \theta Y + Z$; or, it may be written $\theta X + Y + \theta^{-1}Z$.

It follows that the coordinates of any point of the conic satisfy the homogeneous quadratic equation $xz = y^2$, of which, conversely, the parametric expression is obvious. This is generally called the *equation of the conic.* More generally, any conic touching YZ and YX at Z and X has an equation of the form $k^2 xz = y^2$, where k^2 is a fixed number characteristic of that conic; if this conic be determined by a further point of it, of coordinates (ξ, η, ζ), the value of k^2 is $\eta^2/\zeta\xi$; the parametric expression of a point of this conic is then $\theta X + kY + \theta^{-1}Z$.

(18) **The conic in terms of any three points of reference.** Instead of the points X, Y, Z, with their definite relation to the conic, we may refer the points of the plane to three other fixed points, say A, B, C, of general position. The symbols of X, Y, Z will be expressible, in terms of the symbols of A, B, C, by equations

$$X = a_1 A + a_2 B + a_3 C, \quad Y = b_1 A + b_2 B + b_3 C, \quad Z = c_1 A + c_2 B + c_3 C,$$

where $a_1, \ldots, c_3$ are nine constant numbers. Thence the symbol of any point, P, of the form $P = xX + yY + zZ$, will take the form $\xi A + \eta B + \zeta C$, wherein

$$\xi = a_1 x + b_1 y + c_1 z, \quad \eta = a_2 x + b_2 y + c_2 z, \quad \zeta = a_3 x + b_3 y + c_3 z.$$

These are the coordinates of P referred to A, B, C. Now suppose P is a point of the conic; then the coordinates of P, referred to A, B, C, are quadratic polynomials in the parametric number θ, given by

$$\xi = a_1 \theta^2 + b_1 \theta + c_1, \quad \eta = a_2 \theta^2 + b_2 \theta + c_2, \quad \zeta = a_3 \theta^2 + b_3 \theta + c_3.$$

The two points, in particular, where the conic meets the line BC, are then given by the roots of the quadratic equation, in θ, expressed by $a_1 \theta^2 + b_1 \theta + c_1 = 0$. More generally, the two points where the conic meets any line of the plane are likewise given by a quadratic equation in θ. To see this, we have only to regard the symbol of a general point of the line as given in terms of the symbols of two points of the line.

Conversely, any point of symbol $\xi A + \eta B + \zeta C$, where ξ, η, ζ are such quadratic polynomials in a variable number θ as have been put down, describes a conic as θ varies. It is understood that the coefficients, $a_1, \ldots, c_3$, in these quadratic polynomials, are such that the determinant of three rows, with respective elements a_1, b_1, c_1; a_2, b_2, c_2; a_3, b_3, c_3, does not vanish; otherwise, as is easy to see, the point $\xi A + \eta B + \zeta C$ would describe a line.

(19) Some immediate algebraic consequences.

Ex. 1.   Let two points, $P = \theta^2 X + \theta Y + Z$ and $Q = \phi^2 X + \phi Y + Z$, describe two related ranges on a conic. Then the points $\theta X + Y$, $\phi X + Y$, on the tangent YX, obtained by projection of P and Q from the point Z, equally describe related ranges on this line. Thus the parameter ϕ is connected with the parameter θ by an equation of the form $\phi = (a\theta + b)/(c\theta + d)$, where a, b, c, d are numbers whose ratios are constant.

In particular, if P, Q be a pair in an involution of pairs of points on the conic, the relation connecting θ, ϕ is symmetrical in regard to these, and may be written in the form $r\theta\phi - q\,(\theta + \phi) + p = 0$, where p, q, r are numbers whose ratios are fixed. When this is so, the symbols

$$P = \theta^2 X + \theta Y + Z, \quad Q = \phi^2 X + \phi Y + Z,$$

of the points of a pair of the involution, are such that

$$(r\theta - q)\,Q - (r\phi - q)\,P = (\theta - \phi)\,(pX + qY + rZ);$$

this shews that the line PQ, for every pair, passes through the fixed point whose symbol is $pX + qY + rZ$; a result obtained above geometrically.

Further, if θ, ϕ be the symbols of any two points P, Q of the conic, and $xX + yY + zZ$ be the symbol of any point whatever on the chord joining these two points; then x, y, z satisfy the equation $x - y\,(\theta + \phi) + z\theta\phi = 0$. For, the symbol of a point of the chord is of the form $\lambda P + \mu Q$, where λ, μ are numbers; so that the x, y, z are in the respective ratios of $\lambda\theta^2 + \mu\phi^2$, $\lambda\theta + \mu\phi$, $\lambda + \mu$; and these satisfy the linear equation put down. This equation, connecting the coordinates of any point of the chord, is generally called the *equation of the chord*, joining the points θ, ϕ. In particular, putting $\phi = \theta$, the *equation of the tangent* of the conic, at a point θ, is $x - 2y\theta + z\theta^2 = 0$.

Particular consequences are: (i) The points (θ), $(-\theta)$, of the conic, lie on a line through the point Y; (ii) If $xX + yY + zZ$ be any point of the polar line, in regard to the conic, of the point whose symbol is $\xi X + \eta Y + \zeta Z$, then $x\zeta - 2y\eta + z\xi = 0$. For, if (θ), (ϕ) be the two points of the conic whereat the tangents contain the point $\xi X + \eta Y + \zeta Z$, we have $\xi - 2\eta\theta + \zeta\theta^2 = 0$ and $\xi - 2\eta\phi + \zeta\phi^2 = 0$; so that the relation $x\zeta - 2y\eta + z\xi = 0$ is the same as $x - y\,(\theta + \phi) + z\theta\phi = 0$. The condition that the two points $xX + yY + zZ$, $\xi X + \eta Y + \zeta Z$ should be *conjugate* in regard to the conic is therefore $x\zeta - 2y\eta + z\xi = 0$; or, if $(\xi,\ \eta,\ \zeta)$ belong to a given point, this is the *equation of the polar line* of this point.

Ex. 2.   Any two conics in a plane have four points in common, of which two, or three, or all, may coincide. For, taking X, Z to be two arbitrary points of the second conic, and Y the common point of the tangents of this conic at X and Z, the symbols, in terms of the symbols of X, Y, Z, of two general points of the two conics, will respectively be of the forms

$$(a_1\theta^2 + b_1\theta + c_1)\,X + (a_2\theta^2 + b_2\theta + c_2)\,Y + (a_3\theta^2 + b_3\theta + c_3)\,Z,$$
$$\phi^2 X \qquad\qquad + \phi Y \qquad\qquad + Z,$$

where $a_1, \ldots, c_3$ are numbers, and $\theta,\ \phi$ are the parametric numbers determining points of the two conics respectively. For a point common to the two conics, the multipliers of X, Y, Z in the former of these must be proportional to the multipliers of X, Y, Z in the latter. Thus θ is any one of the roots of the *quartic* equation

$$(a_2\theta^2 + b_2\theta + c_2)^2 = (a_1\theta^2 + b_1\theta + c_1)\,(a_3\theta^2 + b_3\theta + c_3);$$

and, when θ is found, ϕ is uniquely determinable.

Ex. 3.   Many of the theorems we have proved geometrically for a conic may be obtained very easily by use of the equations we have given. For instance, Ex. 4 of (16) above.

(20) **The conic as a self-dual construct.** It may be useful to repeat, in part, what we have said about the important relation of pole and polar in regard to a conic. We have seen that, if, from a fixed point T, a variable line be drawn to meet the conic in P and P', and on this line the point $H = (P, P')/T$ be taken, which is the harmonic conjugate of T in regard to P and P', then the locus of the point H, for all lines TPP' through T, is a line. This is the polar line of T in regard to the conic. Moreover, there is only one point which has a given polar line. For, the polar line which is the locus of the point H meets the conic in two points, and the tangents of the conic at these points meet in the point T. This point we call the pole of the given line. The necessary and sufficient condition that the polar of a point T should pass through a point H is that H and T should be harmonic conjugates in regard to the two points in which the line HT meets the conic. Thus, when this is so, the polar of H passes through T. The relation of H and T is reciprocal, and we speak of these as being conjugate in regard to the conic. If, as before, P and P' be two points of the conic such that the line PP' contains T, and the tangents of the conic at P and P' meet in U, the polar of U, which is the line PP', passes through T; so that U lies on the polar of T. More generally, if T and T_1 be any two points whose polars meet in the point L, the polar of L contains both T and T_1, or L is the pole of the line TT_1. Thus, as T_1 describes a range on a given line through T, the polar line of T_1 is one of a pencil of lines, whose centre is the pole of the given line. Likewise, the poles of all the lines of a pencil lie on a line, which is the polar of the centre of the pencil. Again, let HU be a fixed line, whose

pole is the point T, and let U be a variable point of this line; then the polar of U is a variable line, passing through T, which meets the line HU in a point H which is the harmonic conjugate of U in regard to the fixed points in which HU meets the conic. Thus, if H, U be points on a fixed line, which are conjugate to one another in regard to the conic, then H, U are a pair of an involution of points on the fixed line, whose foci are the common points of this line with the conic. Thus, the range described by the variable point U on this line is related to the pencil described by the lines joining H to the pole, T, of the fixed line. In particular, *the polar of a point on the conic is the tangent of the conic at this point.*

Now, consider a fixed point A of the conic; and let the tangent, at a variable point P of the conic, meet the tangent at A in the point T; thus T is the pole of the chord AP. By what we have seen, *the range described by T, as P varies, is related to the pencil described by the lines AP; this range is therefore related to the pencil described by the lines joining any other fixed point of the conic to the variable point P; or, as we have agreed to say, is related to the range of points P on the conic.*

(21) **Algebraic expression of the relation.** We emphasise the important result just obtained by expressing it in terms of the symbols. Take a further fixed point, B,.of the conic, and let the tangent of the conic at B meet the tangent at the original point A, in the point M. Let BP meet this tangent AT in the point U, and BT meet the conic again in Q. As P varies on the conic, the points T, U vary on the tangent at A. We have seen (11) that, regarded as points of the range of points on the conic, the points A and P are harmonic conjugates in regard to Q and B; it follows from this, projecting from B, that on the line AT, the points A, U are harmonic conjugates in regard to T and the point M, where the tangent at B meets the tangent AM. In regard to the symbols of T and M, we can thus write the symbols of A and U in the respective forms $A = T + \lambda M$, $U = T - \lambda M$, where λ is a number. These give $T = A - \lambda M$, and $U = A - 2\lambda M$. Wherefore, B being fixed and P variable, the ranges described by T and U on the tangent AT are related. The range described by U, being given by the lines BP through the fixed point B of the conic, is however related to the range of the point P on the conic. Whence, this range is related to that described by the point T, where the tangent at P meets the tangent AT. This is the result found in (20).

A consequence of the theorem, which is of importance, is that, *if we take the tangents at two fixed points A, A' of the conic, and a tangent at a variable point P of the conic meets these fixed tangents in T and T', respectively, then the ranges described by T and T', on these fixed tangents, are related to one another, and related to the range, on the conic, of the point of contact, P, of the variable tangent.*

(22) **The dual of a conic.** Consider now the dual of our original definition of a conic. In that definition, we had two fixed points, and two related pencils of lines with these points as centres. The variable

point of the conic was the intersection of corresponding rays of the two pencils. Dually, take two fixed lines, l, l', and two related ranges on these lines. There is an aggregate of lines each formed by the join of corresponding points of the two ranges. This is obviously of the same multiplicity (or dimension) as the aggregate of the points of a line, or, of the rays of a pencil. As we speak of a *curve*, regarded as an aggregate of points, so we may give a name to such an aggregate of lines, and call it an *envelope*; and we may call any one of the constituent lines a *tangent* of the envelope. And, just as for the aggregate of points which constituted a conic, so, in this obviously dual figure, and by the dual of the argument used in the other case, each of the lines l, l' is one of the tangents of the envelope. While further, as in the other case, there is, upon each of these lines l, l', say on l, a particular point; namely, the point of the range on l which corresponds to that point of the related range on l' where l meets l'. We may agree to call this point on l the *point of contact* of the tangent l with the envelope; there is a like point on the line l'.

If we compare all this with what is deduced above (21) for a conic, we infer that the aggregate of lines in this dual figure consists in fact of the tangents of a conic, as previously defined. And, to every result for the conic corresponds a dual result for the present figure. Thus also, to every result for the conic corresponds a dual result for the figure formed by the tangent lines of the conic itself. An example of this is the fact, proved for a conic, that just as two points of the conic lie on an arbitrary line, so two tangents of the conic pass through an arbitrary point. More generally, just as there is a tangent line at every point of a conic, so, in the aggregate of lines we have considered here, and called an envelope, there is, upon every one of these lines, a particular point, which we may call the *point of contact belonging to this line*. And, as we shewed that the chord joining two points of a conic should be considered as becoming the tangent line at a point of the conic at which the two points coalesce, so, dually, the point of contact, upon any line of the aggregate, should be considered as arising from the common point of this line with another line of the aggregate which coalesces with the former line. The locus of these points of contact, one upon each line of the aggregate, is the conic whose tangents constitute the aggregate of lines.

That the duality we have explained is important is well illustrated by the interval of time which elapsed between the formulation of Pascal's theorem for six points of a conic (1640), and the recognition of the corresponding theorem for six tangents of a conic (1817), commonly called Brianchon's theorem. Pascal's theorem is that, if six points in order upon a conic be A, B', C, A', B, C', the three intersections $(BC', B'C)$, $(CA', C'A)$, $(AB', A'B)$ lie on a line. Brianchon's theorem is that, if six tangents of a conic, taken in order, be a, b', c, a', b, c', then the three lines joining the respective pairs of intersections (b, c') and (b', c); (c, a') and (c', a); (a, b') and (a', b) meet in a point; or, that the diagonals of the hexagon formed by the six lines meet in a point.

The proof of this can be constructed by steps which are dual, in exact detail, to those taken to prove Pascal's theorem. But the result can be proved directly from the theory of the conic, regarded as a locus; so forming a verification of the validity of the principle of duality as here applied to a conic. It is necessary only to recall that if one point lie on the polar of another point, in regard to a conic, then this other point lies on the polar of the first. For this, denote the points of contact of the six tangents a, b', c, a', b, c', respectively, by A, B', C, A', B, C'. Then the point (b, c') is the pole of the line BC'; whence, the line joining the points (b, c'), (b', c) is the polar of the point $(BC', B'C)$. By Pascal's theorem, the three points such as this last lie upon a line. By the property of polar lines which we have recalled, it follows that the three lines, such as that joining the points (b, c'), (b', c), meet in a point. And, generally, when we are considering the tangents of a conic as the construct dual to that formed by the points of the conic, the dual of any point of the plane is the polar of this point in regard to the conic; and the dual of any line is the pole of this line.

(23) **Examples.** The following examples contain various applications of the duality we have explained. They contain also certain more general results, collected here for subsequent reference.

Ex. 1.   We have, (7), above, used Pascal's theorem to construct the tangent line of a conic which is determined by five points, at any one of these points. Dually, let a, b, c, d, e be five given lines in a plane, of which no three meet in a point. There exists then a definite conic of which these are tangents, the tangent lines of this conic being the joins of corresponding points of ranges on two of these lines which are related to one another by the points where these two lines are, respectively, met by the other three lines. And the point of contact of this conic with any one of the five lines, say with the line e, is found thus: Take the joining line of the two points (a, b), (d, e); also the joining line of the two points (c, d), (a, e). Then the line through the common point of these two joining lines, and through the point (b, c), meets e in the point of contact desired. This fact is a direct consequence of Brianchon's theorem, if we regard the line e as arising by the coalescence of two tangents, e and f, and consider the circumscribed hexagon (a, b, c, d, e, f); the two coalescing tangents are regarded as meeting in the point of contact; just as a tangent is regarded as the chord joining two coalescing points of the conic.

Ex. 2.   We saw that the pencils of lines which join two fixed points of a conic to a varying point of the conic are related to one another; so that we could regard the points of a conic as forming a range, irrespective of the centre of the pencil of lines, and so look upon the conic, like a line, as the seat of related ranges of points. We saw also, dually, that if we take two fixed tangents of the conic, these are met by the tangent, at a variable point of the conic, in ranges which are related to one another, and related also to the range on the conic formed by the point of contact of the variable tangent. Thus, we can speak of *related series of tangents* of the conic; in particular, we can speak of pairs of tangents which form an involution in the series of all the tangents of the conic. Then, as pairs of points, belonging to an involution on the conic, arise from chords passing all through the same point, so the pairs of tangents which form an involution, in the series of all the tangents, are the pairs of tangents drawn from a varying point of a definite line. In fact, the chords joining the points of contact of a pair of tangents of such an involution all pass through the point which is the pole of the definite line.

Ex. 3.　Two points were called *conjugate* in regard to a conic, when the polar of either point contains the other. The condition for this is, that the two points should be harmonic conjugates in regard to the two points in which the conic is met by the line joining them. Dually, two lines are *conjugate* to one another, in regard to the conic, when the pole of either line lies on the other. The condition for this is, that the two lines should be harmonic conjugates in regard to the two tangents which can be drawn to the conic from the common point of the two lines.

Ex. 4.　A conic is the locus of the third angular point of a triangle when the other two angular points lie on fixed lines, and the three sides pass each through a fixed point. Dually, if the three angular points of a triangle move each on a fixed line, and two of its sides pass each through a fixed point, then the third side touches a definite conic.

And again, just as a conic degenerates into two lines when the related pencils, by which it is defined, have the join of the centres of these pencils as a self-corresponding ray, so a conic, regarded as determined by its tangents, becomes the aggregate of the lines of two pencils, when the related ranges, on the two given lines, have the common point of these two lines as a common self-corresponding point of the two given ranges. The two ranges are then in perspective from a point. The tangents of the conic then consist of the pencil of all the lines through this point, together with the pencil of all the lines through the self-corresponding point of intersection of the two lines on which the ranges lie.

A degenerate conic of a locus of points is often spoken of as a *line-pair*; a degenerate conic of an envelope of lines is similarly spoken of as a *point-pair*. In a still more particular case, the two lines of the pair, or the two points of the pair, may coincide with one another.

Ex. 5.　If A, A', B, B' be four points of a conic, and T, U, N be respectively the points of intersection (AA', BB'); $(AB, A'B')$ and $(AB', A'B)$, then each side of the vertex triangle TUN, of the quadrangle formed by the four points, is the polar of the opposite angular point; as we have seen. Dually, if a, a', b, b' be four tangents of a conic, the diagonal triangle of the quadrilateral, formed by these lines, is a self-polar triangle in regard to the conic. Consider the relation of these two triangles when the lines a, a', b, b' are the tangents of the conic at the respective points A, A', B, B'.

Ex. 6.　As we have proved, any line is met in pairs of points belonging to an involution by the conics which can be drawn through four given points; among these conics are the three degenerate conics each formed by a pair of complementary joins of the four points. Dually, the pairs of tangents which can be drawn from an arbitrary point to the conics which touch four given lines form a pencil in involution; three pairs of the rays of the involution are the lines from the arbitrary point, each drawn to a pair of complementary intersections of the four lines.

A connected result of importance is that, if we are given a quadrangle of four points, and an arbitrary point O, there is a corresponding point O', such that O, O' are conjugate points in regard to every conic through the four points. The point O' can be found from O as the common point of the polars of O taken in regard to two of the conics (Chap. vii, Ex. 5). Dually, if we are given a quadrilateral of four lines, then, to an arbitrarily taken line, l, there corresponds another line, l', such that l, l' are conjugate to one another in regard to every conic touching the four lines. The line l' can be constructed from l as the join of the poles of l taken in regard to any two of the conics. We speak of O, O' as conjugate to one another in regard to the quadrangle; and of l, l' as conjugate to one another in regard to the quadrilateral.

Again, if we are given a quadrangle of four points, and an arbitrary line, there are two conics through the four points which touch the line. These are those whose two intersections with the line coincide with one another; namely at the foci of the involution which the conics determine on the line, these being the points of contact of these conics with the line. Dually, given an arbitrary quadrilateral of four lines, and an arbitrary point, there are two conics touching the four lines which pass through the given point; the tangents of these two conics, at this point, are the focal rays of the pencil in involution determined by the pairs of tangents which can be drawn from the point to the various conics touching the lines of the quadrilateral. The two tangents are harmonic conjugates of one another in regard to the tangents drawn from the point to any one of the conics touching the four lines, and are thus conjugate to one another in regard to every one of the conics.

Ex. 7.   There is a property of the pair of points, O, O', conjugate to each other in regard to a given quadrangle, considered in Ex. 6, which is worthy of mention. *Let P, P' be another such pair of points which are conjugate to one another in regard to the quadrangle. Consider the four lines consisting of the two joins of O to P and P', and the two joins of O' to P and P'. These four lines are touched by a conic which also touches the sides of the vertex triangle of the given quadrangle.*

Let T, X, Y, Z denote the points of the given quadrangle, whose vertex triangle is formed by the three intersections $(TX, YZ), (TY, ZX), (TZ, XY)$, which we denote, respectively, by A, B and C. Let the lines $OP, O'P'$ meet in Q, and the lines $OP', O'P$ meet in Q'. By the definition of the conjugate points O, O', the lines CO, CO' are harmonic conjugates in regard to the joins CXY, CZT, which meet in C. This fact we express briefly by $(CO, CO') \sim C(XY, ZT)$. We similarly have $(CP, CP') \sim C(XY, ZT)$. Whence, the lines CXY, CZT are the focal rays of the pencil in involution, with centre at C, which is determined by the two pairs of lines CO, CO' and CP, CP'. But, if we consider the quadrilateral formed by the four lines $OP, OP', O'P, O'P'$, both O, O' and P, P' are a complementary pair of intersections of these four lines. The involutory pencil determined by the pairs of lines CO, CO' and CP, CP' has therefore the lines CQ, CQ' as a pair, for Q, Q' are the third pair of complementary intersections of the lines $OP, OP', O'P, O'P'$. It follows then that $(CQ, CQ') \sim C(XY, ZT)$. And it follows similarly that $(AQ, AQ') \sim A(YZ, XT)$. These two results, together, shew that Q, Q' are a third pair of conjugate points in regard to the original quadrangle $TXYZ$. Now, as CXY, CZT are harmonic conjugates both in regard to the pair CO, CO' and to the pair CP, CP', these lines CXY, CZT are the focal rays of the pencil in involution formed by the pairs of tangents drawn from C to the conics which touch the four lines OP, OP', $O'P, O'P'$; thus a conic can be described, touching these four lines, to touch also any two lines through C which are harmonic conjugates in regard to CXY and CZT; so that a conic exists touching these four lines which also touches the lines CA, CB. As a conic is determined by five of its tangents, and we can use the vertex A just as we have used C, we infer that the conic touching $OP, OP', O'P, O'P'$ which touches CA, not only touches CB but also touches AB. This is the result enunciated.

Ex. 8.   Another property of a pair of points which are conjugate to one another in regard to a given quadrangle, $TXYZ$, is that, *if one of these points, say P, varies on a line, then the conjugate point, say P', varies on a conic, which contains the vertices A, B, C of the vertex triangle. And, the two points common to this conic and the given line are a pair of conjugate points in regard to the quadrangle.*

For, from the harmonic relation expressed in Ex. 7 by

$$(CP, CP') \sim C(XY, ZT),$$

it follows that when P and P' vary, the two pencils, of centre C, namely

$C(P)$ and $C(P')$, are related to one another. Likewise the pencils $B(P)$ and $B(P')$ are related. But, if P varies on a fixed line, the pencils $B(P)$, $C(P)$ are related. Thus also the pencils $B(P')$, $C(P')$ are related. This shews that the locus of P' is a conic, which passes through the points B and C. And, this conic passes through A also; for, by what we have proved, the point A is conjugate, in regard to every conic through T, X, Y, Z, to *any* point on the line BC; so that A is the position of P' when P is at that point of the fixed line which lies on BC. Further, on the given line there are two points which are the foci of the involution determined on this line by all the conics through T, X, Y, Z; these two points are then conjugate to one another in regard to every one of these conics, being harmonic in regard to every pair of the involution. Thus, when P is at one of these foci, then P' is at the other; namely, the conic which is the locus of P' contains both these points. This proves what was stated.

Conversely, if the point P' describes any conic which contains the points A, B, C, then the locus of P is a line meeting the conic in two points which are conjugate to one another in regard to T, X, Y, Z. For, as before, we can infer that the pencils $B(P)$ and $C(P)$ are related. But, in these pencils, the rays $B(C)$, $C(B)$, which are the same line, correspond to one another, since these arise from the rays $B(A)$, $C(A)$ when P' is at A. Thus the locus of P is a line.

Ex. 9. We have remarked that a conic, regarded as a locus of points, may degenerate into two lines; dually, a conic, regarded as the aggregate of its tangents, may degenerate into the lines of two pencils. And as, for conics through four points, the three pairs of complementary joins of these points form three degenerate conics, so, for conics touching four lines, a degenerate conic is formed by two pencils of lines whose centres are a pair of complementary intersections of the four lines. In regard to a conic which degenerates into two lines, a pair of points are conjugate to one another when the two lines which join them to the common point of the two lines are harmonic in regard to these. Dually, in regard to a conic which degenerates into two pencils of lines, two lines are conjugate to one another when they meet the joining line, of the centres of the two pencils, in two points which are harmonic conjugates in regard to these centres. The pole of any line, in regard to such a degenerate conic of two pencils, is the point, on the line joining the centres, which is harmonic, in regard to these centres, to the point where the line meets this joining line. This is the dual of the fact that the polar of a point, in regard to a degenerate conic forming a line-pair, is the harmonic line of the point in regard to the two lines.

Ex. 10. We have proved above, (19), Ex. 2, that two conics, considered as loci of points, have four points in common. Dually, *there are four lines each of which is a tangent of both of two given conics. And, it can be shewn that the diagonal triangle of the quadrilateral formed by the four common tangents of two conics is the same as the vertex triangle of the quadrangle formed by the four common points of the two conics.* In fact, to any point, P, of either one of the two conics, corresponds another point, P', of this conic, which is the harmonic image of P, in regard to an angular point of the vertex triangle taken with the opposite side of this triangle; harmonic images being as defined above (Chap. IV (4), Ex 7). It follows from this, a tangent of a conic being regarded as arising from the chord joining two points of the conic which coalesce with one another, that, to any common tangent of the two conics there corresponds another common tangent, the harmonic image of the former with regard to the same line and point; these tangents meet then on the line. From this, taking the other two angular points of the vertex triangle, each associated with its opposite side, and then forming the harmonic images of the two common tangents spoken of, with regard to these, the result becomes obvious.

A particular corollary is that a conic can be drawn through the four common points of the two conics to contain any one of the pairs of complementary intersections of the four common tangents. Dually, the four common tangents of two conics, taken with a complementary pair of common chords of the two conics, form a set of six lines which are all touched by a conic.

Ex. 11.   *We can prove that the eight points of contact, with two conics, of their four common tangents, lie on another conic.* Also, that *a conic exists having a given triangle as self-polar triangle which also passes through two given points.*

Let UTN be a given triangle; let the operation of passing from one arbitrary point, P, to the point, P_1, which is the harmonic image of P in regard to the point U and the line NT, be denoted by ϑ; so that we may write $P_1 = \vartheta(P)$, and $P = \vartheta(P_1)$. Similarly, when we take the harmonic image in regard to the point T and the line UN, let the resulting point, P_2, be denoted by $P_2 = \phi(P)$. We can then prove that the point $\phi(P_1)$ is the same as $\vartheta(P_2)$; this point, P_3, may then be denoted by $\vartheta\phi(P)$, or by $\phi\vartheta(P)$; it is the harmonic image of P for the point N and the line UT. If we put $P_3 = \psi(P)$, we may say that $\vartheta\phi\psi = 1$, the symbols ϑ, ϕ, ψ being freely commutable, and such that $\vartheta^2 = 1$, $\phi^2 = 1$, $\psi^2 = 1$. With the given triangle UTN, the four points P, P_1, P_2, P_3 may be called, momentarily, an *associated* set. Either of them may be taken first, and the others derived in the way we have followed. By use of the operation ϕ, the join of P and P_1 meets the join of P_2 and P_3 on the line UN, and also the lines $P_3 P$ and $P_1 P_2$ meet on the same line UN. Wherefore T is the pole of UN for any conic which passes through the four points. Likewise for the two other angular points of the triangle UTN; this is therefore a self-polar triangle for any conic through the four points P, P_1, P_2, P_3. It is in fact the vertex triangle of this quadrangle. Now, let Q be any other point. The triangle UTN is then self-polar for the conic defined to pass through P, P_1, P_2, P_3 and Q. By harmonic inversion with the triangle UTN this conic therefore contains also the points Q_1, Q_2, Q_3 associated with Q.

Now, the eight points of contact, with two given conics, of their four common tangents, consist, we have seen, of two sets of four, each forming an associated set, for harmonic inversion with the vertex triangle which is the common self-polar triangle of the two conics. Therefore, these eight points lie on a conic. And this is the conic, having the common self-polar triangle of the two given conics as a self-polar triangle, which passes through one of the four associated points of contact on one conic, and also through one of the four associated points of contact on the other conic.

Dually, the eight tangents of two conics, one to each conic at each of their four common points, touch another conic.

Ex. 12.   *Let T be the pole, in regard to a conic, of the side AB, of a triangle CAB inscribed in the conic; the points L, M, in which the other sides CA, CB are, respectively, met by an arbitrary line drawn through T, are conjugate points in regard to the conic* (Seydewitz's theorem). For, if the arbitrary line drawn through T meets the conic in U and V, then, in the range of points on the conic, these points U, V are harmonic conjugates in regard to A and B, (10), above. Thus, projecting from C, we have L, $M \sim U$, V, which proves the result. The pole X, of the line LM, lies on AB; and the two lines AB, XC are harmonic conjugates in regard to the lines XL and XM.

Dually, *the points where two tangents a, b, of a conic, are met by a third tangent, are joined, to any point on the chord joining the points of contact of a and b, by lines which are conjugate in regard to the conic.*

Ex. 13.   The following result will be useful later: Let the tangents at two points, A, C, of a conic, meet in B. Let P, Q, R be three other points of the conic which are such that the lines PQ, AR meet the tangent at C in two points, say U, U', for which U, $U' \sim B$, C. When this is so, the conic through the five points P, Q, R, B, C touches the line CA at C.

The pair of complementary joins PQ, AR, of the quadrangle A, P, Q, R, meet the tangent CB, respectively, in U and U'; let the pair of complementary joins AP, QR meet CB in L and L' respectively; and also AQ, RP meet CB, respectively, in M and M'. The given conic, and the degenerate conic consisting of the lines PQ, AR, are two conics through the four points A, P, Q, R, whose pairs of intersections with the line CB, namely C, C and U, U', are, in both cases, harmonic conjugates in regard to C and B. Thus C, B are the foci of the involution of pairs of intersections, with CB, of the conics through A, P, Q, R. Thus also L, $L' \sim C$, B and M, $M' \sim C$, B. Whence, the two cross ratios $(C, B/M', L')$ and $(C, B/M, L)$ are equal; it follows that the cross ratio of the pencil $R\,(C, B/P, Q)$, which is equal to $(C, B/M', L')$ and hence equal to $(C, B/M, L)$, is that of the pencil $A\,(C, B/Q, P)$. But this is that of the range $(C, A/Q, P)$ on the conic, and is thus equal to $C\,(C, A/Q, P)$, or $C\,(B, A/Q, P)$, or $C\,(A, B/P, Q)$. Whence we have $R\,(C, B/P, Q) = C\,(A, B/P, Q)$. Now, consider the conic generated by related pencils, whose centres are R and C, to contain the five points R, C, B, P, Q. The equality just found shews that the ray CA of the latter pencil corresponds to the ray RC of the former pencil. This shews that CA is the tangent of this conic at C, as we wished to prove.

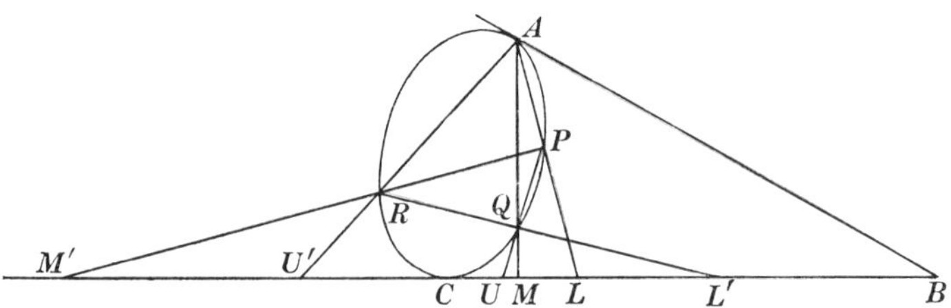

A corollary is, that the conic, determined to contain the four points C, B, P, Q and touch the line CA at C, has, for its fourth intersection with the given conic, a point R with the property that all conics through A, P, Q, R meet the line CB in pairs which are harmonic in regard to C and B. From this it follows, if I, J be any two points on the line CB such that $I, J \sim C, B$, then the conic through I, J and the points A, P, Q, of the original conic, meets this conic again in the point R; this is determined by the fact that AR, PQ meet CB in points U', U for which $U', U \sim C, B$.

Ex. 14.   The following results, for two conics related to one another in a particular way, will subsequently be found to be of interest. Let A, B, C be three points of a conic, and T the pole of the chord AB. On any line drawn through T, take two points D, E which are conjugate to one another in regard to the conic. Denote the intersections (AD, BC) and (BE, AC), respectively, by P and Q. Then

(a) The line PQ contains the point T. For, as in Seydewitz's theorem (Ex. 12), D and E being conjugate points, the lines AD, BE meet on the conic, say in K; and thence, from the inscribed quadrangle $ABKC$, the pole of PQ lies on AB. Hence the pole, T, of AB, lies on PQ.

(b) Consider the conic through the five points C, A, D, E, B. Applying Pascal's theorem to the six points C, C, A, D, E, B, we see that the tangent of this conic at C passes through T. This conic meets the original conic in a further point, C'. We similarly shew that the tangent of this conic at C' also passes through T.

(c) It is clear that the points, D', E', where the line TDE meets the original conic, are conjugate in regard to the second conic (C, A, D, E, B). We can therefore regard this conic as the given one, and construct the original conic as that conic, which passes through A, B, D', E' and has any given point, T, lying on $D'E'$, as pole of AB.

(*d*) The tangent at C, of the original conic, meets AB in the pole of the line CT. Hence, the tangents of the two conics, at the common point C, meet AB in points which are harmonic conjugates in regard to A and B. This can also be deduced from (16), Ex. 9.

(*e*) If any line through T meets the second conic in D_1 and E_1, these points can be shewn to be conjugate in regard to the first conic; so that the intersection (AE_1, BD_1) is a point, H_1, of the first conic, and the intersection (AD_1, BE_1) is a second point, K_1, of this conic. Hence there are two pencils of lines, with centres at the poles, in regard to the two conics, of the common chord AB, such that any line of either pencil meets the two conics in respective pairs of harmonically conjugate points. The result, and its dual, will arise again later.

Ex. 15.   Let T, X, Y, Z be an arbitrary quadrangle of four points, and l an arbitrary line. By these we can define eleven points which we shew to lie on a conic. Three of these are the vertices A, B, C of the quadrangle; two others are the foci of the involution, on the line l, determined by the pairs of points in which l is met by the various conics through T, X, Y, Z; six others lie, respectively, one on each of the six joins of T, X, Y, Z; for instance, if XY meet l in H, the point on XY is $(X, Y)/H$ (cf. Ex. 8).

To prove this, let the pole of the line l, taken in regard to a particular one of the conics through T, X, Y, Z, be denoted by U; and let UK be the polar of H, meeting XY in $K = (X, Y)/H$. The range, on l, of the six points such as H, where l is met by the joins of T, X, Y, Z, form a range related to the pencil, of centre U, of the six polars such as UK. Now take another conic through T, X, Y, Z and let U' be the pole of the line l in regard to this; the six lines such as $U'K$ equally form a pencil related to the range of points H on l. Thus the two pencils are related; therefore, the conic described through U and U' and three of the six points K, contains the remaining three points K. As this conic is determined by any five of the points K, it contains the pole of l in regard to every one of the conics through T, X, Y, Z. In particular, it contains the points of contact with l of the two conics through T, X, Y, Z, which touch this line; and these are the two foci spoken of. The pole of the line l in regard to a degenerate conic, consisting of a complementary pair of joins of T, X, Y, Z, is the intersection of these joins, and is one of the points A, B, C; these points, then, also lie on the conic.

We have thus shewn that *the locus of the poles of l, in regard to all the conics through T, X, Y, Z, is a conic. And have specified eleven points of this conic.*

This conic may be defined by means of its pairs of intersections with the sides of any one of the four triangles TYZ, TZX, TXY, XYZ, say the last of these. Its two intersections with the side XY of this triangle, for example, one of which is the vertex C (or (TZ, XY)), are the harmonic conjugates, in regard to X and Y, of the two points in which XY is met, (*a*) by the given line l, and, (*b*) by another line which may be defined as containing the three points $(Y, Z)/A$, $(Z, X)/B$, $(X, Y)/C$. This latter line is that which is often called *the polar line of the point T in regard to the triangle XYZ* (Chap. iv (4), Ex. 3).

Dually, *the polar lines of a fixed point in regard to all the conics which touch four given lines, touch a conic.* Of this eleven tangents can be specified. And, if these four lines be t, x, y, z, the pairs of tangents to the envelope-conic, drawn from the three points (y, z), (z, x), (x, y), meet the lines x, y, z, respectively, in six points which lie in threes upon two lines (one of these being the line t).

Ex. 16. *If both the triangles ABC and PQR be self-polar in regard to the same conic, the six angular points A, B, C, P, Q, R lie on another conic.*

We have already remarked (in (20)) that if the polar of a point O, in regard to a conic, be the line l, then the pencil of all lines through O is related to the range of points, on the line l, which are the respective poles of the lines of the pencil. In fact, if the polar of a point H, of the range, meets l in H', then H, H' are harmonic conjugates in regard to the two points in which HH' meets the conic.

Now, let the side QR, of one of the self-polar triangles, be met by the sides AC, AB, of the other triangle, respectively in E and F. Then, the points R, Q, E, F are the respective poles of the four lines PQ, PR, PB, PC. Thus, the pencil $P(Q, R, B, C)$ is related to the range (R, Q, E, F), and thus related to the range (Q, R, F, E), of which the points lie, respectively, on the rays of the pencil $A(Q, R, B, C)$. From the fact that the pencils $P(Q, R, B, C)$ and $A(Q, R, B, C)$ are related, it follows that A, B, C and P, Q, R lie on a conic.

We prove similarly that, if any conic through A, B, C, P meet the polar of P, taken in regard to the original conic, in the points Q, R, then these points are conjugate in regard to the original conic. Hence, *if any conic contain the angular points of one self-polar triangle of a given conic, this conic contains also the angular points of an infinite series of self-polar triangles of the original conic; any point of the conic is an angular point of one such triangle.* When this angular point is chosen, its polar, in regard to the original conic, determines the other two angular points, namely by the intersections of this with the second conic.

Ex. 17.   We now prove a result which is related to that given in Ex. 13, but more general than this. Let the tangents at two points A, C, of a conic ϖ, meet in B. Let ρ be any conic drawn through B and C, to touch CA at C; let ρ meet ϖ in the points P, Q, R, besides C. Then let the line BP meet ϖ again in H, and let the intersection (CP, HA) be K. We prove that *the conic, γ, through the five points P, Q, R, K, A touches CA at A.*

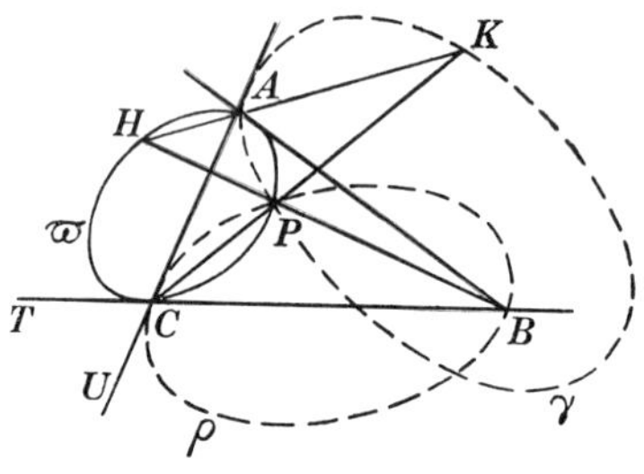

For definiteness of description, take any points T, U, respectively on the sides BC, AC. The two cross ratios $C(U, B/Q, R)$ and $C(A, T/Q, R)$ are the same; so that the range (C, B, Q, R), *on the conic ρ*, has the same cross ratio as the range (A, C, Q, R) on the conic ϖ.

The former of these is $P(C, B/Q, R)$, or $P(K, H/Q, R)$. By considering the intersections of the sides KH, HP, PK, of the triangle KHP, with the line QR, we see that $P(K, H/Q, R)$ is equal to the product

$$K(P, H/Q, R) \cdot H(K, P/Q, R),$$

or to $\qquad\qquad K(P, H/Q, R) \cdot H(A, P/Q, R);$

and, because H, A, C, P, Q, R are on the conic ϖ, this product is

$$K(P, H/Q, R) \cdot C(A, P/Q, R).$$

Likewise, the range (A, C, Q, R), on the conic ϖ, has the cross ratio $P(A, C/Q, R)$, which, considering the triangle PAC, is equal to the product $A(P, C/Q, R) \cdot C(A, P/Q, R)$, of which the latter factor occurs in the preceding equal cross ratio $(C, B/Q, R)$.

Whence we have $K(P, H/Q, R) = A(P, C/Q, R)$, of which the former is $K(P, A/Q, R)$. This shews that the conic, obtained by related pencils whose centres are A and K, by the condition of containing P, Q and R, has the ray AC, of the former, corresponding to the ray KA of the latter. So that the conic, γ, through P, Q, R, K, A has AC for tangent at A; as we wished to prove.

We remark that, as P, H, on the conic ϖ, are harmonic in regard to A and C, it follows that AB meets PK in the point $(P, K)/C$. Thus AB is the polar of C in regard to the conic γ. Thus, both the conics γ and ϖ meet CB in a pair of points which are harmonic in regard to C and B. Wherefore, *a conic can be drawn through the common points A, P, Q, R of the two conics ϖ, γ, to contain any two points I, J, lying on CB, which satisfy I, $J \sim C$, B.*

Ex. 18. Associated with any given conic there is another conic which we call the *conjugate line locus, for two given centres.* It is convenient to collect here various properties of this associated conic.

We have called two lines *conjugate*, in regard to a given conic, when either line contains the pole of the other. Let B and B' be any two fixed points; let their respective polars meet the given conic in A, C and A', C'. The general line, l, through B, has its pole, L, on the line AC; and the line, l', through B' which is conjugate to l, is the line $B'L$. The pencil of lines, l', through B', is thus related to the range of points L, on the line AC, and thus, it is related to the pencil of conjugate lines l through B. Wherefore the common point of the corresponding lines (l, l'), in these two related pencils, describes a conic passing through B and B'. This conic is called *the conjugate line locus, for the centres B and B', in regard to the given conic.* As the line $B'A$ contains the pole A of the line BA, this conic contains the point A; and equally contains C, and the points A', C'; so that the six points B, A, C and B', A', C' lie on a conic, which is the conjugate line locus in question.

Now let AC, $A'C'$ meet in T, which is then the pole of BB' in regard to the original conic. Then the pole of the line BB', of the pencil (B), is T, and $B'T$ is conjugate to BB'; this shews that $B'T$ is the tangent of the conjugate line locus at B'. So BT is the tangent at B; and the polars of T, in regard to the original conic, and in regard to the conjugate line locus, coincide in the line BB'. From this also it follows, if the points on BB' which are harmonic conjugates in regard to B and B', and also conjugate in regard to the original conic, be N and U, then TNU is a common self-polar triangle of the two conics. As the two conics meet in A, C, A', C', the points N, U are the two intersections $(AC', A'C)$ and (AA', CC').

Next, let P be any point of the conjugate line locus; then, as PB, PB' are conjugate lines in regard to the original conic, they are harmonic conjugates in regard to the two tangents which can be drawn from P to the original conic. The locus can then be defined as that of a point P which is such that the tangents to the original conic, from P, are harmonic in regard to the lines PB, PB' joining P to the fixed points B and B'.

This last result shews that quadrangles of four points can be taken on the conjugate line locus, such that the four joins of these points, taken in a certain order, are tangent lines of the original conic; this fact we may express by saying that *the conjugate line locus is quadrilaterally circumscribed to the original conic.* For let the tangents to the original conic, from a point P of the conjugate line locus, meet this locus again in Q and Q'; and let PQ meet BB' in X. The two tangents from Q, to the original conic, are then QP, meeting BB' in X, and a line meeting BB' in the point $(B, B')/X$; this point, however, lies on PQ', because of the property of the tangents PQ, PQ'; denote it by X'. The two tangents from Q', to the original conic, are then $Q'P$, meeting BB' in X', and a line meeting BB' in the point $(B, B')/X'$, which is X, and lies on PQ. The two tangents from Q, Q', to the original conic, other than QP, $Q'P$, thus meet in a point, R, which has the property that the two tangents, drawn to the original conic from this point, meet BB' in points X, X' for which X, $X' \sim B$, B'. This point is therefore on the conjugate line locus; so that P, Q, R, Q' are such a quadrangle as spoken of, with its points on the conjugate line locus, with its sides PQ, QR, RQ', $Q'P$ touching the original conic. This quadrangle, when the two conics are given, is determined by any one of its points. Moreover, as the two complementary joins PQ, RQ', of this quadrangle, and the two complementary joins PQ', RQ, both have their intersections, at X and X', lying on the line BB', it follows that the two further joins, namely, PR and

QQ', both pass through the point T, which is the pole of BB' in regard to the conjugate line locus; it follows also that TXX' is a self-polar triangle in regard to this locus. It has, for one angular point, the point T, which is the pole of BB', and is independent of the quadrangle P, Q, R, Q'. We have seen that the conjugate line locus contains, beside B, the points of contact, A, C, of the tangents from B to the original conic; the four points B, A, B, C form a particular quadrangle P, Q, R, Q'. So, likewise, the four points B', A', B', C' form such a quadrangle.

Conversely, let two conics ω, Ω be such that the tangents of ω, at two of the four common points of these, A and C, meet in a point B of the conic Ω. Let A', C' be the other two common points of these conics, and B' be the common point of the tangents of ω at A' and C'. Then it can be shewn that B' lies on Ω; and that this is the conjugate line locus of ω for the centres B and B'. We have only to consider the conjugate line locus of ω, for the centres B, B', and remark that this contains the five points B, A, C, A', C'; these lie on Ω, which is thus the conjugate line locus in question.

Also, if two conics ω, Ω be such that quadrangles of four points can be inscribed in Ω, of which one point, P, is arbitrary, which are such that four joins, beginning at P, of the points of the quadrangle, taken in order, are tangents of ω, then Ω is the conjugate line locus of ω, for a proper pair of centres. For, if we take P at a common point, A, of the two conics, we deduce that the tangent of ω, at this point, meets Ω in a point B which lies on the tangent of ω at another common point, C, of the two conics; by taking P at one of the two remaining common points of ω and Ω, say at A', we likewise obtain another degenerate quadrangle $A'B'C'B'$. The two points B, B' are the centres, for Ω as a conjugate line locus of ω. The chord BB', of the conic Ω is, by what we have seen, one side of the common self-polar triangle of the two conics (which is determinable as the vertex triangle of the quadrangle formed by the common points of the two conics). The opposite angular point, T, of this triangle, is the intersection of that pair of complementary common chords AC, $A'C'$, of the two conics, whose poles, in regard to ω, are B and B'. As two conics, in general, have three pairs of complementary common chords, we see that there are three different ways in which one of these conics can be conditioned, so as to become the conjugate line locus of the other.

It may increase the definiteness of the ideas if we argue somewhat differently: let two conics Ω, ω be such that quadrangles P, Q, R, Q' can be inscribed in Ω, of which the four joins PQ, QR, RQ', $Q'P$ are tangents of ω, one of the four points, say P, being an arbitrary point of Ω. Then, the point P, of Ω, determines the point R, of this conic, without ambiguity; and, when P is taken where R now is, then R takes the place now occupied by P. Thus, if we think of P, R as determined on Ω by the respective values θ, ϕ of a variable number, either of these determines the other uniquely. If we assume that the construction for R, when P is given, ensures that ϕ is an algebraic function of θ, and conversely, we infer that each of ϕ, θ is a fractional linear function of the other. The points P, R being interchangeable, the relation between ϕ, θ is symmetrical. Wherefore P, R are a pair of an involution on the conic Ω; so that the line PR passes through a fixed point, say T. As we may take P where Q now is, in which case R will be where Q' now is, we infer that the line QQ' also passes through T. The lines PR, QQ' are a pair of complementary joins of the inscribed quadrangle P, Q, R, Q', of Ω; thus, the polar of the fixed point T, in regard to Ω, is the line containing the intersections (PQ, RQ') and (PQ', RQ). But, T is also the common point of two diagonals of the circumscribed quadrilateral of ω given by PQ, QR, RQ', $Q'P$; thus the polar of T in regard to ω is the same line as the polar of T in regard to Ω. We can now complete the argument, and shew that Ω is the conjugate line locus of ω, for two centres B, B', which lie on Ω, and on the common polar line of T.

The conjugate line locus will often arise in what follows. It is a particular case of another locus, called the *harmonic locus* for the conic ω and another conic, say σ. The conjugate line locus arises when σ *degenerates into* two pencils of lines.

Ex. 19. Prove, in general, Maclaurin's theorem (1735) that, if every side of a variable triangle pass through a fixed point, and two of its angular points vary on two curves of orders m and n, then the third angular point describes a curve of order $2mn$; and investigate reductions in this order which may arise when the given curves coincide, or the fixed points lie on a line. (By a curve of order m is meant a curve which is met by a general line in m points.) See Poncelet, *Prop. Proj.*, 1865, t. I, p. 321; and Poncelet's *Introduction* to his volume.

CHAPTER X

ASSIGNMENT OF TWO ABSOLUTE POINTS. PROPERTIES OF CIRCLES

THE geometrical results of elementary metrical geometry are particular cases of theorems which arise when we add to the general figures, so far discussed, certain arbitrarily taken fixed elements. This is a capital *discovery*; which is essentially due to Poncelet.

We take a definite *line*, regarded as determined by real points. This we call the *Absolute* line. It may be any line of the plane, agreed upon; but when taken it remains unaltered. Upon this line we take two definite points, which we call the *Absolute points*; these again are arbitrary points of the line; but, once taken, they remain unaltered. We generally denote these points by I and J.

When we make distinction between real and imaginary elements in the plane, there is convenience in supposing that, upon the Absolute line, there are two real points, X and Y, such that, in terms of the symbols of these, the symbols of I and J have the forms $I = X + iY$ and $J = X - iY$, where $i^2 = -1$. Then I, J may be spoken of as conjugate imaginary points, in regard to X and Y. For X and Y we may take any two real points upon the Absolute line which are harmonic conjugates in regard to I and J.

(1) We use the names common in elementary geometry for figures which are defined with reference to I and J, choosing the names so that the figures have properties which are the same, *in form*, as those arising in elementary geometry.

For instance, we say that two lines in the plane are *parallel* when they meet on the Absolute line. Also, that two lines are *perpendicular* when they meet the Absolute line in two points which are harmonic conjugates in regard to I and J. A consequence of this is that, through an arbitrary point of the plane, not lying on the Absolute line, a single line can be drawn to be perpendicular to a given line; this line is then perpendicular also to all lines which are parallel to the given line.

Also, for brevity of language, if A, B be any two points of the plane, which do not lie on the Absolute line, and the line AB meets the Absolute line in H, we call the point $C = (A, B)/H$, which is the harmonic conjugate of H in regard to A and B, *the mid point of A and B.*

Some particular applications of these definitions may be mentioned at once:

(*a*) The lines joining the angular points, A, B, C, of a triangle, respectively to the mid points of B and C, of C and A, and of A and B, meet in a point. This is an immediate consequence of the definition of harmonic points. If D, E, F be the respective mid points, the lines EF, FD, DE are respectively parallel to BC, CA and AB.

(*b*) The lines, drawn through the angular points A, B, C, of a triangle, respectively perpendicular to BC, CA, AB, meet in a point. For, we have remarked (Chap. vii(4), Ex. 3) that the condition for three lines, drawn through A, B, C, to meet in a point, is that these lines should meet a general line of the plane, which meets BC, CA, AB, respectively, in L, M, N, in such points L', M', N', that the three pairs L, L'; M, M'; N, N' belong to an involution. This condition is satisfied here by the Absolute line, whereon I, J are the foci of the involution.

(*c*) If, as above, D, E, F are the mid points, respectively, of B, C; of C, A and of A, B, then the lines drawn through these points, respectively perpendicular to BC; to CA and to AB, meet in a point. For these lines are the lines, through the angular points D, E, F, of the triangle DEF, which are respectively perpendicular to EF, FD, DE, these being parallel respectively to BC, CA, AB; so that the result follows from (*b*).

Ex.   Let A, B, C, O be any points in a plane, and D, E, F the respective mid points of B, C; of C, A and of A, B, the lines AD, BE, CF meeting in the point G. Further, let the four mid points, respectively of O, A; of O, B; of O, C and of O, G, be denoted by A_1, B_1, C_1, M. Prove that the two lines A_1G and MD are parallel; as are A_1M and GD. Prove also that the four mid points, respectively of D, A_1; of E, B_1; of F, C_1 and of M, G, are the same point.

(2) We have seen (Chap. vii (3)), that if A, A' and B, B' be two pairs of points in a line, there is a pair of points X, X' on the line, for which both A, $A' \sim X$, X' and B, $B' \sim X$, X'; that is, X, X' are harmonic conjugates in regard both to A, A' and to B, B'. Similarly, if a, a' and b, b' be two pairs of lines drawn through a point O, there is a pair of lines x, x' through O, harmonic conjugates both in regard to a, a' and to b, b'. For brevity of language, when b, b' are the lines joining O to the Absolute points I, J, we speak of x, x' as *the bisectors of the angles defined by the lines a, a'*. These bisectors are then the pair of lines drawn through O which have the double property of being perpendicular to one another and also harmonic conjugates in regard to a, a'. This definition will be found to be amply justified by its convenience in describing various geometrical theorems.

When the lines a, a' are real, and the points I, J are conjugate imaginaries, in the sense explained above, these bisectors are always real. This follows from what was proved previously (Chap. vii (3)). The bisectors are also real, as follows from the same reference, when a, a' are conjugate imaginary lines, that is, the point O being real, when a, a' meet the Absolute line XY in points which are conjugate imaginaries in regard to X and Y (where X, Y are as above). It is in fact in order that the bisectors of the angles defined by two real lines should always be real, as in elementary geometry, that we take the points I, J to be conjugate imaginaries.

As an application of this definition of bisectors we prove the theorem: Let A, B, C be real points; take the three pairs of bisectors, respectively, of the angles defined by the lines AB, AC; the angles defined by the lines BC, BA; and the angles defined by the lines CA, CB. These six bisectors meet in threes in four points.

To prove this, let the Absolute line be met by BC, CA and AB, respectively, in L, M, N. On this Absolute line, let X, X' be the points which lie on the bisectors of the angles determined by AB, AC; so that X, X' are the foci of the involution determined by the two pairs of points M, N and I, J; or, we may say, let X, $X' \sim I, J$ and also X, $X' \sim M, N$. Similarly, taking the lines drawn from B, let Y, Y', on the Absolute line, be such that Y, $Y' \sim I, J$ and Y, $Y' \sim N, L$. Finally, let Z, Z', on the Absolute line, be such that Z, $Z' \sim I, J$ and Z, $Z' \sim L, M$. Then, invoking a principle which we have already quoted, the condition for the lines AX, BY, CZ to meet in a point, is that the three pairs X, L; Y, M; Z, N should belong to an involution. This is so if X, Y, Z be properly chosen from the pairs to which they belong, as follows from a property of involutions noted in Example 14 of Chapter VII. The algebraic proof there suggested depends merely on the identity $(a^2 - bc)(b - c) + (b^2 - ca)(c - a) + (c^2 - ab)(a - b) = 0$. Three other such involutions are obtainable if we replace X, Y, Z, respectively, by X, Y', Z'; by X', Y, Z'; and by X', Y', Z. This proves the theorem stated.

The four points so found, each a common point of three of the bisectors, form a quadrangle; of this the original triangle ABC is the vertex triangle. Moreover, if we use the name *orthocentre* of a triangle, for the point in which the lines drawn through the angular points, each perpendicular to the opposite side, have been proved to meet, each of the four points of the quadrangle is the orthocentre of the triangle formed by the other three. And there is a still further remark, subsequently of importance, based on the theorem proved earlier (Chap. VII), that an arbitrary line is met in three pairs of an involution by the pairs of complementary joins of four points; namely, if H, U, V, W be an arbitrary quadrangle, of which A, B, C is the vertex triangle, and we take, on an arbitrary line, the foci of this involution to be the Absolute points of the plane, then the quadrangle H, U, V, W is that formed, as in the theorem, by the bisectors at A, B, C.

(3) *Ex.* 1.   If, in a pencil in involution, of pairs of rays a, a'; b, b'; c, c';..., which all pass through a point, it be the case that the two rays a, a', of one pair, are perpendicular, and this be true also for the rays b, b', of another pair, then the rays c, c', of *every* pair, are perpendicular.

Ex. 2.   Let P, P', Q, Q' be four general points of a plane, the intersections (PP', QQ') and $(PQ, P'Q')$ being denoted, respectively, by T and U; then the mid points of the three pairs of points, P and Q'; P' and Q; T and U, lie in a line. This is the conjugate line of the Absolute line, in regard to the quadrilateral formed by the four lines PP', $P'Q'$, $Q'Q$, QP, according to the definition given (Chap. VII (4), Ex. 8).

Also, let R, R' be the mid points, respectively, of P, Q and of P', Q'; also K, K' the mid points, respectively, of the pairs P, P' and Q, Q'; and V, V' the mid points, respectively, for the pairs P, Q' and P', Q; then the three pairs R, R'; K, K' and V, V' have the same mid point.

Ex. 3.   Let M be the mid point of two points P, Q; and O be a point, not lying on the line PQ, such that OM is perpendicular to PQ; prove that the bisectors of the angles determined by the lines OP, OQ consist of the line OM and the line, through O, parallel to PQ.

(4) A *circle* is defined as a conic which contains the Absolute points I, J. Unless the contrary is stated, it is supposed to be undegenerate. The aggregate of the Absolute line, taken with any other line, is a degenerate circle. So is the aggregate of any two lines of which one passes through I and the other through J. For a general, undegenerate, circle, the *centre* is defined to be the pole of the line IJ, namely, the intersection of the tangents of the circle at I and J.

In general, for any conic, the centre is the pole of the Absolute line in regard to the conic; and is the intersection of the tangents of the conic at the points where it is met by the Absolute line. These two tangents are often called the *asymptotes* of the conic. A line which passes through the centre of a conic is called a *diameter* of the conic; if such line meet the conic in P and P', then the centre is the mid point of P, P'; as follows from the definition of pole and polar. Again, let P, P' and Q, Q' be two pairs of points on a conic which are such that the lines PP' and QQ' are parallel; then the chords PP', QQ' meet in a point, say T, of the Absolute line, and the polar line of T, in regard to the conic, contains the mid points of P, P' and of Q, Q', and contains also the centre of the conic. This again follows from the definition of pole and polar. It follows that the mid points, each of one of a series of parallel chords of a conic, lie on a diameter of the conic; the pole of this diameter is the point of the Absolute line at which the parallel chords meet. This diameter is conjugate to that diameter of the conic which is one of the parallel chords; and these two diameters are symmetrically related to one another. The diameter containing the mid points of a series of parallel chords meets the conic in two points, whereat the tangents of the conic are parallel to the chords; and the polar of any point on this diameter is one of the parallel chords. Conversely, if the polar of a point T, in regard to the conic, meets the conic in P and P', the line joining T to the centre of the conic contains the mid point of P and P'; and this joining line is the polar of the point where the line PP' meets the Absolute line.

In the case of a circle, which contains the points I, J, of the Absolute line, the line joining the centre to the mid point of a chord PP', being the polar of the point, H, where PP' meets the Absolute line, meets the Absolute line in the point $(I, J)/H$. Thus, the line joining the centre to the mid point of a chord of the circle is perpendicular to this chord. Thus, also, two conjugate diameters, through the centre, are perpendicular to one another. Such pairs of conjugate diameters form a pencil in involution, of which the focal rays are the lines joining the centre to I and J; these are the tangents of the circle at I and J. Conversely, if P, P' be any points of the circle, the line, through the mid point of P, P', which is perpendicular to PP', passes through the centre; and the tangent of the circle at any point is perpendicular to the line joining the centre to this point. A particular corollary, and illustration, is that, if A, B, C be any three points of a circle (not at I or J), the three lines drawn, respectively, through the mid points of B, C; of C, A and of A, B, respectively perpendicular to BC, to CA and to AB, meet in a point, namely in the centre of the circle. In general, a conic can be drawn through any five points; thus a circle can be drawn through any three points (not at I, nor J); and its centre is given by the construction we have enunciated.

If D, D' be the points where a line drawn through the centre, O, of a conic, meets the conic, so that O is the mid point of D, D'; and if P be any point of the conic, and the mid point of D, P be M; then it is a consequence of the harmonic relation that the lines $D'P$ and OM meet on the line joining the two points $(D, D')/O$ and $(D, P)/M$; hence OM is parallel to $D'P$. The polar of the point of intersection of OM and $D'P$ contains O, because this point of intersection is on the Absolute line; and this polar contains the mid point, M', of D', P, because this point of intersection is $(D', P)/M'$. Thus, the lines OM, OM' are conjugate in regard to the conic. The lines DP, $D'P$, which join the ends D, D' of a diameter, of a conic, to an arbitrary point P of the conic, are often called *supplementary chords* of the conic; we have proved that the lines drawn through the centre, parallel to a pair of supplementary chords, are conjugate diameters.

In the case of a circle, such conjugate diameters have been proved to be perpendicular. Thus the lines joining an arbitrary point P, of the circle, to the ends D, D', of a diameter, are perpendicular to one another. Conversely, if D, D' be arbitrary points of the plane, and two pencils of lines be taken, with centres at D and D', such that any ray of one pencil is perpendicular to a ray of the other pencil, the two pencils will be related; and the point, P, common to corresponding rays of the two pencils, will describe a conic, passing through D and D'. As the ray through D' which is perpendicular to the ray DI is the ray $D'I$, the conic will contain I; and will likewise contain J; and so, will be a circle. The tangents of this circle at D and D' will both be perpendicular to DD', and the centre of this circle will be the mid point of D, D'.

A circle can be drawn, with centre at a given point, T, to contain another given point, P. For this is only to say that a conic can be drawn to touch the lines TI and TJ at I and J, and to contain another point, P. In particular, let TP be one of the two tangents which can be drawn, from a given point T, to touch a given circle, the point of contact being P; let the circle with centre at T which contains P meet the given circle again in P' (besides I and J and P); then, by what we have proved, the line through the mid point of P, P', which is perpendicular to PP', contains the centre T, of the circle described, as well as the centre, say O, of the given circle. Thus, by the preceding, the line PP' is the polar of T in regard to the given circle. Or, a circle can be drawn, with centre at T, to contain the points P, P', in which the polar of T in regard to a given circle meets this circle. The tangent at P, of the given circle, is the line TP; as T is the centre of the second circle, the tangent at P of the second circle, being perpendicular to PT, is the line PO containing the centre O of the given circle. Thus the tangents at P of the two circles are perpendicular; and the same is true at P'. The two circles are then said to *cut one another perpendicularly, or, to be perpendicular.* And, just as the tangents of the two circles at P (as at P') thus cut the other common chord IJ of the two circles, in two points which are harmonic conjugates in regard to I and J, so it may be shewn that the tangents of the two circles at I (as at J) cut the common chord PP' in two points which are harmonic conjugates in regard to P and P'. It may be shewn also that, if any line be drawn through the centre, O, of one of the circles, the two intersections of this line, with either circle, are harmonic conjugates in regard to the two points in which this line meets the other circle (and similarly for the centre T). These results arise again later. Two circles cutting perpendicularly are in fact such a related pair of conics as was considered in Chap. ix (23), Ex. 14. And, it may be remarked, through two arbitrary points P, P' of a given circle, an infinite series of other circles can be drawn; only one of these cuts the given circle perpendicularly.

If l, m be any two lines which cut the Absolute line respectively in the points L, M, it is very convenient to denote the cross ratio $(L, M/I, J)$ by a symbol. We denote this by $\epsilon\,(l, m)$. For this symbol to have a definite meaning, the order in which the Absolute points I, J are taken must be definite. The order in which the lines l, m are taken is also important; we have in fact $\epsilon\,(m, l) = 1/\epsilon\,(l, m)$. When l, m are perpendicular we have $\epsilon\,(l, m) = -1 = \epsilon\,(m, l)$; and conversely this ensures that the lines are perpendicular. As applications of the notation we may notice:

(a) If l, m, n be any three lines (in a plane) we have

$$\epsilon\,(m, n)\,.\,\epsilon\,(n, l)\,.\,\epsilon\,(l, m) = 1.$$

In particular, if OP, OQ, OR be three lines through a point O, we have $\epsilon\,(OP, OR) = \epsilon\,(OP, OQ)\,.\,\epsilon\,(OQ, OR)$.

(b) If the lines l, l' be perpendicular, $\epsilon\,(l', m) = -\epsilon\,(l, m)$.

(c) If two lines, CA, CB, be perpendicular, then

$$\epsilon\,(AC, AB) = -\epsilon\,(BC, BA).$$

We may agree to abbreviate such an equation by writing

$$\epsilon A\,(C, B) = -\epsilon B\,(C, A).$$

If A, A', P, Q be any four points on a circle (not at I, nor J), the fundamental property of six points of a conic is expressed by

$$\epsilon A\,(P, Q) = \epsilon A'\,(P, Q).$$

Conversely, if P, Q be fixed points, the locus of a point O, when

$\epsilon O\,(P,\,Q)$ remains constant, is a circle, which passes through P and Q. In particular, if $A,\,P,\,Q$ be any three points of a circle, and PT be the tangent at P, we have $\epsilon P\,(T,\,Q)=\epsilon A\,(P,\,Q)$. Also, if $P,\,Q$ be the ends of a diameter of a circle, and R be any point of the circle, then $\epsilon R\,(P,\,Q)=-1$.

The use of the symbol $\epsilon\,(l,\,m)$ enables us to dispense, quite successfully, as will be seen, with the use of the measurement of angle which occurs commonly in elementary geometry. The latter requires the topological notion of the direction of rotation of a line revolving round a fixed point. The symbol $\epsilon\,(l,\,m)$ takes no account of the sense in which either of the lines $l,\,m$ may be supposed to be described. While this leads, in cases, to a loss of topological exactness, it is, in other cases, an advantage. For instance this is so in the descriptive geometrical theorems which occur, in which it can only be said, of two angles, that they are "equal or supplementary", to use the common phraseology.

(5) *Ex.* 1.   Let $D,\,D'$ be the ends of a diameter of a circle, P and Q be any two other points of the circle; also, let $M,\,M'$ be points of the line PQ such that DM and $D'M'$ are perpendicular to PQ. Prove that the mid point of $P,\,Q$ is also the mid point of $M,\,M'$.

    Ex. 2.   Let $H,\,H'$ be points of a circle such that the chord HH' is perpendicular to a diameter DD' of the circle; and let P be any other point of the circle. Prove that the bisectors of the angles determined by the lines $PH,\,PH'$ are the lines $PD,\,PD'$.

    Ex. 3.   Let $A,\,B,\,C$ be three points of a circle, whose centre is O. Prove that

$$\epsilon A\,(B,\,O)=\epsilon B\,(O,\,A)=-\,\epsilon C\,(B,\,A),$$

and that
$$\epsilon O\,(A,\,B)=[\epsilon C\,(A,\,B)]^2.$$

Also, if the lines $AO,\,BO$ meet the circle again respectively in A' and B' and N be the mid point of $A,\,B$, prove that $AB',\,BA',\,NO$ are all perpendicular to AB. Thence the equations follow from $(a),\,(c)$ above.

    Ex. 4.   Let $OX,\,OY$ be the bisectors of the angles determined by any two lines $OP,\,OQ$, the four points $X,\,Y,\,P,\,Q$ being all on the Absolute line, so that $X,\,Y$ are the foci of the involution determined by the two pairs $I,\,J$ and $P,\,Q$. Prove that

$$(P,\,X/I,\,J)=(Q,\,X/J,\,I)=(X,\,Q/I,\,J).$$

Hence $\epsilon O\,(P,\,X)=\epsilon O\,(X,\,Q)$. Similarly we prove that $\epsilon O\,(P,\,Y)=\epsilon O\,(Y,\,Q)$.

If a line which is perpendicular to OX, and therefore parallel to OY, meet $OP,\,OQ,\,OX$ in $G,\,H,\,D$, respectively, then, because $Y=(P,\,Q)/X$, it follows that D is the mid point of $G,\,H$. We similarly prove that

$$\epsilon G\,(H,\,O)=\epsilon H\,(O,\,G).$$

We may illustrate these results by giving here a formal proof of the theorem, which is important in what follows, that, *if a line, drawn through the centre of one of two circles, cut these circles, respectively, in harmonically conjugate pairs of points, then the circles cut perpendicularly.*

    Let AOB be a diameter of a circle of centre O, the points $A,\,B$ being on the circle; let $P,\,Q$ be two points on this diameter such that $P,\,Q\sim A,\,B$, or $Q=(A,\,B)/P$. Let any other circle, containing P and Q, cut the given circle in H and K. Since $HA,\,HB$ are perpendicular, and are harmonic conjugates in regard to HP and HQ, the lines HA, HB are the bisectors of the angles determined by HP, HQ. Whence

$$\epsilon H\,(O,\,P) = \epsilon H\,(O,B)/\epsilon H\,(P,\,B) = \epsilon B\,(H,\,O)/\epsilon H\,(B,\,Q)$$
$$= \epsilon B\,(H,\,Q)/\epsilon H\,(B,\,Q) = \epsilon Q\,(H,\,B) = \epsilon Q\,(H,\,P).$$

Thus $\epsilon H\,(O,\,P)$, equal to $\epsilon Q\,(H,\,P)$, is equal to $\epsilon H\,(H,\,P)$. Thus HO is the tangent, at H, of the circle described through P and Q. This is perpendicular to the tangent, at H, of the original circle. So that the two circles are perpendicular (at their common point H; and similarly at their other common point K).

Ex. 5. We now prove various theorems, of frequent occurrence, in regard to the line which is commonly called the *pedal line, or the Simson line,* of a triangle ABC, in regard to any point O taken on the circle through ABC.

Let A, B, C, O be any points of a circle, and L, N be points on BC, BA, respectively, such that OL and ON are respectively perpendicular to BC and BA. Let H be the point, called the ortho-centre of the triangle ABC, in which the perpendiculars drawn from the angular points, each to the opposite side, have been proved to meet ((1), above). Let AH meet BC in D, and meet the circle again in G. Let OG meet BC in E, and meet LN in K. Let OH meet LN in T.

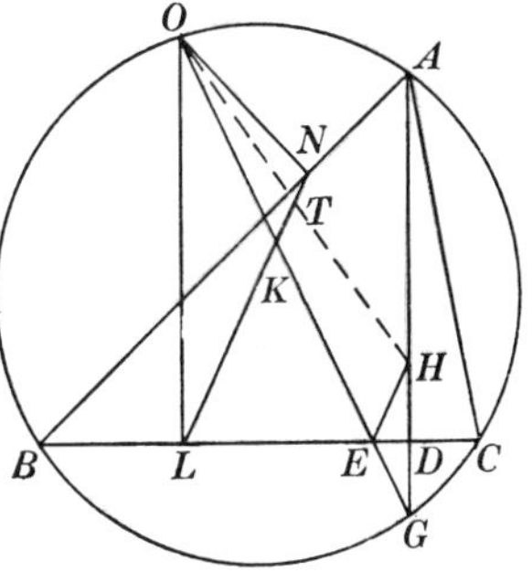

Because CH, CD are respectively perpendicular to AB and AG, we have

$$\epsilon C\,(H,\,D) = \epsilon A\,(B,\,G),$$

of which, as A, B, C, G, are on the circle, the latter is equal to $\epsilon C\,(B,\,G)$, or $\epsilon C\,(D,\,G)$. Thus, by Ex. 4 preceding, D is the mid point of H, G. Thus also $\epsilon E\,(G,\,D) = \epsilon E\,(D,\,H)$. From

$$\epsilon G\,(D,\,E)\,.\,\epsilon D\,(E,\,G)\,.\,\epsilon E\,(G,\,D) = 1 = \epsilon H\,(E,\,D)\,.\,\epsilon E\,(D,\,H)\,.\,\epsilon D\,(H,\,E),$$

wherein $\epsilon D\,(E,\,G) = -1 = \epsilon D\,(H,\,E)$, we infer that $\epsilon G\,(D,\,E) = \epsilon H\,(E,\,D)$, or $\epsilon G\,(H,\,E) = \epsilon H\,(E,\,G)$.

Again, because ON, OL are respectively perpendicular to BN and BL, the circle containing O, B, N contains L. Hence $\epsilon L\,(K,\,O)$, or $\epsilon L\,(N,\,O)$, is equal to $\epsilon B\,(N,\,O)$, or $\epsilon B\,(A,\,O)$; but this is equal to $\epsilon G\,(A,\,O)$, which is equal to $\epsilon O\,(L,\,G)$, or $\epsilon O\,(L,\,K)$, because GA, LO are parallel. From

$$\epsilon L\,(K,\,O) = \epsilon O\,(L,\,K),$$

however, we can infer (Ex. 3, above) that the circle OLE has K for centre, so that K is the mid point of O, E.

Also, because $\epsilon H\,(E,\,G)$ has been shewn equal to $\epsilon G\,(H,\,E)$, or $\epsilon G\,(A,\,O)$, or $\epsilon B\,(A,\,O)$; and this has been shewn equal to $\epsilon L\,(K,\,O)$; we have

$$\epsilon H\,(E,\,G) = \epsilon L\,(K,\,O).$$

Thus we infer that EH is parallel to LK, or LT. And we proved that K is the mid point of O, E; whence T is the mid point of O, H.

As the line joining L to the mid point of O, E thus contains N, it may be inferred that, if this line meets AC in M, then OM is perpendicular to AC.

We have thus proved the theorem, usually associated with the name of Simson (1687–1768), that, *if A, B, C, O be four points of a circle, and OL, OM, ON be the perpendiculars from O, respectively, to BC, CA, AB, then the points L, M, N lie in a line. And, if H be the orthocentre of the triangle ABC, we have proved that this* pedal line, *of O, contains the mid point of O, H.* The theorem is also assigned to W. Wallace.

If OL meets the circle again in L_1, we can prove that AL_1 is parallel to the pedal line. Conversely, if, from the points A, B, C, respectively, three lines be drawn which are parallel to one another, and these meet the circle again, respectively, in L_1, M_1 and N_1; and, then, lines be drawn from L_1, M_1, N_1, respectively perpendicular to BC, CA, AB, these perpendiculars meet in a point of the circle.

Ex. 6. If any four lines be drawn in a plane, any three of these form a triangle. The orthocentres of the four triangles lie in a line. We shall have subsequent proof of this; and of the theorem just proved in Ex. 5. But we add here a proof that these four orthocentres lie in a line, which is founded on the assumption of Pappus' theorem (Chap. III (8)), and illustrates the ideas we are dealing with.

Let the sides BC, CA, AB, of a triangle, meet the Absolute line respectively in D_1, E_1, F_1. Let any other line l meet the Absolute line in U_1, and meet BC, CA, AB, respectively, in D, E, F. On the Absolute line take points D', E', F', defined respectively by $D'=(I, J)/D_1$, $E'=(I, J)/E_1$ and $F'=(I, J)/F_1$. It is sufficient, for our purpose, to prove that three of the orthocentres of the four triangles, formed by four lines, are in a line; we consider then the orthocentres of the three triangles formed by the line l taken in turn with the pairs of the lines BC, CA, AB, namely, the triangles AEF, BFD, CDE. The orthocentre of the triangle AEF is the common point of the lines EF' and FE', which are the lines through E and F, respectively, perpendicular to the opposite sides of the triangle AEF. Similarly, the orthocentres of the triangles BFD and CDE are, respectively, the intersections (FD', DF') and (DE', ED'). As the points D, E, F are on the line l, and the points D', E', F' are on the Absolute line, it follows, by Pappus' theorem, that the three orthocentres are in a line; as we wished to prove. This argument is given in Salmon's *Conics*, 1879, p. 246, as due to De Morgan and W. Burnside.

Ex. 7. Let D, E, F be any points, respectively, on the sides BC, CA, AB of a triangle. We prove that the three circles AEF, BFD and CDE meet in a point (Miquel, *Liouville's Journ.* III, 1838).

The two latter circles, BFD and CDE, meet in D; let O be their other common point (besides I and J). We have, because O, C, D, E are on a circle, $\epsilon O\,(D, E) = \epsilon C\,(D, E)$, and the latter is $\epsilon C\,(B, A)$; also, because O, B, F, D are on a circle, we have $\epsilon O\,(F, D) = \epsilon B\,(F, D)$, of which the latter is $\epsilon B\,(A, C)$. Thus $\epsilon O\,(F, E)$, being equal to the product $\epsilon O\,(F, D) . \epsilon O\,(D, E)$, is equal to $\epsilon B\,(A, C) . \epsilon C\,(B, A)$; and, as for any triangle, this product is equal to $\epsilon A\,(B, C)$, or $\epsilon A\,(F, E)$. From $\epsilon O\,(F, E) = \epsilon A\,(F, E)$, however, we infer that O is on the circle AEF; as we wished to prove.

The three circles have in common the points O, I, J. These are the three points of concurrence of corresponding rays of three related pencils of rays whose centres are the points D, E, F (cf. Chap. VIII (11), Ex. 7).

Ex. 8. A particular case of Ex. 7 arises when the points D, E, F are in a line. In that case we have the theorem that, if four triangles, formed by taking threes of four lines, be considered, then the three circles through the angular points of three of these triangles, respectively, meet in a point. Whence, by symmetry, *the four circles, each from one of the four triangles, meet in a point.*

Ex. 9. It seems proper to mention that the result of Ex. 8 can be deduced from the following theorem of intersection of four planes in space of three dimensions: Let O, A, B, C be four points which are the angular points of a tetrahedron in space; let any line in the plane ABC meet BC, CA, AB, respectively, in L, M, N; and D, E, F be any points arbitrarily taken on the lines OL, OM, ON, respectively. The theorem referred to is that the four planes AEF, BFD, CDE, ABC meet in a point.

A proof of this theorem may be given with the help of the symbols: In terms of the symbols of A, B, C, let the symbols of L, M, N be, respectively, $L=B-\lambda C$, $M=C-\mu A$, $N=A-\nu B$, so that (Chap. IV (4), Ex. 6) the numbers λ, μ, ν are such that we have $\lambda\mu\nu=1$. Let then the symbols of D, E, F, in terms of the symbols of O, A, B, C, be, respectively, $D=pO+B-\lambda C$, $E=qO+C-\mu A$, $F=rO+A-\nu B$, where p, q, r are numbers. If then the numbers x, y, z be, respectively,

$$x=p\lambda^{-1}+q+r\mu,\quad y=p\nu+q\mu^{-1}+r,\quad z=p+q\lambda+r\nu^{-1},$$

it is at once seen that the points of respective symbols,

$$xA+rE-qF,\quad yB+pF-rD,\quad zC+qD-pE,$$

are the same point. This point lies in each of the planes AEF, BFD, CDE, ABC.

We do not enter now into the deduction of the result of Ex. 8 from this theorem of space, beyond remarking that it arises by projection.

Ex. 10.   There is converse of Ex. 8: *If D, E, F, be points, on the sides BC, CA, AB of a triangle, which are such that the common point, proved to exist* (in Ex. 7), *of the circles AEF, BFD, CDE, lies also on the circle ABC, then the points D, E, F are in one line.* For, let EF meet BC in D'. Then, by Ex. 8, the point, besides A, common to the circles AEF, ABC, lies on both the circles BFD' and $CD'E$. By hypothesis this point, say O, also lies on the circles BFD, CDE. The two circles BFD' and BFD thus have the three points B, F, O in common. Thus these circles, supposed undegenerate, coincide with one another. Likewise, so do the circles $CD'E$, CDE. Thus the points D and D' are the same point, which proves the statement made.

This proves again the particular result of Ex. 5, relating to the existence of the pedal line of a point O taken on the circle ABC. For, if OD, OE, OF be the perpendiculars from O to BC, CA, AB, meeting these, respectively, in D, E and F, it is clear that the four points O, E, F, A lie on a circle, as do O, F, D, B and O, D, E, C. Thus the points D, E, F lie in a line.

Ex. 11.   Let us speak of two points, O, O', as being *optical images* of one another, in regard to a line l, when this line l contains the mid point of O, O', and is also perpendicular to OO'. Prove then, from Exx. 5, 6, that if any four lines be given in a plane, there exists a point O such that the four optical images of O, taken in regard to the four lines in turn, lie upon another line.

Ex. 12.   Let A, B, C be any three points in a line, and O be any point, not in this line. Through the three points which are the respective centres of the circles OBC, OCA, OAB there passes a circle. We prove that this circle contains the point O.

Let the centres of the three circles OBC, OCA, OAB be, respectively, D, E, F. As OA is a common chord of the second and third circles, the lines joining F, E to the mid point of O, A are both perpendicular to OA; and are the same line; or, we may say, O and A are optical images in regard to the line EF. Likewise O and C are optical images in regard to the line DE.

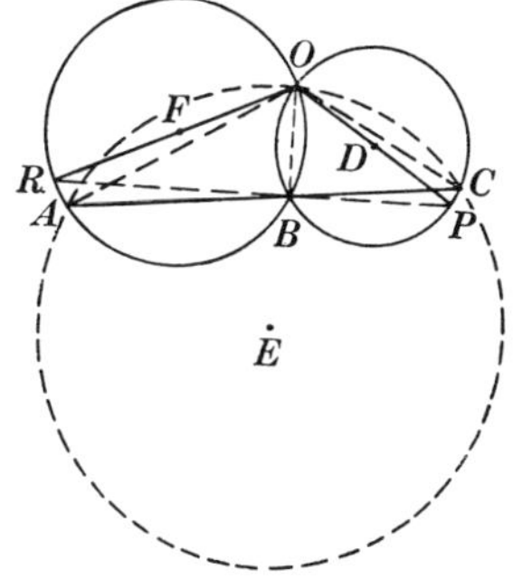

If ODP, OFR be the diameters from O, of the circles (D), (F), the common chord, OB, of these circles, is perpendicular both to BP and to BR. Thus P, B, R are in a line.

We have

$$\epsilon R\,(P,\,O)=\epsilon R\,(B,\,O)=\epsilon A\,(B,\,O)=\epsilon A\,(C,\,O),$$

the points O, R, A, B, being on a circle; and, likewise,

$$\epsilon P\,(O,\,R)=\epsilon P\,(O,\,B)=\epsilon C\,(O,\,B)=\epsilon C\,(O,\,A).$$

Also, from the triangles ORP and OAC, respectively, we have

$$\epsilon O\,(R,\,P).\epsilon R\,(P,\,O).\epsilon P\,(O,\,R)=1=\epsilon O\,(A,\,C).\epsilon A\,(C,\,O).\epsilon C\,(O,\,A).$$

Whence we infer that $\epsilon O\,(F,\,D)$, or $\epsilon O\,(R,\,P)$, is equal to $\epsilon O\,(A,\,C)$; and, because OA, EF are perpendicular, and OC, ED are perpendicular, we have

$\epsilon O\,(A,\,C)=\epsilon E\,(F,\,D)$. Thus we have $\epsilon O\,(F,\,D)=\epsilon E\,(F,\,D)$; this shews that $O,\,F,\,E,\,D$ are on a circle. Which is what we wished to prove.

Now recall Ex. 8. Take four arbitrary lines in a plane. Let the *circumcircles* of the four triangles formed by the threes of these lines meet in a point, O, as is there proved. Then, from the result proved in the present Example, it follows that the centres of the four circumcircles lie on a circle, which passes through O.

Ex. 13.   The point O, of concurrence of the circumcircles of the four triangles formed by four lines in a plane, whose existence is proved in Example 8, is evidently such that its pedal lines, in regard to the four triangles, are the same line. Thus, as (by Ex. 5) the pedal line contains the mid point of O and the orthocentre of any one of the four triangles, we see that the four orthocentres of the triangles lie upon a line; as was otherwise proved in Ex. 6. The point O is thus the point whose optical images, taken in regard to the lines in turn, lie upon a line.

If, in addition to the four lines now considered, we take another line so related to the four that the optical image of O, in regard to this line, lies upon the line containing the optical images of O in regard to the four given lines (and it will appear, in the following chapter, that this is possible in an infinite number of ways), then we shall obtain five circles passing through O, each containing the centres of the four circles circumscribing the triangles formed with a set of four of these lines. It can be shewn that the centres of these five circles lie on another circle, also passing through O.

But a theorem is true for five lines which are all arbitrary: For every four of these there exists a point whose optical images in regard to these four lie upon a line; call this momentarily the optical centre for these four lines. The five optical centres, so obtainable by taking the fours of the lines, lie upon a circle. Calling this, momentarily, the Miquel circle of the five lines, it can be shewn that, if six arbitrary lines be taken, the six Miquel circles, of the six sets of five lines chosen from these, meet in a point. Again, if seven arbitrary lines be taken, the seven points of concurrence, so obtainable from every set of six from the seven lines, lie upon a circle. And so on, indefinitely. (Cf. Note 1, (13), below; and *Principles of Geometry*, IV, pp. 31, 64.)

Ex. 14.   Let DOD' be a diameter of a circle whose centre is O; let T be the common point of the tangents of the circle at two arbitrary points, $P,\,Q$, thereon. Through T draw the line which is perpendicular to the chord DP, and let the chord $D'Q$ meet this perpendicular in L. Prove that OL is parallel to DP.

Ex. 15.   The following result is used below when we are dealing with a parabola. Let $A,\,Y,\,X,\,Z$ be any four points of a circle. Let another circle, which contains A, meet $AY,\,AZ$ respectively in B and C; let P be any point of this second circle, and O any point of the plane. Draw from P the line perpendicular to BP, and draw from O the line perpendicular to YX, and let these lines meet in Q. Draw from P the line perpendicular to CP, and draw from O the line perpendicular to ZX, and let these lines meet in R. We prove that the four points $O,\,P,\,Q,\,R$ lie on a circle.

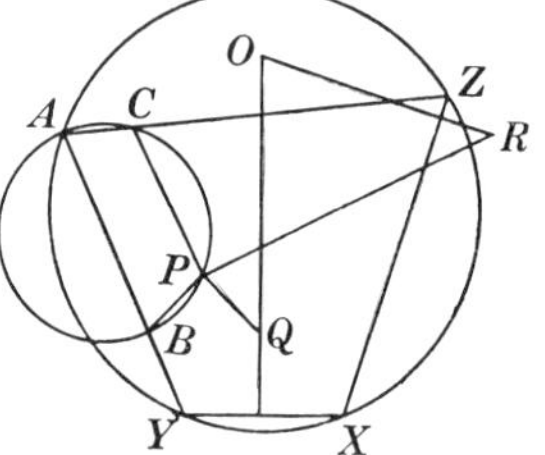

Since $PQ,\,PR$ are, respectively, perpendicular to $PB,\,PC$, we have $\epsilon P\,(Q,\,R)=\epsilon P\,(B,\,C)$, which is equal to $\epsilon A\,(B,\,C)$, or $\epsilon A\,(Y,\,Z)$; and this is equal to $\epsilon X\,(Y,\,Z)$. Since $OQ,\,OR$ are respectively perpendicular to $XY,\,XZ$, we have $\epsilon X\,(Y,\,Z)=\epsilon O\,(Q,\,R)$. From $\epsilon P\,(Q,\,R)=\epsilon O\,(Q,\,R)$, it follows that $O,\,P,\,Q,\,R$ are on a circle.

Ex. 16.   We have shewn above that, if the tangents at points $P,\,Q$ of a circle, whose centre is O, meet in T, then TO is one of the bisectors of the angles determined by the lines TP, TQ. And that the lines OP, OQ are respectively perpendicular to the tangents at P, Q. Conversely, then, if O be any point lying on one of the bisectors of the angles determined by two given lines, a circle exists with centre at O which touches both these lines; it is the circle, with centre at O, which contains the point in which the perpendicular drawn from O, to one of these lines, meets this line.

Now, let ABC be any triangle. Take one of the bisectors of the angles determined by the lines BC, BA; and, also, one of the bisectors of the angles determined by the lines CB, CA; and let these two bisectors meet in O. Then, let OL, OM, ON be the perpendiculars drawn from O respectively to BC, CA, AB. By what has been said, the circle with centre O, which contains L, will touch BC, CA, AB, respectively at L, M, N; and the line OA is one of the bisectors of the angles determined by the lines AB, AC. Incidentally, this proves again a theorem stated above (2). And we see that four circles exist, each of which touches all the sides of a given triangle.

Moreover, as the Absolute points of the plane can be arbitrarily taken, we infer that, through two arbitrary points of a plane, there can be drawn four conics each of which touches all of three arbitrarily given lines of the plane. Dually, four conics can be drawn through three arbitrarily given points of a plane to touch two arbitrarily given lines of the plane. For comparison, we recall that a single conic can be described through five arbitrarily given points of a plane (in general positions), or a single conic to touch five arbitrarily given lines of a plane. And (Chap. IX (23), Ex. 6) that two conics can be drawn through four arbitrarily given points of a plane to touch a given arbitrary line; or two conics through a given point to touch four arbitrary lines. These are all the possible cases in which a conic is required to satisfy five conditions, a condition being that of passing through a given point, or of touching a given line.

Ex. 17.   Recall that the condition that three points, L, M, N, lying respectively on the sides BC, CA, AB, of a triangle, should lie in a line, is the existence of a general point, say D, such that the three pairs of lines DA, DL; DB, DM; DC, DN should belong to a pencil in involution (Ex. 3, Chap. VII (4)). Hence prove that, if a point, O, be joined to three points A, B, C, and the bisectors of the angles determined by the two lines OB, OC meet BC in P and P', while similarly, the bisectors of the angles determined by OC, OA meet CA in Q and Q', and the bisectors of the angles determined by OA, OB meet AB in R and R', *then* the six points P, P', Q, Q', R, R' lie, in threes, upon four lines.

It is also true that, if the bisectors of the angles determined by the two lines AB, AC meet BC in D and D'; and the similar lines through B and C meet CA and AB, respectively, in E, E' and in F, F', then the six points D, D', E, E', F, F' lie in threes upon four lines.

Ex. 18.   Let the bisectors of the angles determined by the lines AB, AC be denoted by l and l', the similar bisectors through B be denoted by m and m', and the similar bisectors through C by n and n'. Let x, y, z be lines, respectively through A, B, and C, which meet in a point, say P. Let the line harmonic conjugate to x, in regard to l, l', or say $(l, l')/x$, be denoted by x'; and similarly let $y' = (m, m')/y$ and $z' = (n, n')/z$. Then the lines x', y', z' meet in a point, say P'.

The points P, P' are called *isogonal conjugates* in regard to the triangle. The lines l, l', m, m', n, n', we have seen, meet in four points. The vertex triangle of the quadrangle formed by these four points is the original triangle ABC. The points P, P' are conjugate points, in a sense previously (Chap. IX (23), Ex. 6) explained, in regard to this quadrangle, being conjugate to one another in regard to every conic passing through the four points of the quadrangle.

It may be proved that, if A', B', C' be the optical images of any point P, in regard to the respective sides BC, CA, AB of a triangle (so that, for instance, BC is perpendicular to PA', and contains the mid point of P, A'), then the isogonal point P', of the point P, in regard to the triangle ABC, is the intersection of lines drawn through A, B, C, respectively perpendicular to $B'C'$, $C'A'$ and $A'B'$; and P' is the centre of the circle $A'B'C'$.

Ex. 19. We have proved that the lines drawn through A, B, C, respectively perpendicular to BC, CA, AB, meet in a point, H, the *orthocentre* of the triangle ABC; and that the lines, respectively through the mid points of B, C; of C, A; and of A, B, which are perpendicular to BC, to CA and to AB, also meet in a point; this we may call the *circumcentre*, O, of the triangle ABC.

The points H, O are isogonal points, in regard to the triangle ABC, in the sense explained in Ex. 18. For, let the line through O, perpendicular to BC, which contains the common point T of the tangents at B, C of the circle ABC, meet this circle in L and M; then, on the circle, the points L, M are harmonic conjugates in regard to B and C; whence the lines AL, AM, which are perpendicular, are the bisectors of the angles determined by the lines AB, AC. The line AH, which is parallel to LM, meets LM in the point $(L, M)/O$; hence the line AH is that expressed by $(AL, AM)/AO$, as in Ex. 18. Similarly for BH, in regard to BO, and for CH, in regard to CO. This shews that H, O are isogonal points.

Ex. 20. We have proved the results, of several of the preceding examples, in a manner similar to that generally followed in elementary geometry, because it is interesting to see that this is possible on the basis of the definitions and axioms which are adopted in this volume. But, as already suggested, the results of Exx. 18, 19 may be approached from a more general point of view. Let T, X, Y, Z be four arbitrary points of a plane; denote the respective intersections (TX, YZ), (TY, ZX), (TZ, XY) by A, B, C. Let I be an arbitrary point of the plane, and J the point conjugate to I in regard to the quadrangle $TXYZ$. By the property of a quadrangle the lines ATX, AYZ are harmonic conjugates in regard to AB, AC. Thus, if we take I, J as the Absolute points, the lines ATX, AYZ are the bisectors of the angles determined by the lines AB, AC. Similarly for lines through B and C. We thus recover the figure, described before, from a different point of view; and, any two points which are conjugate to one another in regard to the quadrangle $TXYZ$ are isogonal points in regard to the vertex triangle ABC, when I, J are the Absolute points.

Ex. 21. Let any conic be taken in a plane, and the pairs of tangents to this conic from the Absolute points I and J be considered. The quadrilateral formed by these four tangents has two pairs of complementary intersections, besides I and J. These are called *the pairs of complementary foci* of the conic.

Recalling, with the ideas explained in Ex. 20, a theorem proved in Chap. IX (23), Ex. 7, we can hence shew that, *for a conic which touches the sides of a triangle, each pair of complementary foci are a pair of isogonal points in regard to the triangle.* As it is clear that, to prescribe a focus of a conic, say S, is to prescribe two tangents, SI and SJ, we can also say that a conic, defined to touch the sides of a triangle and to have a given point S as focus, has its complementary focus at the point which is isogonal to S in regard to the triangle.

For, in the example in Chapter IX referred to, we have shewn that, if I, J and P, P' be two pairs of conjugate points, in regard to a quadrangle $TXYZ$, of which ABC is the vertex triangle, then the seven lines BC, CA, AB, IP, IP', JP, JP' all touch a conic.

Ex. 22. If P, P' be isogonal points in regard to a triangle ABC, we have

$$\epsilon A\,(B, P) = \epsilon A\,(P', C); \quad \epsilon B\,(C, P) = \epsilon B\,(P', A); \quad \epsilon C\,(A, P) = \epsilon C\,(P', B).$$

For, as the lines AP, AP' are harmonic conjugates in regard to the bisectors of the angles determined by AB, AC; and these bisectors are harmonic conjugates both in regard to AB, AC, and in regard to AI, AJ, we see that the bisectors are the focal rays of the pencil in involution of which AB, AC; AP, AP' and AI, AJ are three pairs. Whence $A\,(B, P/I, J) = A\,(C, P'/J, I)$, which is equal to $A\,(P', C/I, J)$. Similarly for the lines through B and C. These prove the results stated; and are obviously sufficient, conversely, to prove that P, P' are isogonal in regard to ABC.

Ex. 23.   Let P be any point in the plane of a triangle ABC, and let the perpendiculars drawn from P to the sides BC, CA, AB meet these, respectively, in D, E, F. The circle containing the points D, E, F is called the *pedal circle* of P in regard to the triangle. Let O be the centre of this circle; let P' be the point, on the line PO, such that O is the mid point of P, P'. Let the perpendiculars from P' to the sides BC, CA, AB meet these, respectively, in D', E', F'. We can shew that these are the second intersections of the pedal circle, respectively, with BC, CA, AB; so that this circle is also the pedal circle of P' in regard to the triangle.

For, because PD, $P'D'$ are parallel, and O is the mid point of P, P', it follows at once that the line, drawn from O to the mid point of D, D', is parallel to PD and to $P'D'$; and is, hence, perpendicular to (BC or) DD'. But also, by what was shewn, the line drawn from the centre O of the circle, perpendicular to DD', will contain the mid point of the two points in which the circle meets the line DD'. Thus, as the circle contains D, it also contains D'. By like reasoning, the circle contains E' and F'.

Ex. 24.   But we can shew further that P and P' are isogonal points in regard to the triangle. For, since $P'E'$ and $P'F'$ are perpendicular, respectively, to AC and AB, a circle contains the four points A, F', P', E'. Thus $\epsilon A\,(P', C)$, or $\epsilon A\,(P', E')$, is equal to $\epsilon F'\,(P', E')$; and, as $F'A$, $F'P'$ are perpendicular to one another, and, therefore, are harmonic conjugates in regard to $F'I$ and $F'J$, we have $\epsilon F'\,(P', E') = -\,\epsilon F'\,(A, E')$, which is equal to $-\,\epsilon F'\,(F, E')$. Again, a circle contains A, F, P, E; so that, similarly,

$$\epsilon A\,(B, P) = \epsilon A\,(F, P) = \epsilon E\,(F, P) = -\,\epsilon E\,(F, A) = -\,\epsilon E\,(F, E').$$

By the preceding example F', F, E', E lie on a circle. Thus

$$\epsilon F'\,(F, E') = \epsilon E\,(F, E').$$

Whence we infer that $\epsilon A\,(P', C) = \epsilon A\,(B, P)$. A precisely similar argument holds in regard to the points B and C. Whence, by Ex. 22, the points P, P' are isogonal in regard to the triangle ABC.

Ex. 25.   Now recall the result proved in Chap. ix (23), Ex. 6, in regard to two points which are conjugate in regard to a quadrangle, that, if one of these points describes an arbitrary line, the conjugate point describes a conic, this conic passing through the three vertices of the quadrangle, and meeting the line in the two conjugate points which lie on this line; and the converse theorem, that, if any conic be taken through the vertices of the quadrangle, the conjugate of a point which describes this conic describes a line. There are upon such a conic two points which are conjugate in regard to the quadrangle; the line is that which joins these points. Recall also that two points which are isogonal points in regard to a triangle are conjugate points in regard to the quadrangle which is formed by the centres of the four circles which touch the sides of the triangle; as we have proved above. We can then infer that, if any point moves on a definite conic which contains the angular points of a given triangle, the point which is isogonal to this, in regard to the triangle, moves on the line joining the unique pair of isogonal points which lie on the conic.

A particular case of this gives a proof of the theorem of the pedal line of a triangle which is given in Ex. 5 above. For, suppose the conic taken through A, B, C to be the circumcircle of the triangle. Upon this, the pair of existing isogonal points are the Absolute points I, J; as we may prove. Therefore, as a point P describes the circumcircle, the isogonal point P' describes the

Absolute line. When this is so, the perpendiculars from P', to the sides BC, CA, AB, meet these in points D', E', F' which lie on the Absolute line. For instance, if BC meet the Absolute line in L, the line $P'D'$ is the line joining P' to the point of the Absolute line which is $(I, J)/L$; and, when P' is on the Absolute line, the joining line coincides with the Absolute line. It follows that the circle which is the pedal circle of P and P', in regard to the triangle, contains the Absolute line as part of itself. The other part is then a line, containing the points D, E, F where the perpendiculars drawn from P to BC, CA, AB meet these sides respectively; this is the pedal, or Simson line, of P in regard to the triangle. The converse is also true, that if the pedal circle of a point P, in regard to a triangle, consists of a line together with the Absolute line, then P lies on the circumcircle.

Ex. 26.   A better approach to the theory of the pedal circle, than that which we have given, is as follows: Let D be any fixed point, and I, J the Absolute points. Let two points, M and N, vary respectively on the lines DI, DJ, describing related ranges on these lines. Consider the point K, common to the lines JM, IN which join J, I, respectively, to corresponding points of the related ranges (M), (N). This point K describes a conic passing through I and J, that is, a circle. Now, let DK meet MN in T. Then T also describes a circle. For JD, JI are fixed lines, and JT, JI are harmonic conjugates in regard to JD and JK (from the quadrangle $DMKN$). Thus the pencil $J(T)$, as T varies, is related to the pencil $J(K)$. This pencil is related to the pencil $I(K)$; and this, in a similar way, is related to the pencil $I(T)$. Thus the pencils $J(T)$ and $I(T)$ are related; which shews that T describes a circle. As MN and DK meet IJ in points which are harmonic conjugates in regard to I and J (from the quadrangle $MIJN$), we can describe T, on MN, as lying on the line drawn from D perpendicular to MN; and T is the midpoint of MN.

The line MN, joining corresponding points of the related ranges (M), (N), touches a conic, of which DI, DJ are tangents. Conversely, if any conic be taken which touches DI, DJ, the variable tangent of this conic meets DI, DJ in points of two related ranges. Of such a conic the point D is a focus, being a common point of a tangent from I with a tangent from J. Thus we may say: *The point where a variable tangent of a conic is met by the perpendicular to this tangent, which is drawn from a focus of the conic, describes a circle.*

Now let D, A, B, C be fixed points. A variable conic through D, A, B, C will meet the lines DI, DJ, respectively, in points M, N, besides D. As the conic varies, these points M, N describe related ranges on DI, DJ. This is clear, in general terms, from the remark that, if M be assigned, the conic through the five points D, A, B, C, M is defined, and so N is determined; and conversely N determines M without ambiguity. If then we assume further, from the nature of the construction, that the parameter which determines M on DI is an algebraic function of the parameter which determines N on DJ, it follows that M, N describe related ranges. The principle is important, and of frequent application. But, in case it is not quite clear, we may argue in detail as follows: Let the lines BC, CA, AB meet DI, respectively, in U, V, W, and AU meet the conic D, A, B, C, M, N again in A'. Then the range (M, U, V, W) is related to the pencil $A (M, A', C, B)$, or to the range (M, A', C, B), taken on the conic. But M, D; A', A; C, B are pairs of an involution on the conic, because the joins meet in U; thus (M, A', C, B), on the conic, is related to (D, A, B, C). Similarly, if BC, CA, AB meet DJ, respectively, in U', V', W', the range (N, U', V', W') on DJ is related to the range (D, A, B, C) on the conic. The ranges (M, U, V, W) and (N, U', V', W') being thus related, and U, V, W, U', V', W' being determined only by D, A, B, C, we can infer that the ranges (M), (N), on DI and DJ, are related; as was stated. Corresponding points of these ranges are determined by a varying conic through D, A, B, C; of such conics three degenerate cases are the line pairs (DA, BC), (DB, CA) and (DC, AB); these give the pairs U, U'; V, V'; W, W', respectively, of corresponding positions of M and N on DI and DJ.

Now, the ranges (M), (N) being related, it follows from what is said above that, if DT be the perpendicular from D to MN, the point T describes a circle. As BC, CA, AB are, as we have remarked, particular positions of the line MN, it follows that *the circle, which is the locus of* T, *is the pedal circle of* D *in regard to the triangle ABC*. The circle is also the locus of the foot of the perpendicular drawn to a tangent of that conic which is the envelope of the lines MN; of this conic, D is a focus. This conic is that touching DI, DJ, BC, CA, AB.

As we have seen, the pedal circle may consist of the Absolute line, with another line; namely, when D is on the circumcircle of ABC.

Ex. 27.   From the definition of the pedal circle reached in Ex. 26, we can prove that, *if* D, A, B, C *be arbitrary points, in general position, the pedal circles of any one of these four points, in regard to the other three, meet in a point.*

For, on the Absolute line, we may consider the involution formed by pairs of points which are harmonic in regard to I and J. This will have one pair in common with the involution which is determined on IJ by all the conics through D, A, B, C. In other words, there is one of these conics in regard to which I, J are conjugate points.

Let this particular conic meet DI, DJ respectively in M and N, and meet the line IJ in X and Y; the pencil, whose centre is the point D of this conic, formed by the four lines DM, DN, DX, DY, has then DX, DY as harmonic conjugates in regard to DM, DN; or the points M, N, *on the conic*, are harmonic conjugates in regard to X and Y. Thus the line MN contains the pole of the chord XY, that is, of the line IJ (Chap. IX (11)); so that MN contains the centre of this conic. If this centre be O, the lines OMN and OD meet IJ in points which are harmonic conjugates in regard to I and J, so that DO is perpendicular to MN; for, if the line DIM meets OJ in I', the polar of I is OJ, and D, M are harmonic conjugates in regard to I and I'. Therefore the pedal circle of D in regard to ABC, which contains the foot, O, of the line drawn from D perpendicular to MN, contains the centre O of this particular conic. In the same way, the pedal circles of A, B, C, each in regard to the other three points of D, A, B, C, all pass through O.

Ex. 28.   If D, D' be two points of the circumcircle, of a triangle ABC, which are on a diameter of this circle, prove that the pedal lines of D, D', in regard to ABC, meet on the pedal circle of the orthocentre H of the triangle ABC, and are perpendicular to one another. Prove further that the pedal circle, of any point of the diameter DD', passes through the common point of these pedal lines.

Ex. 29.   We have spoken of the optical image of a point in regard to a line. We may also define the optical image of a point in regard to a circle, this being the point often called also the *inverse* of the given point in regard to the circle. Two points, P, P', are said to be inverse in regard to a circle when they are conjugate to one another in regard to the circle, and their joining line passes through the centre of the circle. The lines PI, PJ, joining P to the Absolute points I, J, which are on the circle, then meet the circle again in points which lie on the lines $P'J, P'I$. The process of passing from either of the points P, P' to the other is often called *inversion in regard to the circle*. It is frequently useful. The reader may prove the following results:

(*a*) If a point P moves on one circle, its inverse point P', taken in regard to another circle, moves on a third circle, generally called the inverse of the circle on which P moves, in regard to the circle by which the inversion is effected; this being called the *circle of inversion*. In particular, if P describes a line, the inverse point P' describes a circle; this passes through the centre of the circle of inversion. Conversely, the inverse of a circle which passes through the centre of the circle of inversion is a line. In general, if two circles be inverses of one another in regard to another circle, this circle of inversion contains the common points of the two inverse circles.

(b) Let P, P' be any points inverse to one another in regard to a given circle, and Σ denote any circle which passes through P and P'. Then the inverse of any point P of the circle Σ is another point of the same circle; or the circle Σ is its own inverse in regard to the circle of inversion. Further, the circle Σ is perpendicular to the circle of inversion, at both their common points. Conversely, a circle perpendicular to a given circle, cuts any radius of this in points inverse in regard to this.

(c) Let S_1, S_2 be any two circles, and t_1, t_2 the respective tangent lines of these at a common point. Take the inverses of S_1, S_2, in regard to any circle Ω. Let t_1', t_2' be the respective tangents of these inverse circles at their common point which is inverse to the considered common point of S_1 and S_2. With the notation we have used, prove that $\epsilon\,(t_1,\,t_2) = \epsilon\,(t_2',\,t_1')$.

(d) Let U, V be points which are inverses of one another in regard to a circle S. In regard to another circle, Ω, let S' be the inverse circle of S, and U', V' the respectively inverse points of U and V. Prove that U', V' are inverse points in regard to the circle S'.

(e) As an application of these results, we may instance the following: Given any four lines in a plane, we have shewn that the circumcircles of the four triangles, formed by the threes of these lines, meet in a point, say O; and that the centres of these four circles lie in a circle which passes through O (Ex. 12, above). We have also shewn that the optical images of O, taken in regard to the four lines in turn, lie on a line, say p. Now take a circle Ω, with centre at O, as circle of inversion. The four circumcircles spoken of, as passing through O, have, for inverses, four lines of the plane; the centres of the four circumcircles have, as inverse points, points lying on the line p. The four original lines have, as inverses, four circles, which are the circumcircles of the four triangles formed by the lines obtained in the inverse figure.

In regard to the proof of the propositions (a) ... (e), we may give brief indications: The inverse of a point P, in regard to a circle Ω whose centre is O, is the foot of the perpendicular drawn from O to the polar of P taken in regard to Ω. It is clear, from what was said in the preceding chapter in regard to the dual of a conic, that, if P describes a conic, the polar of P, taken in regard to Ω, envelopes another conic. If the first conic be a circle, S, thus passing through the Absolute points I, J, the second conic, therefore, touches the polars of I, J, in regard to Ω, which are the lines OI, OJ. Thus the second conic has O for a focus. We have proved (Ex. 26, above) that the feet of the perpendiculars, drawn from a focus, to the tangents, of a conic, describe a circle. This proves that the inverse of a circle S, in regard to the circle Ω, is another circle, S'. As every point of Ω is inverse to itself, the circle S' passes through the common points of S and Ω. A more direct proof that the inverse of a circle, in regard to a circle, is another circle, is given below (Ex. 22 of (6)).

If S contains two points, P, P', which are inverse in regard to Ω, then S', besides containing the points common to S and Ω, will contain P' and P. Thus S' coincides with S. Thus, also, any line through the centre of Ω meets S in a pair of points which are harmonic conjugates in regard to the pair of points in which this line meets Ω. By considering a line through the centres of both circles, S and Ω, we see that Ω contains two points which are inverse in regard to S. Hence, any line through the centre of S meets Ω in two points which are harmonic conjugates in regard to the two points in which this line meets S. The relation between the circles S, Ω is thus mutual; their tangents at either common point are perpendicular to one another. (Cf. Ex. 4, above.)

Ex. 30. Let A, B, C, D be four points which lie on a circle. Let A', B', C', D' be other four points, satisfying the four conditions that each of the sets of four points

$$A,\, B,\, B',\, A';\quad B,\, C,\, C',\, B';\quad C,\, D,\, D',\, C';\quad D,\, A,\, A',\, D'$$

lie on a circle. We can then prove that the four points A', B', C', D' lie on a circle.

In fact, with the notation employed, we have

$$\epsilon A' (D', A) = \epsilon D (D', A); \quad \epsilon A' (A, B') = \epsilon B (A, B'),$$

and hence $\qquad \epsilon A' (D', B') = \epsilon D (D', A) . \epsilon B (A, B').$

We also have

$$\epsilon C' (C, D') = \epsilon D (C, D'); \quad \epsilon C' (B', C) = \epsilon B (B', C),$$

and hence $\qquad \epsilon C' (B', D') = \epsilon B (B', C) . \epsilon D (C, D').$

Wherefore, the product $\epsilon A' (D', B') . \epsilon C' (B', D')$

is equal to the product

$$\epsilon B (A, B') . \epsilon B (B', C) . \epsilon D (C, D') . \epsilon D (D', A),$$

which is the same as $\qquad \epsilon B (A, C) . \epsilon D (C, A).$

Because A, B, C, D are on a circle, this last product is equal to unity. So, therefore, is the product

$$\epsilon A' (D', B') . \epsilon C' (B', D');$$

whence $\qquad \epsilon A' (D', B') = \epsilon C' (D', B').$

This shews that A', B', C', D' lie on a circle.

The reader may prove that, if we take the inverse of the figure with regard to any circle whose centre is the point D, the theorem becomes that given earlier (Ex. 7), to the effect that if X, Y, Z be arbitrary points on the sides BC, CA, AB of a triangle, the three circles AYZ, BZX, CXY meet in a point, which is associated with the name of Miquel.

Ex. 31. Let L, L'; M, M'; N, N' be the pairs of complementary intersections of four lines in a plane, forming a quadrilateral of which the diagonal triangle is PQR, the lines $LL'QR$; $MM'RP$; $NN'PQ$ being the sides of this triangle. Prove that the three circles having, respectively, as ends of diameters, the points L, L'; the points M, M', and the points N, N'— say, the circles described respectively on LL', MM', NN' as diameters —have two points in common (besides I, J), say the two points O, O'. Prove further that the circle P, Q, R has its centre on the line OO', and is perpendicular to all the three circles.

The former part of this is a particular case of a theorem already proved: if O be one common point of the circles on LL', MM' as diameters, the pencil in involution containing the three pairs of lines OL, OL'; OM, OM'; ON, ON' has two pairs of perpendicular lines, namely OL, OL' and OM, OM'; thus ON, ON' are also perpendicular lines, by a previous remark (Ex. 3 of (3)); thus O lies equally on the circle having NN' as diameter.

For the latter part of this example (31), we may employ what has been said above in regard to inverse circles. Since Q, R are harmonic conjugates in regard to L and L', and are hence inverse points in regard to the circle having LL' as diameter, it follows that the circle PQR cuts this circle perpendicularly; by a similar remark the circle PQR is perpendicular to both the circles having respectively MM' and NN' as diameters. Consider, now, any line through the centre of the circle PQR; and the involution of pairs of points determined on this line by the conics through the four points O, O', I, J; that is, the circles through O and O'. This involution is determined by the two pairs in which the circles having LL' and MM' as diameters meet the line. The foci of the involution are therefore the two points in which the line meets the circle PQR. One conic through O, O', I, J, however, consists of the line OO' and the Absolute line. Wherefore, the line we have drawn, through the centre of the circle PQR, meets the line OO' in the mid point of the two points in which this line meets the circle PQR, namely in the centre of this circle; that is, this centre lies on OO'. This is what we set out to prove. A different proof, independent of the theory of inversion, arises naturally below (see (6), Exx. 14, 15).

Ex. 32.   Let D be any point in the plane of a triangle PQR. Let the bisectors of the angles determined by the two lines DQ, DR meet QR in L and L'; likewise, the bisectors of the angles determined by the lines DR, DP meet RP in M and M'; and the bisectors of DP, DQ meet PQ in N and N'. Then the six points L, L'; M, M'; N, N' lie in threes on four lines, two of these lines passing through each of the six points; so that L, L'; M, M'; N, N' are the complementary pairs of intersections of these four lines, which form a quadrilateral of which PQR is the diagonal triangle. This has already been remarked, and proved, in Ex. 17 above. In terms of the symbols, if DP, DQ, DR meet the Absolute line in points whose symbols are, respectively, $I+pJ$, $I+qJ$, $I+rJ$, then the lines DL, DL' meet the Absolute line in points of symbols $I \pm (qr)^{\frac{1}{2}} J$. And the condition that the three pairs of points

$$(I+pJ,\ I+\epsilon_1\,(qr)^{\frac{1}{2}}\,J),\ \ (I+qJ,\ I+\epsilon_2\,(rp)^{\frac{1}{2}}\,J),\ \ (I+rJ,\ I+\epsilon_3\,(pq)^{\frac{1}{2}}\,J),$$

wherein $\epsilon_1^2=\epsilon_2^2=\epsilon_3^2=1$, should be in involution, may be found to be

$$\epsilon_1\epsilon_2\epsilon_3=-1.$$

It follows, therefore, from Ex. 31, that the three circles L, L', D; M, M', D; N, N', D, which have, respectively, LL', MM' and NN' as diameters, meet in another point X, such that XD passes through the centre of the circumcircle PQR of the diagonal triangle of the quadrilateral spoken of, and that the circle PQR is perpendicular to all the three circles, the points X, D being inverse points in regard to the circle PQR.

Now, let A, B, C be the optical images of D respectively in regard to QR, RP, PQ. Also, let the bisectors of the angles determined by the lines QX, QA meet XA in T and T', so that QT, QT' are perpendicular. The circle having TT' as diameter then passes through Q, and is the locus of a point U which is such that the bisectors of the angles determined by UX, UA pass through T and T'. We may prove that this circle passes through R, and that this circle also passes through the point E defined as the isogonal point of D in regard to the triangle PQR. We shall thus reach the result: *Let D, X be inverse points in regard to a circle PQR; and E the isogonal point of D in regard to the triangle PQR. Then, the inverse point of X, in regard to the circle QRE, is the point A which is the optical image of D in regard to the line QR.* We leave the proof to the reader.

(6) Any two given circles, besides the Absolute points I, J, have two other common points, say H and K. The line HK, joining these points, is called the *radical axis* of the two circles. The two pairs of lines HI, KJ and HJ, KI, which join the common points H, K of the circles respectively to I, J and to J, I, have each a point of intersection. These two intersections are called the *limiting points* of the circles. If we denote these points by S, T, and denote by L the point where the radical axis HK meets IJ, the vertex triangle of the quadrangle formed by the common points, I, H, J, K, of the two circles, is the triangle LST. By a result proved above (Chap. IX (15)), this triangle is self-polar in regard to all conics through the four points I, H, J, K, that is, in regard to all circles passing through H and K. This series of circles, with two common points, is commonly said to form a *coaxal* series. As the poles of the line IJ, in regard to all these circles, lie on the line ST, this line, which is the join of the limiting points, contains the centres of all the circles of the coaxal series; as the lines HK, ST

meet IJ in two points which are harmonic in regard to I and J, the line of centres is perpendicular to the radical axis, and contains the mid point of H, K; as the lines IJ, HK meet ST in two points harmonic in regard to S and T, the radical axis contains the mid point of the two limiting points. It follows that any circle drawn through the limiting points of the coaxal series of circles has its centre on the radical axis of the series; as the limiting points are conjugate points in regard to any circle of the coaxal series, and are on a line through the centre of this circle, they are what we have called *inverse* points in regard to this circle. Whence (Ex. 29 of (5)) *any circle through the limiting points is perpendicular to every circle of the coaxal series.* We are thus led to two coaxal series of circles; one consisting of all circles through the points H, K; the other consisting of all circles through the limiting points S, T of the former series; these circles have the property that every circle of one series is perpendicular to any circle of the other series. The points H, K, common to all the circles of the former series, are evidently the limiting points of the latter series, being the intersections (SI, TJ) and (SJ, TI). Further, by what was seen above (in (5)), any line drawn through the centre of a circle of one series meets the circles of the other series in pairs of points of an involution whose foci are on the first taken circle. We may also prove that if two circles be given, any circle which is perpendicular to both of these passes through·the limiting points of the two circles; for this we may use the criterion for two circles to be perpendicular which we have arrived at, namely that a line through the centre of one of them cuts the two circles in two pairs of points which are harmonic conjugates of one another.

Ex. 1.   Let the line joining the limiting points S, T, of two circles, which is a common diameter of the circles, meet one circle in A, A', and the other circle in B, B'. Then the points S, T, which are conjugate to one another in regard to both circles, are harmonic conjugates both in regard to A, A', and in regard to B, B'. Thus, by what has been proved (Chap. vii (3)), if A, A', B, B' be real points, the limiting points will be real if the points B, B' be *not* separated by the points A, A'. In this case, if, as before, we suppose the Absolute points I, J to be conjugate imaginaries, the common points H, K, of the two circles, which are obtainable by joining I, J to the limiting points, will be conjugate imaginaries; namely, the circles will not meet in real points. Conversely, when the circles meet in real points, the limiting points will be conjugate imaginaries. With the same assumption, as here, for the Absolute points, the line which is the radical axis of the two circles, is real in both cases.

Ex. 2.   A degenerate circle is formed by two lines which join a point O to the Absolute points. Prove that the most general circle which cuts this degenerate circle perpendicularly is an arbitrary circle passing through O. Prove that, as in general, any line through the centre of either of these two circles meets them in harmonically conjugate pairs of points. This is a case of the theorem for a conic with which we have associated the name of Seydewitz (Chap. ix (23), Ex. 12).

Ex. 3.   We have proved that any circle passing through two points, which are *inverse* in regard to a given circle, is perpendicular to this circle.

Prove also that the circle which has, as ends of a diameter, two points which are *conjugate* in regard to a given circle, is perpendicular to this circle.

Ex. 4. We have shewn that an arbitrary line is met by the conics, which pass through four given points, in pairs of points of an involution. Deduce that, if a line touch both of two circles, the lines joining the points of contact to a limiting point of the two circles are perpendicular to one another.

Ex. 5. We have shewn that the diagonal triangle of a quadrilateral is self-polar in regard to any conic touching the four lines (Chap. ix). Moreover, for two conics whose four common points are distinct, there is only one triangle which is self-polar in regard to both. Thus, for the quadrilateral constituted by the four common tangents of two circles, the diagonal triangle is that whose angular points are the two limiting points of the two circles, and the point where the radical axis meets the Absolute line. Wherefore, of the joins of complementary pairs of intersections of the common tangents of two circles, two are constituted by the lines through the limiting points which are perpendicular to the line of centres, and the third by the line of centres. There are then, on this line of centres, two points through each of which there pass two common tangents of the circles. These are called the *centres of similitude* of the two circles. From the figure of the quadrilateral, these two points are harmonic conjugates in regard to the limiting points. Of the involution, on the line of centres, determined by all circles coaxal with the two given circles, the limiting points are the foci, because these are conjugate in regard to any one of the circles; thus there is a circle, coaxal with the two given circles, which contains the centres of similitude of these. This is called the *circle of similitude* of the two circles.

Ex. 6. From the theorem (Chap. ix (23)) that the pairs of tangents, from any point, to the conics which touch four lines, are a pencil in involution, it follows that the tangents to two of these conics, at one of the common points of these conics, are harmonic conjugates in regard to the two lines which join this common point to a complementary pair of intersections of the four common tangents. Whence, for two circles meeting in H and K, the tangents of the circles, at H, are harmonic conjugates in regard to the two lines which join H to the centres of similitude of the two circles. These joining lines are perpendicular, because the circle of similitude, defined in Ex. 5, contains H and K, and has the centres of similitude for ends of a diameter. But, if two lines through H are harmonic in regard to two other lines HC_1, HC_2, through H, which are perpendicular to one another, then the two lines through H, which are respectively perpendicular to the two former lines, are also harmonic in regard to HC_1 and HC_2. As the line joining the centre of a circle to a point H of the circle is perpendicular to the tangent at H, we thus prove that *the centres of the two circles are harmonic conjugates in regard to the centres of similitude.* Likewise, by this argument, the two points which are the poles of the radical axis HK, in regard to the two circles, are harmonic conjugates in regard to the centres of similitude.

We have thus, recalling Ex. 5, found two involutions, on the line of centres of the two circles, for one of which the foci are the limiting points; for the other the centres of similitude are the foci. And *the centres of similitude may be defined as the two points which are the harmonic conjugates both in regard to the centres of the two circles, and in regard to the limiting points.*

We see, therefore, the Absolute points being conjugate imaginaries, as already explained, that, *if the centres of the two circles be real points (or, indeed, if they be conjugate imaginaries), the centres of similitude are real points, whether the two circles cut in two real points, or in two conjugate imaginary points.* As in what precedes, both circles are supposed to contain real points.

Ex. 7. Let O_1, O_2 be the centres of two circles, S and T the limiting points, and C_1, C_2 the centres of similitude. Let P be any point on the circle of similitude (Ex. 5). Because PC_1, PC_2 are perpendicular, and S, $T \sim C_1$, C_2, and also O_1, $O_2 \sim C_1$, C_2, it follows (Chap. VIII (10)) that

$$\epsilon P \,(S,\, C_2) = \epsilon P \,(C_2,\, T), \quad \epsilon P \,(O_1,\, C_2) = \epsilon P \,(C_2,\, O_2), \quad \epsilon P \,(O_1,\, S) = \epsilon P \,(T,\, O_2).$$

Now, let P_1, P_2 be the inverse points of P in regard to the two circles respectively. Then the circle PP_1P_2 is perpendicular to both the given circles, and, hence, contains the points S, T. Whence

$$\epsilon P_2 \,(P_1,\, S) = \epsilon P \,(P_1,\, S) = \epsilon P \,(T,\, P_2) = \epsilon S \,(T,\, P_2);$$

thus the line P_1P_2 is parallel to ST, and perpendicular to the radical axis. Also, the radical axis, containing the mid point of S, T, contains the mid point of P_1, P_2. Thus: *The circle of similitude, of two given circles, is the locus of a point whose inverse points, in regard to the two circles, are optical images of one another in regard to the radical axis of the two given circles.*

Ex. 8. Recurring to the result stated in Ex. 32 of (5), and taking B, C as the optical images of D in regard to RP, PQ, respectively, prove that the three circles of similitude of the pairs of circles (BCD, ABC); (CAD, ABC); (ABD, ABC) all pass through X. Further, that X is also common to the three circles of similitude of the pairs of circles (QRE, PQR); (RPE, PQR) and (PQE, PQR).

Ex. 9. A circle (K), of a coaxal series, being taken, and a circle, (R), which passes through the limiting points of the series, let the common chord of these two circles meet the radical axis of the coaxal series in the point U, and meet the line joining the limiting points of the series in the point V. Prove that U is the pole, in regard to the circle (R), of the line joining the limiting points of the series of circles, and V is the pole, in regard to the circle (K), of the radical axis of the coaxal series.

Ex. 10. We have stated, * as a necessary and sufficient condition for two circles to be perpendicular, that there should be a line, through the centre of one of the circles, whose intersections with either of the circles are harmonic conjugates in regard to the points where the line meets the other circle. Deduce that a circle, having its centre on the radical axis of a series of coaxal circles, which is also perpendicular to one of the circles, passes through the limiting points, and cuts all the circles perpendicularly.

Ex. 11. *From Ex.* 10, *it follows that, if three arbitrary circles be taken, the radical axes, of the three pairs of these circles, meet in a point.* For, denote the circles by S_1, S_2, S_3; let the radical axis, r_{12}, of S_1 and S_2, meet the radical axis, r_{13}, of S_1 and S_3, in the point O. It has been seen that, if the two tangents be drawn from O to the circle S_1, there exists a circle, Σ, with centre at O, passing through the points of contact of these two tangents with S_1, which is perpendicular to S_1. Whence, by Ex. 10, the circle Σ is perpendicular to S_2, and to S_3. We have seen, (6), above, that a circle which is perpendicular to both of two circles has its centre on the radical axis of these. Thus O is on the radical axis, r_{23}, of the circles S_2, S_3.

Ex. 12. Hence we can prove, *if three arbitrary circles be taken, that the six points, formed by the pairs of centres of similitude of the three pairs of these circles, lie in threes upon four lines, of which two pass through each of the points; and, that the diagonal triangle, of the quadrilateral formed by these four lines, has the centres of the three circles for its angular points.* For, the common perpendicular circle, Σ, of the three given circles, found in Ex. 11, with its centre at O, contains the two limiting points of each pair of the three circles. Let the centres of the three circles be denoted by O_1, O_2, O_3. Upon the side O_2O_3, there lie the two limiting points, say S_1 and T_1, of the circles (O_2), (O_3), which lie upon the circle Σ; and also the two centres of similitude of these two circles. These centres of similitude are (Ex. 6) the foci of the involution determined by the two pairs $(O_2,\, O_3)$, $(S_1,\, T_1)$. We can, however, prove that, if a conic, Σ, meets the sides O_2O_3, O_3O_1, O_1O_2 of a triangle, in the respective pairs of points S_1, T_1; S_2, T_2 and S_3, T_3, and we take the foci of the three

---

* Chap. X (5), Ex. 4.

involutions $[(S_1, T_1); (O_2, O_3)]$, $[(S_2, T_2); (O_3, O_1)]$, $[(S_3, T_3); (O_1, O_2)]$, then these foci lie in threes on four lines, which form a quadrilateral of which $O_1 O_2 O_3$ is the diagonal triangle; see Ex. 13. This will prove that the centres of similitude lie on four lines, as stated.

Ex. 13. To prove the theorem assumed in Ex. 12, which has in fact been stated already (Chap. VII (4), Ex. 20), we may proceed as follows, adopting a more familiar notation: *Let the sides BC, CA, AB of a triangle be met by any conic respectively in D_1, D_2; in E_1, E_2 and in F_1, F_2. Let L, L', on BC, be the two points on BC such that L, L' $\sim$ B, C and L, L' $\sim D_1$, D_2; so, on CA, let M, M' be determined from the two pairs C, A and E_1, E_2; and on AB, let N, N' be determined from the two pairs A, B and F_1, F_2. Then, with a proper assignment of notation, as between L and L', M and M', N and N', the four sets, each of three points, L', M', N'; M, N, L'; N, L, M'; L, M, N', lie each in a line, the four lines forming a quadrilateral of which ABC is the diagonal triangle; further, the four sets, each of three lines, AL, BM, CN; AL, BM', CN'; BM, CN', AL'; CN, AL', BM', meet each in a point, the four points so arising forming a quadrangle of which ABC is the vertex triangle.*

For, the conics through the four points E_1, E_2, F_1, F_2 meet the line BC in pairs of points belonging to an involution, of which L, L' are the foci. There is then a conic through E_1, E_2, F_1, F_2 which touches BC at L, and another conic which touches BC at L'. Consider now the conics through E_1, E_2 which touch BC at L; these conics (as passing through four given points, of which two coincide at L) meet the line AB in pairs of points of an involution, of which F_1, F_2 are one pair, and A, B are another pair, since B lies on the tangent at L and A lies on $E_1 E_2$. There is therefore a conic of this series which touches AB at N, and another touching AB at N'. Next consider the conics which touch BC at L, and touch AB at N. These again meet the line CA in pairs of points of an involution, of which E_1, E_2 are one pair, and C, A are another pair (since C lies on the tangent at L, and A lies on the tangent at N). There is therefore a conic of this series which touches CA at M, and another which touches CA at M'. One degenerate conic of this series is formed by the line LN, taken twice over; and this is to be considered as touching CA at the point where this line meets CA. Let this point be named M'; and assume that the conic which touches BC, CA, AB, respectively, at L, M, N is not a degenerate conic. We then have the partial result that N, L, M' lie in a line, and also, as $M = (A, C)/M'$, that the lines AL, CN, BM meet in a point. If, in this argument, we take N' instead of N, and consider the conics which touch BC, AB, respectively, at L and N', we similarly reach the conclusion that one of these conics touches AC at M, and another at M'. As, however, from the first partial result obtained, the line BM meets NL in the point $(N, L)/M'$, we see, from this result, by projecting from M, that ML contains the point N', which is $(A, B)/N$. Thus, one of the two conics of that series, which is defined as touching BC at L and touching AB at N', consists of the line $N'LM$ taken twice over, and there is a conic touching BC, AB and AC, respectively, at L, N' and M'. Proceeding further in the way indicated we reach the complete theorem which was enunciated.

A precisely similar argument will shew that *if X, X' be, on BC, any pair of the involution determined by the pairs B, C and D_1, D_2, while Y, Y' are, likewise, two points on CA such that Y, Y' $\sim$ M, M'; and Z, Z' on AB are such that Z, Z' $\sim$ N, N', then the six points X, X'; Y, Y'; Z, Z' lie on a conic. Further, in the same way, if D_1, D_2 on BC, with E_1, E_2 on CA, and F_1, F_2 on AB, be given points; and L, L' be the foci of the involution $(D_1, D_2; B, C)$, likewise M, M' the foci of the involution $(E_1, E_2; C, A)$, and N, N' the foci of the involution $(F_1, F_2; A, B)$, then, a sufficient condition for D_1, D_2, E_1, E_2, F_1, F_2 to lie on a conic is that the six lines AL, AL', BM, BM', CN, CN' should meet in threes in four points.* The condition for three pairs of points, which lie, respectively, on the sides of a triangle, to be on a conic, is thus deduced from the relations of these pairs with three other pairs, on these sides, which are known to lie on a conic; or is deduced from the behaviour, as to concurrence, of focal rays of involutions.

Ex. 14.   Consider a conic, in a plane; and let P be a point from which the tangents to this conic are perpendicular. Then the lines PI, PJ, joining P to the Absolute points, being harmonic conjugates in regard to the two tangents from P to the conic, are conjugate lines in regard to the conic. If, however, we consider the common point, of a variable line drawn through I, with the line drawn through J, which is conjugate to the former in regard to the conic, the locus of this common point is a conic passing through the points I, J, that is, a circle; for the pencil of lines through I is related to the pencil of the conjugate lines through J. The locus of the common point is that conjugate line locus, previously considered (Chap. IX (23), Ex. 18), with centres at I and J. It is called the *director circle* of the conic; it is the locus of points P, in the plane of the conic, from which the two tangents to the conic are perpendicular to one another.

It follows at once from this definition, if we consider *the system of conics which touch four given lines of the plane*, that *the director circles of this system form a coaxal series*. For, let O be a common point of two of these director circles; the tangents from O to all the conics of the system form a pencil in involution; and this involution has two pairs of rays each consisting of perpendicular lines, namely, the pairs of tangents from O to the two conics whose director circles pass through O. Thus all pairs of this involution are given by two perpendicular lines; and, hence, all the director circles pass through O. Hence these circles also pass all through another point O'.

Of the conics which touch four given lines, there are three which are degenerate. If L, L'; M, M'; N, N' be the pairs of complementary intersections of the four lines, one such degenerate conic consists of the two pencils of lines whose centres are at L and L', taken together; the other two degenerate conics are the pair of pencils with M, M' as centres, and the pair with N, N' as centres. The director circle of the first degenerate conic is the circle having L, L' as ends of a diameter. Thus the three circles described, respectively, on LL'; on MM' and on NN', as diameters, meet in the two points O, O'. These have already been considered above, (5), Ex. 15; and it has been shewn that the circumcircle of the diagonal triangle, PQR, of the quadrilateral formed by the originally given four lines, has its centre on the radical axis, OO', of the three circles, and is perpendicular to all of them. Thus this circle PQR contains the limiting points of the coaxal series of circles formed by the director circles of all the conics touching the four lines ((6), above). The centres of the director circles lie on a line, containing the mid point of O, O', and perpendicular to OO'; this line then contains the three mid points of L, L'; of M, M' and of N, N'; it is the conjugate line of the Absolute line, in regard to the given quadrilateral, in a sense previously explained (Chap. IX (23), Ex. 6). It follows that *the limiting points of the series of coaxal circles, formed by the director circles, are the common points of the circumcircle PQR, of the diagonal triangle of the quadrilateral, with the conjugate line of the Absolute line, taken in regard to the quadrilateral. The two points O, O' may then be constructed as intersections of lines joining the Absolute points to these limiting points.* Further, as the director circle of a conic is the conjugate line locus for two centres, I, J, lying on the Absolute line, it follows, from what was proved (Chap. IX (23), Ex. 18), that the poles of the Absolute line, in regard to the conic and the circle, are the same point; thus the centre of the director circle is at the centre of the conic. Whence, the centres of all the conics touching four given lines lie on the line which is the conjugate line of the Absolute line in regard to the quadrilateral formed by the four lines; this, however, arises from the definition of the conjugate line.

Ex. 15.   From the present point of view, we can at once prove that *the director circle of a conic is perpendicular to the circumcircle of any self-polar triangle of the conic.*

For, we have remarked (Chap. ix (23), Ex. 11) that, if we be given a triangle, and a point, we can construct three other points by taking the harmonic image of the given point in each side of the triangle, in turn, associated with the respectively opposite angular point; and that the aggregate of the four points so arising is a quadrangle of which the given triangle is the vertex triangle. If we have a conic, and a triangle which is self-polar in regard to this, the harmonic image of any point, P, of the conic, in regard to a side and the opposite angular point of the triangle, is another point, P', lying on the conic; and P' can be constructed as the second intersection with the conic of the line joining P to that angular point.

We use the dual of this result. Given any conic, and a self-polar triangle of this, take an arbitrary tangent of the conic. Construct the three other tangents of the conic, each obtained from this one by the process of harmonic inversion. Thus we obtain four tangents of the conic, so situate that the given triangle is the diagonal triangle of the quadrilateral formed by these. Whence, by the preceding example, the circumcircle of the triangle is perpendicular to the director circle of the conic.

Ex. 16. We have defined the line which, for brevity, we call the *polar of a point P in regard to a triangle, ABC.* Namely, if the lines AP, BP, CP meet BC, CA, AB, respectively, in D, E, F, and we take the points D', E', F', respectively, on BC, CA, AB, given by $D' = (B, C)/D$, $E' = (C, A)/E$, $F' = (A, B)/F$, then the polar line in question is the line containing the points D', E', F'. In fact D' is the intersection (BC, EF), and so on.

Now, take any triangle ABC. The *circumcircle* is the circle through A, B, C. We define two other circles associated with the triangle, called respectively the *nine-point circle*, and the *self-conjugate, or self-polar circle* (an illogical, but convenient name). And, we prove that *these three circles have two points in common, or belong to a coaxal series, the radical axis being the polar, in regard to the triangle, of the orthocentre, H, of the triangle.* The orthocentre has already been defined, as the common point of the perpendiculars drawn to the sides, each from the opposite angular point.

The *nine-point circle* is most briefly defined as the pedal circle of the orthocentre, in regard to the triangle. We have proved (Ex. 19, of (5)) that this is also the pedal circle of the circumcentre, O, the centre of the circle ABC. But, from a more general point of view, we have proved (Chap. ix, (23), Ex. 15) that the locus of the poles of a given line, in regard to all conics passing through four arbitrary given points A, B, C, H of a plane, is a conic. We have specified eleven points of this centre-locus, and also the intersections of this locus with the sides BC, CA, AB of the triangle ABC. The eleven points consist of the angular points of the vertex triangle of the quadrangle A, B, C, H; of the two points on the given line which are the foci of the involution determined on this line by the conics through A, B, C, H; and of six points, one on each of the joins of two of the four points A, B, C, H, the harmonic conjugate, on this join, in regard to the two points, of the point where the given line meets this join. The two intersections of the pole locus with the side BC are (*a*) the one of the six points just named which lies on BC, (*b*) the harmonic conjugate, in regard to B and C, of the point where the polar of H, in regard to the triangle ABC, meets BC. If the foci, spoken of, which lie on the given line, be taken as the Absolute points of the plane, then H is the orthocentre of the triangle ABC (three of the conics through A, B, C, H being the line pairs such as AH, BC), and the locus of *the centres* of conics through A, B, C, H becomes a circle. This passes through the three points where AH, BH, CH, respectively, meet BC, CA, AB, these being the angular points of the vertex triangle of the quadrangle A, B, C, H; thus the circle is the pedal circle of H in regard to the triangle ABC. The circle also passes through the three mid points, respectively, of A, H; of B, H; and of

C, H. And it passes through the three mid points, respectively, of B, C; of C, A; and of A, B, and is thus the pedal circle of the circumcentre O, in regard to ABC. The intersections of this *nine-point circle* with the side BC are, as in the general theorem, the harmonic conjugates, in regard to B and C, of the two points where BC is met by the fixed line (now the Absolute line), and where BC is met by the line which is the polar line of H in regard to the triangle ABC.

The so-called *self-polar circle* of the triangle is a circle in regard to which ABC is a self-polar triangle. The existence, and uniqueness, of this circle, follows from the fact (Chap. ix (23), Ex. 11) that there exists a conic, passing through two arbitrary points, which has a given triangle as self-polar triangle. But, in view of the theorem we wish to prove, we reach this circle in another way. Denote the mid points, respectively, of B, C; of C, A; and of A, B by D_1, E_1, F_1; and denote the points where AH, BH, CH, respectively, meet BC, CA, AB by D_2, E_2, F_2. Let the circle proved to contain these six points be denoted by (N). Two involutions of pairs of points on a line have common one pair of points, and the pairs of points on BC, which are harmonic conjugates in regard to B and C, form an involution. Thus there exists on BC a pair of points X, X', belonging to the involution which is determined by the two pairs (B, C) and (D_1, D_2), which are also such that $X, X' \sim B, C$. Let Y, Y' and Z, Z' be the similarly determined pairs, respectively, on CA and AB. By Ex. 13, above, the six points X, X'; Y, Y'; Z, Z' lie on a conic. Denote this conic by (S). Now, let (C) denote the conic which passes through the five points consisting of A and the four points common to the conics (N), (S). The conics through these four points meet the line AB in pairs of points of an involution; of this involution, two pairs are the points F_1, F_2, on (N), and the points Z, Z', on (S); and the points A, B belong to the involution determined by these two pairs. Thus, the conic (C), which contains A, also contains B. By a like argument it contains C.

We can, however, shew that a pair of complementary chords of the two conics (N), (S) consists of the two lines, which are the polars, in regard to the triangle ABC, respectively, of the point where the lines AD_1, BE_1, CF_1 meet, and of the point, H, where the lines AD_2, BE_2, CF_2 meet. For, the necessary and sufficient condition for this is that the two points, in which these lines respectively meet BC, CA, AB, should belong respectively to the involutions determined thereon by the pairs (D_1, D_2), (X, X'); (E_1, E_2), (Y, Y'); and (F_1, F_2), (Z, Z'). These two lines, respectively, meet BC in D_1' and D_2', where $D_1' = (B, C)/D_1$ and $D_2' = (B, C)/D_2$; and the involution determined by the pairs (D_1, D_2), (X, X') is that determined by the pairs (D_1, D_2), (B, C). Of this involution, then, the points (D_1', D_2') are a pair; as we see at once. Similarly on CA and AB.

Wherefore, the three conics, (N), (S), (C), meet the first line in the same two points; and these, lying on (N), we have taken to be the Absolute points of the plane; thus all the three conics are circles. And, the three conics meet the second line, which is the polar of H in regard to the triangle ABC, also in two common points. As the circle (S) meets AB, AC in points, respectively Z, Z' and Y, Y', which are harmonic conjugates of one another respectively in regard to A, B and in regard to A, C, the line BC is the polar of A in regard to the conic (S); so that ABC is a self-polar triangle in regard to the circle (S), while (C) is the circumcircle of the triangle ABC.

The theorem enunciated is thus proved.

We can prove, further, that the pole of either of the two lines considered, in regard to the circle (S), is the pole of the other line in regard to the triangle ABC. Thus the pole of the first line, which we have taken for the Absolute line, in regard to (S), namely the centre of the circle (S), is at the orthocentre, H, of the triangle, and is the pole of the second line in regard to the triangle. The two lines do in fact play symmetrical parts in the figure; and the pole of the second line, in regard to (S), is the pole of the first line in regard to the triangle, say the point G in which the lines AD_1, BE_1, CF_1 meet; when the first line is the Absolute line, G is the point called the *centroid* of the triangle. It can be shewn, and follows from Ex. 17 following, that G lies on the line containing the centres of the coaxal circles (C), (N), (S).

But, for the proof of the statement made in regard to the poles of the two lines in regard to the circle (S), a more direct proof, free from the metrical nomenclature for the triangle, seems simpler. The points X, X', in which (S) meets BC, were defined as being a pair of the involution defined by B, C and D_1, D_2, which are also harmonic in regard to B and C. The points D_1', D_2', in which the first and second lines meet BC, were $D_1' = (B, C)/D_1$ and $D_2' = (B, C)/D_2$. From these relations it can be proved that X, $X' \sim D_1$, D_2', and also X, $X' \sim D_1'$, D_2. Thus, in regard to the circle (S), the point D_1', which is conjugate to A, is also conjugate to D_2, and AD_2 is the polar of D_1'. Whence, considering similarly the sides CA and AB, the point of intersection of AD_2, BE_2, CF_2 is the pole of the line $D_1'E_1'F_1'$. Likewise the point common to AD_1, BE_1, CF_1 is the pole of the line $D_2'E_2'F_2'$. This is the result in question.

Ex. 17.   Referring to Ex. 16, the reader may prove that the centres of similitude of the circumcircle (C) and the nine-point circle (N) are the points H and G. The point H is the orthocentre of the triangle ABC and the pole of the Absolute line in regard to the circle (S); the point G is the centroid of the triangle ABC, and is the pole, in regard to (S), of the common chord of the circles (C), (N) which is complementary to the Absolute line, namely, is the pole of the radical axis of these circles (cf. Ex. 19, below). It is not the case that every three circles which are coaxal have these relations; nor, as is the case here, that the centre of one of the circles, namely (N), is the mid point of the centres of the other two circles (C), (S).

But, in the special case here, the circle having G, H as ends of a diameter, being the circle of similitude of (C) and (N), is coaxal with the three (C), (N), (S). And, it can be shewn, further, that this circle is the conjugate line locus of the circle (S), with G, H as centres.

It may be noticed, too, that all the four coaxal circles, and their two common points, are real under the following condition: that the lines AH, BH meet BC, CA, respectively, in the segments BC, CA containing, respectively, the points D_1', E_1', which lie on the Absolute line; or a similar condition for two other sides of the triangle ABC.

Ex. 18.   It has been seen in Ex. 16, that a unique conic exists, passing through two arbitrary points, to have a given triangle as self-polar triangle. It is suggested that the reader construct a direct proof of this. (See also Chap. IX (23), Ex. 11.)

Ex. 19.   Let any two conics (C), (N), meet in the four points A, B, A', B'. Denote the intersections $(AB', A'B)$ and (AA', BB'), respectively, by X and Y. Let the poles of the chord AB, in regard to the two conics, respectively, be P and Q; these lie in the line XY. Likewise, let P' and Q' be the poles of the chord $A'B'$, which also lie in the line XY.

Prove that the three pairs of points, P, Q; P', Q' and X, Y, belong to an involution; and that the foci, U, V, of this involution, are each a common point of a pair of common tangents of the conics (C), (N). Shew, also, that a conic (S_1) exists, through A, B, A', B', in regard to which U, V are the respective poles of AB and $A'B'$; and a conic (S_2) exists, also through A, B, A', B', in regard to which U, V are the respective poles of $A'B'$ and AB. Further, that there is a conic, containing the six points A, B, A', B', U, V, which is the conjugate line locus, both of (S_1) and of (S_2), with U, V as centres. Essentially, these results have been proved in the preceding examples.

Ex. 20.   The nine-point circle (N), of a triangle ABC, is derivable from the circumcircle (C), by the simple fact that *the mid point of the two points H, P, where H is the orthocentre of ABC and P is any point of the circumcircle, is a point, Q, of the nine-point circle.*

For, consider any conic of which I, J are two points, and let H be any point. Let P be a variable point of the conic, and HP meet the fixed line IJ in the variable point T; take, on HP, the point Q given by $Q = (H, P)/T$. Then, the locus of Q is another conic through I and J. For, in the pencil $I(T, P, Q, H)$, the rays IT, IH are fixed; thus, from the definition of Q, the ray IQ describes a pencil related to the pencil described by IP. Likewise, the ray JQ describes a pencil related to the pencil described by JP. But, as P moves on the given conic, the pencils $I(P), J(P)$ are related; so therefore are the pencils $I(Q), J(Q)$. This shews that Q describes a conic containing I and J. The same conclusion evidently follows if, as P moves on the original conic, Q is taken on HP so that the cross ratio $(Q, T/H, P)$ remains always the same. And, if P_1, P_2, P_3 be three points of the given conic, and Q_1, Q_2, Q_3 be three points such that the cross ratios $(Q_1, T_1/H, P_1)$, $(Q_2, T_2/H, P_2)$, $(Q_3, T_3/H, P_3)$, (where T_1, T_2, T_3 are the points in which HP_1, HP_2, HP_3, respectively, meet IJ) are all equal, then the variable point, Q, of the conic containing the five points Q_1, Q_2, Q_3, I, J, is obtained from P, on the given conic, by the rule given.

The theorem stated, for the deduction of the nine-point circle from the circumcircle, follows, because the nine-point circle, of the triangle ABC, contains the three mid points of H, A; of H, B; and of H, C. We have proved that the nine-point circle is the pedal circle, in regard to the triangle ABC, both of the orthocentre H, and of the circumcentre O; thus the centre of the nine-point circle is the mid point of H, O.

The theorem of this example can be treated with use of the symbols (see Ex. 24).

Ex. 21.   It is also true that *the circles* (C), (N) *are inverses of one another in regard to the circle* (S) (see above, (5), Ex. 29, and Ex. 22, following); namely, the inverse point, P', in regard to the self-conjugate circle (S), of a variable point P of the circumcircle (C), describes the nine-point circle (N).

We prove this by assuming that, if P, P'; Q, Q'; R, R' be three pairs of mutually inverse points, in regard to the circle (S), then the circle PQR is inverse to the circle $P'Q'R'$, in regard to (S), in the sense that the inverse of any point of the circle PQR is a point of the circle $P'Q'R'$. It will follow from this that the three circles have two points in common. We find three points of the circumcircle (C), whose inverse points, in regard to (S), lie on the nine-point circle (N). Incidentally, this furnishes another proof that the three circles are coaxal.

Let the lines HA, HB, HC, joining the orthocentre H to A, B, C, meet BC, CA, AB, respectively, in L, M, N, and meet the circumcircle, again, respectively, in P, Q, R. We have proved (Exx. 4, 5, of (5)) that L, M, N are the mid points, respectively, of H, P; of H, Q and of H, R. Denote the mid points, respectively, of H, A; of H, B; and of H, C by D, E, F. The nine-point circle (N) contains the points L, M, N, D, E, F. Let V denote the point where the line BH meets the Absolute line. The triangle ABC being self-polar in regard to the circle (S), whose centre is H, there is, upon the line BH, an involution of which V, H and M, B are pairs, the foci of this involution being where BH meets (S). From $Q = (V, M)/H$, and $E = (H, B)/V$, it follows, easily, that Q, E are a pair of this involution. These are then inverse points in regard to (S). Likewise, R, F and P, D are two pairs of inverse points in regard to (S). Of these P, Q, R are on the circle ABC, and D, E, F are on (N). The theorem is therefore proved.

Ex. 22.   The assumption made in Ex. 21 is essentially equivalent to the statement made in Ex. 29, of (5), above, that the inverse of a circle, in regard to a given circle, is another circle. The proof there given assumed that the locus of the foot of the perpendicular, drawn from a focus of a conic, to a variable tangent of the conic, is a circle; a result which had been proved. But we now give an alternative proof.

Let two circles meet a given circle of centre O, in the same two points,

L, M. Suppose that upon these circles, respectively, there are two points, *P* and *P'*, which are inverse in regard to the given circle. Let *Q* be any point on the former circle, and *Q'* be the inverse of *Q* in regard to the given circle. Then *Q'* lies on the second circle.

For, consider the circle *P, P', L.* Because *P, P'* are inverse points in regard to the circle *(O)*, the line *OPP'* cuts the circle *(O)* and the circle *PP'L* in two harmonically conjugate pairs of points; therefore ((4), above) the circle *PP'L* is perpendicular to the circle *(O)* and touches the line *OL* at *L.* This proves that

$$\epsilon L\,(O,\,P') = \epsilon P\,(L,\,P').$$

In the same way we prove, for *any* point *M* on *(O)*, that

$$\epsilon M\,(O,\,P') = \epsilon P\,(M,\,P').$$

Whence $\epsilon P\,(L,\,M)$, which is equal to

$$\epsilon P\,(L,\,P')\,.\,\epsilon P\,(P',\,M),$$

is equal to

$$\epsilon L\,(O,\,P')\,.\,\epsilon M\,(P',\,O).$$

We have, however, from the triangles *LOP'*, *MP'O*,

$$\epsilon L\,(O,\,P')\,.\,\epsilon O\,(P',\,L)\,.\,\epsilon P'\,(L,\,O) = 1 = \epsilon M\,(P',O)\,.\,\epsilon O\,(M,\,P')\,.\,\epsilon P'\,(O,\,M);$$

thus

$$\epsilon L\,(O,\,P')\,.\,\epsilon M\,(P',\,O)\,.\,\epsilon P'\,(L,\,M)\,.\,\epsilon O\,(M,\,L) = 1,$$

so that

$$\epsilon P\,(L,\,M)\,.\,\epsilon P'\,(L,\,M) = \epsilon O\,(L,\,M).$$

By a similar argument, because *Q, Q'* are inverse, we have

$$\epsilon Q\,(L,\,M)\,.\,\epsilon Q'\,(L,\,M) = \epsilon O\,(L,\,M).$$

Thus, if *L, M, P, Q* lie on a circle, so that $\epsilon P\,(L,\,M) = \epsilon Q\,(L,\,M)$, it follows that *L, M, P', Q'* lie on a circle. This proves the theorem stated.

Ex. 23. The three coaxal circles *(C)*, *(N)*, *(S)* have here been defined from a given triangle *ABC.* It is, however, the case that if two conics be related as are *(C)* and *(S)*, namely, *if a triangle, ABC, can be inscribed in one conic which is self-polar in regard to the other conic, then an infinite series, of such triangles, A'B'C', can be inscribed in the former conic (C); in more detail, if A' be any point of the conic (C), and the polar of A' in regard to the conic (S) meet the conic (C) in B' and C', then A'B'C' is a self-polar triangle in regard to (S).* This is a corollary from a result formerly proved (Chap. IX (23), Ex. 16) that, if two triangles be both self-polar in regard to a conic, their six angular points lie on a conic.

In the case of the circles, the orthocentre of such a triangle *A'B'C'*, being the centre of the self-polar circle *(S)*, is the same as the orthocentre *H* of the original triangle *ABC*; and the nine-point circle, being the locus of the point which is the mid point of *H* and any point of the circle *(C)*, will coincide with the original circle *(N)*. Thus this circle contains the point where the line *HA'* meets the perpendicular polar *B'C'* of the point *A'*. We thus have another proof of the fact that *(N)* is the inverse of the circle *(C)*, in regard to *(S)*. Or conversely.

Ex. 24. The relation of the two circles *(C)*, *(N)*, with the circle *(S)*, is a particular case of a relation of any two circles with a third circle, which we now consider.

Let *O* be a centre of similitude of any two circles *(p)*, *(q)*, the intersection of a pair of common tangents which meet on the line of centres of the two circles. Let an arbitrary line through *O* meet the circle *(p)* in P_1 and P_2, and meet the circle *(q)* in Q_1 and Q_2. Let this line meet the radical axis of the two circles in *H*, and meet the Absolute line in *K*.

Then, denoting the centres of *(p)*, *(q)* by *P* and *Q*, and by *N* the point where *PQ* meets the Absolute line, we have

$$(P_1,\,Q_2/O,\,H) = (P_2,\,Q_1/O,\,H) = -(P_1,\,Q_1/O,\,K)$$
$$= -(P_2,\,Q_2/O,\,K) = -(P,\,Q/O,\,N),$$

and the value of these cross ratios is independent of the chord drawn through
O. The equations, however, require a proper association of the points of the
pair Q_1, Q_2 with the pair P_1, P_2.

Further, the two given circles (p), (q), are inverses of one another in
regard to the circle of centre O which is coaxal with (p) and (q). Also, the
points P_2, Q_1 are inverses of one another in regard to this third circle, as are
P_1 and Q_2. The three pairs of points, P_1, Q_1; P_2, Q_2 and O, H, are in involution; as are also the three pairs P_1, Q_2; P_2, Q_1 and O, K. And if, by variation
of the transversal drawn from O, the points P_1, P_2, Q_1, Q_2 become, respectively, L_1, L_2, M_1, M_2, then the lines P_1L_1, Q_1M_1 are parallel, that is,
they meet on the Absolute line, as are also the two lines P_2L_2, Q_2M_2. Likewise the two lines P_2L_2, Q_1M_1 meet on the radical axis of (p) and (q), as do
P_1L_1 and Q_2M_2. Further, P_1L_2, Q_1M_2 are parallel, and P_1L_2, Q_2M_1 meet on
the radical axis, etc.

We may state the results for any two conics, of which two complementary
common chords replace what are here called the radical axis and Absolute
line. The point O is still the intersection of two common tangents of the
conics, lying on the common polar line of the point which is common to the
two complementary chords.

The proof may be given geometrically, or, in terms of the symbols. If
I, J be the symbols of the Absolute points, P and Q the symbols of the
centres of the circles (p), (q), the symbols of the corresponding variable
points P_1, Q_1, of the two circles, are of the forms

$$P_1 = P + r\,(\theta I + \theta^{-1}\,J), \quad Q_1 = Q + s\,(\theta I + \theta^{-1}\,J),$$

where r, s are two fixed numbers, and θ is a variable parametric number.
These formulae follow from what is said in Chap. IX (17).

Ex. 25.   State in detail the dual of the general theorem for two conics
which is indicated in Ex. 24, taking two complementary intersections of the
four common tangents of two conics instead of two complementary common
chords.

Ex. 26.   In the figure of Ex. 21 above, if W be any point on the side CA,
prove that the circle which has B, W as ends of a diameter is perpendicular
to the self-polar circle of the triangle ABC, whose centre is H.

Deduce that the line which contains the orthocentres of the four triangles
formed by four lines in a plane is the radical axis of the three circles
described on the (limited) diagonals of the quadrilateral formed by the
four lines, as diameters.

Ex. 27.   Let any line, through the circumcentre O of a triangle ABC,
meet BC, CA, AB in U, V, W, respectively. Let CO, BO meet the circle
again, respectively, in B' and C'. Let $B'W$ and $C'V$ meet in A'. Prove, by
applying Pascal's theorem to the hexagon $ABC'A'B'C$, that A' lies on the
circle ABC. The lines BA', $A'V$ are perpendicular, as are CA', $A'W$; thus
the two circles having BV, CW as diameters meet in A', on the circumcircle.
The lines BV, CW are (limited) diagonals of the quadrilateral formed by the
four lines BC, CA, AB, UVW; hence, the circle having AU as diameter
equally contains A'. The two common points of these three circles are (by
Ex. 26) inverse points in regard to the self-conjugate circle of ABC; and the
nine-point circle is inverse to the circumcircle in regard to the self-polar
circle. Wherefore, the three circles, having AU, BV, CW as diameters,
meet again on the nine-point circle. Hence, if a quadrilateral be formed by
four lines, of which one contains the circumcentre of the triangle formed by
the other three lines, then the two common points of the circles, described
on the (limited) diagonals of the quadrilateral, lie, one of them on the
circumcircle, the other on the nine-point circle, of this triangle (Thébault,
Nouv. Ann. x, 1910; quoted by Coolidge, *The circle and the sphere*, 1916,
p. 113).

Ex. 28.   Prove that a circle can be drawn, perpendicular to a given circle, to pass through two arbitrary points. Thus, if three circles be given, a circle can be drawn coaxal with two of these to be perpendicular to the third. Prove that the three circles, so derivable from three given circles, are themselves coaxal.

Ex. 29. Prove that the circle having the centres of two given circles as ends of a diameter contains four of the six intersections of common tangents of the two given circles.

Ex. 30. Let t, u be two common tangents of two circles which meet in a centre of similitude S; let S' be the other centre of similitude; let O be the point where the common tangent t meets the radical axis. Prove that there is a circle (O), with centre at O, which is perpendicular to both the given circles. Shew that the circle, which is the inverse of the tangent u, in regard to this circle (O), is the nine-point circle of the triangle whose sides are the line t and the two common tangents of the two original circles which meet in the centre S'. (Cf. Exx. 31, 32, following.)

Ex. 31. The result in Ex. 30 depends on this: Let H be one angular point of the diagonal triangle of any quadrilateral, and h the opposite side of this triangle. The four lines of the quadrilateral may be regarded as two pairs, of which those of a pair meet on the line h. Any line through H, meeting h in H', is met, by the two pairs of lines, in two pairs of points defining an involution. The foci of this are H and H'.

Ex. 32. Let a, b, c be any three of the common tangents of two circles, and u the fourth common tangent. Let O be the common point, of the tangent a, and the radical axis of the two circles. Let H be one of the limiting points of the two circles, the opposite side of the common self-polar triangle of the two circles being h. Let c be the common tangent of the two circles which meets the tangent a on h, so that the common tangents b, u also meet on h. Let OH meet h in H'. Then H, H' are harmonic conjugates in regard to the two points where the line OH meets the lines b and u. The line OH is parallel to c, and meets b in the mid point of the two angular points of the triangle (a, b, c) which lie on b. Further, H and H' lie on a circle whose centre is O. Hence we can say: *There is a circle, of centre O, which cuts both the given circles perpendicularly; in regard to this circle, the nine-point circle of the triangle (a, b, c) is the inverse of the fourth common tangent, u, of the two given circles. This shews that this nine-point circle touches both the given circles. Whence it follows that this nine-point circle touches all the four circles which touch the lines a, b, c.*

Ex. 33. Any common tangent of two circles passes through one of the centres of similitude. Let the centre through which any such tangent passes be momentarily spoken of as being *associated* with that tangent. The common tangents of the pairs of three circles form three sets, each of four tangents common to two of the circles. Suppose that the three circles are such that there are three common tangents, one from each set of four, which meet in a point, the three, respectively, associated centres of similitude not being in a line. Then prove that there are three other common tangents, one from each set, which also meet in (another) point, these being the tangents through the same centres as the other three concurrent tangents. A proof is given below in the Miscellaneous Examples (217).

Ex. 34. Let the tangents of two points A, A' of a circle meet in Y; let the line joining A to the mid point of Y, A' meet the circle again in R. Prove that the circle Y, A', R touches $A'A$ at A'.

Ex. 35. Points D, E, F, on the sides BC, CA, AB of a triangle, respectively, are such that $\epsilon\,(DC, DA) = \epsilon\,(EA, EB) = \epsilon\,(FB, FC)$. Prove that the circumcentre of the triangle, formed by the lines AD, BE, CF, is at the orthocentre of the triangle ABC.

Ex. 36. Prove that a circle, perpendicular to the self-conjugate circle of a triangle, which touches the circumcircle, also touches the nine point circle.

Ex. 37. If X, Y, Z, T be the angular points of a tetrahedron, and the four lines XY, YZ, ZT, TX be respectively met by a sphere in P, P'; Q, Q'; R, R'; S, S', such that P, Q, R, S lie in a plane, then P', Q', R', S' also lie in a plane. The planes TXY, XYZ, YZT, ZTX determine four circles on the sphere; and the planes $PQRS$, $P'Q'R'S'$ determine two other circles. If these circles be projected, from P, on to the tangent plane of the sphere at the point diametrically opposite to P, prove that the result obtained is Miquel's theorem ((5), Ex. 7, preceding).

CHAPTER XI

THE PARABOLA

(1) By a *parabola* is meant any general conic which touches the Absolute line. In this chapter, for the sake of brevity, in a plane in which the Absolute points I, J are given, we define a conic from an arbitrarily given point S, and an arbitrary line t, shewing, from the definition, that this conic touches the Absolute line. It will appear (Ex. 1 of (3)) that, conversely, any conic touching the Absolute line can be so defined. We may thus describe the curve we consider, from the start, as a parabola.

Let Z be a variable point on the given line t; join the given point S to Z, and draw from Z the line ZP which is perpendicular to SZ. It is supposed that S does not lie on t. As SZ, ZP are perpendicular, and meet the Absolute line in points which are harmonic conjugates in regard to the Absolute points I, J, the line ZP joins corresponding points of two related ranges, these lying, respectively, on the line t, and on the Absolute line. Thus ZP is a variable tangent of a conic, which touches the Absolute line, and touches the line t. By a previous theory (Chap. ix), the point of contact of this conic, with the line t, is that point of t which corresponds to the point, on the Absolute line, which lies on t; namely the point of contact, A, is the foot of the perpendicular drawn from S to t. Likewise, the point of contact of the conic with the Absolute line is the point thereon which corresponds to the point of t where t meets the Absolute line. This point of contact is, therefore, also on the line AS; we may denote it by A'.

If Z, Z' be any two points of the line t, the lines through Z and Z', respectively perpendicular to ZS and $Z'S$, meet in a point R, which is such that the four points R, Z, Z', S lie on a circle. Thus, in particular, the point of contact, P, of the tangent drawn from Z perpendicular to SZ, may be constructed by taking the circle through S which touches the line t at Z; the line SP is then the diameter of this circle which passes through S. Let the centre of this circle, the mid point of S, P, be O; also let the tangent ZP meet the fixed line SA in T. Then OZ is perpendicular to the tangent of the circle at Z, which is the line t. Thus OZ is parallel to SA. Whence, as O is the mid point of S, P, it follows that Z is the mid point of P, T.

The conic touches the two lines SI, SJ, as well as the side IJ of the triangle SIJ. For, the points where the lines ZS, ZP meet the Absolute line, being harmonic conjugates in regard to I and J, coincide when SZ contains either I or J. The polar of S, in regard to the conic, contains the pole of the line SA; this is the point where the line t meets the Absolute line. Thus this polar is perpendicular to SA. And this polar contains the point, say X, on the line AA', given by $(A, A')/S$.

In other words, the polar of S is perpendicular to SA, and meets this in a point X such that A is the mid point of X, S. This polar of S is called the *directrix* of the curve; the point S is called the *focus*, and the line $XASA'$ is called the *axis*. The circle we have spoken of, containing S, and touching the line t at a point Z, degenerates when SZ contains I, into the two lines which join Z to I and J.

Let the line SZ, joining S to any point Z of the tangent t, meet the directrix in N. Then, because A is the mid point of X, S, and XN is parallel to AZ, the point Z is the mid point of S, N. The tangent ZP, of the conic, is perpendicular to SN; whence, by what we have proved for a circle, the circle with centre P, which passes through S, equally passes through N; and also (Chap. x (5)), we have

$$\epsilon P\,(N,\,Z) = \epsilon P\,(Z,\,S).$$

Again, T, as before, being the point where the tangent PZ meets SA, noticing that Z is the mid point of P, T and SZ is perpendicular to PT, we have $\epsilon P\,(Z,\,S)$, or $\epsilon P\,(T,\,S)$, equal to $\epsilon T\,(S,\,P)$. Thus $\epsilon P(N,T)$, or $\epsilon P\,(N,\,Z)$, is equal to $\epsilon T\,(S,\,P)$, wherein PT and TP are the same line. This shews that PN is parallel to TS; thus PN is perpendicular to the directrix NX; and the circle, through S, with centre at P, touches the directrix at N.

Next, let the tangent PZ meet the directrix in H. Because HZ is perpendicular to SN, and contains the mid point of S, N, it follows, from what we have proved for a circle, that HS is the other tangent, besides HN, drawn from H to the circle, through S, whose centre is P. Thus SP, SH are perpendicular. Whence, if PS meet the conic, which is the locus of P, in the second point P', then HP' is the tangent of the conic at P'; so that the line PSP' is the polar of H in regard to the curve. We see, then, that two perpendicular lines drawn through S, such as SP, SH, are conjugate in regard to the conic. And, it can be shewn that, if the tangent of the curve at P' meet the tangent at A (the line t), in the point Z', then SZ, SZ' are perpendicular; so that the tangents HP, HP', drawn at the ends of the focal chord PSP', are perpendicular. This result arises also below.

(2) More generally, let the tangents of the curve, at any two points P, Q, meet in R; and let the chord of contact PQ meet the directrix in K. The point K, common to the polars of R and S, is the pole of SR; wherefore, the lines SK, SR are conjugate in regard to the curve, and are thus perpendicular, by what is proved above. The same result holds for any conic; namely, if S be a focus of this curve, meaning a point from which the tangents to the curve pass through I and J, and the polar of S be called the corresponding directrix, then the lines joining S to any point R, and to the point where the polar of R meets this directrix, are perpendicular to one another. For, it is a general result for any conic (Chap. ix (23), Ex. 3), that conjugate lines drawn through any point S are harmonic conjugates in regard to the tangents drawn from S to the conic; thus, such conjugate lines, drawn through

a focus S, are perpendicular. But, further, for the parabola, the line
SR, being the polar of K, meets PQ in the point $(P, Q)/K$. Thus SK,
SR are harmonic conjugates, not only in regard to SI, SJ, but also in
regard to SP, SQ; and are the bisectors of the angles determined by
the lines SP, SQ, according to our definition of the bisectors. We have
shewn that this involves $\epsilon S\,(P, R) = \epsilon S\,(R, Q)$. This is clearly also true
if RP, RQ be the tangents from a point R to any conic of which S is
a focus.

(3) In the case of a parabola, but not for any conic, we can prove
further that $\epsilon R\,(P, Q) = \epsilon S\,(R, Q)$. For, we know that four tangents
of a conic meet any two tangents of the conic
in related ranges. Consider the four tangents
of the parabola, RP, RQ, SI, SJ; and the
ranges which these determine, respectively,
on the Absolute line, and on the tangent RQ.
Taking Q for the common point of RQ with
itself (Chap. IX), we thus have

$$\epsilon R\,(P, Q) = \epsilon S\,(R, Q);$$

as stated.

From this it follows that $\epsilon R\,(S, P) = \epsilon Q\,(S, R)$.
For, from

$$\epsilon S\,(R, Q)\,.\,\epsilon R\,(Q, S)\,.\,\epsilon Q\,(S, R) = 1,$$

we have $\qquad\qquad \epsilon R\,(S, Q) = \epsilon S\,(R, Q)\,.\,\epsilon Q\,(S, R),$

which is $\qquad \epsilon R\,(S, P)\,.\,\epsilon R\,(P, Q) = \epsilon S\,(R, Q)\,.\,\epsilon Q\,(S, R),$

wherein $\qquad\qquad\qquad \epsilon R\,(P, Q) = \epsilon S\,(R, Q).$

A particular case of $\epsilon R\,(P, Q) = \epsilon S\,(R, Q)$ is when SR is perpendicular
to SQ. Then, as we have seen, R is on the directrix, and the equation
$\epsilon R\,(P, Q) = -1$ expresses the previous result, that the common point
of two perpendicular tangents of the parabola lies on the directrix. For
a general conic (Chap. X (6), Ex. 14) the common point of two perpendicular tangents describes a circle, the *director circle*, whose centre
is at the centre of the conic. For the parabola, this circle degenerates
into the directrix, taken with the Absolute line. In other words, the
directrix, together with the Absolute line, is the locus of a point H
from which the tangents to the parabola are harmonic conjugates in
regard to HI and HJ, or HI, HJ are conjugate lines. This is the dual of
Seydewitz's theorem (Chap. IX (23), Ex. 12), the directrix being the
polar of the common point of the tangents SI, SJ.

Another corollary is the theorem, subsequently of interest, that if,
on the tangent at a variable point P of the parabola, a point R be
taken so that $\epsilon R\,(S, P)$ is constant, then the locus of R is a line, t,
which touches the parabola, the point of contact being the point Q for
which $\epsilon\,(QS, t)$ has this constant value. A particular case of this is the
property by which we have defined the parabola, namely, when QS
and t are perpendicular. And there is an important consequence of

the result: If, from another point, R_1, of the tangent at Q, another tangent, R_1P_1, be drawn to the curve, and the tangents RP, R_1P_1 meet in V, then the points S, V, R, R_1 lie on a circle; for we have $\epsilon R\,(S,\,P) = \epsilon R_1\,(S,\,P_1)$, which is $\epsilon R\,(S,\,V) = \epsilon R_1\,(S,\,V)$. Thus, *any three tangents of a parabola form a triangle whose circumcircle contains the focus of the curve.*

Ex. 1.   We may prove that any conic which touches the Absolute line has the property by which we have defined the parabola, in the following way: Let the tangents to the conic from I, J, other than IJ, meet in S; let the line joining S to the point A', where the conic touches IJ, meet the curve again in A. Let the tangent at A meet the tangent, at any other point P of the conic, in the point Z, and SZ meet IJ in Y. We are to prove, the line AZ being the tangent at a fixed point A of the curve, that, if the tangent at P meet IJ in X, then X, $Y \sim I$, J, or that SZ, ZP are perpendicular. Let SP meet the curve again in P', and AP, $A'P'$ meet in K. As points of the range on the curve, P and P' are harmonic conjugates in regard to the two points

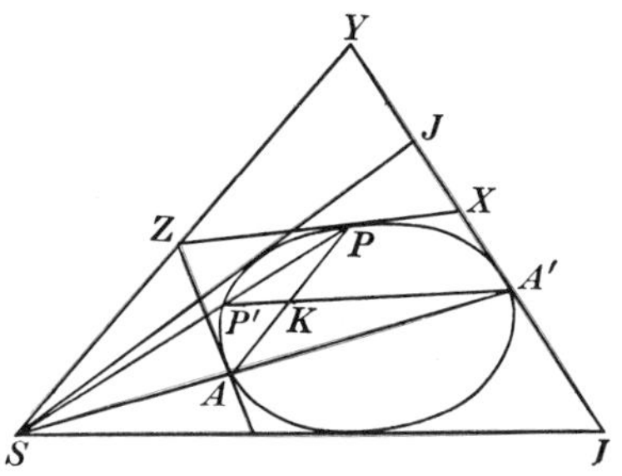

where SI, SJ touch the curve. Thus the tangents at P, P' meet the tangent IJ in points harmonic in regard to I and J. We require then only to shew that the tangent at P' passes through Y. But, from the inscribed quadrangle A, A', P, P', the points K, S are conjugate in regard to the conic; thus the polar of K contains S; and this polar also contains the point Z, which is the pole of AP. Thus the polar of K is SZ, and contains Y; wherefore, the polar of Y contains K, and is hence the line $A'K$, because the tangent at A' contains Y. Thus YP' touches the curve at P'; as we wished to prove.

Ex. 2.   Two of the results given above may be derived from a more general point of view; recalling that, if four tangents of a conic be given, and R be an arbitrary point, then the pairs of lines joining R to the complementary pairs of intersections of the four tangents belong to a pencil in involution, to which belongs also the pair of tangents RP, RQ drawn to the conic from R. Let the four tangents of the conic which are considered be those from the Absolute points I, J; and let S_1, S_2 be another pair of complementary intersections of the tangents, so that S_1, S_2 are a complementary pair of foci of the conic. Then, as RI, RJ; RS_1, RS_2 and RP, RQ are three pairs of a pencil in involution, the two cross ratios $R\,(P,\,S_1/I,\,J)$, $R\,(S_2,\,Q/I,\,J)$ are equal, or $\epsilon R\,(P,\,S_1) = \epsilon R\,(S_2,\,Q)$. This is true, in particular, when R is on the conic, in which case the tangents RP, RQ are the same line. And it is true, for any position of R, when the curve touches the line IJ, say in A'; then the ray RS_2 becomes the ray RA', which is parallel to S_1A', the focus S_1 being the focus which does not lie on IJ.

Whence, for the parabola, we have, (*a*) Let R be a point of the curve, and the tangent at R meet SA' in T. Then we have $\epsilon R\,(T,\,S) = \epsilon R\,(A',\,T)$, where S is the focus; and the latter is $\epsilon T\,(S,\,R)$. This is one result given above. Also, (*b*) when R is not on the parabola, we have $\epsilon R\,(P,\,S) = \epsilon R\,(A',\,Q)$, which is the same as $\epsilon Q\,(A',\,R)$; or, if the tangent at Q meet SA' in U, is the same as $\epsilon U\,(S,\,Q)$. By (*a*) this is $\epsilon Q\,(R,\,S)$. And $\epsilon R\,(P,\,S) = \epsilon Q\,(R,\,S)$ is a result found above.

Ex. 3.   The theorem we have stated above by saying that, if R, R_1, U be the intersections of pairs of three tangents of a parabola, whose focus is

S, then the four points R, R_1, U, S lie on a circle, is the same as saying that, if R, R_1, U and S, I, J be two triangles whose sides touch a conic, then the six angular points of these two triangles lie on another conic. The argument we have employed, using I, J as Absolute points, is quite general. Dually, if the angular points of two triangles lie on a conic, then the six sides of the two triangles touch another conic.

Ex. 4.   We know that a conic can be described to touch five lines. Thus, if we are given a triangle, and a point S, lying on the circumcircle of the triangle, then a conic can be described to touch the sides of the triangle, and touch, also, the lines SI, SJ. By Ex. 3, preceding, this conic will then touch the line IJ. Thus a parabola can be described to touch the sides of a triangle and have its focus at any point of the circumcircle of the triangle.

Wherefore, as the perpendicular from the focus of a parabola, to any tangent of the curve, meets this tangent in a point lying on the tangent at a definite point A of the curve (called the *vertex*), we have the theorem of the pedal line, or Simson line, of a triangle, arising from any point of the circumcircle (Chap. x (5), Ex. 5). If we assume, what was proved (*ibid.*), that the pedal line of a point S, for a triangle whose orthocentre is H, contains the mid point of S, H, and also recall that the vertex A is the mid point of S, X, where X is on the directrix, which is parallel to the tangent at A, and perpendicular to SX, we reach the result that *the orthocentre of a triangle, formed by three tangents of a parabola, lies on the directrix of the curve.*

We can give a proof of this result which is independent of the theory of the pedal line given in the preceding chapter. Consider the quadrilateral formed by three tangents of the parabola, taken with the Absolute line; let the three tangents form the triangle ABC, of which H is the orthocentre. A pair of complementary intersections of the four lines consists of the point A, and the point, say L, where BC meets the Absolute line; the lines HA, HL are perpendicular, and are a pair of a pencil in involution with centre at H, which equally has two other similar pairs of perpendicular rays. Hence, the tangents from H are also perpendicular. Therefore H lies on the directrix, as we were to prove. Conversely, this proves again the theorem given in the preceding chapter, that the pedal line of a triangle, in regard to a point S, lying on the circumcircle of the triangle, contains the mid point of S, H, where H is the orthocentre of the triangle.

And, as a parabola can be found to touch four arbitrary lines, we also have an independent proof of the two results, that the orthocentres of the four triangles formed by four arbitrary lines lie upon a line; and that the circumcircles of these four triangles meet in a point (the focus of the parabola touching the four lines). The feet of the perpendiculars, drawn to the four lines from the common point of the circles, thus lie on a line.

Ex. 5.   Another proof of the fact that the orthocentre of a circumscribed triangle of a parabola lies on the directrix, quoted by Salmon from J. C. Moore (*Conics*, 1879, p. 247), is an application of Brianchon's theorem for six tangents of a conic. Let three tangents, p, q, r, of a parabola, meet, in pairs, in the points A or (q, r), B or (r, p), C or (p, q). Let the directrix meet the tangents q, r respectively in Q and R; the lines through Q and R, on the directrix, respectively perpendicular to q and r, will also touch the parabola; let these be q', r'. Denote the Absolute line by k. We then have a hexagon, q, q', k, r', r, p, of six tangents of the parabola. Of this hexagon, one diagonal, joining the opposite intersection qq' and $r'r$, is the directrix of the parabola. Another diagonal, joining the opposite intersections $q'k$, rp, is the line from B perpendicular to the line q. The third diagonal, joining the opposite intersections kr', pq, is the line from C perpendicular to the line r. The last two perpendiculars meet in the orthocentre H of the triangle ABC; which is thus on the directrix.

Ex. 6.   Let two lines, i, j, be drawn through a point K, on a conic; and let the tangents of the conic at the other two points, where the lines i, j

meet the conic, have D for common point. Prove that, if two points, respectively on i and j, be conjugate in regard to the conic, then the line joining these passes through D. This is the result associated with the name of Seydewitz. It is noticeable that this result is the dual of the theorem that two perpendicular tangents of a parabola meet on the directrix.

On the conic in general, on which K is taken, let P, Q, R be three other points. Let PD, QD, RD meet the conic again, respectively, in L, M, N, and the intersections (KL, QR), (KM, RP), (KN, PQ) be, respectively, P_1, Q_1, R_1. Prove that P_1, Q_1, R_1 lie on a line, which passes through D. This is the dual of the theorem that the orthocentre of a circumscribed triangle of a parabola lies on the directrix; but may be proved directly, either, as in Ex. 4, by considering the involution determined on the line Q_1R_1 by conics which pass through the four points P, Q, R, K; or, as in Ex. 5, by applying Pascal's theorem to the hexagon formed by Q, M, K, N, R, P.

Remark. The theorem, that the circumcircle of a circumscribed triangle of a parabola passes through the focus, was given by Lambert, in 1761. The theorem, that the orthocentre lies on the directrix, was proved, from Pascal's theorem, in 1827, by Steiner (*Werke*, I, pp. 134, 207). This theorem can also be deduced from the theorem (proved in Chap. IX), that the director circles of conics which touch four lines are coaxal. For the four lines, we take three tangents of the parabola, and the Absolute line; three of the director circles are then degenerate, meeting in the orthocentre. (See also Taylor, *Ancient and Modern Geometry of Conics*, 1881, p. 57.)

(4) The theory of poles and polars for a parabola arises easily from the general definition given in Chapter IX, (15). We assume the result there given that, if, from a fixed point O, any line be drawn to meet the parabola in P and Q, then the locus of the harmonic point $R = (P, Q)/O$, taken thereon, is a line. This line is the polar of O in regard to the parabola.

As has been said, we call the line drawn from the focus S, perpendicular to the directrix, the *axis* of the parabola; this meets the curve at A', on the Absolute line, and also at the *vertex A*, which is the mid point of S and the point X where the axis meets the directrix; the tangent of the parabola at A is perpendicular to the axis.

Let a series of parallel chords of the parabola be drawn, all meeting the Absolute line in the same point K. The mid points of these chords lie upon the polar of K, which passes through A', and is parallel to the axis; such a line, containing the mid points of a series of parallel chords, is called a *diameter* of the parabola; the diameter meets the curve also at a point, V, sometimes called the vertex of the diameter, whereat the tangent of the curve is parallel to the chords. The tangents of the curve at the end points, Q and Q', of one of the parallel chords, meet in a point, T, of the diameter; and V is the mid point of T and the mid point, M, of the chord QQ', since $M = (V, A')/T$. Thus, also, the tangent at V contains the mid points of T, Q and of T, Q'; a particular case is when the diameter is the axis of the parabola, which is perpendicular to all the chords whose mid points it contains. Of the series of parallel chords, there is one through the focus S, whose pole, U, is on the directrix; we have proved that this chord is perpendicular to SU. So that, *the diameter of a series of parallel chords of the parabola, and the line through the focus which is perpendicular to these chords, meet on the directrix.*

(5) Now consider a self-polar triangle of the parabola, say THK, where the side HK meets the curve in Q and Q'. We have seen that the pole, T, of the chord QQ' lies on a diameter containing the mid point, M, of Q and Q'; and that this diameter contains the point V of the curve which is the mid point of T and M. Since TK is the polar of H, we have $K=(Q,\,Q')/H$. Conversely, with the same angular point T, any two points $H,\,K$, on the polar QQ' of T, which are such that $H,\,K \sim Q,\,Q'$, make, with T, a self-polar triangle. The line which joins the mid point of $T,\,H$ to the mid point of $T,\,K$ is parallel to HK, and contains the mid point of $T,\,M$; this line is then the tangent of the parabola at the vertex V of the diameter TM. Wherefore, *if $T,\,H,\,K$ be the angular points of any self-polar triangle, the mid points of the pairs $H,\,K$; $T,\,H$; $T,\,K$, say the mid points of the sides of the triangle, form another triangle, whose sides touch the parabola.* In fact, in phraseology used before, the line joining the mid points of $T,\,H$ and $T,\,K$ is the harmonic image of the Absolute line in regard to the point T and the line HK; the curve is its own harmonic image in regard to this angular point and side of a self-polar triangle, and the harmonic image of the tangent at A' is another tangent. So for the other two sides of the new triangle.

Conversely, if we have a quadrilateral formed by three tangents $p,\,q,\,r$ of the parabola, taken with the Absolute line k, the diagonal triangle of this, which is self-polar in regard to the curve, has for sides the three joins of pairs of complementary intersections of the four lines. One of these joins is the line through the point $(q,\,r)$ which is parallel to p; and so on. Thus *the lines through the angular points of a circumscribed triangle of a parabola, each parallel to the opposite side, form a self-polar triangle of the curve.*

It is immediately obvious that the points $(q,\,r)$, $(r,\,p)$, $(p,\,q)$ are the mid points of the sides of the self-polar triangle. The lines drawn through these points, perpendicular to the sides of the self-polar triangle, are respectively perpendicular to the opposite sides of the triangle pqr. These three perpendiculars therefore meet in a point, which is both the orthocentre of the triangle (pqr), and the circumcentre of the self-polar triangle. As the orthocentre of a circumscribed triangle of a parabola lies on the directrix, it follows that *the circumcentre of a self-polar triangle lies on the directrix.* This result is a particular case of one already referred to (Chap. x (6), Ex. 15), that the circumcircle of a self-polar triangle of a conic is perpendicular to the director circle of the conic; for we have seen that the directrix of a parabola arises by degeneration of the director circle.

(6) Conversely to (5), we now prove that *if any circle is described with its centre on the directrix of the parabola, then a triangle can be inscribed in this circle to be self-polar in regard to the parabola, one of the angular points, T, of this triangle, being an arbitrary point of the circle.* This is

equivalent to saying that if the polar of a point T, of the circle, in regard to the parabola, meet the parabola in P, Q, and the circle in H, K, then $H, K \sim P, Q$. Then THK is such a self-polar triangle inscribed in the circle. Let the perpendicular from T, to the directrix, meet this in N, and meet the circle again in T', so that N is the mid point of T, T'. The circles through T and T', which have their centres on the directrix, are a series of conics through four points, and cut the polar line PQ in pairs of points of an involution. We prove that P, Q are the foci of this involution. This involves then $H, K \sim P, Q$, which we wish to prove.

To prove that P, Q are the foci, it is sufficient to shew that P, Q are harmonic conjugates in regard to the intersections of the line PQ with two of the circles through T and T'. This is so for that degenerate circle which consists of the line TT' and the Absolute line; for TT', parallel to the axis, contains the mid point of the points P, Q, on the polar of T. It is also so for that particular circle through T, T' which contains the point, say L, in which PQ meets the directrix. To see this, notice that the line ST, joining the poles of the directrix and of PQ, is the polar of L, and meets PQ in the point, say F, for which $F = (P, Q)/L$; it is sufficient then to shew that T, T', L, F lie on a circle. Now, we have seen, ST, SL are the bisectors of the angles determined by SP, SQ, and are perpendicular; also, as N is the pole of the chord through S which is parallel to PQ, the lines SN, PQ are perpendicular (by (4), above). Wherefore $\epsilon F\,(L, T) = \epsilon S\,(N, L)$, which is equal to $\epsilon T\,(N, L)$, because T, N, L, S are on a circle; and $\epsilon T\,(N, L)$, or $\epsilon T\,(T', L) = \epsilon T'\,(L, T)$, because N is the mid point of T, T', which is perpendicular to NL. The equation $\epsilon F\,(L, T) = \epsilon T'\,(L, T)$, however, proves that T, T', L, F lie on a circle, as we wished to prove. Whence, finally, $H, K \sim P, Q$, and the theorem enunciated is established.

(7) Assuming the theorem, already proved (Chap. x (6), Ex. 15), that the circumcircle of a self-polar triangle of a conic is perpendicular to the director circle of the conic, we can make an inference for any conic which is interesting later for what is called the curvature of the conic. For the parabola in particular a direct proof of this inference can be given. *Let TP, TQ be tangents of a conic, from a point T; let H be the orthocentre of the triangle TPQ. Then H lies on the polar of T taken in regard to the director circle of the conic. For the parabola, this becomes the result that the directrix contains the mid point of T, H.*

Let PH, QH, respectively, meet the opposite sides TQ, TP, of the triangle TPQ, in L and M. As PL, TL are perpendicular, and QM, TM are perpendicular, there is a circle containing T, M, H, L, whose centre is at the mid point of T, H; and there is a circle containing P, M, L, Q, whose centre is at the mid point of P, Q. These two circles are perpendicular; it may, for instance, be shewn that the lines joining L to the centres of these circles, are perpendicular. Therefore, if the former circle, $TMHL$, meet PQ, which is a diameter of the latter circle, in the points U, V, then $V = (P, Q)/U$. Whence, since UV is the polar of T in regard to the conic, which contains P and Q, the triangle TUV is a self-polar triangle in regard to the conic. Hence, assuming the result quoted, the circle $TMHL$ is perpendicular to the director circle of the conic. Thus, the diameter, TH, of this circle, meets the director circle in two points which are harmonically conjugate in regard to T and H. This shews that T and H are conjugate to one another *in regard to the director circle*, which was to be proved. It may be remarked that the line TH, perpendicular to PQ, does not generally contain the centre of the director circle.

Particular cases of this result, for any conic, are: (*a*) when T is on the director circle. Then, the orthocentre H, conjugate to T in regard to the director circle, coincides with T. In this case, the tangents TP, TQ to the conic are perpendicular to one another; (*b*) when T is on the conic. Then the line TH, perpendicular to PQ, is the line through T perpendicular to the tangent at T; which is commonly called the *normal of the conic* at T. In anticipation of a more deliberate consideration given below (Chap. XIII (13)), the reader may prove that, when T becomes a point on the conic, so that P, Q coalesce with one another at T, the point common to the normals of the conic at P and Q becomes a point, say O, which lies on the normal at T, while the orthocentre H, of the triangle TPQ, becomes a point H, equally on the normal at T, such that T is the mid point of H, O. The point H is still conjugate to T in regard to the director circle, and is thus definite when the director circle is given; and hence O is found. The point, O, in which the normals at P, Q meet, when Q coalesces with P at T, is called the *centre of curvature* of the conic at T.

In particular, *for a parabola, let the normal at the point T, of the curve, meet the directrix in U; let H be taken on this normal so that U is the mid point of T, H. Then, the centre of curvature of the parabola at T is the point O, of the normal at T, such that T is the mid point of H, O.*

Ex. Prove that, if T be the pole of a chord PQ of a parabola, the line PQ being the normal of the curve at P, then the directrix contains the mid point of T, P.

(8) Let O be any fixed point. Suppose the perpendicular to be drawn from O to the tangent at a variable point V of the parabola; let this perpendicular meet the diameter through V (which is parallel to the axis ASA'), in the point T. We find the locus of T. If the tangent of the parabola at the vertex A meet the Absolute line in Y, we prove that *the locus of T is a conic containing the three points O, Y and A', touching the axis AA' at A'.*

The pencil $O\,(T)$, of lines from O to the variable point T, is related to the range on the Absolute line determined by the respectively perpendicular tangents of the curve at the points V; and, as the Absolute line is a tangent of the curve, this range is related to the pencil of lines $A'\,(V)$, or $A'\,(T)$. This shews at once that the locus of T is a conic containing the points O and A'. When V is at the vertex, A, the line OT

is parallel to AA', and is the line OA'; the tangent of the locus at A' is therefore that ray of the pencil $A'(V)$ which contains the point of contact V, which is then A; thus $A'A$ is the tangent of the locus at A'. When V is at A', the point T is at Y; for, in general, the lines OT, and the tangent at V, meet the Absolute line in points which are harmonic conjugates in regard to the Absolute points; and the tangent at A' is to be regarded as meeting the Absolute line in the point A' (Chap. IX (22)). This proves the statement made.

We have seen, however ((4), above), that, if the diameter through V meet the directrix in W, and S be the focus, then SW is perpendicular to the tangent at V. Thus *the locus of T contains the point where OS meets the directrix*. The other intersection with the directrix is at Y. We have thus found four points of the locus, and the tangent at one of these; so that the curve is defined. The locus of T, in its nearly identically defined general form, for any conic, was considered by Apollonius (247–205 B.C. See his *Conics*, Book V, Props. 58–63), and will occur again below.

Ex. 1.   The tangent at O, of the locus of T, is the line drawn from O perpendicular to the polar of O with regard to the parabola. In particular, if O be taken on the parabola, the tangent at O is the normal of the parabola at O, which is perpendicular to the tangent at O.

Ex. 2.   If O be taken on the Absolute line, the locus of T degenerates, becoming the Absolute line, together with the line, parallel to the axis of the parabola, which contains the point where OS meets the directrix.

(9) The locus of the point T, considered in (8), besides meeting the parabola at A', will have three other points thereon. Denote these by P, Q, R. When the varying point T is at one of these, say P, the line OP, by the definition of the locus, is perpendicular to the tangent of the parabola at P, this tangent being the polar of P. Thus OP is the normal of the parabola at P. We infer therefore that *there are three points, P, Q, R, of the parabola, whereat the normals pass through an arbitrary point O of the plane*, besides the point A'. For, conversely, any such point lies on the locus of T.

(10) Let Y_1, Y_2 be any two points of the Absolute line which are harmonically conjugate in regard to the point A', where the parabola touches the Absolute line, and the point Y, where the tangent at the vertex A meets the Absolute line. Let P, Q, R be three points of the parabola whereat the normals meet in a point, as in (9). Then *there is a conic through P, Q, R and the vertex, A, which passes through Y_1 and Y_2*.

To prove this, we shew that the involution on the line $A'Y$, determined by conics through the four points P, Q, R, A, has for foci the points A' and Y. This is clear if we find two such conics which meet $A'Y$ each in a pair of points harmonic in regard to A' and Y. And we already have one such conic, namely, the parabola itself, which meets $A'Y$ in two points which coincide at A'. We prove that the degenerate conic, consisting of the lines AR, PQ, is another such, namely, that the lines AR, PQ meet the Absolute line in two points which are harmonic conjugates in regard to A' and Y. This has in fact incidentally appeared already (Chap. IX (23), Ex. 13). But we can give an independent proof thus: let the points P, Q be taken as the Absolute points of the plane. Then the curve so far called a parabola, say (P), is a circle, of which YA, YA' are tangents, touching the curve at A, A', and meeting in Y. The locus (T), of points T, which has been considered, is also a circle, as passing through P and Q; this locus (T) touches $A'A$ at A', which, with the other point at R, makes, with the points P, Q, the four common points of the two circles. Moreover, the circle (T) contains the pole Y, of the chord $A'A$ of the circle (P), taken in regard to this circle. For two circles so related it has been remarked (Chap. X (6), Ex. 34) that the chord AR of the circle (P) contains the mid point of the points A', Y, on the circle (T), this mid point being defined by reference to the Absolute line PQ. So that the line PQ, and the line AR, meet $A'Y$ in two points which are harmonic conjugates in regard to A' and Y. This proves the result required. And, returning to the hypothesis that the curve (P) is a parabola, we see that every conic through P, Q, R, A meets $A'Y$ harmonically in regard to A' and Y.

A particular case is that a conic can be drawn through P, Q, R, A to touch $A'Y$ at Y. Regarding $A'Y$ as the Absolute line, as before, this is another parabola, with axis perpendicular to the axis AA' of the former (its vertex not generally being at A). Another particular case is that the four points P, Q, R, A lie on a circle (for the Absolute points I, J on $A'Y$ are such that I, $J \sim A'$, Y).

(11) If two arbitrary points P, Q be taken on a parabola, the normals of the curve at P and Q meet in a point O, through which passes the normal at a third point R of the curve. Thus R is determined when P, Q are assigned. The result of (10), preceding, gives a construction for R. For since PQ and AR meet $A'Y$ harmonically in regard to A' and Y, it follows at once that, *if the tangent of the parabola which is parallel to PQ meets the axis in T, then AR is parallel to the other tangent which can be drawn to the curve from T.* A particular corollary of this is that, if a series of parallel chords, PQ, of the parabola be taken, the point of intersection of the normals of the curve at P and Q describes a line, namely, the normal of the curve at the point R; the line AR is in fact the same for all the parallel chords PQ. Further, if the tangents from T to the parabola, one of which is parallel to PQ, touch the curve in V and V', the diameters from V and V', parallel to the axis, contain respectively the mid point of P, Q and of A, R; while the mid point of V, V' is on the axis. Wherefore, considering the mid points of P, Q and of A, R, the point which is the mid point of these mid points lies on the axis. So that *the centroid of the three points P, Q, R of the parabola, whereat the normals meet in a point, lies on the axis* (the centroid of three points P, Q, R, already explained, is the point common to three lines such as that joining R to the mid point of P, Q).

(12) We have (Chap. IX (23), Ex. 18) defined the conjugate line locus, for any conic, with assigned centres T, T', as being the locus of the common point of two lines, drawn respectively through T and T', so as to be conjugate to one another in regard to the given conic. This locus is a conic, passing through T and T', meeting the given conic on the polars of T and T'; and the pole of TT', in regard to the given conic, is equally the pole of TT' in regard to the locus.

For a parabola, let two points T, T' be taken, such that the focus S is the mid point of T, T'; let the tangents to the parabola, from T, be TP, TQ, and, from T', be $T'P'$, $T'Q'$. We prove that *the conjugate line locus, for the parabola, with centres at T and T', is a circle.* This is definable, therefore, as the circle TPQ, or as the circle $T'P'Q'$. For, if I, J be the Absolute points, the tangents from I to the parabola consist of the line IS, and the Absolute line; as S is the mid point of T, T', the lines IT, IT' are harmonic conjugates in regard to these tangents from I; thus IT, IT' are conjugate in regard to the parabola. This shews that the conjugate line locus, with T, T' as centres, contains the point I. Likewise this locus contains J; and is, therefore, a circle.

The centre of this circle lies on the line, perpendicular to TT', through the mid point S, of the chord TT'. This is the line joining S to the point, H, on the directrix which is common to the chords PQ, $P'Q'$ of the parabola. The point H is the pole of TT' in regard to the parabola; it is thus also the pole of TT' in regard to the circle; so that HT, HT' touch the circle at T and T', respectively.

The common self-polar triangle of the parabola and the circle has, therefore, H for one angular point, the other two angular points being those points on TT' which are both conjugate in regard to the parabola and harmonic conjugates in regard to T and T'.

Ex. 1. Let the tangents at two points, P, Q, of the parabola meet the axis in L and M, respectively, and meet one another in T. Then, the tangent at T of the circle TLM contains the focus, S, of the parabola.

For we have proved ((3), above) that $\epsilon T(Q, S) = \epsilon P(T, S)$; and also that the latter, being $\epsilon P(L, S)$, is equal to $\epsilon L(S, P)$. If K be any point of the circle, we have $\epsilon L(S, P) = \epsilon L(M, T) = \epsilon K(M, T)$, which is the same as $\epsilon T(M, T)$, or $\epsilon T(Q, T)$. From $\epsilon T(Q, S) = \epsilon T(Q, T)$, however, it follows that TS is the tangent of the circle at T.

Now, let the lines through L, M, which are respectively perpendicular to the tangents LP, MQ, meet in H, so that TH is a diameter of the circle TLM, and is perpendicular to TS; and let the normals of the parabola at P and Q meet in O. Then, because the line from S, drawn perpendicular to PL, contains the mid point of P, L, and is parallel to LH and PO, with a similar statement for QM, we infer that the line HO contains S, and that S is the mid point of H, O.

Thus, *if the normals at the points P, Q, R, of a parabola, meet in O, and H be taken on OS so that S is the mid point of O, H, then the circle having S, H as ends of a diameter is the circumcircle of the triangle formed by the tangents of the parabola at P, Q, R.*

Ex. 2. Any chord PP' of a parabola is conjugate to the line drawn parallel to the axis through the mid point of P, P'. Hence shew that if a variable chord of a parabola, PP', pass through a fixed point B of the axis, the locus of the mid point of P, P' is another parabola, having B for vertex, with the same axis.

Ex. 3. Prove that the four common points of two parabolas, whose axes are perpendicular, lie on a circle.

Ex. 4. For two parabolas having a common focus, the bisectors of the angles determined by their directrices are a pair of common chords.

Ex. 5. For two parabolas having a common directrix, and foci at S and S', the line through the mid point of S, S', perpendicular to SS', contains a common point of the two curves.

Ex. 6. Let P, Q be two given points, on two given lines TP, TQ. Prove that a parabola exists touching TP and TQ, at P and Q, respectively. Construct the vertex, the focus, the directrix, and a general point of the curve. The focus is the further intersection of the circle through T and P, which touches TQ at T, with the circle through T and Q which touches TP at P.

Ex. 7. Let an arbitrary line and four points, O, A, B, C, be given. We know that the complementary joins OA, BC; OB, CA; OC, AB determine on the line three pairs of points belonging to an involution. Prove that, *if P, P' be any pair of this involution, the conic through A, B, C, and P, P', contains O.*

As a particular case of the dual of this result, prove that the orthocentre, of a circumscribed triangle of a conic, of which no two sides are perpendicular, does not lie on the director circle of the conic unless the conic is a parabola (in which case the director circle degenerates).

An application of the theorem enunciated is that, if A, B, C and O, P, P' be two triangles which are both self-polar in regard to a conic, then their six angular points lie on another conic. (This is obtained by taking PP' for the arbitrary line considered.)

We may compare the result: *If P, P' be a pair of the involution determined on a line by the complementary joins of four points O, A, B, C, there exists a conic for which ABC and OPP' are both self-polar triangles.* For, if I, J be the foci of the involution, we know (Chap. IX (23), Ex. 11) that there exists a conic through I, J for which ABC is a self-polar triangle; and it can be proved that OPP' is self-polar in regard to this. The so-called self-polar circle of a given triangle (considered above, Chap. X (6), Ex. 16) is an example.

We see then that, if ABC and OPP' be two inscribed triangles of a conic, there exists another conic in regard to which both these are self-polar. In general there is only one such conic.

Ex. 8. A parabola touches the sides of a triangle PQR, its point of contact with QR being the mid point of Q, R. A fourth tangent of the parabola, and the line through P which is parallel to this fourth tangent, meet QR in H and K. Prove that H, $K \sim Q$, R.

Ex. 9. Let a parabola touch the sides BC, CA, AB, of a triangle, respectively at X, Y, Z. We know that the circles AYZ, BZX, CXY have a common point, say O; and that the circle ABC contains the focus S of the curve. Let P, Q, R be the respective centres of the circles AYZ, BZX, CXY. We can prove that the circle PQR contains the focus S.

Because BZ and BX are tangents of the parabola, the line SB is perpendicular to SQ (by (12) above). Likewise SC is perpendicular to SR. Also, S and O are not the same point, or SB would be tangent to the circle BXZ at S; and so on. Again, OY is perpendicular to the join, RP, of the centres of the two circles having O, Y for common points; likewise OZ is perpendicular to the join, QP, of the two circles having O, Z for common points. Whence

$$\epsilon S\,(Q,\,R) = \epsilon S\,(B,\,C) = \epsilon A\,(B,\,C) = \epsilon A\,(Z,\,Y) = \epsilon O\,(Z,\,Y) = \epsilon P\,(Q,\,R);$$

and this proves the result.

Ex. 10. The proof given above (in (3)), that the circumcircle of a circumscribed triangle of a parabola contains the focus, establishes the theorem

that the six angular points, of any two circumscribed triangles, of any conic,
lie upon another conic. We may repeat the argument in general terms: Let
a, b, c and a', b', c' be the sides of two circumscribed triangles of a conic, the
respectively opposite angular points being A, B, C and A', B', C' (so that
A is the intersection (b, c), etc.). Then the ranges of points, determined on
the tangents a, a', respectively, by the tangents c, b, c', b', are related. The
points of these ranges lie, respectively, on the pencils A' (B, C, B', C'),
A (B, C, B', C'). That these pencils are related proves that the six angular
points lie on a conic. Dually, if two triangles have their six angular points
on a conic, their six sides touch a conic.

Now, let three tangents, of one conic, ω, form a triangle PQR; let Ω be
any conic passing through P, Q, R; and P' be any point of this conic Ω. Let
the pair of tangents to ω, drawn from P', meet Ω again in Q' and R'. By the
dual theorem referred to, there exists a conic touching the six sides of the
two triangles PQR and $P'Q'R'$; and this, as touching five tangents of the
conic, ω, coincides with ω. Thus $Q'R'$ touches ω whatever point of Ω is taken
for P'; and there is an infinite series of triangles whose angular points lie on
Ω, whose sides touch ω.

A conic Ω which contains the angular points of a triangle whose sides
touch a conic ω is said to be *triangularly circumscribed* to ω. The argument
shews that a sufficient condition, for the re-
lation between the conics Ω, ω, is the existence
of a single triangle inscribed in Ω and circumscribed to ω. Thus, taking one angular point P,
of this triangle, to be at a common point of the
two conics, we see that a sufficient condition
is that *the tangent of the inscribed conic, ω, at
a point common to the two conics, should meet
the circumscribed conic, Ω, at a further point,
whereat the tangent is also a tangent of the inscribed conic.* A common point of the two
conics being thus associated with one of the four
common tangents, every other common point
is associated with a definite common tangent.

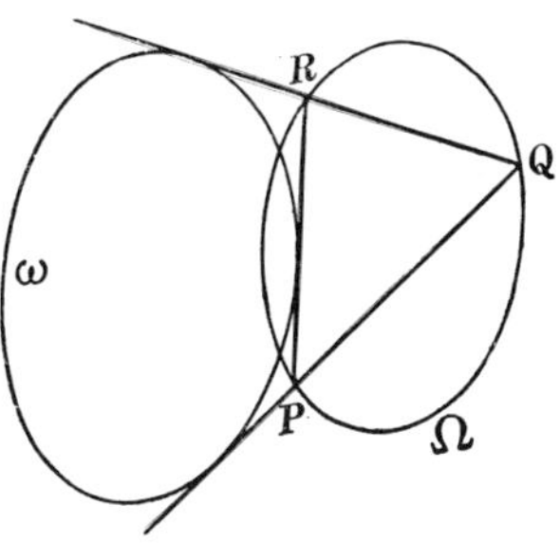

In particular, any circle, drawn through the focus of a parabola, is triangularly circumscribed to the parabola. Two tangents of the parabola,
drawn from a point of the circle, meet the circle again in two points which
lie on a tangent of the parabola.

Ex. 11.   We can prove, in connexion with the theory of the parabola, a
theorem which is proved in general terms below (Chap. xiv (21), Ex. 20). *On a
given conic, ω, let O be a given point; and let A, B, C, be given points not lying
on ω. Through the four points, O, A, B, C, a series of conics can be drawn.
Each of these determines on ω, besides the point O, three points, forming an
inscribed triangle of ω. The theorem is that the sides of all these triangles touch
another conic.* Three particular conics of the series arise by taking, through
$OABC$, such a degenerate conic as the line-pair OA, BC; the inscribed triangle of ω then arising has its angular points where BC meets ω, and at the
further point P where OA meets ω. The conic spoken of thus touches BC,
and touches the lines joining P to the points of ω which lie on BC. We first
prove then that there is a conic touching the nine lines so arising.

Denote the further intersections with ω of the lines OA, OB, OC, respectively, by P, Q' and S. Let BC, CA, AB respectively meet ω in Q, R;
in R', P' and in I, J. The nine lines spoken of consist then of QR, RP, PQ,
IJ, SI, SJ and $Q'R'$, $R'P'$, $P'Q'$. Take the points I, J for the Absolute
points. Then the conic ω is a circle, of which PQR and SIJ are inscribed
triangles. There exists then a conic touching the six sides of the two triangles,
namely, a parabola whose focus is S. We are to prove that the lines $Q'R'$,
$R'P'$, $P'Q'$, which are the remaining three of the nine specified lines, **are**

tangents of this parabola. Take, first, $R'P'$. We proved that if H, H_1 be points on a tangent of a parabola of focus S, and the other tangents from these points be respectively t, t_1, then

$$\epsilon\,(HS,\,t) = \epsilon\,(H_1 S,\,t_1).$$

Because CP' is parallel to OP, we have

$$\epsilon C\,(S,\,P') = \epsilon O\,(S,\,P) = \epsilon Q\,(S,\,P);$$

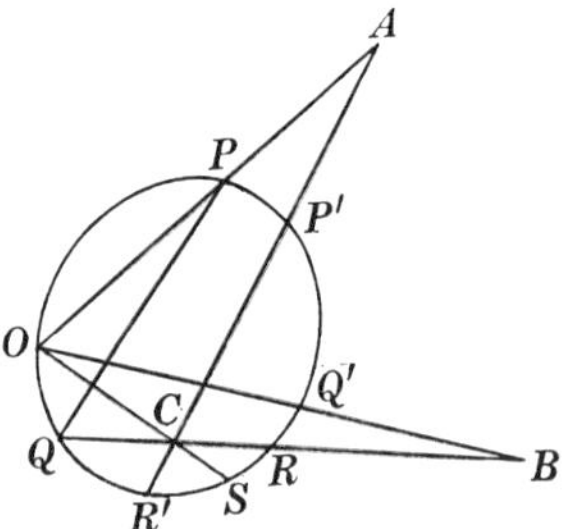

also, Q, C are two points on the tangent QR of the parabola, and QP is a tangent from Q. Thus CP', drawn from C, is also a tangent. That is, $R'P'$ touches the parabola. Next take $R'Q'$. Let $R'Q'$ meet QR in H. Then, OQ' being parallel to QR, we have $\epsilon H\,(C, R') = \epsilon Q'\,(O, R') = \epsilon S\,(O, R')$. Thus the points C, R', S, H lie on a circle. Whence, as CH, CR' are tangents of a parabola whose focus is S, it follows that $R'H$, or $R'Q'$, also touches the parabola. Then, finally, as S is on the circumcircle of $P'Q'R'$, and $R'P'$, $R'Q'$ touch the parabola, so also does $P'Q'$. Thus the nine lines in question touch a conic.

To prove that, if any conic through O, A, B, C meet the circle further in P_1, Q_1, R_1, then the lines $Q_1 R_1$, $R_1 P_1$, $P_1 Q_1$ touch the parabola, it will be sufficient to prove that, if Q_1, R_1 be two points of the circle such that $Q_1 R_1$ touches the parabola, then a conic contains the six points Q_1, R_1, O, C, B, A; for a conic is determined by five points. For this, by the converse of Pascal's theorem, it is sufficient to prove that the three intersections $(Q_1 R_1, CB)$, $(R_1 O, BA)$, (OC, AQ_1) lie in a line. Let K, L denote, respectively, $(Q_1 R_1, CB)$ and (OC, AQ_1), so that LQ_1 is parallel to OP and to $R'P'$. To prove that the join of these two points K, L contains the point $(R_1 O, BA)$ is to prove that KL is parallel to $R_1 O$.

But, because KQ_1 and QP are tangents of the parabola, which meet the tangent QR, respectively, in K and Q, we have

$$\epsilon K\,(S, Q_1) = \epsilon Q\,(S, P) = \epsilon O\,(S, P), = \epsilon L\,(S, Q_1).$$

Hence L, K, S, Q_1 lie on a circle, and, therefore,

$$\epsilon K\,(L, Q_1) = \epsilon S\,(L, Q_1) = \epsilon S\,(O, Q_1) = \epsilon R_1\,(O, Q_1),$$

so that $R_1 O$ is parallel to KL, as desired.

The whole theorem is therefore proved. But the general proof of Chapter xiv below may be preferred.

Ex. 12. We deduce, from Ex. 11, that, if PQR, $P'Q'R'$ be two circumscribed triangles of a parabola, then the second common point of the circumcircles PQR, $P'Q'R'$, besides the focus S, is a point which lies on the conic which (Ex. 10) contains P, Q, R, P', Q', R'.

Denote the circles PQR, $P'Q'R'$ respectively by ω and ω'; and the conic $PQRP'Q'R'$ by Ω. Let M denote the fourth common point of ω and Ω. By Ex. 11, the conics through S, I, J and M meet Ω in the angular points of triangles whose sides touch a conic, say σ. One such triangle is PQR, inscribed in ω. Moreover, the conic σ touches the sides of the triangle SIJ. Thus σ is a conic which touches six tangents of the original parabola, and coincides with this. The particular conic through S, I, J, M which contains P', on the conic Ω, thus contains the points Q', R' of Ω, which lie on the tangents $P'Q'$, $P'R'$ drawn from P' to the parabola. This particular conic, containing the points S, I, J, P', Q', R' of the circle ω', thus coincides with ω'. Wherefore ω' passes through M. This establishes the theorem.

In general terms, *considering three circumscribed triangles of any conic, and the three conics which each contain the angular points of two of these triangles, these three conics have a common point.*

Remark. The results of Exx. 11, 12 are found in Steiner's posthumous lectures on Conics (ed. Schröter). They may be proved by the theory of plane cubic curves (cf. *Principles of Geometry*, II, p. 245), which was also used by Steiner to obtain properties of conics. But the result of Ex. 11, that, if O, A, B, C be given points, of which O lies on a given conic Ω, then the conics through these four points determine inscribed triangles of Ω whose sides touch another conic, is an easy consequence of the use of the symbols, and the notion of the equation of a conic, to which we have (Chap. IX (18)) briefly referred (cf. also the subsequent chapter XVI on the equation of a conic).

For, if θ denote the parameter by which the points of the conic Ω are given, it can be shewn that the parameters of the three points, in which a variable conic through O, A, B, C meets Ω, are the roots of a cubic equation of the form $u + \lambda v = 0$, wherein u, v are definite cubic polynomials in θ, and λ is a variable parameter, depending on the conic which is drawn through O, A, B, C. If two roots of this cubic equation be θ_1, θ_2, the two equations $u_1 + \lambda v_1 = 0$, $u_2 + \lambda v_2 = 0$ (where u_1 is the result of putting θ_1 for θ in u, etc.) lead to $u_1 v_2 - u_2 v_1 = 0$; after division by $\theta_1 - \theta_2$, we thus obtain a polynomial relation connecting $(\theta_1 \theta_2,$ $\theta_1 + \theta_2, 1)$, homogeneously quadratic in these three numbers. Now if, as is possible in an infinite variety of ways, we suppose the equation of the conic Ω to be put into the form $xz - y^2 = 0$, the chord of the conic which joins the points (θ_1), (θ_2) has the equation

$$x - y\,(\theta_1 + \theta_2) + z\theta_1\theta_2 = 0$$

(as we have seen, Chap. IX (19), Ex. 1). The existence of a homogeneous quadratic polynomial relation connecting the coefficients u, v, w, in the equation of a line, $ux + vy + wz = 0$, can, however, be proved to secure that the line touches a definite conic. This establishes the result of Ex. 11.

CHAPTER XII

THE RECTANGULAR HYPERBOLA

(1) A *rectangular hyperbola* is a conic which meets the Absolute line in two points which are harmonic conjugates in regard to the Absolute points, or, is any conic for which the Absolute points are conjugate points. In general, the tangents of a conic at the points where the curve meets the Absolute line (which meet in the centre when the curve is not a parabola) are called the *Asymptotes* of the conic. Thus, also, a rectangular hyperbola is a conic whose asymptotes are perpendicular. The parabolas which touch the Absolute line at the Absolute points are particular rectangular hyperbolas.

An immediate consequence of the definition is that, if PCP' and QCQ' are perpendicular chords of the rectangular hyperbola, meeting in the centre C of the curve, then the four lines which join P and P' to Q and Q' pass in pairs through the Absolute points I, J. For, let the intersections $(PQ, P'Q')$ and $(PQ', P'Q)$ be, respectively, U and V. Then the vertex triangle of the quadrangle P, Q, P', Q', inscribed in the curve, which is the triangle UCV, is (by general theorems) self-polar in regard to the curve. Thus U, V are on the Absolute line, and are conjugate in regard to the curve; being also harmonic conjugates in regard to the points where the perpendicular lines PP', QQ' meet the Absolute line, they are in fact the Absolute points I, J.

The relation of Q, Q' to P, P' is thus the same as that of the common points of a series of coaxal circles to the limiting points of the series. This relation is often expressed by saying that Q, Q' are the *antipoints* of P, P' (or P, P' the antipoints of Q, Q'). In the case usually taken, in which the Absolute points I, J are conjugate imaginaries, if one of the two pairs P, P' and Q, Q' consist of real points, the other pair are conjugate imaginary points.

(2) Conics through four given points, of general position, meet the Absolute line in pairs of points belonging to an involution; this will have one pair which belongs also to the involution of which I, J are the foci. Thus a *single rectangular hyperbola can be drawn through any four points of general position*. If, however, the four points be such that one of them is the orthocentre of the other three (and, therefore, any one of the four points has this property), it will appear, immediately, that any rectangular hyperbola through three of the points contains the fourth point. In this case, then, a rectangular hyperbola can be drawn through the four points to contain any other general point.

Four points, of which any one is the orthocentre of the other three, are in fact such that every one of the three complementary pairs of joins, of the four points, is a degenerate rectangular hyperbola. The four points may be defined as the common points of any two of these degenerate conics. And, it is true that every conic passing through the four common points of two rectangular hyperbolas is itself a rectangular hyperbola; for the points in which this conic meets the Absolute line are a pair of the involution determined thereon by the two rectangular hyperbolas, and are therefore harmonic conjugates in regard to I and J. Thus, *every conic through three points A, B, C, and the orthocentre, H, of these, is a rectangular hyperbola.*

Conversely, every rectangular hyperbola passing through three points, A, B, C, contains also the orthocentre, H, of these. Indeed, *if we are given three points A, B, C, and also an involution on a given line, the conics, of which one is defined as containing A, B, C and a pair of points of this involution, have another common point.* For take two of the conics; by their intersections with the given line, these determine the involution; let D be the fourth common point of these conics, besides A, B, C. Then, a conic through A, B, C, D described to contain one point of a pair of the involution, contains the other point of the pair. This is then one of the conics defined as containing the five points consisting of A, B, C and a pair of the involution, and can be any one of these. The argument assumes that a definite conic is determined by five points, provided no four of these lie in a line; and it is assumed that the given line does not coincide with any one of BC, CA, AB. Thus, in particular, any rectangular hyperbola through three points contains also the orthocentre of these.

(3) Another proof of this last result, which uses Pascal's theorem, has historical interest (Brianchon and Poncelet, *Gergonne's Ann.* XI (1821), p. 205). Let H be the orthocentre of A, B, C; let a rectangular hyperbola through A, B, C meet the Absolute line in E and F. Let EH meet AB in K; also let AF meet CH in M, and BC meet the Absolute line in U. The lines KH, MA, containing respectively E, F, for which $E, F \sim I, J$, are perpendicular to one another. And AK, CH are perpendicular. Thus K is the orthocentre of the triangle HAM; so that MK is perpendicular to AH, and contains the point U. In the hexagon B, C, H, E, F, A, the points of intersection of the pairs of opposite sides are U, M, K, which are in line. Whence the rectangular hyperbola containing the five points A, B, C, E, F, by the converse of Pascal's theorem, contains also H; as we wished to prove.

(4) By using what is in fact the dual of the theorem we have proved, namely, of the theorem that the conics defined as passing through three points, and the various pairs of the points of an involution, have all a further common point, we can connect what has been said with a previous result, namely, that the directrix of a parabola touching the sides of a triangle passes through the orthocentre of the triangle.

Suppose we have a rectangular hyperbola passing through three points A, B, C, whose orthocentre is H. Consider the circle of centre H for which ABC is a self-polar triangle (Chap. X (6), Ex. 15). Take the particular dual figure obtained by replacing any point of the given figure by its polar line taken in regard to this circle. Any line of the given figure, defined as containing two points, will then be replaced by the pole of this line, in regard to the self-polar circle. A conic, regarded as a locus of points, will be replaced by an envelope of which two tangents pass through an arbitrary point; this we assume to be another conic (cf. Chap. ix (22)). In particular, the conic of the new figure which replaces the given rectangular hyperbola through A, B, C will have the polar lines BC, CA, AB as tangents. The Absolute points I, J, lying on the self-conjugate circle, whereat HI, HJ are the tangents of this circle, will be replaced by the lines HI, HJ; and the Absolute line replaced by the point H. As the rectangular hyperbola meets the Absolute line in points which are harmonic conjugates in regard to I and J, the new conic will have tangents, from H, which are harmonic in regard to HI and HJ, and so are perpendicular to one another. The new conic is thus one of a system, having three given tangents BC, CA, AB, of which each conic touches the lines of a pair of a pencil in involution, whose centre is at H. By the dual of the theorem referred to, all conics of this system have a common tangent (besides BC, CA, AB). The focal lines of the pencil in involution are the lines HI, HJ; also, one of the three degenerate conics of the system consists of two pencils of lines, one with centre at A, the other with centre at the point where BC meets the Absolute line. All the degenerate conics thus have the Absolute line for a tangent. This line is then the common tangent of the system of conics. Thus the new conic, which replaces the given rectangular hyperbola, is a parabola. And as the tangents to this, drawn from H, are perpendicular to one another, the directrix of the parabola passes through H.

(5) For a conic which is not a parabola, a pair of *conjugate diameters*, that is, a pair of lines, drawn through the centre, which are conjugate to one another in regard to the conic, are harmonic conjugates in regard to the tangents of the conic drawn from the centre, that is, in regard to the asymptotes. In *the case of a rectangular hyperbola, the asymptotes are perpendicular. They are, therefore, the bisectors of the angles determined by any pair of conjugate diameters.*

If X, X' denote the points where the rectangular hyperbola meets the Absolute line; and p, p' be one pair of conjugate diameters, meeting the Absolute line, respectively, in P, P', while another pair of conjugate diameters q, q' meets the Absolute line in Q, Q', respectively— then, the pair X, X' are harmonic conjugates in regard to P, P', also in regard to Q, Q', and in regard to the Absolute points I, J, because the curve is a rectangular hyperbola. Whence we have

$$(P,\, Q/I,\, J) = (P',\, Q'/J,\, I) = (Q',\, P'/I,\, J).$$

And therefore $\epsilon\,(p,\, q) = \epsilon\,(q',\, p')$.

Now, any chord of the curve, and the line joining the centre to the mid point of this chord, are parallel to two conjugate diameters, as we have seen in considering the theory of polars for a conic (Chap. ix (15)). Hence we can prove that, *if P, Q, R be three points of a rect-*

angular hyperbola, whose centre is C, and D, E, F be the respective mid points of Q, R; of R, P and of P, Q, then the circle DEF contains the centre C. For, by what has just been said, we have $\epsilon C\,(E, F) = \epsilon P\,(F, E)$, which, because DE, DF are respectively parallel to PF, PE, is the same as $\epsilon D\,(E, F)$. This shews that the four points D, E, F, C lie on a circle.

The circle containing the mid points of the sides of a triangle is called the nine-point circle of the triangle, and has already been considered (Chap. x (6), Ex. 15). We have then proved that *the centres of all rectangular hyperbolas, containing the angular points of a triangle, lie on the nine-point circle of the triangle.* The fact has already been proved. For we have shewn above that these hyperbolas all pass through the orthocentre of the triangle; the locus of the poles of the Absolute line in regard to these conics is thus a conic of which eleven points can be specified (Chap. ix (23), Ex. 15).

Another simple application of the relation $\epsilon\,(p, q) = \epsilon\,(q', p')$ is the fact that if the tangents, at any two points P, Q of a rectangular hyperbola whose centre is C, meet in T, then $\epsilon T\,(P, Q) = \epsilon C\,(Q, P)$. This may be compared with the corresponding result for a circle, which is $\epsilon T\,(P, Q) = \epsilon C\,(P, Q)$.

(6) Again, *if PCP' be a diameter of a rectangular hyperbola, the centre C being the mid point of P, P', and H, K be any two points of the curve, then $\epsilon P\,(H, K) = \epsilon P'\,(K, H)$.*

In fact, the lines HP, HP', of which the latter is parallel to the line joining C to the mid point of P, H, are parallel to conjugate diameters; as also are the lines KP, KP'. The result then follows from $\epsilon\,(p, q) = \epsilon\,(q', p')$.

The result is often expressed by saying that any chord, HK, of a rectangular hyperbola, subtends equal or supplementary angles at the ends of a diameter. Using notions of topology into which we do not now enter, and supposing P, C, P', H, K to be real points, the Absolute points being conjugate imaginaries, as previously explained, it can be said that the angles subtended by the chord HK at P and P' are *equal* when the line HK and the Absolute line meet the diameter PP' in two points which are separated by P and P'; and these angles are *supplementary* in the other case.

The result has several interesting applications, (a), (b), (c):

(a) If P, P' be any two fixed points, and a point H moves so that the product $\epsilon P\,(H, P').\,\epsilon P'\,(H, P)$ is constant, then H describes a rectangular hyperbola, of which the centre is the mid point of P, P'. We can, in fact, replace $\epsilon P\,(H, K) = \epsilon P'\,(K, H)$ by

$$\epsilon P\,(H, P').\,\epsilon P\,(P', K) = \epsilon P'\,(K, P).\,\epsilon P'\,(P, H),$$

which is $\quad \epsilon P\,(H, P').\,\epsilon P'\,(H, P) = \epsilon P\,(K, P').\,\epsilon P'\,(K, P).$

The equation $\epsilon P\,(H, P').\,\epsilon P'\,(H, P) = \text{constant}$ is to be compared with the equation $\epsilon P\,(H, P').\,\epsilon P'\,(P, H) = \text{constant}$, that is, $\epsilon H\,(P, P') = \text{constant}$, which holds for a variable point H on a *circle*, of which P, P' are any two points.

(*b*) From the relation $\epsilon P\,(H, K) = \epsilon P'\,(K, H)$, if we suppose H to coalesce with P, and take any point T on the tangent at P, we infer, if K be any point of the curve, that $\epsilon P\,(T, K) = \epsilon P'\,(K, P)$. This also is to be compared with a well-known result for a circle. In particular, *if the normal to the rectangular hyperbola, at a point P, meet the curve again in K, we infer that the circle, having PK for diameter, contains the other extremity, P', of the diameter PCP'.*

Also, if T be that point of the tangent of the hyperbola, at P, which is the pole of the chord joining P, to any point K of the curve, so that CT, containing the mid point of P, K, is parallel to $P'K$, we infer that $\epsilon P\,(T, K)$, as being equal to $\epsilon P'\,(K, P)$, is equal to $\epsilon C\,(T, P)$. Thus, the circle PCT touches the chord PK at P. Likewise the circle KCT touches KP at K. Whence, *the involution determined on the chord PK, by circles drawn through C and through the pole T of the chord PK, has P and K for foci.*

(*c*) If PH be perpendicular to PK, so that

$$\epsilon P\,(H, K) = -1, \quad = \epsilon P'\,(K, H) = \epsilon P'\,(H, K),$$

then, also, $P'H$ is perpendicular to $P'K$. Wherefore, *the circle having any chord HK of the hyperbola for diameter, meets the hyperbola again at the ends P, P' of a diameter of the hyperbola.*

(7) In fact, if the circle, having a chord HK of a rectangular hyperbola for diameter, meets the hyperbola again in P and P', the chord HK is perpendicular to the parallel tangents of the hyperbola at P and P'. For, regarding P as a fixed point, let a variable pair of perpendicular lines be drawn through P, meeting the curve again respectively in U and V. Then the points U, V are a pair of an involution of pairs of points on the curve. Thus all the lines UV meet in a point. By supposing U to coalesce with P, we see that this point lies on the normal of the hyperbola at P. By supposing the pair of perpendicular lines PU, PV to be those to the points where the hyperbola meets the Absolute line, that is, to be parallel to the asymptotes, we see that the point through which all the chords UV pass is on the Absolute line. Thus, the lines UV are all perpendicular to the tangent at P; and thus also perpendicular to the parallel tangent at P'.

We see then that, *if a series of parallel chords HK of a rectangular hyperbola be taken, the circles described on HK, as diameter, are a coaxal series, whose common points are those two points P, P' of the hyperbola whereat the tangents are perpendicular to the parallel chords.* By a remark made earlier, the limiting points, of this coaxal series of circles, are the points Q, Q' of the hyperbola which lie on the diameter perpendicular to the diameter PP'. These points Q, Q' are then the common points of the associated coaxal series of circles which have, as diameters, the chords of the hyperbola perpendicular to the tangents at Q, Q'.

We may remark another proof of the same result: let the tangent at a point P, of the rectangular hyperbola, be perpendicular to a chord

HK of the curve. Consider the four points consisting of *H*, *K*, and the point *P* regarded as the coalescence of two points of the curve. These four points are common to two rectangular hyperbolas, namely the given hyperbola, and the degenerate rectangular hyperbola which consists of the line *HK* and the tangent at *P* of the given curve. All conics through these four points are then rectangular hyperbolas. In particular, the two lines *PH*, *PK* are such a conic; these are therefore perpendicular to one another. In fact, any conic through *P*, *H*, *K* which touches the given hyperbola at *P* is a rectangular hyperbola. Regarding then *P* as a fixed point of the curve, we may take for *HK* any one of the appropriate series of parallel chords; and all the circles having such a chord *HK*, for diameter, pass through *P*.

(8) We have already used, for a parabola, a process known to Apollonius, for drawing lines normal to the curve, through an arbitrary point *O* (Chap. xi (8)). A very slight change is necessary to obtain the corresponding result *for any conic,* whose centre is *C*. Let *CP*, *CV* be a variable pair of conjugate lines, through *C*, the points *P*, *V* being on the conic. These pairs of conjugate diameters belong to a pencil in involution, whose focal rays are the asymptotes of the conic. Draw, from the fixed point *O*, the line perpendicular to *CP*; and let this meet *CV* in *T*. As the tangent at *V* is parallel to *CP*, the line *OT* is perpendicular to the tangent at *V*. The pencil *O* (*T*), formed by the lines *OT*, is therefore related to the pencil of respectively perpendicular lines *C* (*P*); and this, because *CP*, *CV* belong to a pencil in involution, is related to the pencil *C* (*V*), or *C* (*T*). The locus of *T* is thus a conic containing *O* and *C*. The tangent of this conic, at *O*, is the line through *O* which arises when *CV* passes through *O*; it is thus the line, through *O*, perpendicular to the polar of *O* in regard to the given conic. The tangent of the locus of *T*, at the point *C*, is the line through *C* which is conjugate to the diameter perpendicular to *CO*. Further, there is one pair of conjugate lines *CP*, *CV*, of the pencil in involution, which consists of perpendicular lines, namely the lines called the *axes of the conic.* The locus of *T* thus contains the two points where these axes meet the Absolute line; these points are harmonic conjugates in regard to the Absolute points. The *locus of T is thus a rectangular hyperbola.* We may call it the *Apollonius hyperbola of O in regard to the original conic.*

Evidently, a point, *N*, of the original conic at which the normal is the line *NO*, is a position for the point *T*; this then coincides with *V* in the description above, the conjugate diameter *CP* being parallel to the tangent at *N*. Conversely, any one of the four points in which the Apollonius hyperbola meets the original conic is a point whereat the normal passes through *O*. *Four normals of the original conic can therefore be drawn, in general, through an arbitrary point O.* In the case of the parabola, one of these normals is parallel to the axis, *SA'*.

Ex. 1. Let *S* be a focus of the original conic, of which *C* is the centre. The directrix corresponding to *S*, namely the polar of *S* in regard to the conic, is parallel to one of the axes of the conic, and meets the Absolute line in a point whose polar line is the other axis, SC. Thus the Apollonius hyperbola meets the directrix in the point where this meets the Absolute line. We can prove that *its second intersection with the directrix lies on the line OS*. For, let OS meet this directrix in H. The polar of H is the line, l, through S, which is parallel to the diameter CP of the conic that is conjugate to CH. This line, l, is therefore conjugate to SH, in regard to the conic, and is therefore perpendicular to SH; for it is the harmonic conjugate of SH in regard to the tangents, SI, SJ, of the conic. Whence, H is the point where the perpendicular drawn from O to CP meets the conjugate diameter CH. Thus H is a point of the Apollonius hyperbola.

Incidentally we notice that, if two conjugate diameters of a conic meet a directrix in H and H', the corresponding focus is the orthocentre of the triangle CHH'.

Ex. 2.   If O be on the given conic, the tangent at O, of the Apollonius hyperbola, is the normal of the conic at O; this being the perpendicular from O to the polar of O in regard to the conic. If the pole, in regard to the given conic, of the chord of the conic which is the normal at O, be Q, the diameter CQ is conjugate to that diameter of the conic which is parallel to the normal. Thus Q *lies on the Apollonius hyperbola of O.*

Ex. 3.   If O be on one of the axes of the given conic, the *Apollonius hyperbola of O degenerates into two lines.* For, this axis then contains three points of the hyperbola, namely the points C, O, and the point where the axis meets the Absolute line. The hyperbola thus breaks up into this axis, and another line, which, as the hyperbola is rectangular, is perpendicular to this axis. By Ex. 1, this other line meets a directrix of the conic, which is parallel to the axis on which O lies, on the line joining O to the focus corresponding to this directrix.

Ex. 4.   From the theorem of (7), *the two points, say H, K, where the Apollonius hyperbola of a general point O, in regard to a conic, meets the polar of O, in regard to the conic, are such that OH, OK are perpendicular*; and also, any conic through O, H, K which touches the Apollonius hyperbola at O is a rectangular hyperbola. This follows because the Apollonius curve is a rectangular hyperbola touching, at O, the perpendicular from O to the polar line HK.

Ex. 5.   *If the normals at four points of a conic meet in O, the other points of these normals which lie on the conic, lie on a rectangular hyperbola, which touches the Apollonius hyperbola of O at the point O, its other two intersections with this hyperbola being on the polar of O in regard to the original conic.*

For, denote this polar line by l; and take the harmonic image of the Apollonius hyperbola, and of the original conic, with O as centre, and l as axis. The original conic is then its own image; the image of the Apollonius hyperbola is another conic meeting the Apollonius hyperbola on the line l, and touching it at O. This new conic is therefore also a rectangular hyperbola (Ex. 4). That the harmonic image of a conic is another conic is most briefly obvious because its points can be given by the same parametric number as that which gives the points of the original conic, while two of its points lie on an arbitrary line.

Ex. 6.   *The centre of a rectangular hyperbola, which passes through points A, B, C, whose orthocentre is H, and meets the circumcircle ABC in the further point D, is the mid point of D, H.*

For, we have proved that the hyperbola contains H; let H' be the other extremity of the diameter of the hyperbola through H. Then, by (6) above, $\epsilon H'\,(B, C) = \epsilon H\,(C, B)$. But, as HC, AB are perpendicular, and HB, AC are perpendicular, $\epsilon H\,(C, B) = \epsilon A\,(B, C)$. Because, then, $\epsilon H'(B, C) = \epsilon A\,(B, C)$, it follows that H' is on the circumcircle ABC.

Ex. 7.   *Let the normals at four points P, Q, R, D of a conic (C), whose centre* *is C, meet in a point. Let D' be the other point of (C) such that C is the mid point of D, D'. Then, the four points A, B, C, D' lie on a circle.*

We prove, in two ways, a more general result, including this. The first is an immediate use of Ex. 6. Namely, *let two points, L, M, of a conic (R), be conjugate in regard to a conic (C); let K be the pole of the line LM in regard to the conic (C); let P, Q, R, D be the common points of the two conics; and let the line KD meet the conic (C) again in D'. Then the lines PQ, RD' meet LM in two points which are harmonic in regard to L and M.* For, at first regard L, M as the Absolute points. Then the conic (C) is a rectangular hyperbola, and the conic (R) is a circle meeting this in P, Q, R, D. The point K is the centre of the hyperbola, and D' is the other extremity of the diameter of this which contains D. Thus, by Ex. 6, the point D' is the orthocentre of the triangle PQR. Thus PQ, RD' meet the present Absolute line LM in two points which are harmonic in regard to L, M. This proves the general result enunciated.

Another proof involves more detail: Let LM meet the conic (C) in X, Y; and let LM meet the common chords PQ, RD, of the two conics, respectively in U and V, and meet RD' in O. Then, because DD' contains the pole K of the chord XY, in regard to (C), the points D, D', as points of the range of points of the conic (C), are harmonic conjugates in regard to X and Y. Whence, considering the pencil whose centre is R, we have $O, V \sim X, Y$. Through the four points P, Q, R, D there pass the conics (R) and (C) and also the degenerate line-pair PQ, RD; thus the pairs, L, M; X, Y and U, V, on the line LM, are in involution; and $L, M \sim X, Y$ is also true. Wherefore, by an easily verified property of involutions, the points $(L, M)/U$ and $(X, Y)/V$ are the same point. The latter is O, which is thus $(L, M)/U$. As O is on RD', and U is on PQ, the result enunciated is proved; namely, PQ and RD' meet LM harmonically in regard to L, M.

Now, the points P, Q, R, D' are on the conic (C), which meets LM in two points which are harmonic in regard to L and M; wherefore a conic can be described through P, Q, R, D' to contain any two assigned points of the line LM which are harmonic in regard to L and M.

Let then (C) be any conic, of centre C, meeting the Absolute line in X and Y, and CL, CM be *the axes of this conic.* Let (R) be any rectangular hyperbola containing the points L, M where these axes meet the Absolute line; and let I, J be the Absolute points, which are harmonic in regard to L and M, by the definition of the axes. Finally, let P, Q, R, D be the common points of these conics, and D' the other end of the diameter DC of the conic (C). Then, a circle contains P, Q, R, D'.

This includes the original theorem of this Example (7), the hyperbola (R) being the Apollonius hyperbola, of some point, in regard to the conic (C). But we have had no need to utilise the fact that the Apollonius hyperbola contains the centre C.

Ex. 8. It is clear that, if two points P, Q be taken on the conic (C), the Apollonius hyperbola of that point, O, which is common to the normals of (C) at P and Q, determines the other two points, R and D, whereat the normals pass through O. And the Apollonius hyperbola may be defined as the conic which contains the angular points of the fixed self-polar triangle CLM, and also the points P, Q. That P, Q, R, D' lie on a circle does not determine R, D when P, Q are given. This may be taken to correspond to the fact that the proof, that P, Q, R, D' lie on a circle, does not need the assumption that the Apollonius hyperbola contains the centre C. A geometrical construction for the points R, D, whereat the normals of the conic (C) pass through the common point of the normals at two given points P, Q, is as follows: Let CL, CM be the axes of (C), the points L, M being on the Absolute line. Consider the conic C, L, M, P, Q, or (R). Then RD is the common chord of the conics $(C), (R)$ which is complementary to the chord

PQ. We may construct the line *RD* by taking, on each side of the triangle *CLM*, the involution, thereon, determined by the two conics (*C*), (*R*), and then finding, on this side, the point, conjugate, in this involution, to the point where *PQ* meets this side. The three points so found are on the line *RD*. See also Ex. 10, following.

Ex. 9. Let four points of a rectangular hyperbola be such that the normals of this hyperbola at these points meet in a point. Prove that each of the four points is the orthocentre of the other three. But this condition is not sufficient to ensure that the normals at these points meet in a point; it must be supplemented by the condition of the involution on one of the two axes *CL*, *CM*, of the hyperbola, as explained in Ex. 8.

Ex. 10. If the chord *PQ*, of a conic whose centre is *C*, meets the axis *CL* in *T*; and the polar of *T* meets the axis in *N*, while *N'* is the point of this axis for which *C* is the mid point of *N'*, *N*, then the chord *RD* of the conic (for which the normals at *P*, *Q*, *R*, *D* meet in a point) contains the point *N'*. *Using this for both axes, we construct the line RD.*

Ex. 11. *Let A, B, C, P be four points of a rectangular hyperbola. Then the pedal circle of P, in regard to ABC, contains the centre of the hyperbola.* For we have seen (Chap. x (5), Ex. 26) that, if the lines, joining *P* to the Absolute points of the plane, are met by a variable conic, which passes through *A*, *B*, *C*, *P*, in the respective points *M* and *N*, and the line *PL*, drawn perpendicular to *MN*, meets *MN* in *L*, then the pedal circle in question is the locus of *L*. We have proved, however, in (1), that, if the conic taken be the given rectangular hyperbola, the points *M*, *N* are the extremities of that diameter of the hyperbola which is perpendicular to *PK*, where *K* is the centre of the hyperbola. Thus *K* is a possible position for the point *L*.

As a rectangular hyperbola can be drawn through four arbitrary points, it follows that, if four arbitrary points be taken, the pedal circles of any one of these, in regard to the other three, are four circles which meet in a point. Through this point there pass also, by (5), the nine-point circles of the four sets of three points contained in the four points. The eight circles coincide when any one of the four points is the orthocentre of the other three.

When the four points *A*, *B*, *C*, *P* lie on a circle, the pedal lines, of any one in regard to the other three, meet in a point. This point, by Ex. 6 above, is the mid point of any one of the four given points, and the orthocentre of the other three.

Ex. 12. *The director circle of a rectangular hyperbola is degenerate, consisting of the lines joining its centre to the Absolute points.* For, if *CIJ* be a self-polar triangle in regard to a conic, and *P* be a point such that *PI*, *PJ* are conjugate lines in regard to the conic, it follows, because the pole of *PI* is the intersection of *CJ* with the polar of *P*, that the polar of *P* meets *PJ* on *CJ*. Thus, unless *P* is on *CJ*, the polar of *P* contains *J*, and then *P* lies on *CI*.

Also, *if PQR be a self-polar triangle in regard to the rectangular hyperbola, the circumcircle of PQR contains the centre C.* For, if *PQR* and *P'Q'R'* be any two self-polar triangles in regard to any conic, the pencils *P* (*Q*, *R*, *Q'*, *R'*) and *P'* (*R*, *Q*, *R'*, *Q'*) are such that corresponding rays are conjugate lines in regard to the conic, so that the former pencil is related to the latter, and hence related to the pencil *P'* (*Q*, *R*, *Q'*, *R'*). This shews that *P*, *Q*, *R*, *P'*, *Q'*, *R'* lie on a conic. In particular, then, as *C*, *I*, *J* is a self-polar triangle in regard to the hyperbola, the circle *P*, *Q*, *R*, *I*, *J* contains the centre *C*.

The theorem that *P*, *Q*, *R*, *C* lie on a circle, when the conic is a rectangular hyperbola, is a particular case of a previous result (Chap. x (5) Ex. 15), that, for any conic, the circumcircle of a self-polar triangle is perpendicular to the director circle of the conic.

We have shewn above, in (6(*b*)), that, if *T* be any point on any circle through the centre *C* of the rectangular hyperbola, the polar of *T* meets the circle in two points, U, V, which are conjugate to one another in regard to the hyperbola; so that TUV is a self-polar triangle in regard thereto. Thus, *the necessary and sufficient condition for a circle to be such that triangles can be inscribed therein, which are self-polar in regard to the rectangular hyperbola, is that the circle contains the centre of the rectangular hyperbola. Of such a triangle, one angular point is an arbitrary point of the circle.*

Ex. 13. The vertex triangle of the quadrangle formed by four points, in general position, is known to be self-polar in regard to any conic through the four points. It is, therefore, self-polar in regard to the rectangular hyperbola which can be drawn through the four points; and the circumcircle of this triangle therefore contains the centre of the hyperbola. This centre is, however, the pole of the Absolute line, in regard to the hyperbola, and so lies on the eleven-point conic which (Chap. ix (23), Ex. 15) is the locus of the centres of all conics through the four points of the quadrangle. This eleven-point conic contains the angular points of the vertex triangle. Thus, *if the possible rectangular hyperbola be described through four arbitrary points, the centre of this hyperbola is the fourth point common to the circumcircle of the vertex triangle of the four points and the conic which is the locus of the centres of all conics through the four points.*

In general terms, let X, Y, Z be the vertices of the quadrangle, also I, J the Absolute points, and P_1, P_2 the points where the axes of the two parabolas through the four points meet the Absolute line; then, the circumcircle is the conic X, Y, Z, I, J; the eleven-point locus is the conic X, Y, Z, P_1, P_2; and the rectangular hyperbola is the conic through the four given points which contains the points U, V of the Absolute line for which U, $V \sim I$, J and also U, $V \sim P_1$, P_2. This hyperbola contains, also, the orthocentres of the four triangles formed by taking threes of the four given points. These orthocentres form another quadrangle inscribed in the rectangular hyperbola.

Ex. 14.   *The centroid of the four points, common to a rectangular hyperbola and a circle, is the mid point of the centres of the two curves.*

Let A, B, C, P be these four common points; let O, H be the circumcentre and orthocentre, respectively, of ABC. Also let D, L, Q be the respective mid points of B, C; of A, P and of H, P. Then, as was proved above, Q is the centre of the rectangular hyperbola. What we prove is that, the mid point of O, Q is the mid point of D, L. By symmetry, this will then equally be the mid point of E, M and of F, N, if E, F be the respective mid points of C, A and of A, B; and M, N be the respective mid points of P, B and of P, C. It is this common point of the joins, of the mid points of the three complementary pairs of A, B, C, P, which we call the centroid of these points.

To prove that the mid points of O, Q and of D, L are the same point, it is sufficient, as we easily see, to prove that OD, LQ are parallel lines, and also OL, DQ are parallel. The former is clear because LQ is parallel to AH, which is perpendicular to BC and therefore parallel to OD. To prove that OL, DQ are parallel, remark that L, D are the respective mid points of the chords PA, BC of the hyperbola; thus, Q being the centre of this, the lines QL, AP are parallel to conjugate diameters, as are QD and BC. Thus, because the asymptotes of a rectangular hyperbola are the bisectors of the angles determined by any pair of conjugate diameters, by (5), we have

$$\epsilon O\,(D,\,L) = \epsilon\,(BC,\,PA) = \epsilon Q\,(L,\,D).$$

As, then, OD and QL are parallel, it is true that OL and QD are parallel; and the theorem is established.

We may remark, in regard to the figure, that the centre of the nine-point circle of the triangle ABC is the mid point of O, H, this circle being the pedal circle both of O and H in regard to the triangle. If U be the mid point of A, H, so that DU is a diameter of the nine-point circle, the centre of this circle is also the mid point of D, U. Hence we can prove that the lines OU, DH are parallel.

Ex. 15. Let five points be taken on a circle. Through every four of these a rectangular hyperbola can be described. The five centres of these hyperbolas lie on another circle.

We prove this result simply by use of the symbols. Let U be any point not lying on the Absolute line I, J. We can then express the symbol, P, of any point of the plane which does not lie on the Absolute line, in terms of the symbols of U, I, J, in the form $P = U + \xi I + \eta J$, where ξ, η are numbers. Then the line joining two points, P, P', meets the Absolute line in the point of symbol $P - P'$, and the symbol of the mid point of P, P' is $\frac{1}{2}(P + P')$, the multiplier $\frac{1}{2}$ being introduced so that the symbol of the point may be of the form $U + \lambda I + \mu J$, with 1 as the multiplier of the symbol U; such a form we may speak of as *reduced*. If P, P', P'' be any three points, the point whose reduced symbol is $\frac{1}{3}(P + P' + P'')$ evidently lies on the line joining any one of P, P', P'' to the mid point of the other two. This point is the centroid of P, P', P''. Likewise, for four points, the point whose reduced symbol is $\frac{1}{4}(P + P' + P'' + P''')$ evidently lies on the line joining any one of these points to the centroid of the other three, and is the centroid of the four points. And so on, for a set of any number of points. If these points be divided into two batches, the line joining the centroid of one batch to the centroid of the other batch contains the centroid of the whole set.

With this understanding, the result of Ex. 14 preceding may be expressed by saying that, if A, B, C, P be four points of a circle of centre O, and Q be the centre of the rectangular hyperbola containing these four points, then, in terms of the reduced symbols of O and Q, the reduced symbol of the centroid of A, B, C, P is $\frac{1}{2}(O + Q)$.

Now, take five points, say $A_1, \ldots, A_5$, on the circle. Let Q_5 denote the centre of the rectangular hyperbola which contains $A_1, \ldots, A_4$; and so on. Also, let G be the centroid of the five points. Using reduced symbols for the points, we have the two relations connecting the symbols

$$G = \tfrac{1}{5}(A_1 + \ldots + A_5), \quad \tfrac{1}{4}(A_1 + \ldots + A_4) = \tfrac{1}{2}(O + Q_5).$$

Thus
$$A_5 + 2Q_5 = 3G',$$

where G' is such that
$$2O + 3G' = 5G.$$

Thus G' is the symbol of a definite point on the line OG, and the points A_5, Q_5 lie on a line through this point. Denote now by K the point where the line joining A_5, Q_5 meets the Absolute line, whose symbol is $K = A_5 - Q_5$. From $A_5 + 2Q_5 = 3G'$, we thus have

$$A_5 = G' + \tfrac{2}{3}K, \quad Q_5 = G' - \tfrac{1}{3}K,$$

so that the cross ratio $(Q_5, A_5/G', K)$ is $-\frac{1}{2}$. This shews (Chap. x (6), Ex. 24) that, as A_5 takes various positions on the original circle of centre O, the corresponding positions of Q_5 lie on another circle, the point G' being a centre of similitude of the two circles. And, if a point O' be taken on the definite line OG', with symbol given by $O + 2O' = 3G'$, and L, of symbol $O - O'$, be the point where the line OG' meets the Absolute line, we have,
from

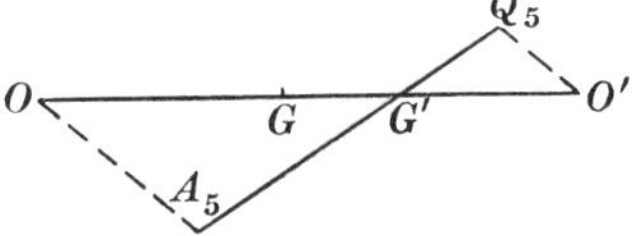

$$O - O' = L, \quad O + 2O' = 3G',$$

the equations
$$O = G' + \tfrac{2}{3}L, \quad O' = G' - \tfrac{1}{3}L,$$

so that the cross ratio $(O', O/G', L)$ is $-\frac{1}{2}$. The line $O'Q_5$ is parallel to OA_5, because $(Q_5, A_5/G', K) = (O', O/G', L)$.

We infer, therefore, that, as the point A describes a circle, of centre O, with $A_1, \ldots, A_5$ as particular positions, there is a corresponding point Q, on another circle (whose centre is at O'), of which $Q_1, \ldots, Q_5$ are particular positions.

The equations are consistent with the symbols of the varying points A, Q, which are expressed (Chap. IX (17)) by

$$A = O + r\,(\theta I + \theta^{-1}J), \quad Q = O' - \tfrac{1}{2}r\,(\theta I + \theta^{-1}J),$$

where r is a fixed number, and θ is a varying numerical parameter. As follows from what was said earlier, these are the appropriate expressions for points on two circles of centres O and O' (Chap. IX (17)); but we have avoided here the use of this fact. These lead to

$$A + 2Q = O + 2O' = 3G',$$
$$A - Q = O - O' + \tfrac{3}{2}r\,(\theta I + \theta^{-1}J), = L + \tfrac{3}{2}r\,(\theta I + \theta^{-1}J).$$

(Cf. also Chapter XV below, on the Theory of Length.)

Ex. 16. In connexion with the preceding example, we may remark further that, *the centre of the conic which is the locus of the centres of all conics through four given points, is the centroid of the four points.*

Let the four given points be A, B, C, D. Let U, V be the respective mid points of A, B and of C, D; and let U', V' be the respective mid points of A, D and of B, C. The locus of the centres of conics through A, B, C, D contains the four points U, V, U', V'. It is a property of the quadrangle that the mid point of U, V coincides with the mid point of U', V', in the point which is the centroid of the four points. The pole of the point which is the mid point of U, V, taken in regard to the locus of centres, is on the Absolute line; and similarly for U', V'. Thus the common point of the lines UV, $U'V'$ is the centre of the locus of centres; and it is also the centroid of the four points.

In the particular case when A, B, C, D lie on a circle, we have already seen (Ex. 14) that the centroid of the four points is the mid point of the two points constituted by the centre of the circle, and the centre of the rectangular hyperbola which passes through the four points. A more particular case, when D is the orthocentre of A, B, C, is that the centre of the nine-point circle of ABC is the centroid of A, B, C and D. The centre of the nine-point circle is also the mid point of D and the circumcentre of ABC; the position of the centroid of ABC, on the line joining the orthocentre and circumcentre of ABC, is thus determined.

Ex. 17. In the immediately preceding examples we have, for simplicity, spoken of the centroid of three or more points. This is constructed, by taking harmonic conjugates, from the Absolute line. This line, however, is arbitrary. We may thus formulate results having a more general expression, starting from an arbitrary line. For instance, instead of the centroid of a triangle of three points, we may speak of the point which we have previously called the pole of the arbitrary line in regard to the triangle. And, for four points, we have the result that, if we take the poles of an arbitrary line in turn in regard to each of the four triangles formed by the threes of four points, then the lines joining these poles each to the remaining point, are four lines which meet in a point. This point we may call the pole of the arbitrary line in regard to the quadrangle formed by the four points. And there is, similarly, a pole of this line in regard to any number of points. And, dually, if any number of lines be given, there is a *polar*, in regard to these, of an arbitrary point of the plane.

APPENDIX TO CHAPTER XII.
FEUERBACH'S THEOREM

We associate with the name of Feuerbach (1822) the theorem that the four circles which touch the sides of a triangle are all touched by the circle which contains the mid points of the sides of the triangle, that is, by the nine-point circle of the triangle. We have already remarked (Chap. x (6), Ex. 32) that this follows because the nine-point circle is the inverse of a common tangent of two of the four circles, the inversion being so taken that these two circles are both unaltered thereby. In this Appendix, we give two other proofs of the theorem. One of these is a direct proof which is very simple, though perhaps a little intricate to describe. The other is in connexion with the theory of a rectangular hyperbola; and establishes a result of which Feuerbach's theorem is a particular case. Further general results in regard to circles in a plane are contained in Note 1.

(1) Let D, E, F be any points, which lie respectively on the sides BC, CA, AB of a triangle, and are such that the lines AD, BE, CF meet in a point. From the converse of Pascal's theorem it can be proved that a conic exists touching BC, CA, AB, respectively, at D, E, F. Denote this conic by Ω. Also, let D', E', F' be other points, respectively on BC, CA, AB, for which AD', BE', CF' meet in a point; and let Ω' denote the conic touching BC, CA, AB, respectively, at D', E', F'. The two conics Ω, Ω' will have another common tangent, whose points of contact with Ω, Ω' we denote respectively by U and U'. Let PQR be the diagonal triangle of the quadrilateral formed by the four common tangents of the two conics, which is self-polar in regard to both conics. We suppose the notation taken so that the pair of complementary intersections of the four common tangents, namely the points (EE', FF'), or A, and (BC, UU'), lie on the line QR; likewise the pair (FF', DD'), or B, and (CA, UU'), lie on RP; and the third pair (DD', EE'), or C, and (AB, UU'), lie on PQ. The lines DU, EF, which are the polars, in regard to the conic Ω, respectively of the points (BC, UU') and A, which are on QR, thus meet in P, as do, similarly, the lines $D'U'$ and $E'F'$. Likewise, the intersections (EU, FD) and $(E'U', F'D')$ coincide at Q, and the intersections (FU, DE) and $(F'U', D'E')$ coincide in R. Thus *P can be defined as the intersection* $(EF, E'F')$; *likewise Q as* $(FD, F'D')$, *and R as* $(DE, D'E')$; *and, then, U is the common point of DP, EQ, FR, while U' is the common point of D'P, E'Q and F'R*. So that, given the points of contact of Ω, Ω' with three of their common tangents, their common self-polar triangle, and the points of contact with their fourth common tangent, are found by linear constructions.

Now consider, in particular, the line $E'F'$. Let this meet the common tangents BC, UU', of the conics Ω, Ω', respectively in X and X', also meet the side QR of the common self-polar triangle in P', and meet the conic Ω in H and K. From the figure, the tangents BC, UU' are harmonic conjugates in regard to the two lines consisting of QR, which contains the point (BC, UU'), and the line joining this point to P; and P lies on $E'F'$. Thus, on this line $E'F'$, we have $X, X' \sim P, P'$. But, also, QR, AP are harmonic conjugates in regard to AB, AC; so that $P, P' \sim E', F'$. Further, as P is the pole of QR, in regard to Ω (and to Ω'), we have $P, P' \sim H, K$. Therefore, on the line $E'F'$ there is an involution, of which P, P' are the foci, in which three pairs are X, X' and H, K and E', F'. This involution may be defined by the two pairs H, K and E', F'.

Introduce then another conic, say ω, defined as containing D', E', F' and touching Ω at U. Consider also the chord common to ω and Ω which is complementary to the tangent at U. These two common chords, and the conics ω, Ω, determine pairs of points of an involution on the line $E'F'$; of this involution H, K, on Ω, and E', F', on ω, are pairs; and the tangent at U meets $E'F'$ in X'. Thus, the complementary chord, spoken of, passes through X, which is the common point of BC and $E'F'$. By parity of reasoning, the common chord of ω, Ω, which is complementary to the common tangent at U, contains the three points $(BC, E'F')$, $(CA, F'D')$ and $(AB, D'E')$. This chord is thus the line sometimes called the polar, in *regard to the triangle ABC,* of the common point of AD', BE' and CF'.

Finally, let the conic Ω, touching BC, CA, AB respectively at D, E, F, be a circle; and let D', E', F' be the respective mid points of B, C; of C, A; and of A, B. The polar line of the point (AD', BE', CF') is then the Absolute line. As this has been shewn to be a common chord of the conics ω, Ω, it follows that the conic ω is a circle, containing the points I, J where Ω meets the Absolute line. As ω contains D', E', F', it is the nine-point circle of the triangle ABC. We have thus proved that this *nine-point circle touches the circle Ω, which is any circle touching BC, CA, AB, at the point U.*

We have seen that P, Q, R may be respectively defined as the points $(EF, E'F')$, $(FD, F'D')$, $(DE, D'E')$; and then the point of contact U as the common point of (DP, EQ, FR). Or, we may obtain U, on the circle Ω, as the point of contact, with Ω, of the fourth common tangent of Ω with the conic Ω', which, in the case considered, touches BC, CA, AB at their respective mid points D', E', F'.

Ex. 1. The point U is obtainable in a third way as follows: *Through the point D, where the circle Ω touches BC, draw the line parallel to the chord of contact EF; let this line meet Ω again in the point U_1. Then U is the second point of Ω lying on the line $D'U_1$.*

In regard to the point U_1, here used, we remark that, if, respectively on AC, AB we take the points $E_1' = (C, A)/E'$ and $F_1' = (A, B)/F'$, then AD', BE_1', CF_1' meet in a point, and $E_1'F_1'$ is the Absolute line; and, also, that U_1 is the position, on the conic Ω, which would be taken by U, if, instead of E', F', we took, respectively, E_1' and F_1'. If we did so, then in place of P, which is $(EF, E'F')$, we should have P_1, defined as $(EF, E_1'F_1')$. Then, as before, U_1 would lie on DP_1; and this is parallel to EF.

To prove that $D'U_1U$ lie on a line, we may reason thus: Consider a point T, which varies on $D'A$. Let the points (BT, AC) and (CT, AB) be, respectively, N and M; and the points (EF, MN), (FD, ND'), (DE, MD') be, respectively, P', Q' and R'. As before, the lines DP', EQ', FR' meet in a point of the conic Ω, say in V; and the conic Ω is determined solely by D, E and F. When T is at the point of $D'A$ which is common to BE' and CF', then V is at U; when T is at the point of $D'A$ which is common to BE_1' and CF_1', then V is at U_1. In general, the range of points V, *on the conic* Ω, is related to the range of points T on the line $D'A$, as follows from the construction of V from T. Now consider the pairs of positions of T on $D'A$ which form the involution of which D' and A are the foci; these give pairs of points on Ω, forming an involution thereon, whose joining line, therefore, passes through a definite point. This point is the pole of the chord of Ω which joins the two points thereon corresponding to the foci D', A. We can shew that, when T is at D', then V, on Ω, is at D; and, when T is at A, the point V is at the point of contact of the second tangent which can be drawn to Ω from D', which we may denote by V_0. Then U_1 and U will be harmonic conjugates, on Ω, in regard to D and V_0, and the line U_1U will pass through D', as we wish to prove. To shew that D and V_0 are the positions of V, as stated, we remark, first, that when T is at D', the line MN coincides with BC, and meets EF in a position of P' which lies on BC; the corresponding position of V is the second intersection of DP' with the conic Ω, namely, is D. Next, when T is at A, both M and N are at A, and both Q' and R' are on $D'A$; this line is thus the polar of P', in regard to Ω. Whence, the polar of D' contains P', and is the line DP'. The position of V is the second intersection of DP' with Ω, and is thus the point V_0, the point of contact with Ω of the second tangent from D'. The result stated is thus wholly proved. (Cf. Lachlan, *Modern Geometry*, 1893, p. 74.)

Ex. 2.   If the tangent of the circle Ω, at the point U_1, meets BC in L, prove that A and L are the centres of similitudes of Ω and another of the circles touching BC, CA, AB. And, that the tangent U_1L is the inverse of the nine-point circle in regard to the circle of centre D' which contains D (cf. Chap. x (6), Ex. 32). Remark also that the tangent of Ω at U meets BC in $(D, D')/L$.

Ex. 3.   Prove that the common tangents, other than BC, CA, AB, of the four circles which touch BC, CA, AB, consist of three pairs of parallel lines. And that the lines of one pair are parallel to the tangent at A of the circumcircle ABC, and meet BC in points which are harmonic conjugates in regard to B and C.

Ex. 4.   On the circle Ω, the point of contact U, and the Absolute points I, J, form a triad which is *apolar to* the triad D, E, F. (Cf. *Principles of Geometry*, Vol. ii, pp. 59, 117; also Gibbins, *Math. Gazette*, xix (1935), p. 34.)

Ex. 5.   A proof of Feuerbach's theorem may be given involving the *lengths* of common tangents of circles involved. See p. 366 below.

(2) The theorem that the nine-point circle is touched by the *four* circles which touch the sides of the triangle can be generalised. We shew how to define an infinite series of circles (including the four spoken of), all of which touch the nine-point circle; the points of contact are all the points of this circle, taken in order.

We have explained what is meant by a pair of isogonal points in regard to a triangle; and proved that, if one of these describes a line, the other describes a conic, which passes through the angular points of the triangle. The conic meets the line in the only pair of points on the line which are isogonal to one another (Chap. x (5), Ex. 25). In particular, if the line described by one of the isogonal points contains the circumcentre of the triangle, then the conic described by the other point passes through the point isogonal to the circumcentre, namely, the orthocentre of the triangle. The conic is then a rectangular hyperbola. The two common points of the line and hyperbola, being isogonal to one another, have then the same pedal circle in regard to the triangle. *Every diameter of the circumcircle thus gives rise to a rectangular hyperbola, and a pedal circle.* We prove that *this circle touches the nine-point circle, the point of contact being the centre of the rectangular hyperbola.* There are four diameters upon each of which the two isogonal points are coincident, the corresponding pedal circles being the four which touch the sides of the triangle.

Denote the circumcentre and the orthocentre of the triangle ABC, respectively, by O and H; and the centre of the nine-point circle, which is the mid point of O, H, by N. For a diameter of the circumcircle, we denote by U, V the isogonal points in which this is met by the isogonally corresponding rectangular hyperbola, this hyperbola meeting the circumcircle in the point D besides A, B, C. The centre of the hyperbola, K, which lies on the nine-point circle, is the mid point of H, D (Chap. XII (8), Ex. 6). The centre of the pedal circle of U, or V, is the mid point, M, of U, V, lying on the diameter considered. We have proved that this pedal circle passes through the centre, K, of the rectangular hyperbola (Chap. XII (8), Ex. 11). If then we prove that MK contains the centre, N, of the nine-point circle, the pedal circle will touch the nine-point circle, at K, which is what we desire to see. As N, K are the mid points respectively of H, O and of H, D, and NK is parallel to OD, we shall prove that MK contains N if we prove that MK is parallel to OD.

Now, the diameter KM of the hyperbola contains the mid point of the points U, V lying on the curve. Thus KM and UV are parallel to conjugate diameters. We have thus to prove that the rectangular hyperbola, corresponding to the diameter OM considered, meets the circumcircle in a point D such that OM, OD are parallel to conjugate diameters of the hyperbola; which is achieved if we prove that the bisectors of the angles determined by these lines OM, OD are parallel to the asymptotes of the hyperbola. Let the diameter OM meet the circumcircle in X' and Y'. Then the bisectors of the angles determined by OM, OD are parallel to DX', DY', from the properties of a circle. We have thus only to prove that DX', DY' *are the lines joining D to the points where the hyperbola meets the Absolute line.*

Notice now that *the conic isogonally corresponding to the Absolute line, in regard to the triangle ABC, is the circumcircle ABC, containing the Absolute points I, J.* These points I, J are isogonal points, in regard to the triangle ABC, because they are conjugate to one another in regard to the three degenerate conics which consist of the two bisectors of the angles determined either by AB, AC, or by BC, BA, or by CA, CB. The conic isogonal to the Absolute line is then the conic through

A, B, C, I, J. From this fact it follows that the isogonal of the point X', common to the line OM and the circumcircle, is a point, X, where the hyperbola, isogonal to OM, meets the Absolute line, isogonal to the circumcircle. Similarly, the isogonal of the other point, Y', where OM meets the circumcircle, is the other point, Y, where the hyperbola meets the Absolute line. Also, the isogonal of the point D, common to the hyperbola and the circumcircle, is the point, D', where OM meets the Absolute line. It follows then, from the relation of isogonal points to the bisectors of the angles determined by AB, AC, that

$$\epsilon A\,(D',\,X) = \epsilon A\,(X',\,D);$$

and this, because Y' is on the circumcircle, is equal to $\epsilon Y'\,(X',\,D)$. Now AD' is parallel to $Y'X'$; thus AX is parallel to $Y'D$; and X is on the Absolute line. Thus DX is parallel to DY', or the lines DX, DY' are the same. So the lines DY, DX' are the same. And this is the result we wished to obtain. If we assume that X, Y are respectively isogonal to X', Y', we can state the proof in more summary form, as in Ex. 2, below.

We have thus shewn: *Let any rectangular hyperbola through A, B, C meet the circumcircle ABC again in D. Let the lines through D, parallel to the asymptotes of the hyperbola, meet the circumcircle again, at the ends of a diameter, in X' and Y'. Then the points U, V, in which the hyperbola meets the diameter $X'Y'$, have the same pedal circle in regard to the triangle ABC; and this pedal circle touches the nine-point circle, of the triangle ABC, at the centre of the hyperbola.* In particular, there are four positions of D for which the points U, V coincide; then the pedal circle becomes a circle touching the sides of the triangle. (Fontené, 1905.)

It may in fact be proved that, if we take different diameters $X'Y'$, the locus of the points U, V is a cubic curve, which contains A, B, C and the circumcentre O. From a point O of such a curve, four tangents of the curve can be drawn. The points of contact, in this case, are the centres of the circles which touch the sides BC, CA, AB. Thus also, from what is known of such a curve, the cubic curve touches, at O, the conic containing O and the four centres of the circles which touch BC, CA, AB.

Ex. 1. The pedal circle, in regard to ABC, of any point of the diameter $X'Y'$, being equally the pedal circle of some point of the hyperbola, passes through the centre, K, of the hyperbola; this lies in the nine-point circle, and is defined by the diameter alone. In particular, the pedal lines of the two points X', Y' meet in K (and are perpendicular to one another).

Ex. 2. Let the diameter $X'Y'$ meet the sides BC, CA, AB respectively in A', B', C'; these are respectively isogonal to the points A, B, C. The range X', Y', A', B', C', D', O, on the diameter, is thus related to the range X, Y, A, B, C, D, H, on the hyperbola. In particular, the pencil $D\,(A, B, C, X)$ is related to the range (A', B', C', X'). It is, however, a property of any conic, that, if a chord $X'Y'$ meet the sides BC, CA, AB of a triangle inscribed in the conic, respectively, in A', B', C', then the range (A', B', C', X') is related to the pencil $D\,(A, B, C, Y')$, subtended at any point D of the conic. If the conic be the circumcircle, it follows that $D\,(A,\,B,\,C,\,Y')$ is related to $D\,(A,\,B,\,C,\,X)$. Thus $X,\,D,\,Y'$ are in line; as proved above.

Ex. 3. Another series of circles which all touch the nine-point circle consists of the director circles of conics which touch the sides of the triangle and pass through the circumcentre. Each of these circles also touches the circumcircle. To prove this we may first shew that a circle described, with centre on a conic, to touch the director circle, is triangularly circumscribed to the conic.

Ex. 4. The line which joins the point of contact of the nine-point circle, and the inscribed circle (centre I), of a triangle ABC, to the mid point of A and I, meets the inscribed circle again on the diameter of this circle which is perpendicular to the side BC.

CHAPTER XIII

SOME ELEMENTARY PROPERTIES OF CONICS

(1) We repeat the definitions of the more frequent points and lines associated with a conic to which names are usually given.

The *centre* of the conic is the pole of the Absolute line. When the conic touches the Absolute line, it is called a *parabola*; the centre is then the point of contact. Unless said, we suppose here that the conic is not a parabola. The *axes* of the conic are the two lines through the centre which are both conjugate to one another in regard to the conic and perpendicular to one another. The *asymptotes* of the conic are the lines joining the centre to the two points where the conic meets the Absolute line. They are the tangents to the conic drawn from the centre, and are the focal rays of the pencil in involution which is formed by the pairs of conjugate lines drawn through the centre. Any line which passes through the centre of the conic is called a *diameter*.

It is assumed that the conic, like a real line, contains an infinite series of real points. Also, it is generally assumed (as explained above, Chap. x) that the Absolute points are conjugate imaginary points. Thus the conic may meet the Absolute line in two real points; in this case it is called a *hyperbola*; otherwise the conic meets the Absolute line in two *conjugate imaginary* points, in which case it is called an *ellipse*. In either case the axes are real lines (Chap. VII (3)). Thus also, any pair of conjugate diameters, being harmonic conjugates in regard to the asymptotes, are either both real or both imaginary. But, in regard to the intersections of these diameters with the curve, there is a difference between the cases of the hyperbola and ellipse. In the case of the ellipse, both of two real conjugate diameters meet the curve in two real points. In the case of the hyperbola, if one of these diameters meets the curve in two real points, the conjugate diameter meets the curve in two points which are conjugate imaginaries of one another. This statement may be proved, if we remark that the centre, taken with the two points where a pair of conjugate diameters meet the Absolute line, form a self-polar triangle in regard to the conic, by proving the more general statement that, if *three real lines be the sides of a self-polar triangle of a real conic, just one of these lines meets the conic in conjugate imaginary points*. This, in turn, may be proved by remarking that, if the pairs of the points of the conic which lie on the sides of this triangle be, respectively, P, P'; Q, Q' and R, R', then, in the range of points on the conic, any one of these pairs consists of points which are harmonic conjugates in regard to both the other two pairs, say $P, P' \sim Q, Q'$ and $P, P' \sim R, R'$. When three pairs of points on a real line have this relation, one pair always consists of conjugate imaginary points (cf. the chapter on involutions, preceding, Chap. VII (3)). The same is therefore true for the conic.

The pairs of tangents drawn to the conic, still supposed to contain an infinite series of real points, from the Absolute points, form a circumscribed quadrilateral of the conic; the diagonals of this consist of the Absolute line and the axes of the conic. On each of the axes there is then a pair of complementary intersections of the tangents, which are a pair of complementary *foci* of the conic. The polar lines of the foci, in regard to the conic, are the *directrices* of the conic. Let the tangents to the conic from one Absolute point touch the conic in U, V, so that UV is the polar of this Absolute point; and let the tangents to the conic from the other Absolute point touch the curve in U', V'. The lines UV, $U'V'$ thus meet in the pole of the Absolute line, namely in the centre of the conic. We assume that these lines are conjugate imaginaries. Thus U is conjugate imaginary to one of U', V', say to U'; likewise, then, V is conjugate imaginary to V'. Wherefore, both the lines UU' and VV' are real; and the foci of the conic which are the poles of these lines are both real. The other pair of complementary foci are conjugate imaginaries. The notions underlying these statements, as that the line joining two conjugate imaginary points is a real line, may be made explicit by use of the symbols; it is clear that, if the symbols, in terms of three real points A, B, C, of two conjugate imaginary points be $P = xA + yB + zC$ and $P = x'A + y'B + z'C$, where x', y', z' are numbers respectively conjugate imaginary to the numbers x, y, z, then two conjugate imaginary numbers λ, λ' give the symbol $\lambda P + \lambda' P'$ of a real point of the line joining P and P'. And so on.

In the case of a hyperbola, one of the axes meets the curve in real points, and the other axis meets the curve in conjugate imaginary points; this is a particular case of a statement made above. From the quadrilateral, of the four tangents drawn from the Absolute points, we see that, on either axis, there is an involution, wherein one pair consists of the centre and the point where the axis meets the Absolute line, and another pair consists of a focus and the point where the corresponding directrix meets this axis. The foci of this involution are the points where this axis meets the curve. Hence we may prove that the two real foci of the hyperbola are on that axis which meets the curve in real points. Further, the centre is the mid point of the two foci of the curve which lie on either axis, and of the two points where this axis meets the curve; and, the directrices corresponding to the foci on either axis are perpendicular to this axis. This last statement is also true for an ellipse, which is met in two real points by both axes. And, in both cases, the imaginary foci of the curve on one axis are the *antipoints* of the real foci on the other axis, the lines joining any focus, to both the foci of the other pair, passing through the Absolute points.

The *director* circle of the conic, which may be regarded as the locus of a point from which the tangents, drawn to the curve, are perpendicular, passes through the four points U, V, U', V', spoken of above, which are points of the curve lying on the four directrices (two perpendicular directrices meeting in every one of these points). Indeed, a point of the conic may be regarded as the common point of two coincident lines which both touch the curve at this point, and two coincident lines, both passing through one of the Absolute points, satisfy the condition of being perpendicular to one another. But the director circle is the conjugate line locus (Chap. ix (23), Ex. 18) for the conic, with centres at the Absolute points; and this gives the more natural proof that the director circle contains the points U, V, U', V'.

The *auxiliary* circle of the conic, is the circle, with centre at the centre of the conic, which contains the so-called *vertices*, A, A', of the conic, where it is met by the axis which contains the real foci, S, S'. As the directrix which corresponds to S meets the axis AA' in a point X such that $X = (A, A')/S$, we see, considering the tangents of the auxiliary circle at the two points, U_1, U_1', where the auxiliary circle meets the directrix, that these tangents meet in S; or, that S and the directrix are pole and polar, both in regard to the conic and in regard to the auxiliary circle. The tangents of the auxiliary circle at U_1 and U_1' are perpendicular to the lines joining the centre C to U_1 and U_1', respectively; it can be shewn that the joining lines, CU_1, CU_1', are the asymptotes of the conic. In fact, the line drawn through a focus of a conic, to be perpendicular to an asymptote, meets the asymptote on the directrix which corresponds to the focus. This is a particular application of the result previously referred to that, if the tangent at any point P, of a conic, meets the directrix corresponding to a focus S, in the point Z, then the lines SZ, SP, being conjugate to one another, and, therefore, harmonic in regard to the tangents from S, are perpendicular, which is obtained by taking P on the Absolute line. Hence also, the directrices corresponding to the real foci of a hyperbola meet the auxiliary circle in real points; but, for an ellipse, the directrices meet the auxiliary circle in imaginary points. In particular, a rectangular hyperbola meets a directrix in points which lie on the lines joining the centre to the Absolute points; and, for this curve, the point X, where a directrix, which is perpendicular to the axis of real foci, meets this axis, is the mid point of the centre C and a focus S.

(2) We define now, for a conic with two real complementary foci, S and S', a circle, with centre at one of these foci, which we call *the subdirector circle* with that centre. It has been noticed above that the director circle of the conic meets the directrix which corresponds to the focus S' in points which lie on the conic. The subdirector circle of centre S also passes through these two points. The subdirector circle has S' for one of the limiting points of itself and the director circle; also S' is a centre of similitude for the subdirector circle and the auxiliary circle.

Let I, J be the Absolute points. Let the tangent at a variable point P of the conic meet $S'I$, $S'J$ in L, M, respectively. Let IM, JL meet in Q; and let $S'Q$ meet the tangent at P in R.

The ranges on the tangents $S'I$, $S'J$, determined by the varying tangent at P, are related to the range described by P on the conic, and to one another. Thus the pencils $J(Q)$, $I(Q)$, with centres at J and I, respectively, described by the lines JL and IM, are related. This shews that Q describes a conic, passing through I and J, namely, a circle. When L is at I, then P is the point of contact of IS

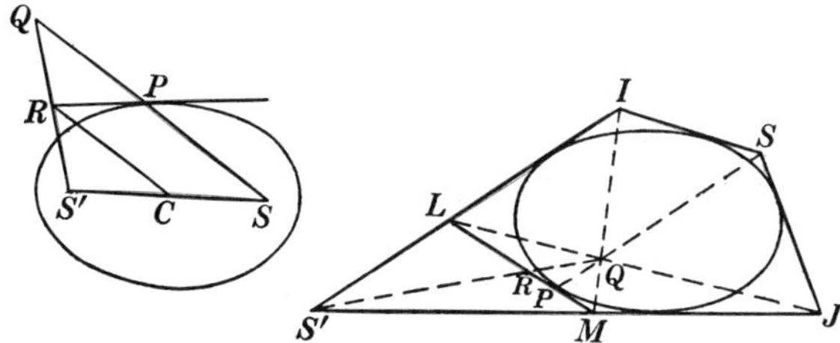

with the conic, and M is the point where IS meets $S'J$. Thus the ray JI, of the pencil $J(Q)$, corresponds to the ray IS of the pencil $I(Q)$; whence, IS is the tangent at the point I, of the circle which is the locus of Q. In the same way JS is the tangent at J. Thus S is the centre of the circle. Considering the quadrangle $S'MQL$, of which the vertex triangle is RJI, the point R is the harmonic conjugate, in regard to S' and Q, of the point where $S'Q$ meets the Absolute line; so that R is the mid point of S', Q. Again, RS', RP, or QS', LM, are harmonic conjugates in regard to RI and RJ; thus $S'R$ is perpendicular to the tangent at P. Further, considering the five tangents of the conic, SI, SJ, LI, MJ, LM, Brianchon's theorem shews that SQ contains P. Wherefore, the point Q may be defined as the common point of SP with the perpendicular, $S'R$, drawn from S' to the tangent at P. Equally, as the point on the line $S'R$ for which R is the mid point of S' and Q. Incidentally, also, we see that the tangent at P is one of the bisectors of the angles determined by the lines PS, PS'. This, however, is a particular case of the fact that, if P be any point of the plane, the three pairs of lines PS, PS'; PI, PJ; and the tangents from P to the conic, form three pairs of a pencil in involution.

It is the locus of Q which we call *the subdirector circle of centre S*; there is another such circle of centre S'.

If the line $S'Q$ meet the Absolute line in K, it is true that $(R, Q/S', K) = \frac{1}{2}$. For, using the symbols of the points, because R, K are harmonic conjugates in regard to S' and Q, we can put, for the symbols, $R = Q + \lambda S'$ and $K = Q - \lambda S'$, where λ is a number; hence $Q = K + \lambda S'$, which gives $R = K + 2\lambda S'$. Therefore (see Chap. x (6), Ex. 24), the locus of R is a circle, and S' is a centre of similitude of this circle and the subdirector circle. The centre of the circle, which is the locus of R, is that position of R which is given by the relation $(R, Q/S', K) = \frac{1}{2}$ when Q is taken at the centre S of the subdirector circle (see the reference given). Whence, *the perpendicular drawn from a focus of the conic to the tangent at a variable point of the conic meets the tangent in a point whose locus is a circle, with centre at the centre of the conic.* By taking the tangents at the points, A, A', of the conic which are on the axis SS', we see that this circle, the locus of R, is the auxiliary circle, as defined above.

That the subdirector circle passes through the points where the given conic meets the directrix, which corresponds to S', is clear by taking the variable point P of the conic, in turn, at the points of contact with the conic of the tangents $S'I$ and $S'J$. We have already remarked that the director circle of the given conic passes through these two points of contact; as the joins of S', to these two common points of the two circles, pass through I and J, it follows that S' is a limiting point of the two circles.

Ex. 1.   The subdirector circle is so related to the given conic that an infinite series of triangles can be inscribed in the circle whose sides touch the given conic. And all these triangles have the same nine-point circle, which is the auxiliary circle of the given conic.

Let the tangent at the point P, of the conic, meet the subdirector circle in V and W, and the line QRS', referred to above, meet this circle again in U. We can then prove that S' is the orthocentre of the triangle UVW. For we have $\epsilon U\,(Q,\,W) = \epsilon V\,(Q,\,W)$; and this is equal to $\epsilon V\,(W,\,S')$, because QS' is perpendicular to VW, and R is the mid point of $Q,\,S'$. Hence, if VS' meet WU in R', we have

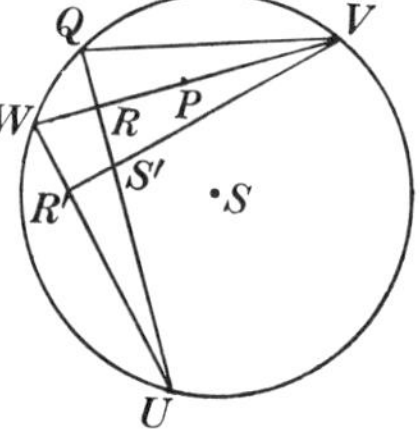

$$\epsilon U\,(R,\,R') = \epsilon V\,(R,\,R').$$

Thus a circle contains $VRR'U$; and VR' is perpendicular to UW. This shews that S' is the orthocentre of the triangle UVW. Of this triangle, also, S is the circumcentre. Whence (Chap. x (5), Ex. 19) a conic exists, with S and S' as foci, touching the sides VW, WU, UV; and this, as having $S,\,S'$ as foci and touching VW, is the original conic. Thus, the triangle UVW is circumscribed to the original conic; and an infinite series of such triangles is possible, by taking different tangents VW of the original conic; all these are inscribed in the subdirector circle. The nine-point circle, of the triangle UVW, has its centre at the mid point of the orthocentre and circumcentre of this triangle, that is, the mid point of $S,\,S'$, that is, the centre C of the original conic; and it passes through R and R'. It is therefore the auxiliary circle of the original conic.

It has been remarked (Chap. xi (12), Ex. 10) that a sufficient condition, for a conic to be triangularly inscribed to a circle, is, that the tangent of the conic, at a common point of the two curves, should meet the circle again at the point of contact, with this, of a common tangent of the two curves. This condition is satisfied here; for, a common point of the two curves is the point where $S'J$ touches the conic; and this tangent, $S'J$, contains the point J, which is the point of contact with the circle of the common tangent SJ.

We have proved that the self-polar circle of a triangle has its centre at the orthocentre of the triangle, and passes through the common points of the circumcircle and nine-point circle; the radical axis of the circles is the polar line, in regard to the triangle, of the orthocentre. Thus all the triangles UVW are self-polar in regard to this same circle (see Chap. x (6), Ex. 16). We have also proved (Chap. x (6), Ex. 15) that the circumcircle of a triangle, which is self-polar in regard to a conic, is perpendicular to the director circle; it will be shewn below to follow from this that the director circles of all conics which touch the sides of a triangle are perpendicular to the self-polar circle of the triangle (Chap. xiv (16)). The circle, spoken of above, with centre at S', in regard to which all the triangles UVW are self-polar, may, therefore, be defined, as the circle, of centre S', which is perpendicular to the director circle of the given conic. The reader may prove, further, that the other tangent of the given conic which is parallel to UV, contains the mid point of S' and U.

Ex. 2.   On the directrix of the original conic which corresponds to the focus S', take a point T; prove that the polar of T, in regard to the subdirector circle whose centre is the other focus, S, meets the directrix on the polar line of T in regard to the conic; this is a focal chord, $PS'Q$, passing through S', meeting the conic, say, in P and Q. Also that the polar line of T, in regard to the circle, meets this in two points lying, respectively, on SP and SQ.

Ex. 3.   We can construct the original conic when the subdirector circle, whose centre is S, is given, by the following theorem: Let a circle of centre S, and also a line, be given. Let S' be one of the limiting points of the line and circle (regarded as one of a coaxal series of circles of which the line is the radical axis). Let the tangent at a variable point U of the circle meet the line in T; and the line from T perpendicular to $S'U$ meet the radius SU in P. Then, the locus of P is a conic, whereof S, S' are foci, and the given line is the directrix corresponding to S'.

(3) We have proved in (2) that, if a perpendicular be drawn from a focus of a conic to meet a variable tangent, the point of meeting describes a circle, the auxiliary circle of the conic. We give a more direct proof of this. The figure forms a natural introduction to the theory of the important envelope which is called *the harmonic envelope* of two conics, considered in (4).

Let I, J be the Absolute points; S and S' a complementary pair of real foci of a conic; and TP, TP' be the tangents from any point T, touching the conic in P and P'. It is a fundamental theorem that the pairs of rays TI, TJ; TS, TS'; TP, TP' are in involution. From this it follows that $\epsilon T\,(P, S) = \epsilon T\,(S', P')$. Wherefore, if BC, CA, AB be three tangents of the conic, we have $\epsilon B\,(A, S') = \epsilon B\,(S, C)$; also $\epsilon C\,(B, S') = \epsilon C\,(S, A)$; and $\epsilon A\,(C, S) = \epsilon A\,(S', B)$, of which the two former involve the last. Thus, if S, S' be two points for which the two former hold, it follows that the conic with foci at S and S' which touches BC, equally touches AB and AC. Remark, also, that if OA, OB be two perpendicular lines, and OA', OB' be two other perpendicular lines through the same point O, then

$$\epsilon O\,(A, A') = \epsilon O\,(B, B').$$

Now suppose that A, A' are the ends of a diameter of a circle, whose centre is C; also that S, S' are two points on this diameter, such that C is the mid point of S and S'. Let P be any point of this circle; and let PQ, drawn perpendicular to PS, meet the circle again in Q. It is then easy to see that QP is perpendicular to QS'. Let PQ meet the tangents of the circle, at A and A', respectively in X and X'. With appropriate notation, it may then be seen that (SX', AQ), $(SX, A'Q)$, $(S'X, A'P)$, $(S'X', AP)$ are four pairs of parallel lines.

For, because QX, QS' are perpendicular, and AX, AS' are perpendicular, a circle contains X, Q, S', A, so that

$$\epsilon X\,(Q,\,S') = \epsilon A\,(Q,\,S') = \epsilon A\,(Q,\,A') = \epsilon P\,(Q,\,A'),$$

which shews that XS', PA' are parallel. And so on.

Next, with fixed S, S', allow P, Q to vary on the circle. Then the pencils $A'(P)$, $A(P)$ are related; therefore the pencils $S'(X)$, $S'(X')$ are related. Whence, the line PQ, giving related ranges on the lines AX, $A'X'$, touches a conic of which these are tangents. By considering the triangle formed by AX, $A'X'$ and XX', and using

$$\epsilon X\,(S,\,A) = \epsilon P\,(S,\,A) = \epsilon X\,(Q,\,S'),$$

with $\qquad\qquad \epsilon X'\,(A',\,S') = \epsilon Q\,(A',\,S') = \epsilon X'\,(S,\,P),$

we see, from what was said above, that a conic exists, with S, S' as complementary foci, which touches AX, $A'X'$ and PQ. This conic is identified by its two foci and one of the fixed lines XA, $X'A'$. Wherefore, the variable line PQ envelopes this conic. And we can prove that it touches XA, $X'A'$ at A and A', respectively.

Conversely, if the conic be given, with vertices at A, A' and foci at S, S', the lines drawn from S, S', perpendicular to a tangent of the conic, meet the tangent in points, P, Q, which lie on the circle of which AA' is a diameter. This is the auxiliary circle of the conic.

(4) *Suppose two circles to be given, with centres at O, Q, respectively. Let a line meet the former of these circles in A, A', and meet the other circle in B, B', these points being so related that A, A' are harmonic conjugates in regard to B, B'; or, as we write, such that A, $A' \sim B$, B'. We prove that the line is a tangent of a definite conic.* As the tangent, of one of the circles, at a point common to the circles, is met by the two circles in harmonically conjugate pairs of points, the conic, if existing, touches the tangents of the two circles drawn at their common points; in particular, the conic touches the tangents of the circles at the Absolute points; it will thus have the centres, O, Q, of the two circles, for a pair of complementary foci.

Let P be any point of the line $ABA'B'$; then we may easily prove that the two points $(A, A')/P$ and $(B, B')/P$ belong to the involution, on this line, which is determined by the two pairs of points, A, A' and B, B'. If then M be the mid point of A and A', and N be the mid point of B and B', the points M, N are a pair of this involution.

Hence, a circle can be drawn, coaxal with the two given circles, to pass through M and N. The centre of this circle lies on the line of centres, OQ, of the two given circles; this centre also lies on the line drawn, perpendicular to MN, through the mid point of M and N. As OM, QN are both perpendicular to the line $AMA'BNB'$, the centre of this circle is at the mid point of O, Q. Therefore, having a fixed centre, and being coaxal with the given circles, this third circle is the same whatever be the line $ABA'B'$ originally taken. Wherefore, by (3) preceding, this line MN is the tangent of a definite conic, with

O, Q as foci; and the auxiliary circle of this conic is the circle, coaxal with the two given circles, whose centre is the mid point of O, Q. This proves the theorem stated.

(5) It is thus true, for any two conics whatever, *that a variable line, which meets the conics in harmonically conjugate pairs of points, envelopes another conic.* This is called the *harmonic envelope* of the two conics. Another proof of the theorem, not using the particular phraseology here employed, is given later (Chap. xiv (6)). The dual theorem is therefore true: *A variable point, from which the pairs of tangents, to two given conics, are harmonically conjugate to one another, describes another conic.* And, as dual of what was said above, this conic passes through the eight points of contact of the four common tangents of the two conics, with these. This conic is called the *harmonic locus* of the two conics. If one of the two given conics, regarded as an envelope, be degenerate, and consist of the lines of two pencils, with centres at X and Y, the theorem becomes: *If a conic, and two points X, Y be given, the locus of a point P such that PX, PY are harmonic conjugates in regard to the tangents from P to the given conic, or, what is the same, such that PX, PY are conjugate lines in regard to the given conic, is a conic.* This is the conjugate line locus of the given conic, with centres at X and Y, previously considered (Chap. ix (23), Ex. 18).

(6) When neither of the two given conics is degenerate, but they are related to one another in a suitable way, the harmonic envelope, and the harmonic locus, are degenerate. And it is the case that, when one of these is degenerate, so is the other. *The necessary and sufficient condition, for the harmonic envelope and the harmonic locus, of two conics, to be degenerate, is, that the conics should meet one side of their common self-polar triangle in two harmonically conjugate pairs of points. In this case, the dual condition is also satisfied; that the pairs of tangents to these conics, from the opposite angular point of this triangle, form two harmonically conjugate pairs of rays.* When this is so, the eight tangents of the two conics, at their four common points, meet, *in fours*, in two points, lying on the side of the self-polar triangle referred to; the harmonic envelope then degenerates into the two pencils having these two points as centres. And, also, the eight points of contact with the conics, of their four common tangents, lie, *in fours*, on two lines, both passing through the angular point of the self-polar triangle referred to; the harmonic locus then degenerates into these two lines.

The matter is referred to again below, with another treatment of the harmonic locus (Chap. xiv (7)). But, all the facts are really contained in the statements that, for two circles which are perpendicular, the harmonic envelope consists of the two pencils of lines whose centres are the centres of the circles; and the harmonic locus consists of the two lines which are the polars of the two centres of similitude, taken in regard to the circles, each of these polar lines arising twice over. A single sufficient condition for two circles to be perpendicular is the existence of a line, passing through the centre of one of the circles, but not through an Absolute point, whose points of meeting with the two circles consist of two harmonically conjugate pairs of points (Chap. x (5), Ex. 29; etc.). When the circles meet in real points, the harmonic locus consists of two real lines.

Ex. Let the line drawn from a point P of an ellipse, whose centre is C, to be perpendicular to the axis of real foci SS', meet the auxiliary circle in Q and Q'. Let the normal to the ellipse, at P, meet the diameters CQ, CQ' in R and R' respectively. Prove that P is the mid point of R, R'; also that, as P describes the ellipse, the points R, R' describe circles whose centres are at C. A hyperbola can be drawn through P having the same (four) foci as the ellipse. Prove that CR, CR' are the asymptotes of this hyperbola. Prove also that R, R' are conjugate to one another in regard to all conics passing through the four foci.

(7) The theorem of the auxiliary circle, proved in (2) and (3), is a particular case of another theorem, of which we give a proof that illustrates the fundamental methods of this volume. The theorem is connected with a certain theorem for any circle, as will appear in (8).

Let P be a variable point of a conic, of which S is a focus. Let a line be drawn from S, to meet the tangent at P in the point Q, such that $\epsilon Q\,(P, S)$ is constant. The theorem is that the locus of Q is a circle. This degenerates when the given conic is a parabola; we suppose this is not so.

The tangents to the conic, from S, pass through the Absolute points, I, J. Let the tangent at P meet the Absolute line in T, and meet the tangents SI, SJ in X and Y, respectively. Let SQ meet the Absolute

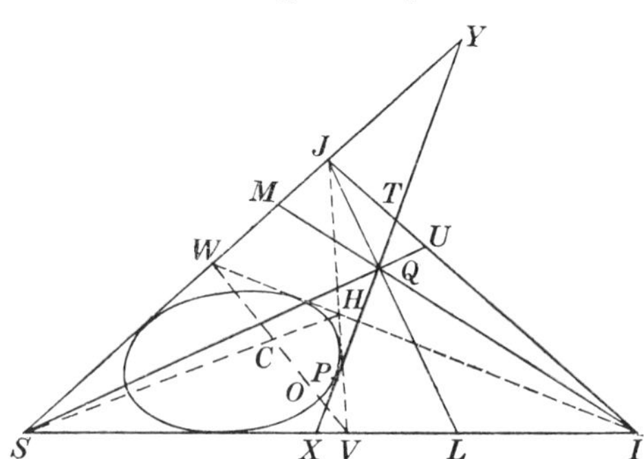

line in U; and JQ, IQ meet the tangents SI, SJ in L and M, respectively. Let the tangents to the conic from I, J, other than SI, SJ, meet in H; and JH, IH meet SI, SJ in V and W, respectively.

We are given that, with a constant number, m, the cross ratio $(T, U/I, J)$ is equal to m. By projection from Q, on to SI and SJ, we thence infer that $(X, S/I, L) = (Y, S/M, J) = m$. From these (Chap. VIII (1)) we have

$$(X, L/S, I) = (m-1)/m, \quad (Y, M/S, J) = 1-m.$$

Now use the symbols of the points, putting

$$X = S + \xi I, \quad L = S + \lambda I, \quad Y = S + \eta J, \quad M = S + \mu J$$

where ξ, λ, η, μ are numbers. We have then $\xi/\lambda=(m-1)/m$, and $\eta/\mu=1-m$. We shall denote the inverses of these, respectively $m/(m-1)$ and $1/(1-m)$, by e and f; so that $e+f=1$.

As P varies on the conic, the points X, Y, on the tangents SI, SJ, respectively, describe ranges related to one another, and to the range of points P on the conic. Thus, as P varies, and consequently the numbers ξ, η vary, there exists an equation $\eta=(a\xi+b)/(c\xi+d)$, where a, b, c, d are numbers whose ratios are fixed. Replacing ξ, η by λ/e and μ/f, this gives $\mu=f(a\lambda+be)/(c\lambda+de)$, in which a, be, c, de and f are fixed, but μ, λ vary together. This shews that, as Q varies, the ranges described by M and L are related. Hence the pencils $I\,(Q)$ and $J\,(Q)$ are related; which proves *that Q describes a conic passing through I and J*; and this is the proposition to be proved. When the original conic is a parabola, the locus of Q breaks up into the line IJ, and another line tangent to the parabola (Chap. XI (3)).

We now examine various particular properties of the locus of Q:

(*a*) When L is at I, or $\lambda=\infty$, then $\mu=fa/c$, and M is the point whose symbol is $cS+faJ$. This gives the ray of the pencil $I\,(Q)$ corresponding to the ray JI of the pencil $J\,(Q)$, namely, the tangent at I of the locus of Q. Similarly, when M is at J, or $\mu=\infty$, then $\lambda=-ed/c$, and L is the point whose symbol is $cS-edI$; this gives the tangent at J of the locus of Q. It follows that the centre of the circle which is the locus of Q, is the point O of symbol

$$O=cS-edI+faJ.$$

(*b*) When Y is at J, or $\eta=\infty$, then $\xi=-d/c$, and X is the point whose symbol is $cS-dI$; this is the point V. Likewise, when X is at I, or $\xi=\infty$, then $\eta=a/c$, so that Y is then the point whose symbol is $cS+aJ$; this is the point W. Thus the point H, where JV meets IW, which is the focus of the original conic that is complementary to S, is of symbol

$$H=cS-dI+aJ.$$

(*c*) The line VW contains the point whose symbol is $eV+fW$, namely, $e\,(cS-dI)+f\,(cS+aJ)$; because $e+f=1$, this is $cS-edI+faJ$. Thus the centre O of the circle which is the locus of Q, lies on the line VW.

(*d*) The centre, C, of the original conic, being the point common to VW and SH, is then seen to be the point whose symbol is $2cS-dI+aJ$; this is both $cS+H$ and $V+W$.

(*e*) When SQ is perpendicular to the tangent PQ, or $m=-1$, we have $e=f=\frac{1}{2}$. Then the centres of the original conic, and the circle locus of Q, coincide; in the point whose symbol is $V+W$.

Ex. 1.   The centre of the circle locus of Q being O, as above, let the line drawn through S, to be perpendicular to SO, meet VW in D. Shew that the symbol of D is $-eV+fW$. Hence $D=(V,\,W)/O$. Prove also that the circle touches the original conic at the two points of contact of the tangents from D.

The point O lies on the axis of the original conic which is perpendicular to SH; its position may be defined by $\epsilon O\,(C,\,S)=m=\epsilon Q\,(P,\,S)$. If N be the point where the perpendicular from S meets the tangent at the variable point P, prove that

$$\epsilon S\,(C,\,N)=\epsilon S\,(O,\,Q).$$

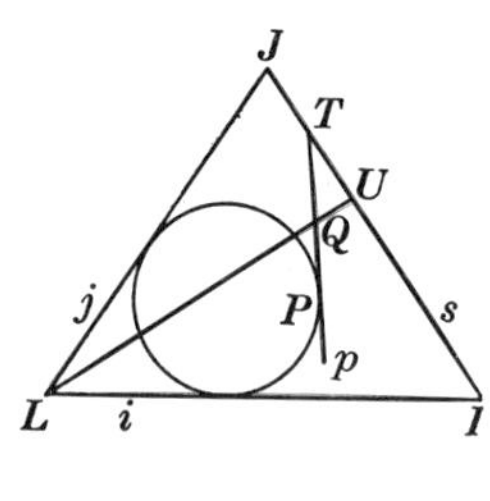

And infer, from the fact that N describes a circle of centre C, the fact that Q describes a circle of centre O. A point of contact, Q_0, of the circle and conic, is such that $\epsilon Q_0\,(D,\,S)=m$.

Ex. 2. A particular case of the theorem for the locus of Q is that, if t, t' be two parallel tangents of the original conic, and u, u' be another pair of parallel tangents, and a line LSL', through a focus S, meet t, t', respectively, in L and L', another line MSM' meet u, u' in M, M', respectively, and a third line NSN' meet u, u' in N, N', respectively, in such a way that

$$\epsilon\,(t,\,LS)=\epsilon\,(u,\,MS)=\epsilon\,(t',\,L'S)=\epsilon\,(u',\,M'S),$$

and

$$\epsilon\,(t,\,LS)=\epsilon\,(NS,\,u)=\epsilon\,(t',\,L'S)=\epsilon\,(N'S,\,u'),$$

then L, M, L', M' lie on a circle, and L, N, L', N' lie on another circle.

(8) We consider now the converse of the theorem of (7). *Let P be a variable point of a fixed circle, of centre O, and S an arbitrary fixed point. Let a line, q, be drawn from P so that $\epsilon\,(q,\,PS)$ is constant, $=m$, say.*

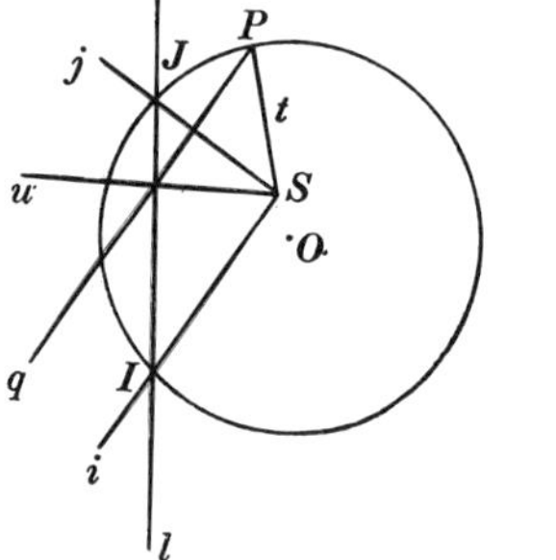

Then q envelopes a conic of which S is one focus. We have already considered the case when $m=-1$ ((3), above). The centre, C, of this conic is such that $\epsilon O\,(C,\,S)=m$. That this is the converse of (7) is clear. But this converse is also a dual of the original theorem; the original conic and a variable tangent of this are dual of a circle and a variable point of this.

In detail, we may say: In the original theorem, the fixed tangents, i, j, of a given conic, meet in a point L, and meet a fixed line s respectively in the points I, J. A variable tangent, p, of this conic, meets the line s in the point T. Then the point U is taken on the line s, such that $(T,\,U/I,\,J)$ is constant. The theorem is that the point Q, common to the tangent p and the line LU, describes a conic passing through I and J. In the present theorem, two fixed points I, J, of a given circle, are joined by a line l; and i, j are the lines joining I, J, respectively, to a fixed point S. Then, the line joining S to a variable point P, of the circle, is t. Another line u is drawn through S, such that $(t, u/i, j)$ is constant. The theorem is that the line q, which joins P to the common point of l and u, envelopes a conic which touches the lines i and j.

Remark. There is another point of view from which the preceding results may be considered. And, in particular, this leads to another proof of the theorem of the auxiliary circle.

Let two conics, which we denote by ω and ψ, touch one another at two points, X and Z, the common pole of XZ being Y. Let the tangent, at a variable point, P, of the conic ω, meet the conic ψ in Q and Q'. It can then be shewn that the ranges (Q), (Q'), on ψ, are related; and the common corresponding points of these ranges are at X and Z. Thus the cross ratio, on the conic ψ, expressed by $(Q, Q'/X, Z)$, is constant as P varies on ω.

This is most readily proved, perhaps, by use of the symbols. In terms of the symbols of X, Z, Y, we can suppose the symbol of P to be $\theta X + \theta^{-1}Z + k^{-1}Y$ (Chap. ix (17)), where k is a fixed number, and θ is a variable number; and, at the same time, the symbols of Q, Q', on the other conic ψ, to be respectively $\phi X + \phi^{-1}Z + Y$ and $\phi_1 X + \phi_1^{-1}Z + Y$, where ϕ, ϕ_1 are variable numbers. Then, expressing that the two points of the line QQ' which lie on the conic ω coalesce with one another, we find $\phi/\phi_1 = (k+h)^2$, where $h = (k^2 - 1)^{\frac{1}{2}}$. This proves that Q, Q' describe related ranges on the conic ψ. From this, if an arbitrary fixed point S be taken on the line XZ, and QS meet the conic ψ, again, in R, it follows that RQ' meets XZ in a point C, which is also fixed. This point C is given by $(S, C/X, Z) = (Q, Q'/X, Z)$. An essentially equivalent theorem is that, if two circles, ω, ψ, have a common centre, and a variable tangent of ω meet ψ in Q and Q', then, for any point, O, of ψ, the number $\epsilon O\,(Q, Q')$ is constant as Q, Q' vary.

If, in particular, ψ be the auxiliary circle of the conic ω, and S be any focus of ω, it can be shewn, by taking a particular position of P, that C is the common centre of ω and ψ; so that QS, QP are perpendicular. For, let P be taken at the point of contact of any tangent drawn from S to the conic ω; this tangent passes through an Absolute point, I, which lies on the auxiliary circle ψ. Then R, Q' coincide in I; and the line RQ' is the tangent of ψ at I, and passes through the centre of the circle.

(9) The following results may also be given. Let S be a focus of a conic, and X be the point of the corresponding directrix lying on the perpendicular drawn from S. Let the tangent, at a point P of the conic, meet this directrix in K, so that SK, SP are perpendicular. Let PM, perpendicular to the directrix, meet this in M; and SN, parallel to the directrix, meet PM in N. Also, let ST, drawn from S per-

pendicular to the tangent at P, meet this tangent in T. Denote the Absolute points by I, J. Then

(a) The points X, T, N are in line.

(b) The four pairs of lines SP, SX; SN, SK; ST, SM; SI, SJ are pairs of a pencil in involution.

(c) The bisectors of the angles determined by the lines TS, TX contain the vertices, A, A', of the conic.

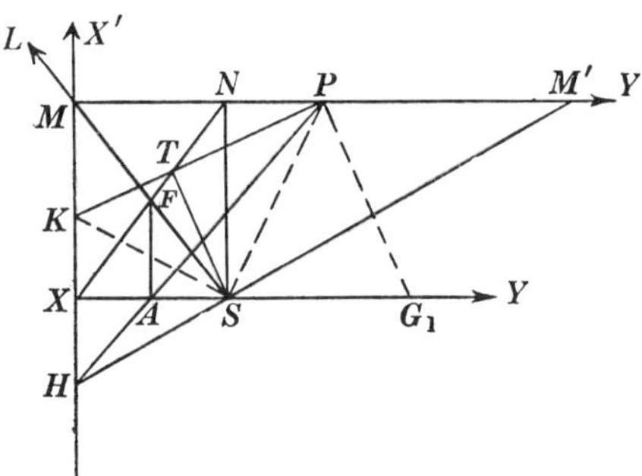

(d) The line, CP, joining the centre C of the conic to P, and the conjugate diameter of this, meet the directrix in points V and V', such that S is the orthocentre of the triangle CVV'.

(e) If the tangents at the vertices A, A' of the conic meet SM respectively in F and F', then PF, PF' are the bisectors of the angles determined by the lines PS, PM.

(f) Let the tangents of the conic at two points P, Q meet in O. Let CO meet the directrix in V; let the chord PQ meet the axis SX in W. Then VW is parallel to SO. Again, let the chord PQ meet the directrix in Z; and let the line through S, parallel to PQ, meet CZ in U. Then OU is parallel to the directrix. Thus Z is the orthocentre of the triangle SVW. Also, let the line from O, which is perpendicular to PQ, meet the axis AA' in G. Then U is the orthocentre of the triangle GOS.

To prove (a): By the definition of T, a circle contains S, P, N, T, and a circle contains S, X, K, T. Wherefore, $\epsilon T\,(P, N) = \epsilon S\,(P, N)$. Because SP, SK are perpendicular, and SN, SX are perpendicular, the latter is $\epsilon S\,(K, X)$, which is equal to $\epsilon T\,(K, X)$. Hence, as TP, TK are the same line, so are TN, TX the same line.

For (b): As a circle contains S, P, M, K, and SX is parallel to PM, we have
$$\epsilon S\,(X, M) = \epsilon M\,(P, S) = \epsilon K\,(P, S) = \epsilon S\,(T, P),$$
which is
$$S\,(X, M/I, J) = S\,(T, P/I, J) = S\,(P, T/J, I).$$
This shews that SP, SX; ST, SM; SI, SJ are three pairs of a pencil in involution. But, SK, SN are, respectively, harmonic conjugates of SP, SX in regard to SI, SJ. Thus SK, SN are another pair of the pencil in involution.

The focal rays of the involution are the common bisectors of the angles determined by the pairs of lines ST, SM; SN, SK and SP, SX.

The bisectors for the last pair are the lines joining S to the point H, where PA meets the directrix, and to the point, where the tangent at P meets the tangent at A.

If the normal of the curve at P meet the axis of the conic in G_1, it is part of the theorem that $\epsilon P\,(S,\,G_1)$, or $\epsilon S\,(P,\,T)$, $=\epsilon M\,(S,\,P)$.

For (c): From the theorem of the auxiliary circle, T lies on the circle of which $A,\,A'$ is a diameter. Thus $TA,\,TA'$ are perpendicular. Also $X=(A,\,A')/S$, so that $TA,\,TA'$ are harmonic conjugates in regard to $TS,\,TX$. Wherefore $TA,\,TA'$ are the bisectors spoken of, for these are harmonic conjugates in regard to TI and TJ.

For (d): Noticing that CV' is parallel to the tangent PT, it is sufficient to prove that ST and CP meet on the directrix. Now, conjugate lines through S are perpendicular, because they are harmonic conjugates in regard to the tangents $SI,\,SJ$. Thus, the pole of the line drawn through S parallel to the tangent at P, lying on ST and the directrix, is the point where ST meets the directrix. And the line joining C to this pole is the line CP.

For (e): Let MP meet HS in M', and meet the Absolute line in Y. Let SM, and the directrix, respectively, meet the Absolute line in L and X'. We have then

$$(M',\,M/P,\,Y)=H\,(S,\,X/A,\,Y)=X'\,(S,\,X/A,\,Y)=(S,\,M/F,\,L).$$

Thus the two related ranges $M',\,M,\,P,\,Y$ and $S,\,M,\,F,\,L$ have the point M in common. Hence $M'S$ and PF meet on the Absolute YL, and are parallel. We know, however, that SH, or $M'S$, is one of the bisectors of the angles determined by $SP,\,SA$. Whence, PF is one of the bisectors of the angles determined by PS and PM.

For (f): It is sufficient to shew that SO is parallel to WV, and OU is parallel to the directrix. We assume, from what has been said, that

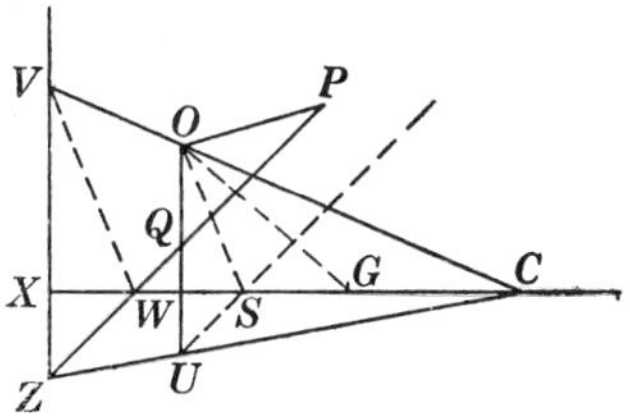

SV is perpendicular to PQ, and SO is perpendicular to the conjugate line SZ. And we assume that, if the polars of any three points, $A,\,B,\,C$, form a triangle $A'B'C'$, then the lines AA',BB',CC' meet in a point (Chap. xiv (1)). Since the line $ZWQP$ is perpendicular to SV, and VZ is perpendicular to SW, it follows that SZ is perpendicular to VW, which is there-

fore parallel to OS. Again, if we join $S,\,C,\,O$ to the respective poles of $CO,\,OS,\,SC$, we prove that OU is parallel to the directrix.

Ex. 1. If the line through X, which is parallel to SM, meet SP in R, prove that R lies on the circle PKX. (Here, as before, M is the foot of the perpendicular from P to the directrix, and K is the point of the directrix lying on the tangent at P.) Prove, further, that the locus of R, as P varies, is a circle with centre at S, which passes through the points where SN meets the conic. The line through S, parallel to the directrix, is called the *latus rectum* of the conic.

Ex. 2. On the tangents drawn to the conic from any point T, point

P, Q are respectively taken, of arbitrary positions. The other tangents from P, Q are drawn; and these meet in T'. Prove that $\epsilon S\,(T,\,P) = \epsilon S\,(Q,\,T')$.

Ex. 3. Let P, P' be any two points of the conic, such that PP' contains the focus S. Let the normals at P, P' meet in O. From O a line is drawn parallel to the axis of the conic to meet PP' in E; and also a line is drawn perpendicular to PP', from O, to meet PP' in N. Prove that E is the mid point of PP', and is equally the mid point of NS. Prove, further, that, for different chords PSP', the locus of N is a conic, having the same centre, C, and the same asymptotes as the given conic, the vertices of this conic being at S and at the complementary focus S'.

To prove this, let the tangents at P, P' meet in U, on the directrix; also, let the perpendiculars from P, P' to the directrix meet this in M, M', respectively; and the perpendiculars from S to PU, $P'U$ meet these, respectively, in T and T'; also, PP' meet the directrix in Y; also, as before, let SX, perpendicular to the directrix, contain the vertex A. As in the involution in (*b*) above, SM, SM' are paired respectively with ST, ST', and $M, M' \sim X, Y$, we see that ST, ST' are harmonic conjugates in regard to SX, SY. Taking the intersections of these four lines with the Absolute line, and the pencil with centre O which contains these four intersections, we infer that OP, OP' are harmonic conjugates in regard to OE and the line from O parallel to PP'. This shews that E is the mid point of P, P'. Since OU is a diameter of the circle $OPUP'$, it follows (Chap. x) that E is also the mid point of S, N, the line SU being parallel to ON. If AL, parallel to PP', meet the conic again in L, it follows that CE contains the mid point of A, L, and that CL contains N.

But, if a point, L, on a conic, be joined to the vertices A, A', and SN, parallel to AL, meet CL in N, it follows, if we consider the respective intersections, A_0, L_0, A_0, of CA, CL, CA', with the Absolute line, that $(S,\,A_0/C,\,A) = (N,\,L_0/C,\,L)$; and hence, if S' be the focus complementary to S, so that $(S',\,A_0/C,\,A') = (S,\,A_0/C,\,A)$, it follows that NS' is parallel to LA'. Hence, noticing that when L coincides with L_0, so does N, the result enunciated is established.

(10) The system of conics which touch the tangents drawn to a conic from the Absolute points I, J is called a system of *confocal* conics. The diagonals, other than IJ, of the quadrilateral formed by the tangents, contain each a pair of complementary foci, say S, S' and H, H'. If SS' and HH', which we suppose to be real, meet the Absolute line in X and Y, we suppose, as before, that the symbols of I, J are of the forms $X \pm iY$.

The pairs of tangents drawn to the conics of the confocal system, from an arbitrary point, P, form a pencil in involution, in which PI, PJ; PS, PS'; and PH, PH' are pairs; and there are two conics of the confocal system passing through P, of which the respective tangents at P are the focal rays of the involution. These tangents are perpendicular; they meet the axis SS' in two points which are harmonic conjugates in regard to S and S'; and they are the bisectors of the angles determined by the lines PS, PS'.

We can shew that one of the conics, of the confocal system, which pass through P, is an ellipse, and the other is a hyperbola. For, let S, S' be real, the axes SS', HH' meeting the Absolute line, as stated, respectively in points whose symbols are X and Y. Also, let the tangents at P meet the Absolute line in the points of respective symbols $X + mY$, $X + m'Y$, where m, m' are numbers. As these points are harmonic conjugates in regard to the Absolute points, whose symbols we have supposed to be $X + iY$ and $X - iY$, we prove at once that $mm' = -1$. Let the line CP, joining the centre C to the point P, meet the Absolute line in the point of symbol $X + pY$. The asymptotes of a conic are harmonic in regard to the axes, and in regard to the conjugate diameters which consist of CP and the line through the centre parallel to the tangent at P. The points of the Absolute line which are harmonic conjugates, in regard to the points whose symbols are X, Y and $X + pY$, $X + mY$, are those whose symbols are $X \pm (pm)^{\frac{1}{2}} Y$. Therefore, the asymptotes of one of the two conics through P, belonging to the confocal system, meet the Absolute line in the points whose symbols are $X \pm (pm)^{\frac{1}{2}} Y$, while the asymptotes of the other conic meet the Absolute line in the points whose symbols are $X \pm (-p/m)^{\frac{1}{2}} Y$. One of these pairs of points is real, and the other consists of conjugate imaginary points. The former conic is therefore a hyperbola, and the latter is an ellipse.

The two conics of the confocal system which pass through P will have three other common points. We have remarked (Chap. IX) that a conic is its own harmonic image in regard to the point and line constituted by an angular point of a self-polar triangle and the opposite side; and the triangle CXY is self-polar in regard to all the confocal conics. Wherefore, the other three common points of the two confocals through P are obtained by taking harmonic images in regard to CX, CY; and one of these three points is the point P' such that C is the mid point of P, P'.

If we consider the pencil of lines with centre at Y, and the poles of these lines in regard to the confocal hyperbola through P; and add an assumption derived from the topology of the plane, which we have already incidentally used (Chap. VI), namely, that two (real) points, which are harmonically conjugate in regard to two points A_1, A_1', are *separated* by A_1, A_1', unless they coincide at one of these, we can infer, since the pole of YX is at C, that *the tangents of the hyperbola, at all its real points, meet the axis SS' in the segment SS' which contains C*. If we recall, further ((1), above), that the auxiliary circle of the hyperbola, described on the axis A_1A_1' as diameter, meets the directrix corresponding to S in real points, we infer also, from this, that the vertices A_1, A_1' lie in this segment SS' which contains C. For the ellipse which passes through P, *the tangents at real points, including those at the vertices A, A', meet SS' in the segment SS' which does not contain C*.

Ex. If P, Q be the common points of two confocal conics, such that PQ is perpendicular to the axis SS', then PQ meets the asymptotes of either conic in two points which lie on the auxiliary circle of the other conic.

(11) We have seen (Chap. IX) that the polars of a point, in regard to the conics through four points, all have a common point. Dually, the poles of a line, in regard to the conics touching four lines, lie on a line.

There will be one of the system of these latter conics which touches the given line; the locus of the poles of this line passes through the point of contact. If the system of conics with four given tangents be a confocal system, and the given line meet the Absolute line in K, the line which is the locus of poles of the given line will pass through the point $(I, J)/K$, and will thus be perpendicular to the given line. Thus, *the locus of the poles of a given line, in regard to a system of confocal conics, is the normal, at the point of contact, of the conic of the system which touches the given line.* Whence, also, *the tangent and normal, at a point of a conic, are conjugate lines in regard to all conics confocal with the given conic.*

As a particular application we prove the following: *Let the normal at a point, P, of a conic, meet this conic again in P', and meet another conic, confocal with the given conic, in Q and Q'. Then, the normals of the second conic at Q and Q' meet on the tangent at P' of the first conic.* Let the tangents at the points P, P' of the original conic, ω, meet in the point T, and the tangents at the points Q, Q' of the confocal conic, ψ, meet in U. Then U, being the pole of QQ' in regard to ψ, lies on PT, which is the normal at P of that other conic, ω', of the confocal system, which passes through P. From U draw the tangent, other than UP, to the conic ω; and let this meet $P'T$ in N. By a theorem already remarked (Chap. IX (23), Ex 12), lines which are conjugate to another in regard to the conic ω, drawn through U and N, meet on PP'. Thus QN, conjugate to QU, is the normal to the conic ψ at the point Q where QU touches ψ, likewise $Q'N$ is the normal at Q'. This proves the result stated.

A related theorem may be added: *Let the two tangents from a point P, of a conic ω, drawn to touch a confocal conic ψ, meet ω again in R and R'. Then RR' passes through the point, T, which is the pole, in regard to ω, of the normal of this conic at P.* For, the pairs of tangents, drawn from P, to all the conics (ψ) of the system confocal with ω, form a pencil in involution; thus the points (R), (R'), where such a pair of tangents meet the conic ω, form a pair of an involution on this conic; therefore, the chord RR' passes through a point T, independent of the particular conic ψ. In particular, when ψ coincides with ω, the points R, R' coincide at P, and the chord which joins them is the tangent to the conic ω at P; so that T lies on this tangent. When ψ is the other conic, ω', through P, of the confocal system, the points R, R' coincide in the point P' where the normal at P, of the conic ω, meets this conic again; in this case the chord RR' is the tangent of the conic ω at the point P'. Thus T is the pole, in regard to ω, of the chord PP', which is normal to ω at P; as was stated.

The reader may prove, further, that, if the tangent at a point P, of a conic, meet a complementary pair of directrices in H and H', then the pole of the normal at P, of the conic, is the point T of the tangent at P which is given by $(H, H')/P$. Thus T and P are conjugate points, not only in regard to the conic, but also in regard to the degenerate conic formed by this pair of directrices. These points are, therefore, conjugate to one another in regard to every conic passing through the four common points of these two conics. In particular, *T and P are conjugate to one another in regard to the director circle of the conic, which we have shewn* ((1), above) *to contain the common points of the conic with the directrices.* Other properties of the point *T* arise below.

(12) Since each of two confocal conics is the harmonic image of itself in regard to the axes, the four common points of these, being harmonic images in pairs, lie upon a circle, whose centre is the common centre of the conics. *Let the two pairs of tangents be drawn, from any point of this circle, to the two conics. Then, either tangent of one of these pairs is perpendicular to a particular tangent of the other pair.* The circle in question is thus a generalisation of the director circle of a single conic. We may call it the *co-director* circle of the two confocals.

We prove this result by considering a dual figure. Any two conics whatever may be regarded as being two conics of a confocal system, namely, by taking a pair of complementary intersections of their four common tangents as the Absolute points of the plane. Dually, then, we consider any two conics; and with them we consider that conic which touches the four common tangents of these two conics, and also touches a pair of complementary common chords, c, c', of these conics. This further conic then, in the dual figure, represents the circle spoken of for the original two confocals. We are to prove that, if any tangent of this further conic meets the two given conics, respectively, in P, P' and in Q, Q', then Q is the harmonic conjugate of P (or of P'), in regard to the two points where the tangent meets the common chords c, c'. It will then be sufficient to regard the two given conics as circles, the complementary common chords c, c' being the Absolute line, in the new figure, and the radical axis of the two circles. The theorem then becomes: *Given any two circles, there exists a parabola touching the four common tangents of the circles, and touching their radical axis; and, if any tangent of this parabola meets the circles, respectively, in A, A' and in B, B', and meets the radical axis in C, then C is the mid point of A, B, and of A', B'; or, C is the mid point of A, B', and of A', B.*

To prove this, let O, Q be the centres of the two given circles, and S the mid point of O, Q. Let C be any point of the radical axis. Through C draw the line perpendicular to SC, meeting the circles, respectively, in A, A' and in B, B'; let M be the mid point of A, A' and N be the mid point of B, B'; let the line $ABA'B'$ meet the Absolute line in D. By the definition of a parabola, as C varies on the radical axis, this line envelops a parabola, having S for focus, and the radical axis for a tangent. Also, if UV be a common tangent of the two circles, touching them respectively in U and V, and meeting the radical axis in H, then H is the mid point of U, V; and, because OU, QV are perpendicular to UV, and S is the mid point of O, Q, therefore SH is perpendicular to UV. Thus, by taking C to be at the points, in turn, in which the common tangents of the two circles meet the radical axis, we see that the parabola touches the four common tangents of the two circles. And it touches a complementary pair of common chords of these. It is thus the parabola spoken of. To prove that, for any position of C on the radical axis, this point is the mid point of A, B, and of A', B'; or is the mid point of A, B', and of A', B, we require an elementary property of involutions, which has already been enunciated (Chap. VII (4)), but may be reproved here. The pairs A, A'; B, B'; C, D belong to an involution; and the points M, N are given by $M=(A, A')/D$ and $N=(B, B')/D$. As OM, QN are perpendicular to the line $ABA'B'$, and S is the mid point of O, Q, the point C is the mid point of M, N, or $C=(M, N)/D$. Now, expressing the symbols of the pairs A, A', B, B' in terms of the symbols of the pair C, D of the involution, in the forms $A=C+aD$, $A'=C+a'D$, $B=C+bD$, $B'=C+b'D$, where a, a', b, b' are numbers, the property of the involution involves that $aa'=bb', =m$, say. From $M=(A, A')/D$, and $N=(B, B')/D$, we can express the symbols of M, N in the forms

$$M=C+\tfrac{1}{2}(a+a')\,D \text{ and } N=C+\tfrac{1}{2}(b+b')\,D.$$

Whence, from $C=(M, N)/D$, it follows that $a+a'+b+b'=0$. Now this is $a+b+m(a^{-1}+b^{-1})=0$, or $(a+b)(m+ab)=0$, or $(a+b)(a+b')=0$. When $a+b=0$, then C is the mid point of A, B, and of A', B'; when $a+b'=0$, then C is the mid point of A, B', and of A', B.

This proves the theorem for the two circles, and hence also that for the two confocal conics.

Ex. 1. In the theorem given for two circles, besides the six specified tangents for the parabola, there are the two joining S to the Absolute points I, J. The theorem can be enunciated for any two conics, of which one pair of complementary common chords is selected; if one of these be IJ, the points O, Q are the poles of IJ in regard to the conics, and S is the harmonic conjugate, in regard to O and Q, of the point where the line OQ meets IJ. We may equally consider the complementary common chord.

Ex. 2. The general dual theorem may also be enunciated: For any two given conics, there exists a conic passing through the common points of the two conics, and also through a specified pair of complementary intersections of the common tangents of the two conics, say L and M. Four other points of this conic can be described: take the harmonic line of L in regard to the polar lines of L with respect to the two conics; this harmonic line meets the common tangents of the two conics, which meet in L, in two points which lie on this third conic. And two other points of this conic are similarly determined from M. Of the two tangents drawn from an arbitrary point, P, of the third conic, to one of the two original conics, either is the harmonic conjugate, with regard to PL and PM, of a proper one of the two tangents drawn from P to the other conic.

One particular application, which has already arisen (Chap. X (6), Ex. 5), is when L and M are the centres of similitude of two circles. The third conic is then the coaxal *circle of similitude* of the two circles, passing through L and M. The other intersections of this circle with the tangents from L to the two given circles then lie on the harmonic line of L, taken in regard to the two lines which are the lines of contact of the tangents from L to the two circles; and similarly for M.

Ex. 3. If a tangent at P, of one conic, meet a perpendicular tangent, at Q, of a confocal conic, in T, which lies on the circle, say γ, containing the common points of the two conics, prove that the circle which has PQ for diameter touches the circle γ at T; thus the line TC, joining T to the centre, C, of the confocals, contains the mid point of P, Q. Prove, also, that, as T varies on γ, the line PQ touches a third conic confocal with the two given conics. Notice too that, if one of the two original conics be given, and also a circle γ, having the same centre, be given arbitrarily, the second confocal conic can be deduced. Also that the auxiliary circle of a conic is the co-director circle of this conic and the degenerate confocal which consists of the two pencils whose centres are at the real foci of the given conic. What are the corresponding results for the circle of similitude of two circles?

(13) **The circle of curvature at a point of a conic.** If any circle be taken meeting a given conic, and also a pair of complementary common chords of the two curves, then the bisectors of the angles determined by these two lines are parallel to the axes of the conic. For, on the Absolute line, the three pairs of points determined, respectively, by the given conic, by the circle, and by the pair of complementary chords, are in involution; and, by definition, the axes of the conic determine the foci of this involution.

Let, then, P be any point of a given conic; and H be a further point of the conic which is so taken that the bisectors of the angles determined by the tangent at P, and the chord PH, are parallel to the axes of the conic. Then, let Q, R be two further points of the conic such that the chord QR is parallel to PH. By what we have said, it follows that the circle PQR touches the conic at P. Supposing now that the chord QR coalesces with PH, we infer that *the circle drawn through H, to touch the conic at P, has P for its fourth intersection with the conic.* This circle is called *the circle of curvature of the conic* at P. It appears also that, if P, P', P'' be three points of the conic, and P', P'' come to coincidence with P, then the circle $PP'P''$ becomes the circle of curvature at P; and the remaining intersection, H, of this circle with the conic, is such that the bisectors of the angles determined by the line PH and the tangent at P are parallel to the axes of the conic. We assume, further, as a consequence of this, that if the normals of the conic at points P, P' meet in O, then, when P' comes to coincidence with P, the point O becomes the centre of the circle of curvature at P; this is briefly called the *centre of curvature* at P. Another result is also required: let the tangent at P', of the conic, meet the tangent at P in the point T; consider the orthocentre, K, of the triangle PTP', which is the common point of the line through P' parallel to the normal at P, with the line through P parallel to the normal at P'. When P' comes to coincidence with P, the point K takes a definite position, O_1, on the normal at P. As PK, OP' are parallel, and $P'K, OP$ are parallel, we assume that P is the mid point of O, O_1.

But, further, P' being as yet distinct from P, let $P'T$ meet PK in N', and PT meet $P'K$ in N. The points P, P', being vertices of the quadrangle K, N', T, N, are conjugate in regard to the circle containing these four points; thus, P, P' are harmonic in regard to the points, say X and Y, in which this circle meets the line PP'. As P, P'

are on the given conic, it follows that TXY is a self-polar triangle in regard to the conic. Hence, the circle K, N', T, N, containing T, X, Y, is circumscribed to a self-polar triangle of the conic, and so (Chap. x (6), Ex. 15) is perpendicular to the director circle of the conic. Wherefore, the line TK, which contains the centre of this circle, is met, by the circle K, N', T, N, and the director circle of the conic, in two harmonically conjugate pairs of points. Thus T, K are conjugate points in regard to the director circle of the conic (not, generally, inverse points, the line TK not, usually, containing the centre of the director circle). If, now, P' come to coincidence with P, and K, as we have said, becomes the point O_1, of the normal OP, we infer that P and O_1 are conjugate in regard to the director circle. *We have thus the following construction for the centre of curvature, O, of a conic, at a point P: on the normal at P take the point, O_1, which is conjugate to P in regard to the director circle of the conic; then, on this normal, take the point, O, such that P is the mid point of O, O_1. The point O is the centre of curvature.* The point O_1 may be constructed by taking the point, M, inverse to P in regard to the director circle (lying on the line PC, joining P to the centre C), and then drawing MO_1, through M, perpendicular to CP, to meet the normal of the conic at P in the point O_1.

Particular applications of the theorem are:

(*a*) To the parabola: If PM be drawn parallel to the axis, so that the mid point of P, M is on the directrix (M being then the inverse point of P in regard to the degenerate director circle), then the line through M, parallel to the directrix, meets the normal at P in the point O_1. By a result referred to above (in (11)), the line O_1M contains the pole of the chord of the parabola which is normal at P.

(*b*) To the rectangular hyperbola: The inverse point, M, of the point P, in regard to the director circle, then coincides with the centre, C; and O_1 is the point where the normal at P is met by the line drawn through C perpendicular to CP. If the normal at P meet the curve again in Q, and P' be the other extremity of the diameter PC of the hyperbola, we know (Chap. xi (6)) that QP' is perpendicular to PP'. Thus O_1 is the mid point of P, Q; also, the pole of the normal chord PQ lies on O_1C.

The figure has many properties, such as those numbered here (i)–(xv), which we leave to the verification of the reader.

Let P be any point on a conic whose centre is C; let the normal at P meet the conic again in Q, and T be the pole of PQ. A line, PP_1, is drawn through P, such that PQ, PT are the bisectors of the angles determined by PC and PP_1; the line from T perpendicular to PP_1

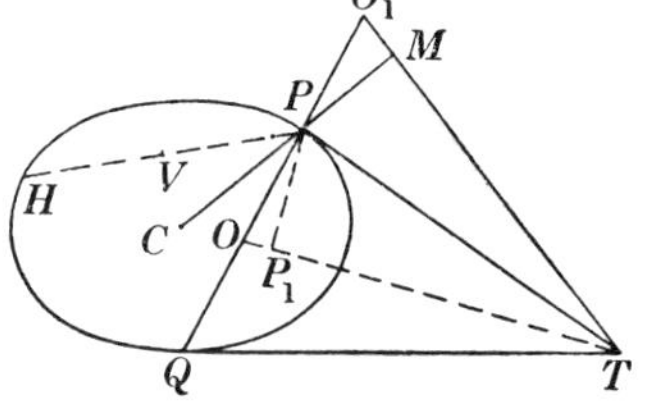

meets this in P_1 and meets the normal at P in O. The point O_1, on the normal at P, is such that P is the mid point of O, O_1; and CP, which is perpendicular to O_1T, meets this in M. The common chord of the circle of curvature of the conic at P, with the conic itself, is PH; and V is the mid point of P, H. We have, then,

(i) O is the centre of curvature of the conic at P.

(ii) Denoting by ω the other conic, confocal with the given one, which passes through P and touches PQ at P, the points P, O are harmonic conjugates in regard to the two points in which the normal PQ meets the directrices of the conic ω.

(iii) The points P, T are harmonic conjugates in regard to the two points in which PT meets the directrices of the original conic.

(iv) The point T is the centre of curvature at P of the conic ω. And, if PT meet this conic again in Q', the tangent at Q', of this conic, contains O.

(v) The line TM is the polar of P in regard to the director circle of the original conic; so that P, T are conjugate points in regard to this director circle.

(vi) The circle having PT as diameter, which contains P_1 and M, is perpendicular to this director circle; as is also the circle having O_1P as diameter, which contains M.

(vii) If a point, P', be taken on the normal PQ, and the tangents be drawn from P' to the original conic, the orthocentre of the triangle formed by P' and the points of contact of these tangents, takes the position O_1, when P' coalesces with P.

(viii) In the circle PO_1M, having O_1P as diameter, an infinite series of triangles can be inscribed all self-polar in regard to the given conic (later, we shall express this by saying that the circle PO_1M is *outpolar* to the given conic. See Chap. xiv (16)).

(ix) If, as in (vii), a point P' be taken on the normal PQ, and tangents be drawn from P' to the given conic, the circle containing P' and the points of contact of these tangents, becomes the circle POP_1, when P' coalesces with P.

(x) If tangents PU, PU' be drawn from P to any conic confocal with the given conic, touching this in U and U', the circles PUU', for different confocals, have all another point in common, besides P; this was essentially proved. This other point is P_1; it is conjugate to P in regard to all conics through the four foci of the confocal conics.* Any such circle PUU' is the conjugate line locus, defined above (Chap. ix (23), Ex. 18), for the particular confocal of which PU, PU' are the tangents, with centres at P and P_1. In particular, the conjugate line locus for the original conic, with centres at P and P_1, is the circle P_1OP, having PO as diameter. In general, as has been proved above, if PU, PU' meet the original conic again in W and W', the chord WW' contains T.

(xi) The point P_1, conjugate to P in regard to all conics through the four foci of the confocals, is obtainable by drawing the perpendicular PP_1, to the line OT, meeting this in P_1. The line OT is the join of the centres of curvature at P, of the original conic and the confocal, ω,

---

* This can easily be proved.

through P. Of conics confocal with the original conic, there is a single one which touches the joining line PP_1; the asymptotes of this are CP and CP_1.

(xii) The Apollonius rectangular hyperbola of the centre of curvature O, in regard to the original conic (Chap. XII (8)), touches this conic at P. The centre of the hyperbola is the mid point of P, V. The circle O_1PM, having O_1P as diameter, is the circle of curvature, at P, of this hyperbola.

(xiii) Tangents PU, PU' being drawn from P to a confocal conic, touching this at U and U', the polar line UU', of P, in regard to this confocal, and the normals of this at U and U', touch a parabola which is the same whatever the confocal chosen. This parabola touches the axes of the confocals, and touches PT and PO, respectively, at T and O. The focus of this parabola is at P_1, and PC is the directrix.

(xiv) If, as before, PU, PU' be the tangents from P to any conic confocal with the given conics, the orthocentre of the triangle PUU' describes a rectangular hyperbola. This hyperbola contains P and O_1, and has one asymptote parallel to CP; and touches the line PP_1 at P. If O_2 be the point on the tangent PT such that P is the mid point of T and O_2, this hyperbola also contains O_2. We may prove this by re-marking that the orthocentre of any triangle PUU' is the common point of a line through P, perpendicular to UU', with the polar line of P in regard to the director circle of the confocal touched by PU, and PU'. These polar lines form a pencil, with centre on the Absolute line, which is related to the pencil of lines drawn from P, perpendicular to the lines UU'.

(xv) The four normals are drawn from a point K to a conic, and the tangents to this conic at the feet of these normals. Find the focus of the parabola which touches these four tangents. And, if the four normals be drawn to the conic from this focus, and the tangents at the feet of these normals, find the focus of the parabola which touches this second set of four tangents.

The theorem for the geometrical construction of the centre of curvature of a conic is given by Steiner (1845), *Gesamm. Werke*, II, p. 341.

CHAPTER XIV

GENERAL THEOREMS. APPLICATIONS

(1) *If, in regard to a conic, the angular points A, B, C, of one triangle, be the respective poles of the sides $B'C'$, $C'A'$, $A'B'$, of another triangle, the joining lines, AA', BB', CC', meet in a point. The relation of the triangles is reciprocal, the angular points A', B', C' being the respective poles of BC, CA, AB.* We give two proofs of the theorem:

(a) We have remarked (Chap. VII (4), Ex. 3) that the condition that three lines, AA', BB', CC', drawn respectively through A, B, C, should meet in a point, is the existence of a line met by the pairs of lines, (AA', BC), (BB', CA), (CC', AB), in three pairs of points in involution. We can shew that $B'C'$ is such a line. Let CA, the polar of B', meet $B'C'$ in B_1; which is thus conjugate to B' in regard to the conic. Let AB, the polar of C', meet $B'C'$ in C_1; which is thus conjugate to C' in regard to the conic. The point X, where BC meets $B'C'$, is the pole of the line AA', joining the poles of $B'C'$ and BC; thus X is conjugate to the point X_1 where AA' meets $B'C'$. Wherefore B', B_1; C', C_1; X, X_1 are three pairs in involution on $B'C'$. And these are respectively the points in which $B'C'$ is met by the lines BB', CA; CC', AB and BC, AA'.

(b) Let the points $(BB', C'A')$, $(CC', A'B')$ be, respectively, M and N; and O be the point $(MN, B'C')$. Consider the conjugate line locus of the given conic, with centres at B' and C'. The line $B'M$, containing B, the pole of $C'A'$, is conjugate to $C'A'$, or $C'M$; thus M is on this locus. So N is on this locus. As then B', C', M and N are on the conjugate line locus, the polar of O in regard to this locus is the line joining the points $(B'M, C'N)$, $(B'N, C'M)$, of which the latter point is A'. By the properties of the locus, the pole of $B'C'$, in regard to the locus, coincides with the pole of $B'C'$, in regard to the original conic, which is A. Thus the polar of O, on $B'C'$, in regard to the locus, contains the point A. Whence the lines $B'M$, $C'N$ meet on $A'A$. But these are $B'B$, $C'C$. This again proves the result.

A particular consequence of the theorem is that, if a conic, which is not degenerate, touches BC, CA, AB respectively in D, E and F, then AD, BE, CF meet in a point.

Ex. Prove the converse theorem, that, if two triangles ABC, $A'B'C'$ be such that AA', BB', CC' meet in a point, then a conic exists for which the angular points of either triangle are the poles of the corresponding sides of the other triangle.

(2) *Let a conic, ω, meet the sides BC, CA, AB of a triangle respectively in D, D'; in E, E'; and in F, F'. On BC, let P, P' be any pair of points of the involution (B, C), (D, D'). So, on CA, let Q, Q' be any pair of*

points of the involution (C, A), (E, E'); and, on AB, let R, R' be any pair of points of the involution (A, B), (F, F'). Then the six points P, P', Q, Q', R, R' lie on a conic.

For, the conics through E, E', F, F' meet BC in pairs of points of an involution. Two such conics are the conic, ω, and the line pair AB, AC. Whence there exists a conic, say ω_1, containing E, E', F, F' and P, P'. Again, the conics through F, F', P, P' meet CA in pairs of points of an involution. Two such conics are the conic ω_1, and the line pair BC, BA, of which the former meets CA in E, E'. Whence, there exists a conic containing F, F', P, P', Q, Q', say ω_2. Lastly, the conics through P, P', Q, Q' meet AB in pairs of points of an involution. Two such conics are the line-pair CA, CB, and the conic ω_2, which meets AB in the points F, F'. It follows that P, P', Q, Q' and R, R' lie on a conic. As we wished to prove.

Ex. If this last conic be ω', prove that the seven points, which consist of A, B, C and the four common points of ω and ω', lie on a conic.

(3) On the lines BC, CA, AB, respectively, let D, D'; E, E' and F, F' be three pairs of points. Denote the foci of the involution, on BC, determined by (B, C) and (D, D'), by U, U'. Likewise, let V, V' be the foci of the involution (C, A), (E, E'); and W, W' the foci of the involution (A, B), (F, F'). Thus we have $U' = (B, C)/U$; $V' = (C, A)/V$; $W' = (A, B)/W$; so that it can be proved that, in the case when AU, BV, CW meet in a point, so also do AU, BV', CW', and also AU', BV, CW'; and AU', BV', CW. The four points then arising form a quadrangle of which ABC is the vertex triangle.

We prove now that *a necessary and sufficient condition, for the six points D, D', E, E', F, F' to lie on a conic, is that the six lines AU, AU', BV, BV', CW, CW' should meet in threes in four points; which is secured if the three lines consisting of one of AU, AU', taken with one of BV, BV', and one of CW, CW' meet in a point.*

This theorem follows from (1) and (2) preceding. To prove that the condition is necessary, suppose that a conic contains D, D'; E, E' and F, F'. Then, because the point U, taken twice over, and also the point U', taken twice over, belong to the involution (B, C), (D, D'), with a similar statement for the points on CA and AB, it follows, from (2), that, with a proper choice of notation, a conic touches BC, CA, AB, respectively, in U, V, W. If this conic is undegenerate, it follows, from (1), that the lines AU, BV, CW meet in a point. There is, however, the possibility, that the points U, V, W should lie in a line, in which case the conic which touches BC, CA, AB at these points would consist of this line taken twice over. Allowing for such a possibility, we may prove that, by proper choice of the notation, the consequence of the existence of a conic containing D, D', E, E', F, F' is that AU, BV, CW, also AU, BV', CW', and two other sets named above, are three lines each meeting in a point; there are then four undegenerate conics touching BC, CA, AB at U, V, W; at U, V', W', and so on; and there are four degenerate conics, each consisting of a line taken twice over, namely, the lines $U'VW$, $UV'W$, UVW', $U'V'W'$. To prove that the condition stated is sufficient, suppose that AU, BV, CW meet in a point, and therefore also AU, BV', CW', and two other sets. Then, because D, D' belong to the involution determined by the two pairs which consist of B, C and the point U taken twice over, with a similar statement on CA and AB, it follows from (2) that D, D', E, E', F, F' lie on a conic.

If we use the symbols of the points, taking for D, D', E, E', F, F' the respective symbols $B+dC$, $B+d'C$, $C+eA$, $C+e'A$, $A+fB$, $A+f'B$, where d, d', e, e', f, f' are numbers, then the symbols of U, U', V, V', W, W' are, respectively (Chap. VII (3)), $B\pm pC$, $C\pm qA$, $A\pm rB$, where $p^2=dd'$, $q^2=ee'$, $r^2=ff'$. The necessary and sufficient condition, that D, D', E, E', F, F' lie on a conic, is then that $pqr=1$ or $pqr=-1$, namely, $dd'ee'ff'=1$; for we have shewn that, for the lines joining A, B, C, respectively, to $B+pC$, $C+qA$, $A+rB$, to meet in a point, the condition is $pqr=1$. The condition for the points $B+pC$, $C+qA$, $A+rB$ to lie on a line is $pqr=-1$ (Chap. IV (4), Ex. 6).

(4) The dual of the theorem in (3) may be enunciated: Let the lines BC, CA, AB be denoted by a, b, c, respectively. Let l, l' be two lines through A; also m, m' two lines through B, and n, n' two lines through C. Let u, u' be the focal rays of the pencil in involution determined by the two pairs b, c and l, l'; with v, v' and w, w' for the focal rays respectively of the pencils in involution (c, a), (m, m') and (a, b), (n, n'). Then, a necessary and sufficient condition, that the six lines l, l', m, m', n, n' should touch a conic, is that the six points of intersection (a, u), (a, u'), (b, v), (b, v'), (c, w), (c, w') should lie in threes on four lines. Denoting these points, respectively, by U, U', V, V', W, W', this is secured if there is a line containing one of U, U', and one of V, V', and also one of W, W'. If for instance these points, on a line, be U', V', W', then the lines u, v, w meet in a point. In general the four lines form a quadrilateral of which ABC is the diagonal triangle.

But, denoting the six points (a, l), (a, l'), (b, m), (b, m'), (c, n), (c, n'), respectively, by L, L', M, M', N, N', the points U, U' are the foci of the involution (B, C), (L, L'), and so on. Thus we see that *a necessary and sufficient condition, that l, l', m, m', n, n' should touch a conic, is that L, L', M, M', N, N' should lie on a conic.* This is the most convenient way of regarding the dual theorem.

Ex. 1.   On BC, CA, AB, respectively, let D, D' and E, E' and F, F' be pairs of points. Let AD, BE, CF meet in a point, and also AD', BE', CF' meet in a point. Then D, D', E, E', F, F' lie on a conic. Also, if these points lie on a conic, and AD, BE, CF meet in a point, then also AD', BE', CF' meet in a point.

Ex. 2.   With points D, D' on BC, etc., such that AD, AD', BE, BE', CF, CF' touch a conic Ω, and the points D, D', E, E', F, F' lie on a conic ω, prove that there exists a conic through the four common points of Ω and ω which touches BC, CA, AB. And, dually, a conic through A, B, C, which touches the four common tangents of Ω and ω.

Ex. 3.   In Ex. 2, regard D, D' as the Absolute points of the plane, and A as a focus of the conic Ω. The theorem of the existence of the conic ω, when Ω is given, then becomes the theorem that, *if S be a focus of a conic Ω, and t a variable tangent of this conic, and on t be taken a point E such that $\epsilon\,(t,\,ES)$ is a constant number, then the locus of E is a circle*; this was proved in Chapter XIII, in (7).

Ex. 4.   Let A, B, C be the vertex triangle of a quadrangle, and P, P' be conjugate points in regard to the quadrangle, and likewise Q, Q' and R, R' be two other pairs of conjugate points (Chap. IX (23), Ex. 6). Prove that the six lines AP, AP', BQ, BQ', CR, CR' touch a conic.

Ex. 5.   Let B, D, C, L' be points of a line, and U, U' the foci of the involution determined by the pairs $(B,\,C)$ and $(D,\,L')$, while $D'=(B,\,C)/L'$, prove that U, U'; B, C; D, D' are three pairs of an involution. This is most easily obvious by interpreting the points as lying on a conic; then the theorem becomes the familiar one that, triangles can be inscribed in a conic whose sides pass respectively through the angular points of a self-polar triangle of the conic. But if, using the symbols, we take $D=B+dC$, $L'=B+l'C$, where d, l' are numbers, then we have $U=B+(dl')^{\frac12}\,C$, $U'=B-(dl')^{\frac12}\,C$, and $D'=B-l'C$. From these the result is obvious.

Now, let D, E, F be points respectively on the sides BC, CA, AB of a triangle, such that AD, BE, CF meet in a point. Also let L', M', N' be points, also on these sides respectively, which lie in a line; so that, if $D'=(B,\,C)/L'$, and $E'=(C,\,A)/M'$, and $F'=(A,\,B)/N'$, then AD', BE', CF' meet in a point. Then a conic contains the six points D, D', E, E', F, F'. Next, let U, U' be the foci of the involution determined by the pairs $(B,\,C)$ and $(D,\,L')$; likewise, V, V' the foci of the involution $(C,\,A)$, $(E,\,M')$; and W, W' the foci of the involution $(A,\,B)$, $(F,\,N')$. By what we have shewn, the pairs $(B,\,C)$, $(D,\,D')$, $(U,\,U')$ are in involution; so, likewise, are the pairs $(C,\,A)$, $(E,\,E')$, $(V,\,V')$, and the pairs $(A,\,B)$, $(F,\,F')$, $(W,\,W')$. Wherefore, as D, D', E, E', F, F' lie on a conic, it follows, from (2), that the points U, U', V, V', W, W' lie on a conic.

One important corollary of this is that, *if ABC be any triangle, O any point, and p' any line, then a conic exists having ABC as a self-polar triangle for which O is the pole of the line p'.* For, let AO, BO, CO meet BC, CA, AB, respectively, in D, E, F, and let p' meet BC, CA, AB, respectively, in L', M', N'. Let U, U' be the foci of the involution $(B,\,C)$, $(D,\,L')$; likewise V, V' of the involution $(C,\,A)$, $(E,\,M')$, and W, W' of the involution $(A,\,B)$, $(F,\,N')$. We have proved that U, U', V, V', W, W' lie on a conic. Since $B=(W,\,W')/A$ and $C=(V,\,V')/A$, the point A is the pole of BC in regard to this conic; and so on; so that ABC is a self-polar triangle for this conic. Next, from $L'=(U,\,U')/D$, the polar of L', in regard to this conic contains D; and this polar contains A because L' lies on BC. Thus AO is the polar of L'; and so on; thus O is the pole of the line $L'M'N'$, or p'. This proves the result.

(5) *Let ABC be a triangle inscribed in a given conic, whose sides BC, CA, AB are met by a line, p', respectively in L', M', N'. Draw through L', M', N' any lines, meeting the conic, respectively, in X, X', in Y, Y', and in Z, Z'. Then the six lines AX, AX', BY, BY', CZ, CZ' touch a conic.*

Let these pairs of lines meet BC, CA, AB, respectively, in D, D'; in E, E'; and in F, F'. It is then sufficient to shew that D, D', E, E', F, F' lie on a conic (by (4)). Let the tangents of the given conic at A, B, C, respectively, meet BC, CA, AB in L, M, N. By the dual of (1), the points L, M, N lie on a line. Denote this line by p. Now consider the degenerate conic consisting of the two lines p', p. Its pairs of intersections with BC, CA, AB, respectively, are L', L; M', M; N', N. If we can shew that (B, C); (L', L); (D, D'), which lie on BC, are in involution; and also that, similarly, (C, A), (M', M), (E, E') are in involution, and (A, B), (N', N), (F, F') are in involution, it will follow, by (2), that the six points D, D', E, E', F, F' lie on a conic. Now consider the four points which consist of X, X', and the two points of the original conic which coincide at A, on the tangent at A. The conics through these four points meet BC in pairs of points of an involution; one such conic is the original conic, giving the points B, C; another such conic is the degenerate conic constituted by the line $L'XX'$ and the tangent of the original conic at A, and this gives the pair L', L on BC; and a third such conic is the degenerate conic consisting of the lines AX, AX'; which gives the points D, D'. Thus the three pairs B, C; L', L; D, D' are in involution. Similarly on CA and AB. Thus the proposition is established.

Ex. 1. Let points L', M', N', respectively, on the sides BC, CA, AB of a triangle inscribed in a conic, lie on a line. Let the polars of these points, in regard to the conic, meet the conic, respectively, in P, P'; in Q, Q'; and in R, R'; and these polars meet BC, CA, AB, respectively, in U, V, W. Also let D, E, F be the respective poles of BC, CA, AB. Prove that AU, BV, CW meet in a point. Also that AP, AP' are the focal rays of the pencil in involution determined by the two pairs of lines (AB, AC), (AD, AU). Hence shew that the six lines AP, AP', BQ, BQ', CR, CR' meet in threes in four points.

(6) *Let two conics be given. Let a variable line meet one conic in a pair of points which are conjugate in regard to the other conic (so that this is then true if the conics be interchanged). We prove that the line envelops a third conic. This is that called the harmonic envelope of the two given conics.* We have already given a proof of the theorem when the two given conics are circles (Chap. XIII (4)).

Let A, B, C be three points of a conic ω. Let the polars of A, B, C, taken in regard to another conic Ω, meet the original conic ω, respectively, in X, X'; in Y, Y'; and in Z, Z'. The triangle whose sides are the polars of A, B, C in regard to Ω is in perspective with ABC, (1); thus these triangles have an axis of perspectivity; so that the points, L, M, N, where XX', YY', ZZ', respectively, meet BC, CA, AB, lie in a line.

As the polar of B, in regard to Ω, passes through Y, the polar of Y, in regard to Ω, passes through B; let this polar meet the first conic ω again in D. By (5) preceding, the lines AX, AX', BY, BY', CZ, CZ' touch a conic. Denote this conic by ψ. But, consider the other triangle CAY inscribed in ω. The polars, in regard to Ω, of C, A, Y, meet ω, respectively, in Z, Z'; in X, X'; and in B, D. Whence, as before, the lines CZ, CZ', AX, AX', YB, YD touch a conic. Of these, however, the first five are tangents of the conic ψ, with which this second conic then coincides. Wherefore, ψ may be defined as the conic touching the five lines BY, BY', CZ, CZ', YD. These, however, are defined solely by the points B, C, and are the same whatever A may be. We have therefore proved that the line joining an arbitrary point A, on the conic ω, to either of the points X, X', on ω, which are conjugate to A in regard to a second conic Ω, touches a definite conic ψ, the same for all positions of A. Which is the theorem enunciated. The argument given is taken from Steiner (Vorträge, *Theorie der Kegelschnitte*, 1867, p. 438, ed. Schröter).

The dual theorem is that *the locus of a point from which the tangents to one given conic are harmonic conjugates, in regard to the tangents to another given conic, is a third conic.* This is called the *harmonic locus* of the two given conics. The conjugate line locus of a given conic, with two given centres (Chap. IX (23), Ex. 18), is a particular case, that namely in which one of the two given conics degenerates into two pencils.

(7) In a previous chapter (XIII), we have given conditions for the degeneration of the harmonic envelope, and of the harmonic locus of two given conics. We shall incidentally recur to this now. We prove first that, *if two conics ω, Ω be in such general position that they have a common self-polar triangle, then this is also a self-polar triangle for the harmonic locus, Φ, and for the harmonic envelope, ψ, of the two conics.* Take the case of the harmonic locus Φ. Let U be the common point of a pair of common tangents of the two given conics ω, Ω. This point U lies on one side, say a, of the common self-polar triangle of the conics ω, Ω; denote by A the opposite angular point of this triangle. Each conic, of ω, Ω, is its own harmonic inverse in regard to the point A and the line a. Consider the quadrangle formed by the four points of contact of the common tangents from U to the two conics ω, Ω. The vertex triangle of this has U for one angular point; the other two angular points consist of A, and a point, V, lying on the line a; the vertex triangle is self-polar in regard to any conic through the four points of contact of the common tangents from U to ω and Ω. By the definition of Φ, this conic contains the eight points of contact of the four common tangents of ω and Ω; and, in particular, Φ contains the four points of the quadrangle under consideration. Wherefore, A is the pole of the line a in regard to the conic Φ. A similar argument is applicable to every angular point of the self-polar triangle of ω and Ω. This triangle is then self-polar in regard to Φ. Likewise, it is self-polar in regard to ψ.

A corollary from this concerns the case when ω, Ω are so related that the harmonic locus Φ breaks up into two lines, it being supposed that ω, Ω still have a definite common self-polar triangle. We assume that, if a definite triangle is self-polar in regard to a conic which degenerates into two lines, then one angular point, say A, of this triangle, is at the common point of the two lines. Then, as A is on the harmonic locus of the two conics, ω, Ω, the tangents from A, to the two conics, are two harmonically conjugate pairs of lines. And the sides, of the self-conjugate triangle of ω and Ω, which meet in A, are, as in general, harmonic conjugates in regard to both these pairs of tangents. Now, the tangents from A, to ω, Ω, touch these conics in the points where these respectively meet the side a. Thus, when Φ breaks into two lines intersecting in the angular point A of the self-polar triangle of ω and Ω, the conics meet the opposite side a in harmonically conjugate pairs of points, the angular points, of their self-polar triangle, which lie on a, being harmonically conjugate in regard to both conics. This shews also that the side a is a line of the harmonic envelope ψ of ω and Ω. Thus, by the dual converse of the assumption already made, when Φ degenerates into two lines, ψ degenerates into two pencils, whose centres are on a side of the common self-polar triangle of ω and Ω.

(8) *The polar reciprocal of one conic, ω, in regard to another conic, ω_0.* Suppose the two given conics to be undegenerate. Let E, A be two points lying on the conic ω; and e, a the polars of these, respectively, taken in regard to the conic ω_0. Then, the common point (e, a) is the pole, also in regard to ω_0, of the line EA. Thus, if E, A, B, C, D be points lying on the conic ω, the pencil $E\,(A, B, C, D)$ is related to the range, $e\,(a, b, c, d)$, of points where e is met by the respective polars a, b, c, d, of A, B, C, D, taken in regard to ω_0. Wherefore, the aggregate of the lines which are the polars, in regard to ω_0, of the points of the conic ω, consists of the tangents of another conic, ω'. And any point of this conic ω', which may be regarded as the common point of two coalescing tangents, is, conversely, the pole, in regard to ω_0, of a tangent of the conic ω. Thus the conic ω can be derived from ω' just as ω' was derived from ω. The conics ω, ω' are said to be *polar reciprocals* of one another in regard to the conic ω_0. Evidently, the angular points of any triangle which is self-polar in regard to ω are the poles, in regard to ω_0, of the sides of a triangle which is self-polar in regard to ω'; the former is the vertex triangle of a quadrangle of four points lying on ω, the latter is the diagonal triangle of a quadrilateral of four corresponding tangents of ω'. In particular if, as is general, the conics ω, ω_0 have a common self-polar triangle, this is the common self-polar triangle of ω' and ω_0.

Ex. 1.   Prove that the polar reciprocal of a circle ω, of centre C, in regard to another circle ω_0, of centre C_0, is a conic ω' of which C_0 is a focus, the corresponding directrix of ω' being the polar of C, taken in regard to ω_0. In general, if P and p be any point and its polar line in regard to a conic ω; and the polar of P in regard to a conic ω_0 be p', while the pole of p, also in regard to ω_0, be P'; then P' is the pole of p' in regard to the conic ω', which is the polar reciprocal of ω in regard to ω_0. When ω, ω_0 are circles, the asymptotes of the conic ω' are perpendicular to the tangents drawn from C_0 to touch ω. Thus ω' is an ellipse, or a hyperbola, according as these tangents are imaginary, or are real; and ω' is a parabola when C_0 is a point of ω.

Many theorems, concerning a conic and one of its foci, may be derived from properties of circles, by reciprocation in regard to a circle whose centre is at the focus.

Ex. 2.   If the circles of a coaxal series be reciprocated, in regard to a circle whose centre is at one limiting point S of the series, prove that the reciprocal conics are a confocal system, of which S is one focus. The complementary focus is the reciprocal of the radical axis of the series of circles.

Ex. 3.   If the conic ω be a parabola, and the reciprocating conic ω_0 be a circle, which passes through the focus of the parabola and has its centre C_0 on the directrix, prove that the conic ω', obtained from ω by reciprocation in regard to ω_0, is a rectangular hyperbola, touching the directrix at C_0, whose centre is at the focus, its asymptotes being parallel to the tangents drawn to the parabola from C_0.

Ex. 4.   Let the tangents at points P, P' of a parabola, whose focus is S, meet in T. Let the line through T parallel to the axis meet the directrix in H. Prove that there is a conic, having S, H as foci, which touches the tangents TP and TP'. By taking a series of parallel chords PP' of the parabola, for all of which the point H is the same, we thus have a confocal system of conics, each having two such tangents as TP, TP' common with the parabola.

Conversely, let ω, ω' be any two undegenerate conics having four distinct common tangents, and, therefore, a common self-polar triangle. Let O be one of the common points, and p be one of the common tangents, of these two conics. By Ex. 5 of (4) above, there exists a conic, ω_0, having the common self-polar triangle of ω and ω' for a self-polar triangle, and O for the pole of the line p. Bearing in mind that a conic is its own harmonic image, in regard to any angular point of a self-polar triangle and the opposite side, it is thence easy to see that the conics ω, ω' are polar reciprocals of one another in regard to the conic ω_0. Any one of the common points of ω and ω' is then the pole, in regard to ω_0, of a definite one of the common tangents of ω and ω'. Thus, *when two conics ω, ω' are in such general relation that they have a definite common self-polar triangle, there are four conics, ω_0, in regard to which ω, ω' are polar reciprocals; for we may choose the common tangent p, to associate with the assigned common point, O, in four ways.*

When two, or more, of the common points, or tangents, of ω and ω' coincide, and the common self-polar triangle becomes specialised, it is still true that there is a conic ω_0, and there may be an infinite series of such, in regard to which ω and ω' are polar reciprocals. (Cf. *Principles of Geometry*, Vol. II, p. 147.)

Ex. 5.   When, in Ex. 4, the conics ω, ω' are circles with a common centre, we may take, for a conic ω_0 in regard to which ω, ω' are polar reciprocals, a properly chosen circle with the same centre. But we may also take, as reciprocating conic, a rectangular hyperbola, whose vertices are the end points of any diameter of the concentric reciprocating circle ω_0.

(9) *The angular points of any two triangles, whose sides touch the same conic, lie on another conic.* Let the triangles be ABC and $A'B'C'$, the sides respectively opposite to the angular points being a, b, c, a', b', c'. Recall that four tangents of a conic meet any two other tangents in related ranges. Then, the pencil $A\,(B', C', B, C)$ meets the tangent a' where this is met by the tangents c', b', c, b, respectively. This range on a' is related to the range on the tangent a where this is met by the same four tangents; the points of this latter range lie on the rays of the pencil $A'\,(B', C', B, C)$. As the two pencils are thus related, the six angular points lie on a conic.

Dually, if the angular points of two triangles lie on a conic, their six sides touch a conic.

A corollary is: *If two conics ω, ω_0 be such that a single triangle exists* *whose angular points A, B, C lie on ω, while its sides touch ω_0, then there is an infinite series of such triangles.* For, let any tangent of ω_0 meet ω in B' and C'; draw from each of B' and C' the other possible tangent of ω_0, these two tangents meeting in P. By what is proved, there exists a conic containing the six points A, B, C, P, B', C'; this conic, however, as containing five points of ω, coincides with ω. Wherefore, P lies on ω, and $PB'C'$ is another triangle, with angular points on ω, whose sides touch ω_0. This proves the statement made. The conic ω is then said to be *triangularly circumscribed* to ω_0, and ω_0 to be *triangularly inscribed* to ω.

If, for two conics in this relation, ABC be a triangle inscribed in ω whose sides touch ω_0, and the point A come to coincidence with a common point of ω and ω_0, then the sides AB, AC come to coincidence with the tangent of ω_0 at this point, while the side BC, which touches ω_0, becomes also a tangent of ω, the points B and C coalescing. We infer therefore that *a sufficient condition, for ω to be triangularly circumscribed to ω_0, is that the tangent of ω_0, at one of the common points of the two conics, should contain the point of contact, with ω, of one of the common tangents of ω and ω_0.* When this is so, it follows from the theory that the tangent of ω_0, at *every* one of the common points of the two conics, contains the point of contact with ω of a certain one of the four common tangents. Thus, when the two conics ω, ω_0 have four distinct common points, and four distinct common tangents, there are four ways in which ω can be triangularly circumscribed to ω_0.

Ex. 1. If the two conics ω, ω_0 be circles, a sufficient condition for the relation considered, is that an asymptote of ω_0 should contain one of the four points of contact, with ω, of a common tangent.

Ex. 2. For a degenerate conic, of two lines, ω, to be triangularly circumscribed to a conic ω_0, one of the two lines must touch ω_0.

Ex. 3. For a conic ω to be triangularly circumscribed to a degenerate conic, of two pencils, ω_0, the conic ω must contain the centre of one of the two pencils.

(10) *If two triangles be both self-polar in regard to the same conic, their six angular points lie on a conic.* Let ABC and $A'B'C'$ be two such triangles. The rays of the pencil $A\,(B, C, B', C')$ are, respectively, conjugate to the rays of the pencil $A'\,(C, B, C', B')$; so that these pencils are related. Hence the pencils $A\,(B, C, B', C')$ and $A'\,(B, C, B', C')$ are related. This shews that the six angular points lie on a conic.

It follows that *if Ω, ω be two conics such that a single self-polar triangle of ω is inscribed in Ω, then there exists an infinite series of triangles inscribed in Ω, all of which are self-polar in regard to ω.* For, let ABC be a triangle, inscribed in Ω, which is self-polar in regard to ω. Let the polar of an arbitrary point A' of the conic Ω, taken in regard to ω, meet Ω in B' and C'; and let the polar, also in regard to ω, of the point B', which passes through A', meet $B'C'$ in D. Then $A'B'D$ is a self-polar triangle in regard to ω. Thus, there is a conic containing the six points A, B, C, A', B', D; of these, five lie on Ω; so that this conic coincides with Ω, and, therefore, D coincides with C'. This proves the result. Of two conics Ω, ω, in this relation, Ω is said to be *outpolar* to ω. When this is so, it is also true that triangles exist, in infinite number, self-polar in regard to Ω, whose sides touch ω. This fact justifies the practice of saying that, when Ω is outpolar to ω, then ω is also *inpolar* to Ω. In the general case, in which there exists a conic ω_0 in regard to which Ω, ω are polar reciprocals, the fact, of the self-polar triangles of Ω circumscribed to ω, follows, by reciprocation in regard to ω_0, from the existence of the self-polar triangles of ω inscribed in Ω. A fuller treatment of outpolar and inpolar conics, given below (in (16)), provides another proof, independent of the hypothesis of the reciprocating conic ω_0, that when Ω is outpolar to ω, then ω is inpolar to Ω.

(11) *Let one conic, ω, be triangularly circumscribed to another conic ω_0. Then, all the triangles inscribed in ω, whose sides touch ω_0, are self-polar in regard to a third conic.* First, we prove, without reference to the conic ω_0, that if two triangles be inscribed in a conic ω, then there exists a conic Ω in regard to which both these triangles are self-polar. We have shewn, denoting these triangles by ABC and $A'B'C'$, that there exists a conic, Ω, for which $A'B'C'$ is self-polar, and A is the pole of BC ((4), Ex. 5). If we prove that B, C are conjugate points in regard to this conic Ω, then ABC will also be self-polar in regard to Ω, and the required result obtained. Let AB', AC' meet BC, respectively, in X' and Y', and $A'B'$, $A'C'$ meet BC, respectively, in Y and X. Then, in regard to Ω, the point X, common to the polars of A and B', is the pole of AB', and is thus conjugate to the point X', which lies on AB'. Also the point Y, common to the polars of A and C', is the pole of AC', and is thus conjugate to the point Y', which lies on AC'. Whence, the conic Ω meets BC in two points, U, V, which are the foci of the involution determined by the pairs X, X' and Y, Y'. The conics through the four points A, A', B', C' meet BC in pairs of an involution; and two such conics are the line pairs $(A'C', AB')$ and $(A'B', AC')$, which give, respectively, the pairs X, X' and Y, Y'; a third such conic is the conic ω. Thus B, C belong to this involution. They are thus harmonic conjugates in regard to the foci U, V; and these lie on Ω. Thus B, C are conjugate in regard to Ω. Wherefore ABC, as well as $A'B'C'$, is self-conjugate in regard to Ω.

Now suppose that the two triangles ABC, $A'B'C'$, inscribed in ω, are chosen so that their sides touch the conic ω_0, which, by hypothesis is triangularly inscribed in ω. If we take the polar reciprocal of ω, in regard to Ω, the conic so obtained will touch the sides of the triangles ABC, $A'B'C'$, which are the polars of the angular points in regard to Ω. This reciprocal conic is thus identified with ω_0; so that Ω is one of the conics in regard to which ω, ω_0 are polar reciprocals of one another. By the construction of Ω, however, the conic ω is outpolar to Ω; there are, therefore, triangles inscribed in ω, with one angular point arbitrary, which are self-polar in regard to Ω; and, as ω_0 is the polar reciprocal of ω in regard to Ω, the sides of all these triangles touch ω_0.

These are then the triangles inscribed in ω, with sides touching ω_0, each of which is determined by the assignment of one angular point. All these triangles are then self-polar in regard to Ω; as we wished to prove.

In the general case, there are four conics Ω in regard to which the given conics ω, ω_0 are polar reciprocals of one another; and there are four ways in which ω can be triangularly circumscribed to ω_0, as we have seen.

Also, we remark, *if a conic ω be outpolar to a conic Ω, the harmonic envelope of ω and Ω is that conic, ω_0, which is the polar reciprocal of ω in regard to Ω; and this is triangularly inscribed to ω.*

Ex. If the conic ω be triangularly circumscribed to the conic ω_0, then one of the conics Ω, by which ω is reciprocated to ω_0, passes through the common points of ω with the harmonic locus of ω and ω_0.

(12) We have considered three possible relations of a conic ω to a conic ω_0, namely, (a), ω may be triangularly circumscribed to ω_0; (b), ω may be outpolar to ω_0; (c), ω may be inpolar to ω_0. Two of these relations may exist together, and then the third also exists, and ω is then also triangularly inscribed to ω_0. Also, when all three relations exist, there is a third conic, such that the three conics form a symmetrical system, the polar reciprocal of any one, in regard to any other, being the third of the conics.

We examine this statement with nomenclature already employed. We *suppose first that ω is both triangularly circumscribed to ω_0 and outpolar to ω_0; and prove that ω is then also inpolar to ω_0.* Let S, I, J be three points lying on ω, forming a triangle whose sides touch ω_0. Take I, J for Absolute points. Then ω_0 is a parabola of which S is the focus; and ω is a circle passing through S. This is because ω is triangularly circumscribed to ω_0. We prove further, as a consequence of the fact that ω is outpolar to ω_0, that the centre of the circle ω lies on the directrix of the parabola ω_0; and that the tangent of ω, at S, which is the polar of this centre in regard to the parabola, meets the parabola in two points which lie on the tangents of the circle at the two points where it meets the directrix. We can then infer that the circle is inpolar to the parabola.

Let N be one of the points where the circle ω meets the directrix of the parabola ω_0. As ω is outpolar to ω_0, the polar of N, in regard to the parabola, which is the line through S perpendicular to NS, is met by the parabola and circle respectively in two harmonically conjugate pairs of points; let these points consist of H, K on the parabola, and S, M on the circle. Then, from $M = (H, K)/S$, it follows that M is on the directrix; hence, as NS, SM are perpendicular, the centre, O, of the circle is on the directrix, at the mid point of N, M. Whence, the line joining O to the mid point of S, M is perpendicular to SM, and touches the parabola at a point P which lies on the tangent of the circle at M. Likewise, the line joining O to the mid point of S, N touches the parabola at a point, Q, which lies on the tangent of the circle at N. Thus MS is the polar of P in regard to the circle. Hence PHK is a self-polar triangle, in regard to the circle, whose angular points lie on the parabola. Or, the parabola is outpolar to the circle. From this, by (10), the circle is inpolar to the parabola, as well as outpolar to it. The proof is in reality quite general, though expressed in terms of a particular nomenclature.

We notice further that the triangle whose angular points are P, Q and the point where the lines NQ, MP, parallel to the axis, meet the Absolute line, is inscribed to the parabola and also is circumscribed to the circle. Thus ω is not only *triangularly circumscribed*, but also *triangularly inscribed* to ω_0.

Moreover, as each of ω and ω_0 is outpolar to the other, the polar reciprocal of either, in regard to the other, is the same conic, Ω, which is the harmonic envelope of the two curves. The polar reciprocal of the parabola, in regard to the circle, is seen without difficulty to be the rectangular hyperbola, with centre at S, having SM, SN for asymptotes, which touches the directrix at O. This hyperbola, Ω, is triangularly circumscribed to the parabola. For, the points where OP, OQ meet the Absolute line are on the hyperbola, as is O; and the joins of these points touch the parabola. Further, the parabola is inpolar to the hyperbola; for the parabola touches the lines SI, SJ, joining S to the Absolute points, and also the Absolute line; and these three lines form a self-polar triangle in regard to the hyperbola. Whence, by a statement we have made, the hyperbola is outpolar to the parabola. The reader may prove that the circle and the parabola are polar reciprocals of one another in regard to the hyperbola. On the whole then, the three conics form a set having entirely symmetrical relations with one another.

We have considered two conics of which one is outpolar to the other and also triangularly circumscribed to this. We may also begin by considering two conics of which one is both outpolar and inpolar to the other, and obtain the third conic of the set.

As will be seen below, Chapter xvi, the algebraic expression of the relations between the conics is very simple.

Ex. 1.   It is known that the tangents of a conic, at the angular points A, B, C of a triangle inscribed in the conic, meet the respectively opposite sides in points L, M, N which lie on a line. Prove that any conic, which touches the four lines consisting of BC, CA, AB and the line LMN, is both outpolar and inpolar to the original conic. Conversely, if two conics be in this reciprocal relation to one another, then there exist triangles ABC, inscribed in one of these conics, such that the other conic touches BC, CA, AB and the corresponding line LMN. Prove that, then, the polar reciprocal of either conic, in regard to the other, is a conic in regard to which all the triangles ABC are self-polar.

Ex. 2.   If the orthocentre of a triangle PQR, which is inscribed to a parabola, lie on the directrix of the parabola, prove that the parabola and the self-conjugate circle of the triangle PQR are two conics of which either is both outpolar and inpolar to the other.

Ex. 3.   Prove that the general condition for a parabola to be outpolar to a circle is, that the diameter of the circle which is parallel to the directrix of the parabola should meet the curves in two harmonically conjugate pairs of points.

Ex. 4.   The normals at three variable points P, Q, R of a conic meet in a point which lies on the normal at a fixed point of the conic. Prove that the lines QR, RP, PQ are tangents of a fixed conic; so that the poles of these lines, in regard to the original conic, describe a third fixed conic. And prove that these three conics are such that any two are polar reciprocals in regard to the third.

(13) *Any conic passing through the common points of two conics, which are both outpolar to the same third conic, is also outpolar to this.* For, let A be a common point of two conics ω, ω', which are both outpolar to a conic Ω. Let the polar of A, in regard to Ω, meet ω, ω', respectively, in P, Q and P', Q', and meet Ω in U, V. Then U, V are harmonic conjugates both in regard to P, Q, and in regard to P', Q'. Thus U, V are likewise harmonic conjugates in regard to the points, P'', Q'', in which this polar line is met by any conic, ω'', which contains the common points of ω and ω'; for P'', Q'' are a pair of the involution determined by the pairs P, Q and P', Q'. Wherefore ω'' contains the angular points, A, P'', Q'', of a triangle which is self-polar in regard to Ω. Thus ω'' is outpolar to Ω. Dually, any conic touching the common tangents of two conics, which are both inpolar to another conic, is also inpolar to this.

(14) *Two lines which are conjugate to one another, in regard to a conic Ω, form a degenerate conic which is outpolar to Ω.* Let the lines meet in O, and meet the polar of O, in regard to Ω, in D and C. Then the polar of C, in regard to Ω, is the line OD. Let A, B be any two points on OD, which are conjugate to one another in regard to Ω. Then the triangle ABC, whose angular points lie on the degenerate conic formed by the two lines, is self-polar in regard to Ω. Another such triangle, self-polar in regard to Ω, is formed by D and any two points, on OC, which are conjugate in regard to Ω. Conversely, if a conic outpolar to Ω degenerate into two lines, these are conjugate in regard to Ω.

The result in (13) therefore leads to the statement that, if ω, ω' be two conics both outpolar to a conic Ω, then each of the pairs of complementary common chords of ω and ω' consists of two lines which are conjugate to one another in regard to Ω. And, hence, leads also to the statement that, *if two pairs of lines be taken, each consisting of lines which are conjugate to one another in regard to a conic Ω, then, any conic passing through the four points in which one pair of the lines meets the other pair, is outpolar to Ω. In particular, the third pair of lines containing these four intersections consists of two lines which are conjugate to one another in regard to Ω.*

We may give a direct proof of this last statement: Let A, B, C, D be four points, which are the common points of two pairs of lines AB, CD and AC, BD. Let A', B', C' be the poles, in regard to a conic Ω, respectively, of BC, CA, AB. We know, by (1) above, that AA', BB', CC' meet in a point. But, if AB, CD be conjugate lines, in regard to

Ω, then CD contains C'; and, if AC, BD be conjugate, in regard to Ω, then BD contains B'. Whence AA' contains D. So that BC, AD are conjugate lines in regard to Ω.

For convenience, a quadrangle, such that each of the three pairs of complementary joins, of the points of the quadrangle, consists of lines which are conjugate to one another, in regard to a conic Ω, may be called *self-conjugate* in regard to Ω.

(15) It follows then, from (14), that any conic ω, which passes through the four points of a quadrangle which is self-conjugate, in regard to a conic Ω, is outpolar to Ω. And, then, that the four points, common to ω and any pair of lines which are conjugate in regard to Ω, form a self-conjugate quadrangle in regard to Ω. Thus ω, or any conic outpolar to Ω, contains the points of an infinite series of quadrangles which are self-conjugate in regard to Ω. And, if a triangle ABC, inscribed in ω, be self-polar in regard to Ω, and D be any point of ω, then the line AD, as containing the pole, A, of BC, is conjugate to BC; so that $ABCD$ is a special quadrangle, self-conjugate in regard to Ω, which is inscribed in ω.

(16) We have defined a conic ω as outpolar to a conic Ω when a triangle, which is self-polar in regard to Ω, can be inscribed in ω. But it now appears, from (15), that we may define a conic ω as outpolar to a conic Ω, when ω contains the four points of a quadrangle which is self-conjugate in regard to Ω. We can then shew that *there are quadrilaterals, circumscribed to Ω, with the property that the three pairs of complementary intersections of the lines of the quadrilateral each consist of two points which are conjugate in regard to ω.*

For, consider any quadrangle inscribed in ω, which is self-conjugate in regard to Ω; let any pair of complementary joins of the points of this quadrangle, which are conjugate lines in regard to Ω, meet in the point E; and another such pair meet in F. The tangents drawn to Ω from E are then harmonic conjugates in regard to the lines, conjugate in regard to Ω, which meet in E; and the tangents to Ω from F are likewise harmonic conjugates in regard to the complementary joins of the quadrangle which meet in F. Let one of the tangents to Ω, from E, meet one of the tangents to Ω, from F, in the point L; and the other tangent from E meet the other tangent from F in the point L'. Then, the points L, L' are conjugate to one another in regard to two, and therefore in regard to all, of the conics which contain the points of the quadrangle; they are a pair of points such as we have previously called conjugate in regard to this quadrangle. Wherefore, the quadrilateral formed by the two tangents to Ω, from E, taken with the two tangents to Ω, from F, is such that every pair of complementary intersections, of the lines of the quadrilateral, consists of two points which are conjugate to one another in regard to ω. Thus, *we may speak of Ω as being inscribed in a quadrilateral which is self-conjugate in regard to ω, just as ω was circumscribed to a quadrangle self-conjugate in regard to Ω.*

Now, we have shewn, (13), that any conic through the common points of two conics which are outpolar to a conic Ω, is itself outpolar to Ω, as being circumscribed to a triangle self-polar in regard to Ω. And, the two conics outpolar to Ω may be taken to consist each of two lines conjugate to one another in regard to Ω. Wherefore, dually, any conic Ω inscribed to a quadrilateral which is self-conjugate in regard to ω, is inscribed to a triangle which is self-polar in regard to ω. Thus, also, *if a conic ω be circumscribed to a self-polar triangle of a conic Ω, then Ω is inscribed to a self-polar triangle of ω. This justifies the usage we have adopted, of saying, when ω is outpolar to Ω, that Ω is inpolar to ω*; and makes no appeal to the theory of polar reciprocal conics.

The theorem that, if two pairs of complementary intersections, of the lines of a quadrilateral, consist each of points which are conjugate to one another in regard to a conic, then the third pair has the same property, was given by Hesse (*Crelle's Journ.* xx (1840), p. 301). An interesting form of the theorem is as follows: Let A, B, C, D be four points such that AB, CD are parallel, and also AD, BC are parallel, while both of AB, CD are perpendicular to both of AD and BC; so that the four points A, B, C, D form what is called a *rectangle*. Let P, P' be points of the diagonal AC such that $P' = (A, C)/P$, and Q, Q' be points of the diagonal BD such that $Q' = (B, D)/Q$. Then the four points P, P', Q, Q' lie on a circle.

Ex. Let D, D' be points conjugate in regard to a conic ω; let Ω be the degenerate conic consisting of the two pencils whose centres are at D and D'. Construct a triangle, self-polar in regard to ω, whose sides are lines of the conic Ω; so that Ω is inpolar to ω. Construct, also, a triangle, inscribed in ω, of which every two sides are conjugate in regard to Ω; so that ω is outpolar to Ω. In particular, ω *is outpolar to the pencil of lines taken twice over, whose centre is any point of ω. Dually, any tangent of a conic, taken twice over, is a degenerate conic outpolar to the given conic.*
Thus, taking any two points, O, O', which are conjugate to one another in regard to a given quadrangle, any conic through the four points of the quadrangle is outpolar to the degenerate conic which consists of the two pencils of which O, O' are the centres. And, dually, any conic touching the four lines of a quadrilateral is inpolar to the line pair formed by two lines which are conjugate to one another in regard to the quadrilateral.

(17) We return to the condition that the harmonic envelope, and the harmonic locus, of two conics, should be degenerate, already dealt with (Chap. x, and (7), above). Let two conics, ϕ, ψ, have a common self-polar triangle. Let C be one angular point of this, and c the opposite side. There are two common chords of the two conics, say l_1 and l_2, which meet in C; and there are two points lying on the side c, say L_1 and L_2, through each of which there pass two common tangents of the two conics. The poles, P_1, Q_1, of the line l_1, taken in regard to ϕ and ψ, both lie on the side c; as do the poles of l_2, say P_2, Q_2. Likewise, the polars p_1, q_1, of the point L_1, taken in regard to ϕ and ψ, both pass through C; as do the polars of L_2, say p_2 and q_2. Suppose, now, that the points where ϕ meets the side c, are harmonic conjugates in regard

to the points where ψ meets this line. Then the tangents drawn to ϕ from C are harmonic conjugates in regard to the tangents from C to ψ. In this case, the points P_1, Q_1, P_2, Q_2, coincide in pairs: say P_1 and Q_2 coincide, and also Q_1 and P_2. The harmonic envelope, of ϕ and ψ, consists then of two pencils whose centres are at these points of coincidence. Likewise, dually, the four lines p_1, q_1, p_2, q_2 coincide in pairs: say p_1 and q_2 coincide, as do q_1 and p_2. The harmonic locus, of ϕ and ψ, then degenerates into these two lines of coincidence. The special conditions may arise in three ways, according to the side of the common self-polar triangle of ϕ and ψ which is considered.

We have spoken of the case of two circles, when the side of the common self-polar triangle, which is considered, is the line of centres of the two circles. Then the circles are perpendicular to one another. But, if we take the case when the side of the common self-polar triangle, which is considered, is that through the limiting point, H, of the two circles, then the lines l_1, l_2 are those which join the other limiting point to the Absolute points of the plane. The condition of degeneration, of the harmonic locus and envelope, in this case, is that the side of the common self-polar triangle, which contains H, should meet the circles in two harmonically conjugate pairs of points. *It can be proved that this is equivalent to saying that the limiting point H is the mid point of the centres of the two circles.* In general, upon any side of the common self-polar triangle of the two conics ϕ, ψ, there are two related ranges of points (E), (E'), defined by the fact that E, E' are conjugate to the same point of the side, in regard to ϕ and ψ, respectively. The condition for the degeneration under consideration is that E, E' should be harmonic conjugates in regard to the two angular points of the self-polar triangle which lie on this side. For instance, when we consider a side of the common self-polar triangle of two circles which contains a limiting point H, the related ranges (E), (E') are defined by the fact that the lines, joining E, E' to the centres of the two circles, are parallel to one another. If E, E' are harmonic in regard to the vertices of the triangle which lie on the side through H, then H is the mid point of E, E'. It is therefore the mid point of the centres of the two circles.

Ex. 1.   Through four arbitrary points, lying on a conic, ω, there can be drawn three other conics ω_1, ω_2, ω_3, so related to ω that the harmonic envelope and locus, of the pair consisting of ω and any one of ω_1, ω_2, ω_3, are degenerate.

This result can be deduced from the following: Take four arbitrary points, and a complementary pair of joins of these four points; take also an arbitrary line, p. Then, there is one conic, through the four given points, for which the pole of the line p is on one of the joins spoken of; and another conic, through the four given points, for which the pole of p is on the complementary join. For these two conics the harmonic locus and envelope degenerate, as is easy to see. Let C be the common point of the complementary joins selected, and A, B be the other two angular points of the vertex triangle of the quadrangle formed by the four given points. The harmonic locus of the two conics consists of the two lines through C which are harmonic conjugates both in regard to the two complementary joins selected and in regard to the lines CA, CB. And it is thus independent of the line p. The harmonic envelope of the two conics consists of two pencils whose centres are, the point where the line p meets AB, and the harmonic conjugate of this point in regard to A and B. This formulation includes the case of two circles which are perpendicular, the line p being the line, through the centre of one of the circles, which is perpendicular to the line of centres.

A particular case is that of two conics whose centres lie, one in each of a pair of complementary common chords of the two conics. For instance, the conics may be a rectangular hyperbola, and a circle described on a chord of this, as diameter. For, it can be proved that the complementary common chord of the circle and hyperbola contains the centre of the hyperbola. In this case, any line, perpendicular to the diameter of the circle, cuts the two curves in harmonically conjugate pairs of points.

Ex. 2.　When the harmonic locus of two conics breaks up into two lines, the range of four points determined by the two conics on one of these lines is related to the similar range on the other line. And there is a dual theorem in regard to the rays of the two pencils of the degenerate harmonic envelope.

Ex. 3.　If two conics have their four common points coincident, the harmonic locus and envelope of the two conics coincide, and have the same four coincident common points with the two given conics.

(18) *Let ω be a given conic, and ω_1, ω_2, ... be a system of conics with four common tangents. The harmonic loci, of ω and every one of the conics ω_1, ω_2, ... in turn, form a series of conics with four common points.* For let A be a common point of the harmonic locus of ω and ω_1, with the harmonic locus of ω and ω_2. Then the two tangents from A to ω are harmonic conjugates both in regard to the two tangents drawn from A to ω_1 and in regard to the two tangents drawn from A to ω_2. The two tangents from A to ω are thus the focal rays of the involution of pairs of tangents drawn from A to all the conics which touch the common tangents of ω_1 and ω_2. So that A is likewise a point of the harmonic locus of ω and any conic of the system ω_1, ω_2, ω_3, ... And the same is true for all the four points common to the harmonic locus of ω and ω_1 with the harmonic locus of ω and ω_2.

Dually, the harmonic envelopes of a particular conic and every one in turn of a series of conics which have four points in common, form a system of conics having four common tangents.

In the system of conics touching the four common tangents of ω_1 and ω_2 are three degenerate conics, each consisting of two pencils of lines. The harmonic locus for a conic ω and a degenerate conic, of two pencils, with centres at O and O', is the conic we have called the conjugate line locus of ω, with centres at O, O'. Hence, the two harmonic loci, of ω and ω_1, ω_2 in turn, meet in the four points common to two conjugate line loci of ω; and, evidently, the pairs of centres, for these two conjugate line loci, determine the common tangents of the system (ω_1, ω_2).

If ω itself consists of two pencils, with centres, say, at I and J, the harmonic loci of ω, taken in turn with all the conics of the system (ω_1, ω_2), is still a series of conics with four common points. Thus, if we take the conjugate line loci of all conics of a system (ω_1, ω_2), with four common tangents, the centres being the same points I, J in every case, we obtain a series of conics, with four common points; and I, J are two of these four points. This result has been proved already, with the enunciation: *The director circles of conics having four common tangents are coaxal.*

For a further particular case, consider the harmonic loci of a conic ω and all the conics in turn of a system (ω_1, ω_2), with four common tangents, the conic ω being one of this system. Then, regarding this system as a confocal system, it may be proved that the harmonic loci of ω and the other conics of this system, in turn, all pass through the common points of ω with its directrices; thus the director circle of ω belongs to the series of harmonic loci.

(19) A series of conics having four common points may, for a simple reason which will appear, be called a *linear* series. Thus the harmonic loci, of a conic and all the conics of a system with four common tangents, taken in turn, form such a *linear* series. Of this series there is just one conic which passes through an arbitrary point. We consider now the conics which are outpolar to a given conic. We prove that one such conic passes through four arbitrary points; more generally, if r be one of the positive integers from 0 to 5, and $s = 5 - r$, we prove that there is one conic which is outpolar to r given arbitrary conics and passes through s arbitrary points ($0 \leqslant r \leqslant 5$, $s = 5 - r$). And we speak of the conics outpolar to one or more conics as forming a *linear series*; and of the *linearity of the condition* for a conic to be outpolar to a given one. The matter is very simple when we use the *equation of a conic* (see Chap. XVI, below); but there is distinct advantage in a preliminary descriptive account.

We prove first (for the case $s = 4$), that there is one conic through four arbitrary points which is outpolar to a given conic Ω. Let A be one of the four points; take the polar line of A in regard to Ω. The conics through the four points determine on this polar line an involution, of which one pair, say U, V, consists of points which are conjugate in regard to Ω. The triangle AUV is then self-polar in regard to Ω, and the conic through the four points, which contains U and V, is the conic required, outpolar to Ω; and this is evidently the only such conic.

Next, consider the conics through three points A, B, C, which are outpolar to Ω. Take an arbitrary line, p; through A, B, C, and an arbitrary point X of this line p, there is one conic outpolar to Ω; let this meet the line p again in X'. There is only one conic, through A, B, C and X', which is outpolar to Ω; this is then the conic, through A, B, C, X, which determined X'. We infer that the pairs of points X, X', on the line p, of which each thus determines the other without ambiguity, form an involution on this line. Or, *the conics through three points, outpolar to Ω, determine an involution of pairs of points on an arbitrary line.* Now take a second conic, Φ, and, for the line, take the polar of the point A in regard to Φ. There is then a single conic, through the three points A, B, C, which is outpolar to Ω and meets the polar line of A, taken in regard to Φ, in two points U, V which are conjugate in regard to Φ. The triangle AUV is self-polar in regard to Φ; and the conic in question is that conic through A, B, C which is outpolar in regard both to Ω and Φ. This proves the theorem for the case $r=2$, $s=3$.

By a similar argument, the conics, through two points A, B, which are outpolar both to Ω and Φ, determine an involution on an arbitrary line; and there is thence a single conic through A, B which is outpolar to Ω, Φ and a third conic Ψ ($r=3$, $s=2$).

Likewise, there is a conic through a single point, A, which is outpolar to four conics Ω, Φ, Ψ, Σ; and, thence, the conics outpolar to these four determine an involution on an arbitrary line. We can prove, however, that *if a series of conics be such as to determine an involution on every arbitrary line, then these conics have four points in common*. For, the involution on the line is determined by two of its pairs, and, thence, by two of the conics. Take the line to contain one of the common points of two of the given conics; then any pair of the involution on this line has this point as one of its two points. Hence, all the given conics pass through all the common points of any two of the given conics. Wherefore, the conics outpolar to four conics Ω, Φ, Ψ, Σ have four points in common; so that one such conic can be drawn through an arbitrary point. Whence it can be proved, as before, that a conic can be drawn outpolar to five arbitrary conics.

To these theorems there are duals, relating to conics which are inpolar to given conics and touch given lines.

Incidentally, we see that the limitation in the freedom of a conic, involved in the condition that the conic shall meet a given line in a pair of points of a given involution on this line, is the same as that imposed by the condition that the conic shall contain a given point.

(20) The conics which pass through four points may, temporarily, be called *a punctual series of conics, founded on two conics*; and the conics which touch four lines may be called *a tangential system of conics, founded on two conics*. We may consider a punctual series of conics founded on more than two conics; and the applications are important. Thus we may have a punctual series, founded on three conics, say ω_1, ω_2, ω_3. This is often called a *net* of conics; just as the conics through four points are called a *pencil*. The net (ω_1, ω_2, ω_3) is defined by taking any general conic, ω, of the pencil (ω_2, ω_3), and then the pencil (ω_1, ω). It can be shewn, (a) that any conic of the net is so obtainable in three ways, corresponding to the three pairs from ω_1, ω_2, ω_3; (b) that the net can be founded equally on *any* three conics of the net, taken in place of ω_1, ω_2, ω_3, provided these do not belong to a pencil; and (c) that a conic of the net exists passing through any two given points of the plane. For this last reason, the conics of the net

are often said to be an ∞^2 aggregate; in the same way, the conics of a pencil are an ∞^1 aggregate.

As a particular example we may consider the punctual net of conics, founded on three *circles*, these not being coaxal. The general conic of the pencil determined by two of the circles will then be a circle (coaxal with these two), and the general conic of the pencil, determined by this and the third given circle, will then also be a circle. So that all the conics of the net are circles. Now, the three original circles are all perpendicular to a definite circle; and, a circle which is perpendicular to both of two circles is perpendicular to any circle coaxal with these. Thus all the circles of the net are perpendicular to a definite circle. Conversely, all the circles which are perpendicular to a given circle may be seen to be a net, founded on three circles; and for these three circles we may take any three circles which are perpendicular to the given one and are not coaxal.

Or again, if ω_1, ω_2, ω_3 be three *circles* which are all outpolar to a given conic Ω, it follows, by the definition of a net (see (13) above), that there is a net, founded on ω_1, ω_2, ω_3, of which all the circles are outpolar to Ω. Conversely, the aggregate of all *circles* outpolar to a given conic is a net, founded on three of these circles.

Among the circles outpolar to a given conic are line-pairs; the common point of such a pair then lies on the director circle of the conic. Moreover, it can be seen that, the pair of lines joining any point of a circle to the Absolute points, is met by a line, through the centre of the circle, in a pair of points which are harmonic conjugates in regard to the points in which this line meets the circle (this is a case of Seydewitz's theorem). Thus, *the director circle of a conic is perpendicular to the degenerate circles which are outpolar to the conic* (these degenerate circles consisting of lines meeting on the director circle). By what we have said, the aggregate of all circles outpolar to the conic can be founded on three such degenerate circles; wherefore, *the director circle of a conic is perpendicular to every circle outpolar to the conic.* This has been proved already (Chap. x (6), Ex. 15). Conversely, *any circle which is perpendicular to the director circle of a conic is outpolar to the conic,* being a member of the net of such circles. This latter statement can be proved independently by remarking that a common chord of the conic and the director circle which contains the centre of both curves, meets the circle, say σ, supposed to be perpendicular to the director circle, in two points which are conjugate in regard to the director circle and therefore conjugate in regard to the conic; while the pole of this chord, in regard to the conic, lies on this circle σ, being at one of the Absolute points. Thus the circle σ is circumscribed to a triangle which is self-polar in regard to the conic.

The theorem that the circumcircle of the diagonal triangle of a quadrilateral $ABCD$, is perpendicular to the circles of which AC, or BD, are diameters, was proved algebraically by Gaskin, *The construction of a conic section...by the properties of involution and anharmonic ratio,*

Cambridge, 1852. An argument quite similar to the above is applicable to an outpolar sphere of a quadric surface, and the director sphere of this; or, indeed, to the analogous result in higher space.

Ex. 1.   The circumcircle of a triangle is outpolar to the self-polar circle of the triangle. Find a triangle, circumscribing the self-polar circle, which is self-polar in regard to the circumcircle.

Ex. 2.   Two lines, x and p, meet the sides BC, CA, AB of a triangle respectively in L, L'; in M, M'; and in N, N'. On BC take the points D, D' given respectively by $D = (B, C)/L, D' = (B, C)/L'$; with similar pairs, E, E' and F, F', on CA and AB, respectively. Prove that the six points $D, D',$ E, E', F, F' lie on a conic. We may speak of the conic, and the line pair (x, p), as being *harmonically conjugate in regard to the triangle*. The conic has the property that AD, BE, CF meet in a point, as do AD', BE' and CF'.

Ex. 3.   We have proved that the locus of the poles of a line x, in regard to all conics containing four points A, B, C, P, is a conic. Let p be the line which is called the *polar of the point P in regard to the triangle*, meeting BC in L such that the lines AL, AP are harmonic in regard to AB, AC, and meeting CA, AB in correspondingly defined points, M and N, respectively. Prove that the locus of the poles of the line x, in regard to the conics through A, B, C, P, is the harmonic conjugate of the line pair (x, p), in regard to the triangle ABC. Denote this locus by (N).

Ex. 4.   Let the common points of the pole locus (N), of Ex. 3, with the line x, be denoted by I and J; and the common points of this locus with the line p be denoted by U and V. The points I, J are the foci of the involution, determined on the line x by the conics through A, B, C, P. Prove, (a) that the seven points A, B, C, I, J, U, V lie on a conic. This we denote by Ω; (b) that there exists a conic through the four points I, J, U, V in regard to which the triangle ABC is self-polar. This conic we denote by (S). Further, if O be the common point of the lines x, p, and the tangent at A, of the conic Ω, meet BC in X, prove that the lines OX, OA are harmonic conjugates in regard to x and p. There is a similar statement for the points Y, Z, where the tangents of Ω at B and C, respectively, meet CA and AB. The points X, Y, Z lie on a line; which we denote by h.

Ex. 5.   Now consider the conics touching the four lines BC, CA, AB and h. Prove that the locus of the poles of the line x, in regard to these conics, is the line p. Also, if Q_1, Q_2 be the points (IV, JU) and (IU, JV), prove that the conjugate line locus, for any conic touching BC, CA, AB, h, with centres at I and J, passes through Q_1 and Q_2.

Ex. 6.   If now, in the preceding examples, the points I, J be taken as the Absolute points, and, therefore, x as the Absolute line, the point P is the orthocentre of the triangle ABC; and the conics through A, B, C, P are rectangular hyperbolas. The locus of the centres of these, that is, of the poles of x, is the nine-point circle of the triangle ABC. The line p is now the polar of the orthocentre in regard to the triangle ABC; and the nine-point circle meets p in the points U, V; as in a previous account. Through these points there passes also the self-conjugate circle of the triangle (Chap. x (6), Ex. 16). The line p contains the mid point of A and X (with similar statements for B, Y and C, Z). As A and X are a complementary pair of intersections of the four lines BC, CA, AB, h, the locus of centres of conics touching these four lines is the line p, as in Ex. 5; and this line contains the centres of the director circles of this system of conics. These are a coaxal series of circles, with Q_1, Q_2 as common points (besides I and J). And Q_1, Q_2 are the limiting points of the coaxal circles through U and V.

Ex. 7.   Consider a rectangular hyperbola touching the sides BC, CA, AB of a triangle. If its centre is on the circumcircle, or, also, if its centre is on the nine-point circle of the triangle, it can be shewn that the hyperbola contains the circumcentre of the triangle.

The rectangular hyperbola is inpolar to the self-conjugate circle of the triangle; this is then outpolar to the hyperbola, and so contains the centre of the hyperbola. This centre, therefore, in both the cases spoken of, is one of the common points, U, V, of the three circles named. Wherefore, the circumcircle, as containing the centre of the rectangular hyperbola, is outpolar to this. But, the circumcircle is also triangularly circumscribed to the hyperbola. Whence, by what was shewn above (in (16)), the hyperbola is outpolar to the circumcircle. Now, the Absolute line meets the hyperbola and the circumcircle in two harmonically conjugate pairs of points. Therefore, the hyperbola, being outpolar to the circumcircle, contains the centre of this circle. This is the result stated.

More generally, for any conic touching the sides of a triangle to contain the circumcentre, it is sufficient that its director circle should touch the circumcircle. And then, the self-polar circle being perpendicular to the director circle, this circle also touches the nine-point circle.

Ex. 8. Let ω, ω' be two conics which are polar reciprocals in regard to a conic Ω. Let σ be the harmonic envelope of Ω and ω; and σ' be the harmonic envelope of Ω and ω'. Prove that Ω is the harmonic locus of σ and σ'.

Ex. 9. Let ω be a degenerate conic, of two lines, which meet in O, and are not conjugate in regard to a certain other conic Ω. Let U, V be the points of contact with Ω of the two tangents drawn thereto from O. Let σ be the harmonic envelope of Ω and ω. Prove that Ω is the conjugate line locus of σ, with centres at U and V.

Ex. 10. We have seen (Chap. xiii (2), Ex. 1) that the subdirector circle of a conic, with centre at one focus, S, of this conic, is triangularly circumscribed to the conic; and that the triangles in the circle, circumscribed to the conic, are all self-polar in regard to the circle, with centre at the complementary focus S' of the conic, which is perpendicular to the director circle of the conic. Prove that the subdirector circle is the polar reciprocal of the original conic in regard to the latter circle, of centre S'.

Ex. 11. *Through four given points there pass two conics which are triangularly circumscribed to a given conic.* Denote the given points by A, B, C, D, and the given conic by ω. Let the tangents to the conic ω, from the point D, be t and t'. Through an arbitrary point P, taken on the tangent t, a conic can be drawn containing A, B, C, D; this conic will meet the tangent t' in a point P', besides D. Thus, the series of conics through A, B, C, D determine two related ranges (P), (P'), on t and t', respectively. Wherefore the line PP' envelopes a conic, touching t and t'. This conic has then two other tangents which are also tangents of ω. Let these two lines meet t in P_1, P_2, and, respectively, meet t' in P_1', P_2'. The conic, say Ω_1, containing A, B, C, D, P_1, P_1', is then triangularly circumscribed to ω; and so is the conic, say Ω_2, containing A, B, C, D, P_2, P_2'. Conversely, any conic through A, B, C, D which is triangularly circumscribed to ω, must be Ω_1 or Ω_2. This proves the statement made.

It will appear below (Chap. xvi), that the condition, for a conic Ω to be triangularly circumscribed to a conic ω, is the vanishing of a certain polynomial in the six coefficients which occur in the equation of the conic Ω, this polynomial being homogeneously of the second degree in these coefficients. This would furnish an immediate proof of the theorem stated.

Dually, there are two conics triangularly inscribed to a given conic, in a system of conics which touch four lines. In another phraseology, two parabolas exist which touch the sides of a given triangle and have the focus on a given circle. In fact, this circle has two points lying on the circumcircle of the triangle formed by the given tangents.

Ex. 12. Let four tangents of a parabola, whose focus is S, be taken; and Ω be a circle passing through S. If four other parabolas be constructed, each to touch three of the four given tangents, and have its focus on the circle Ω, these parabolas have a common tangent (besides the Absolute line).

Ex. 13. Of the conics of a confocal system, there are two which are triangularly circumscribed to an assigned conic of the system.

(21) Repeating in part what has been said (Chap. IX (23), Ex. 18), to enter into somewhat more detail, we remark that, if Ω *be the conjugate line locus of a conic ω, for centres I, J, it is possible to find four points A, B, C, D, on the conic Ω, such that the four joining lines AB, BC, CD, DA all touch ω; and, of such sets of four points, there is an infinite series, one of the four points being an arbitrary point of Ω.* This is clear by regarding I, J as the Absolute points, and considering perpendicular tangents of ω. Conversely, if two conics Ω, ω be such that four points can be taken on Ω such that the four joins AB, BC, CD, DA touch ω, then Ω is the conjugate line locus of ω for two properly chosen centres. For let E, F be respectively the points (AB, DC) and (AD, BC); let Ω meet EF in I and J; and let P be any point lying on Ω. Then E, F are conjugate points in regard to Ω; thus DC, DA (or DE, DF) are harmonic conjugates in regard to DI, DJ. Whence, as points of the range on the conic Ω, the points A, C are harmonic conjugates in regard to I and J. By similar reasoning, the points B, D are also harmonic conjugates in regard to I and J. Therefore, PI, PJ are the focal rays of the pencil in involution determined by the pairs of rays PA, PC and PB, PD. But A, C and B, D are two pairs of complementary intersections of the four tangents, of the conic ω, AB, BC, CD, DA; thus PI, PJ are the focal rays of the pencil in involution formed by the pairs of tangents, from P, to the conics which touch these four lines. In particular, the lines PI, PJ are conjugate in regard to the conic ω. This shews that Ω is the conjugate line locus of ω, for centres I, J. And thus, also, an infinite series of sets of four points exists, lying on Ω, whose four joins, when the points are taken in order, are tangents of ω.

The point where AC, BD meet is one angular point of the vertex triangle of the quadrangle $ABCD$, and is also one angular point of the diagonal triangle of the quadrilateral formed by AB, BC, CD, DA. Thus this point, and the line EF, are pole and polar in regard to both the conics ω and Ω. And if, on EF, we take the pair of points which are harmonic conjugates in regard to I, J, and are also conjugate in regard to ω, these two points form, with the common point of AC, BD, the common self-polar triangle of ω and Ω. The point (AC, BD) is the common point of a complementary pair of common chords of the two conics, the poles of these chords, in regard to ω, being the points I, J. For we have seen that the conjugate line locus of a conic ω, for centres I, J, contains the points of contact of the tangents drawn to ω from these points. There are thus three ways in which it may happen that, of two conics Ω, ω which have a common self-polar triangle, Ω is the conjugate line locus of ω. The centres are the points where Ω meets a selected side of the common self-polar triangle. And a sufficient condition for the relation is, that Ω should contain the pole, in regard to ω, of one of the common chords of the two conics; then, Ω also contains the pole, in regard to ω, of the complementary common chord.

As an example, take the case of two circles. When the selected side of the common self-polar triangle is the line of centres of the two circles, then *the circle Ω, if it is the conjugate line locus of the circle ω, must contain the pole of the Absolute line in regard to ω, that is, the centre of ω*; Ω will then also contain the pole of the radical axis in regard to ω. These two points are thus the centres for which Ω is the conjugate line locus of ω. Suppose, however, that the selected side, of the common self-polar triangle of the two circles, is the line, h, drawn through the limiting point, K, perpendicular to the line of centres. Then, the pair of common chords of the two circles which meet in the other limiting point, H, are the lines through H to the Absolute points, which contain the common points of the two circles. The pole, in regard to ω, of one of these chords, lies on the line h, and also on one of the asymptotes

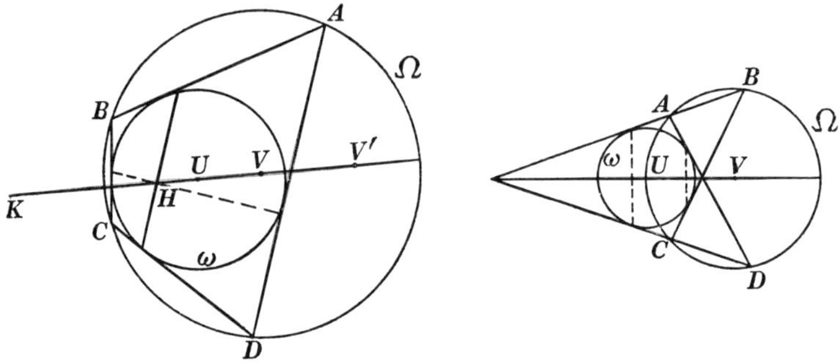

of ω. The necessary condition, for the circle Ω to be a conjugate line locus of the circle ω, is, therefore, that one of the common points of Ω and the line h should lie on an asymptote of ω; the other common point will then lie on the other asymptote of ω; and it will be with these two points as centres that Ω will be the conjugate line locus of ω. In other words, *the radical axis of the two circles (one degenerate) which are constituted by ω and the two asymptotes of the circle ω, must be the side h of the common self-polar triangle of Ω and ω.* If the circles are real, the radical axis, of ω and the "point circle" spoken of, is real; therefore the line h is real. So that the circles Ω, ω have no real common points. The reader may verify that, *if U, V be the centres of the circles ω, Ω, respectively, and V' be the inverse point of V in regard to ω, then the condition stated is the same as that the mid point of K, V' coincides with the mid point of H, U.*

Ex. 1.   When A, B, C, D are points of the conic Ω such that AB, BC, CD, DA touch the conic ω, the line joining the points of contact of AB, CD meets the line joining the points of contact of AD, BC in the selected angular point of the common self-polar triangle of Ω and ω. These joining lines are harmonic conjugates in regard to the common chords of the two conics which meet in this angular point. Also the lines AC, BD pass through this point. Thus, in the second case considered for two circles, these joins of opposite points of contact meet in the limiting point H, and are perpendicular to one another.

Ex. 2.   There exists a conic, independent of the set A, B, C, D, in regard to which AB, CD are conjugate lines, and also BC, AD are conjugate lines.

Ex. 3.   As the relation of Ω to ω is the dual of the relation of ω to Ω, and the tangents to ω from any point of Ω are harmonic conjugates in regard to the lines joining this point to the two intersections of Ω with a certain side of the common self-polar triangle of Ω and ω, it follows that, the two points, where Ω is met by any tangent of ω, are harmonic conjugates in regard to the two points where this tangent meets the tangents to ω from the opposite angular point of the self-polar triangle.

A particular case is that, any tangent of a conic meets the director circle and the asymptotes of the conic in two harmonically conjugate pairs of points. Hence, any fixed tangent of a conic is met, by variable pairs of parallel tangents, in pairs of points of an involution, whose foci are on the asymptotes of the conic. Also, any pair of points of the involution lie on a pair of conjugate diameters of the conic.

Ex. 4.   Let Ω be a given circle, and C a given point. Let P, P' be two variable points of the circle such that CP, CP' are perpendicular. Prove that PP' envelopes a conic, ω, for which Ω is the conjugate line locus, with centres at the two points of Ω whereat the tangents pass through C. The polar reciprocal of the conic ω, in regard to the circle Ω, is another circle, which is a conjugate line locus for the circle Ω.

Ex. 5.   If Ω be a conjugate line locus, for a conic ω, and the tangents, to ω, from a variable point P of Ω, meet Ω again in Q and R, prove that QR passes through a fixed point. This is one of the angular points of the common self-polar triangle of Ω and ω.

When Ω, ω are any two general conics, prove that *the line QR envelopes another conic, ψ. Shew that this conic ψ contains the four common points of Ω and ω; also that ψ touches the four tangents of Ω drawn at the points where the tangents of ω, at the four common points of Ω and ω, meet Ω again.*

For Ω to be *triangularly* circumscribed to ω, the conic ψ must coincide with ω. Already ψ contains the four points of ω which lie on Ω; thus ψ will coincide with ω if it contain another point of ω. For instance, assuming Ω, ω to have a common self-polar triangle, it would be sufficient for ψ to contain one (and therefore the other) of the points where ω meets one side of this triangle. When ψ coincides with ω, the four tangents of Ω named as tangents of ψ will be tangents of ω. Thus we recover a condition already given (Chap. XI (12), Ex. 10), for Ω to be triangularly circumscribed to ω.

For Ω to be a conjugate line locus of ω, the tangents of the conic ψ must consist of two pencils, which coincide with one another. This follows from the result stated at the beginning of this example. But the four tangents specified for ψ, in general, being tangents of Ω, cannot all pass through a point (Ω being undegenerate), unless they coincide in pairs. There are, then, two common points of Ω and ω whereat the tangents of ω meet on Ω, the tangents of ω at the other two common points of Ω and ω equally meeting on Ω. The tangents of ψ are then the repeated pencil of lines whose centre is the common point of the tangents of Ω, at the two points of Ω thus arising; each of these tangents counts twice among the four tangents of ψ specified in the general case. This is the relation of Ω and ω when Ω is a conjugate line locus of ω.

These considerations become more definite when approached algebraically (see Chap. XVI, below). But the preliminary descriptive approach gives greater content to the algebraic results.

Ex. 6.   Let a conic Ω be the conjugate line locus of a conic ω, for centres I, J. Let U, V be any two points of Ω. The four points, common to the pairs of tangents drawn to ω from U and V, form a quadrangle in regard to which I, J are conjugate points (that is, I and J are conjugate in regard to all conics through these four points). And Ω is a conjugate line locus not only of ω, but also of all conics touching these four tangents.

Thus, if we are given a conic ω, and also three points I, U, V, there is a unique conic Ω, passing through I, U, V, which is the conjugate line locus of ω for two centres of which I is one. This is the conic, through I, U, V, which contains the points of contact with ω, of the two tangents drawn thereto from I. The other centre, J, for this conjugate line locus, is the point conjugate to I, in regard to the quadrangle formed by the intersections of the pairs of tangents drawn to ω from U and V.

Ex. 7. A circle with centre at one focus, S, of a conic, is a conjugate line locus of the conic, if this circle meets the directrix corresponding to the complementary focus, S', at two points lying on the circle of which SS' is a diameter. The point S', and its corresponding directrix, are then the common pole and polar for the two curves.

Prove that any circle containing the complementary foci S, S', of a conic, is the conjugate line locus of the conic, for centres at the two points where this circle meets the axis of the conic which is perpendicular to SS'. Thus, any quadrangle inscribed in the circle, of which two pairs of complementary joins touch the conic, has the third pair of complementary joins parallel to SS'.

Ex. 8. If a conic Ω, which is not degenerate, be a conjugate line locus for the conic which is the harmonic envelope of Ω and another conic ω, then ω consists of two lines.

Ex. 9. If A, B, C, D be four points lying on a conic Ω, there is an infinite series of pairs of lines such that the harmonic envelope, of Ω and such a pair of lines, is tangent to the lines AB, BC, CD, DA. Such a pair of lines consists of two which pass through the common point of AC and BD, and are harmonically conjugate in regard to these.

Ex. 10. If a conic, which is a conjugate line locus of a conic Ω, degenerate into two lines, then, either, these lines are conjugate with respect to Ω, and thus form a conic outpolar to Ω, or else, one of these lines touches the conic Ω, and then the lines form a degenerate conic triangularly circumscribed to Ω.

Thus also, as has been remarked already, the conjugate line locus of a conic, for centres E, E' which lie on a tangent of the conic, consists of the tangent EE' together with the chord QQ', which joins the points of contact of the other tangents from E, E' to the conic. (This is the dual of the so-called Seydewitz's theorem.)

Ex. 11. Considering the degenerate conic consisting of the two pencils of lines passing through the Absolute points I, J, the conjugate line locus of this, with centres at points A, A', is the circle of which AA' is a diameter. When A is on the Absolute line, and B is the point of the Absolute line given by $B = (I, J)/A$, the locus degenerates into the Absolute line taken with the line BA'. If A' is also on the Absolute line, the locus becomes the Absolute line taken twice, unless $A' = (I, J)/A$. In this case, every point of the plane satisfies the condition for being on the conjugate line locus.

Ex. 12. In order to place a set of well-known theorems in proper perspective, we state the theory of circles which are conjugate line loci for a given conic as follows:

Let S, H', H, S' be any quadrangle, the intersections of complementary joins $S'S$, HH', and HS', $H'S$, being respectively I and J. Let E be any point, and F the conjugate point in regard to this quadrangle, which is conjugate to E in regard to all conics through S, H', H, S'. As I, J are equally conjugate in regard to the quadrangle $SH'HS'$, it follows (Hesse's theorem, (16), above), that if the intersections (IE, JF) and (IF, JE) be, respectively, L and M, then L, M are also conjugate in regard to this quadrangle.

Further, considering the quadrangle E, M, F, L, of which two pairs of complementary joins meet in I, J, respectively, the lines IS, IH are harmonic conjugates in regard to IE and IF, and the lines JS, JH are harmonic conjugates in regard to JE and JF. Thus, S, H are conjugate points in regard to the quadrangle E, M, F, L; as also are S', H'. When I, J are taken as the Absolute points, the lines $SH, S'H'$ are perpendicular, and their common point, C, is the mid point both of S, H and of S', H'. Similarly EF, LM are perpendicular, and their common point, P, is the mid point both of E, F and of L, M. Also, as E, F are conjugate in regard to $SH'HS'$, the lines CE, CF are harmonic conjugates in regard to CS, CS'; as also are

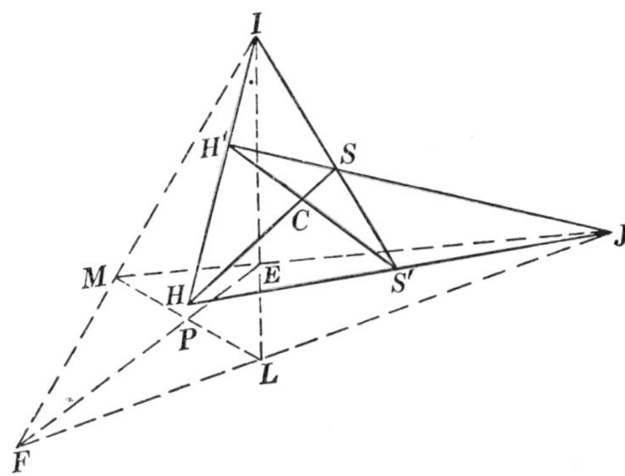

CL, CM. Thus CS, CS' are the focal rays of the involution pencil of pairs of tangents from C to all conics touching EM, MF, FL, LE. In a similar way, PE, PL are the focal rays of the involution pencil of pairs of tangents from P to all conics touching $SH', H'H, HS', S'S$. When I, J are the Absolute points, these latter conics are a confocal system, with S, H as complementary foci, and S', H' as complementary foci, of which C is the centre. There is a conic of this confocal system which touches EF at P. For this conic, the diameter conjugate to CP is the line joining C to the intersection (EF, IJ), this point being $(E, F)/P$. Also, as CS, CS' are conjugate diameters for the confocal conics, the lines CE, CF are harmonic conjugates in regard to *two* pairs of conjugate diameters of the confocal which touches EF at P, so that CE, CF are the asymptotes of this conic. Likewise CL, CM are the asymptotes of the confocal which touches LM at P.

The conjugate line locus of any one of the confocals, with E, F as centres, passes through I, J; and is a circle. Thus, if all the confocals be taken, we obtain a series of coaxal circles, with E, F as common points. The limiting points of this series are at L and M; for IE, JF meet in L and IF, JE meet in M.

The poles of the line EF, with respect to these confocals, lie on the line LM, this being the normal at P of the confocal touched by EF. We know, however, that the pole of EF, with respect to one of the conjugate line loci, defined with E, F as centres, coincides with the pole of EF in regard to that confocal for which we consider the conjugate line locus. Thus the tangents of the conjugate line locus, at E and F, meet at the point of LM which is the pole of EF for the confocal considered.

A particular one of the system of confocals is constituted by the two pencils of lines whose centres are S and H. The corresponding conjugate line locus is a circle containing S and H, these being the points of contact of the tangents drawn from E and F to this particular confocal. The tangents of this circle at E and F pass through the pole of EF with respect to this degenerate confocal, namely the point of SH which is harmonic, in regard to

S and H, to the intersection (EF, SH). As PE, PL are harmonic in regard to PS, PH, this point is on LM. Thus, in the range of points on this circle, the points E, F are harmonically conjugate in regard to S and H; the tangents of this circle at S, H, therefore, meet on EF; therefore these meet at $(EF, S'H')$; for $S'H'$ is perpendicular to SH, and contains the mid point of S, H.

Many particular results can be stated for the figure:

(*a*) The points L, M are the *antipoints* of E, F; for the lines joining L, or M, to E and F, pass through I and J. Thus, with the usual conventions of reality, if E, F be real, lying each on one asymptote of a hyperbola, then L, M are conjugate imaginaries, lying each on one asymptote of a confocal ellipse.

(*b*) We may take the point E arbitrarily. Then F is obtained as the point on the circle ESH, containing E and the complementary foci S, H, which is the harmonic conjugate of E, on this circle, in regard to S and H. Thus, if EE', parallel to the axis SH, meet this circle again in E', then F is on $E'C$. Or, if E_1 be the inverse point of E in regard to the circle of which SH is a diameter, then F is the optical image of E_1 in regard to the line SH. It is also true (Chap. XIII, (13)) that, if O, Q be the centres of curvature at E for the two confocals which pass through E, then F is on OQ, and EF is perpendicular to OQ.

Hence, also, CS is one of the bisectors of the angles determined by CE and CF; and $\epsilon E\,(F, S) = \epsilon E'\,(F, S)$, ($E'$ being on FC), $= \epsilon E\,(H, C)$. It is common to speak of two triangles ABC, $A'B'C'$ as being *similar* when $\epsilon A\,(B, C) = \epsilon A'\,(B', C')$, $\epsilon B\,(C, A) = \epsilon B'\,(C', A')$, and $\epsilon C\,(A, B) = \epsilon C'\,(A', B')$. The two triangles SCE, FCS are then similar; as are the triangles EPS, HPE (P being the mid point of E, F). Further, the tangents, at E, of the confocals which pass through E, are the bisectors of the angles determined by EC, EF.

(*c*) The asymptotes, CE, CF, of the hyperbola, meet the auxiliary circle of the confocal ellipse through P, in two points, lying on a line, perpendicular to SH, which contains P. If P varies on this ellipse, the points E, F describe circles with centres at C.

(*d*) Let D, D' be a pair of the involution of which L, M are the foci, on the tangent at P of the ellipse; so that CD, CD' are conjugate diameters of the ellipse. Consider the point $(DS, D'H)$, as D varies on the tangent. This point lies on the circle, through S, H, whose centre is the point $(EF, S'H')$. A pair of this involution consists of the foot of the perpendicular from H to the tangent at P, taken with the point of this tangent which lies on the directrix of the ellipse corresponding to S. A general pair of the involution is given by the points of the tangent lying on any pair of parallel tangents of the ellipse. The involution is thus the dual of that obtained, on a conic, by joining a fixed point of the conic to the points where the conic is met by the rays of a pencil of lines.

(*e*) The circles $EFSH$, $EFS'H'$ are perpendicular. The former is a conjugate line locus for the confocals. The circle CEF is triangularly circumscribed to the confocal hyperbola which touches EF at P.

(*f*) Consider the conic, Ω, which is the envelope of the polars of E in regard to all the confocals (touching SH', $H'H$, HS', $S'S$). This is the dual of an eleven-point conic previously considered (Chap. IX). The conic Ω is a parabola, having CE as directrix and F as focus. The conic Ω is symmetrical in regard to the four lines SH', $H'H$, HS', $S'S$; but, in associating F with E, as the point conjugate to E in regard to all the conics through the points S, H', H, S', we have selected a particular pair of complementary intersections of the four lines SH', $H'H$, HS', $S'S$. By taking the other two pairs of complementary intersections, we obtain, instead of F, two other points, say F_1 and F_2. The three points F, F_1, F_2, being on the polar of E in regard to Ω, are in line. From this more general point of view, we are led to the consideration of three pencils of conics: (*a*) the circles *EFSH*; (*b*) the circles *EFS'H'*; (*c*) the rectangular hyperbolas through *S, H, S', H'*.

Ex. 13.   Consider a series of conics, ω, which have four points in common; and another conic Ω. We have seen ((18), above) that the harmonic envelope of Ω and a conic ω of the series, belongs to a system of conics which have four common tangents. Conversely, if we are given a conic Ω, and four lines in general position, then the conics ω which are such that the harmonic envelope of Ω and ω touches the four lines, form a series of conics with four common points. In this series there are three degenerate conics, each a pair of lines. Thus, *given a conic Ω and four lines a, b, c, d, in general position, there exist three pairs of lines, such that every one of a, b, c, d is met, by the two lines of one of the three pairs, in points which are conjugate in regard to Ω.*

Dually, given a conic, σ, and four points *A, B, C, D*, in general position, there exist three pairs of points, such that, the two lines joining any one of *A, B, C, D* to the points of each one of the three pairs of points, are a pair of conjugate lines in regard to σ. Thus, given four points *A, B, C, D*, it is possible in three ways to find a pair of points such that the conjugate line locus, of a given conic σ, with the points of a pair as centres, shall pass through the given points *A, B, C, D*. Or, *through four points of general position, there are three conics which are conjugate line loci for a given conic.*

This result follows at once algebraically from the formulae given in the subsequent chapter on the equation of a conic (Chap. xvi).

As remarked in Exx. 8, 9, above, if the four given lines, or points, in the dual theorems here stated, have special relations with the given conic, the number of determined conics may be infinite.

Ex. 14.   Let the complementary pairs of intersections of four tangents of a conic, ω, be *I, J; S, H* and *S', H'*. Let the conjugate line loci of ω, for the respective pair of centres *I, J; S, H* and *S', H'*, be $\omega_1, \omega_2, \omega_3$. Prove that there are three conics, touching the given tangents of ω, for which ω is a conjugate line locus, namely, the polar reciprocals, in regard to ω, respectively, of $\omega_1, \omega_2, \omega_3$.

Ex. 15.   As we have said, the pole of a line which meets the sides *BC, CA, AB* of a triangle, respectively, in *L, M, N, in regard to the triangle,* is the point where the lines *AD, BE, CF* meet, if $D = (B, C)/L$, $E = (C, A)/M$, and $F = (A, B)/N$. Prove that this pole, for a tangent of a conic which touches the sides of the triangle, lies on the axis of perspective of the two triangles formed by *A, B, C* and by the three points of contact of the given conic with *BC, CA, AB*.

If a conic Ω be triangularly circumscribed to a conic ω, the pole of a fixed tangent of the conic ω, in regard to a variable triangle inscribed in Ω and circumscribed about ω, describes a line.

Ex. 16.   Let pairs of parallel tangents be drawn to the conics of a confocal system from a point *E* of the Absolute line. Let $F = (I, J)/E$. Prove that the points of contact of the tangents lie on a rectangular hyperbola, passing through the four foci of the system. And that this is the conjugate line locus of every one of the confocals, for the centres *E, F*. If *A, B, C, D* be four points of the hyperbola such that *AB, BC, CD, DA* touch a particular one of the confocals, prove that *AB, CD* are parallel, as are *AD, BC*.

The theorem can be stated in general terms, for any system of conics which have four common tangents. Of the general theorem, another case is the following: The tangents to a system of confocal conics, drawn from a point *E* on the axis *SH*, have their points of contact on a circle, namely the circle of which *EF* is a diameter, where $F = (S, H)/E$.

And the dual theorem relates to a series of conics which have four points in common. In particular: If a fixed line be taken parallel to the radical axis of a series of coaxal circles, the tangents of the circles at the points, where the circles meet the line, touch a conic. This touches the given line; and has the limiting points of the series of circles for foci.

Ex. 17.   Considering a series of conics with four points in common, let O be a fixed point on a common chord of the conics. The points of contact of the tangents drawn from O, to the conics, lie on another conic. This conic meets the common chord in points which are conjugate in regard to the conics of the series. The conic also contains the two common points, of the conics of the series, which lie on the chord complementary to the original chord, this complementary chord being the polar of O in regard to the locus. And this locus contains two angular points of the common self-polar triangle of the series of conics.

In particular, for a series of coaxal circles, the points of contact of a pencil of parallel tangents lie on a rectangular hyperbola; this hyperbola contains the common points, and also the limiting points of the series of circles. We have already seen, conversely (Chap. XII (7)), that the circles described on parallel chords of a rectangular hyperbola, as diameters, are coaxal.

The theorem has a dual. Of this an example is: If a fixed line be drawn through a focus S, of a system of confocal conics, the tangents of these conics at the points where they are met by the line, touch a parabola. The focus of the parabola is the focus, H, of the confocals, which is complementary to S; and the parabola touches the axis $S'H'$ of the confocals which is conjugate to SH.

Ex. 18.   As before, let F be the point conjugate to an arbitrary point E, in regard to the quadrangle of the foci, S, H, S', H', of a system of confocal conics, whose centre is C. Prove that the points of contact of tangents drawn from E to these confocals, lie on a (cubic) curve. This curve may be described as the pedal of E in regard to that parabola, already met with, which has F as focus and EC as directrix. (The pedal of a curve, in regard to a point, being the locus of the foot of the perpendicular drawn, from the point, to the variable tangent of the curve.) This curve contains the four foci of the confocals; and the Absolute points. Also, if the axes, $SH, S'H'$, of the confocals, meet the Absolute line in X and Y, the curve contains the points where EX, EY, EC, respectively, meet YC, CX, XY. The cubic curve crosses itself at E, with tangents which are those of the two confocals which pass through E.

Ex. 19.   Prove that the inverse of the cubic curve of Ex. 18, with respect to a circle with centre at E, is a rectangular hyperbola. Also, that this hyperbola is the polar reciprocal of the parabola spoken of which has F as focus and CE as directrix, in regard to a circle whose centre is E. In particular, if this circle contains the focus S, the centre of the hyperbola is at S, and the hyperbola touches CE at E. (Cf. Ex. 3 of (8) above.)

To place the cubic curve of Ex. 18 in proper perspective, we may remark that, if CXY be the diagonal triangle of the quadrilateral formed by the lines $ISS', IHH', JS'H, JSH'$, while the lines joining an arbitrary point E to the points C, X, Y, respectively, meet the opposite sides of this triangle in N, L and M, then, through the nine points constituted by N, L, M and the six intersections of the lines of the quadrilateral, there pass an infinite series of cubic curves (though in general a cubic curve is determined by nine general points). Of these cubic curves, the one which contains E has a double point at E, at which the tangents are the focal rays of the pencil in involution which is formed by joining E to the complementary pairs of intersections of the lines of the quadrilateral.

Ex. 20.   We have proved in Chapter XI ((12), Ex. 11), with phraseology related to a parabola, a theorem of which we now give a general proof: *Let* Ω *be any given conic, on which* O *is a given point; and* P, Q, R *be three other given points of the plane. Let a conic be drawn through* O, P, Q, R, *which meets* Ω, *besides in* O, *in the points* A, B, C. *The triangle* ABC, *so obtained, is circumscribed to another conic which is determined by* O, P, Q, R.

Denote the intersections (AB, RP), (AB, PQ), (OC, RP), (OC, PQ), respectively, by M, N, H, K. Let the fixed points where Ω is met by QR, RP, PQ, respectively, be X, Y; W, T; U, V.

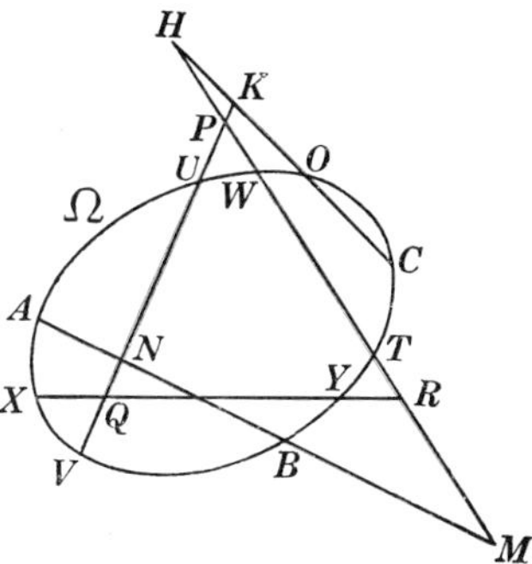

The conics through O, A, B, C meet PQ in pairs of points of an involution, which is determined by the two pairs U, V and P, Q, lying respectively on the conic Ω and that conic we have supposed drawn through O, P, Q, R (which contains O, A, B, C). To this involution belong the pair K, N, lying on the lines OC, AB. These same conics through O, A, B, C, likewise determine an involution on PR, containing the pairs W, T and P, R, and the pair H, M, lying on the lines OC, AB. As the conic taken through O, P, Q, R varies, the points A, B, C vary on Ω, but U, V, P, Q, W, T, P, R are fixed.

Thus N, K are a variable pair of a fixed involution on PQ, and M, H are a variable pair of a fixed involution on PR. Thus on PQ, the ranges (N), (K) are related; and, on PR, the ranges (M), (H) are related. The ranges (K), (H) are in perspective from the fixed point O, and are thus related. Wherefore the ranges (N), (M) are related. This shews that the variable line NM, or AB, envelopes a conic, of which PQ, PR are tangents. As one conic through O, P, Q, R is the line-pair OP, QR, we infer that the envelope also touches QR. Further, if OP meet the conic Ω again in P', the envelope also touches the lines joining P' to the two points X, Y, where QR meets Ω; likewise, if OQ and OR meet Ω again in Q', R', respectively, the envelope touches $Q'W, Q'T, R'U, R'V$. Thus, given Ω, and O, P, Q, R, we have found nine tangents of the envelope, which are sides of three triangles inscribed in Ω; of these nine tangents, the sides QR, RP, PQ are three. The conic so identified is that enveloped by any side AB of the variable triangle ABC, and so touches BC and CA.

We have remarked, in the place referred to (Chap. xi (12), Ex. 11), on a briefer proof of the result, obtained by using the parametric expression of the points of the conic Ω.

Ex. 21.   The preceding example (20) determines a conic, ω, triangularly inscribed to a given conic, Ω, having three given tangents, QR, RP, PQ. There is an infinite series of such conics, according to the position of O on Ω.

Such a conic is determined by two arbitrary triangles inscribed in Ω. Let $Q'WT$, $R'UV$ be two such triangles. We can construct an appropriate circumscribed triangle, PQR, of the conic ω, as follows: Take an arbitrary point O, on Ω; let P be the point (WT, UV), and Q be the point (OQ', UV), and R the point (OR', WT). Then, OQ', WT and OR', UV are two degenerate conics containing O, P, Q, R. Thus, by Ex. 20, if A, B, C, O be the intersections with Ω of any conic through O, P, Q, R, the sides BC, CA, AB touch the conic ω which touches the sides of the triangles $Q'WT, R'UV$. In particular, if OP meet Ω again in P', and QR meet Ω in X and Y, the sides of the triangle $P'XY$ touch the conic ω.

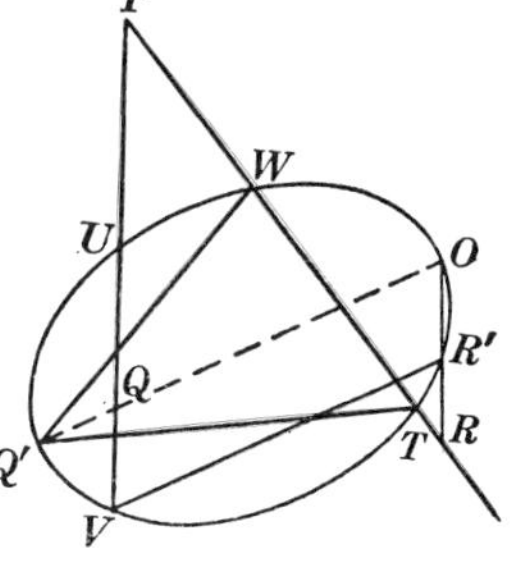

Evidently, by taking O variously on Ω, the side QR takes various positions. Also, we may recall ((19), above), there are two conics, triangularly inscribed to a given conic, which touch four given lines.

Ex. 22. Deduce from Ex. 21 that if three triangles be circumscribed to the same conic, then the three conics, each of which contains the angular points of two of these conics, have a common point.

Ex. 23. As in Chap. x (5), Ex. 26, let the lines joining a point D of a fixed conic to the Absolute points I, J meet the conic again in M, N. Prove that the pedal circles of D in regard to all triangles ABC inscribed in the conic pass through the foot of the perpendicular drawn to MN from D. If $A_1B_1C_1$, $A_2B_2C_2$ be two such triangles, and ω the conic touching the sides of these, and t the fourth common tangent of ω and the conic which touches DI, DJ, B_1C_1, C_1A_1, A_1B_1, prove (see Ex. 22, preceding) that the pedal circle of D in regard to the triangle $A_2B_2C_2$ passes through the foot of the perpendicular drawn from D to the line t. Thus, if ω remain fixed, the pedal circles of D with regard to all triangles $A_2B_2C_2$ are a coaxal system (J. H. Grace). Prove also that, if a parabola be triangularly inscribed to a conic Ω, the circumcircles of the triangles inscribed in Ω whose sides touch the parabola are a coaxal system.

Ex. 24. Let A, B, C, D be any four points; let d be the polar of D, and A', B', C' the respective poles of BC, CA, AB in regard to any conic. Prove that if AD, BD, CD meet d in L, M, N respectively, then $A'L$, $B'M$, $C'N$ meet in a point.

CHAPTER XV

LENGTH, AND DISTANCE, FOR REAL POINTS. LINEAR TRANSFORMATIONS

In the present chapter we shew, by the detailed consideration of examples, how we can, by means of the symbols we have used for points, define a *distance* between two real points, and the length of the segment of the line joining these points. This depends upon two quite arbitrary *Absolute points*, and is thus more general than distance as used in ordinary metrical geometry. But it appears that the metrical theorems of ordinary geometry remain true with the new definitions. Many such metrical theorems have been investigated; we give examples of these. The distance of real points, which we consider, is the positive distance, as in Euclid's theory. The *signed* distance, by which, for example, the distance AB is considered as the negative of the distance BA, can be superimposed upon this; and is not considered here.

(1) We assume two points, I, J, to be given, as *Absolute points* of the plane. These are arbitrary points; but it is supposed, as has several times been explained, that upon the (real) Absolute line which contains the Absolute points, there are two points, X, Y, which we regard as real, such that the symbols, I, J, of the Absolute points are expressible in terms of the symbols of X and Y, in the form $I = X + iY$, $J = X - iY$. The most important consequence of this convention is that, as we have seen, the bisectors of the angles determined by two real lines, are necessarily real. When this convention is adopted, we may speak of the Absolute points as *conjugate imaginaries*. It is assumed that the numerical multipliers implicit in the symbols I, J are definitely chosen. In general, the points of the plane considered are supposed not to lie on the Absolute line unless this is specifically mentioned.

Let O be any definite point of the plane, not on the Absolute line. We can suppose the ratios of the numerical multipliers implicit in the symbols O, I, J to be determined by the assignment of the point whose symbol is $O + I + J$. The symbol of any point P of the plane, *not on the Absolute line*, can then be supposed to be $P = O + \xi I + \eta J$, where ξ, η are definite numbers. Moreover, with the convention referred to, that I, J are conjugate imaginaries, it follows that, if O and P be real points, the numbers ξ, η are conjugate imaginaries. If then

$$P' = O + \xi' I + \eta' J$$

be the symbol of another real point, also not on the Absolute line, the product $(\xi' - \xi)(\eta' - \eta)$ will be real and positive. *By the distance PP', or the length PP', we mean the positive square root of $(\xi' - \xi)(\eta' - \eta)$.*

This distance is independent of the point O. For, if O_1 be any other real point (not on the Absolute line), with a symbol

$$O_1 = O + aI + bJ,$$

the symbols of P and P' take the forms

$$P = O_1 + (\xi - a)\, I + (\eta - b)\, J, \quad P' = O_1 + (\xi' - a)\, I + (\eta' - b)J,$$

and the distance of P, P', expressed by O_1 with I, J, is independent of the numbers a, b.

Also, if, instead of the symbols I, J we use symbols I_1, J_1 for the Absolute points, such that $I = mI_1$, $J = nJ_1$, then distance, as defined by means of O_1, I_1, J_1, will be distance, as already defined, multiplied by $(mn)^{\frac{1}{2}}$, which is real and positive because m, n are conjugate imaginary numbers. This factor affects then the distances of all pairs of points.

The distance of a point P from a point P' increases indefinitely as P' approaches to the Absolute line; for then, in the general form of the symbol of the point P', namely $mO + p'I + q'J$, in which m, p', q' are numbers, the number m approaches to zero, and one at least of $\xi' = p'/m$, $\eta' = q'/m$ becomes infinite. We do not define the distance of two points which lie on the Absolute line; and, as has been said, of points, P, not lying on this line, we consider only real points, for which ξ, η are conjugate imaginary numbers. We shall often denote the distance of two points P, P' by $\overline{PP'}$.

(2) Let R be a real point lying on the line joining the real points P, Q, with symbol $R = aP + bQ$, in terms of the symbols of P and Q, wherein a, b are real numbers; no one of the three points lying on the Absolute line. Let the symbols of P, Q be taken in the reduced forms, explained in (1) namely $P = O + \xi I + \eta J$, $Q = O + \xi'I + \eta'J$. The point, where the line PQ meets the Absolute line, has then the symbol $P - Q$. The reduced form of the symbol of R is then

$$R = O + (\lambda\xi + \mu\xi')\, I + (\lambda\eta + \mu\eta')J,$$

where $\lambda = a/(a+b)$, $\mu = b/(a+b)$, in which $a + b$ is not zero; and $\lambda + \mu = 1$. The distances $\overline{PR}$, $\overline{QR}$ will then respectively be

$$\overline{PR} = [(\lambda - 1)\,\xi + \mu\xi']^{\frac{1}{2}}\, [(\lambda - 1)\,\eta + \mu\eta']^{\frac{1}{2}},$$

which is
$$[\mu\,(\xi' - \xi)]^{\frac{1}{2}}\, [\mu\,(\eta' - \eta)]^{\frac{1}{2}};$$

and
$$\overline{QR} = [\lambda\xi + (\mu - 1)\,\xi']^{\frac{1}{2}}\, [\lambda\eta + (\mu - 1)\,\eta']^{\frac{1}{2}},$$

which is
$$[\lambda\,(\xi - \xi')]^{\frac{1}{2}}\, [\lambda\,(\eta - \eta')]^{\frac{1}{2}}.$$

When R is in that segment PQ which does not contain the point where PQ meets the Absolute line, we have proved, in a preceding chapter (VI), that a, b, and hence also λ, μ, may be regarded as both positive. In this case then $\overline{PR} = \mu\overline{PQ}$, $\overline{QR} = \lambda\overline{PQ}$, $\overline{PR} + \overline{QR} = \overline{PQ}$. But, when R is in the other segment PQ, the numbers λ, μ are of opposite sign; for instance, suppose λ to be positive and μ negative. Then $\overline{PR} = -\mu\overline{PQ}$, $\overline{QR} = \lambda\overline{PQ}$, $\overline{QR} - \overline{PR} = \overline{PQ}$. In the former case, the mid point of P, Q, being the harmonic conjugate of the point whose symbol is $P-Q$, has the reduced symbol $C=\frac{1}{2}P+\frac{1}{2}Q$; and $\overline{PC}=\frac{1}{2}\overline{PQ}=\overline{CQ}$.

In general, it is easy to see that the results obtained, in Chapter VI, for the values of θ in the symbol $P+\theta Q$, of a real point of the line PQ, are expressible in terms of distance, as here defined. But, we recall, to obtain those results, we employed topological assumptions not inherent in the symbols as originally defined.

Ex. 1.   If Z be the point, of symbol $P-Q$, where the line PQ meets the Absolute line, and R be a point of symbol $R=P+\theta Q$, we have

$$(R, Z/P, Q)=-\theta.$$

We have found that if R be in that segment PQ which does not contain Z, then $\overline{RP}/\overline{RQ}=\theta$.

Ex. 2.   Let A, B, A', B' be four real points on a line OU, for which O, A, B, A', B', U are in order; let the symbols be such that

$$A=O+aU, \quad B=O+bU, \quad A'=O+a'U, \quad B'=O+b'U,$$

so that

$$(a-a')\,B=(b-a')\,A+(a-b)\,A', \quad (a-a')\,B'=(b'-a')\,A+(a-b')\,A'.$$

These give $$\qquad (B, B'/A, A')=\frac{a-b}{b-a'}\Big/\frac{a-b'}{b'-a'}.$$

Hence, with the order supposed for the four points, using the positive distances defined, we have

$$(B, B'/A, A')=-\frac{\overline{BA}}{\overline{BA'}}\Big/\frac{\overline{B'A}}{\overline{B'A'}}.$$

In particular, if B', B be harmonic conjugates in regard to A, A', we have

$$\overline{BA}/\overline{BA'}=\overline{B'A}/\overline{B'A'},$$

which leads to $$\qquad 2/\overline{AA'}=1/\overline{AB}+1/\overline{AB'}.$$

Ex. 3.   Let four points of a plane, in order, denoted by 1, 2, 3, 4, be such that the distances $\overline{12}, \overline{23}, \overline{34}, \overline{41}$ are $c_{12}, c_{23}, c_{34}, c_{41}$. Let the lengths of the perpendiculars, drawn from a point O, to the respective joining lines, be $p_{12}, p_{23}, p_{34}, p_{41}$. Also, let the lines joining O to the four points meet the Absolute line respectively in P_1, P_2, P_3, P_4. Prove that, save for sign,

$$\frac{p_{12}\cdot p_{34}}{p_{41}\cdot p_{23}}=(P_1, P_3/P_2, P_4)\Big/\frac{c_{12}\cdot c_{34}}{c_{41}\cdot c_{23}}.$$

In the Greek metrical geometry, the theorem that the locus of a point O for which the left side of this equation is constant, is a conic, was so far conspicuous as to elicit a proof from Newton (*Locus ad quatuor lineas*; Pappus, *Collectio*, Liber VII; Newton, *Principia*, Liber I, sect V, lemmas 17–19). The equation we have stated shews that the theorem is identical with the cross-ratio definition of a conic.

Ex. 4.   Let A, B', C, A', B, C' be six points of a conic. Let I, J be the points of the conic lying on the Pascal line which contains the three points $(BC', B'C), (CA', C'A), (AB', A'B)$. Prove that, if I, J be taken as the Absolute points, the lenths $\overline{AA'}, \overline{BB'}, \overline{CC'}$ are equal.

Thus, *any three chords of a conic may be regarded as chords of a circle which have equal lengths.* Also, there are conics, having double contact with a given conic, which touch the sides of an arbitrary triangle (in fact, four such).

(3) Let OA, OB be real perpendicular lines which meet the Absolute line in the real points A, B; let the symbols of the Absolute points, in terms of the symbols of these points, be $I = \frac{1}{2}(A + iB)$ and $J = \frac{1}{2}(A - iB)$, so that $A = I + J$, $iB = I - J$. Consider a real point P, not on IJ, whose symbol, expressed in terms of the symbols of O, A, B, is $P = O + xA + yB$. This is $P = O + \xi I + \eta J$, if $\xi = x - iy$, $\eta = x + iy$. The numbers x, y are real, ξ and η are conjugate imaginaries. Let BP meet OA in M, and AP meet OB in N; thus PM, PN are the lines through P which are perpendicular to OA, OB, respectively. From the identity

$$P + \tfrac{1}{2}i\,(\eta - \xi)\,B = O + \tfrac{1}{2}(\xi + \eta)\,A,$$

as M is the common point of OA and PB, we see that we may take, for the reduced symbol of M, the form $M = O + \frac{1}{2}(\xi + \eta)(I + J)$, or $M = O + x(I + J)$. Likewise, from

$$P - \tfrac{1}{2}(\xi + \eta)\,A = O + \tfrac{1}{2}i\,(\xi - \eta)\,B,$$

we have, for the reduced symbol of N, the form $N = O + \frac{1}{2}(\xi - \eta)(I - J)$, or $N = O - iy\,(I - J)$. From these, by the definition, the distance $\overline{OM}$ is x, or $-x$, according as x is positive or negative, and the distance $\overline{ON}$, or $(-iy \cdot iy)^{\frac{1}{2}}$, is y, or $-y$, according as y is positive or negative.

Also, because $P = O + \xi I + \eta J$, and $M = O + \frac{1}{2}(\xi + \eta)(I + J)$, the distance $\overline{PM}$, being $[\frac{1}{2}(\xi - \eta) \cdot \frac{1}{2}(\eta - \xi)]^{\frac{1}{2}}$, or $(-iy \cdot iy)^{\frac{1}{2}}$, is y, or $-y$, according as y is positive or negative, *and is equal to* $\overline{ON}$. While, likewise, because $P = O + \xi I + \eta J$, and $N = O + \frac{1}{2}(\xi - \eta)(I - J)$, the distance $\overline{PN}$, being $[\frac{1}{2}(\xi + \eta) \cdot \frac{1}{2}(\xi + \eta)]^{\frac{1}{2}}$, or $(x \cdot x)^{\frac{1}{2}}$, is x, or $-x$, according as x is positive or negative, *and is equal to* $\overline{OM}$. *Thus OMPN is the figure called a rectangle, having opposite sides equal and parallel, and contiguous sides perpendicular.*

Ex. If points N_1, P_1 be taken on NP, of respective symbols $N + nA$, $P + pA$, prove, for $p \geqslant n$, $x > 0$, that $\overline{N_1 P_1} = \overline{OM}$ requires $p = n$. And that, then, the lines ON_1, MP_1 are parallel, and of equal length.

(4) Let OM, ON be perpendicular lines, meeting the Absolute line respectively in A and B, and, as in (3), let the lines through M and N, respectively parallel to ON and OM, meet in P, with the symbol $P = O + \xi I + \eta J$. Then, from

$$M = O + \tfrac{1}{2}(\xi + \eta)(I + J) \text{ and } N = O + \tfrac{1}{2}(\xi - \eta)(I - J),$$

the length $\quad\quad \overline{MN}, = (\eta \xi)^{\frac{1}{2}} = [(x + iy)(x - iy)]^{\frac{1}{2}}.$

Thus $\quad\quad\quad\quad (\overline{MN})^2 = (\overline{OM})^2 + (\overline{ON})^2.$

This is the generalised form of *the result known by the name of Pythagoras*, for Euclid's distances, which was, and remains, a very great discovery.

(5) *All real points of a circle are at the same distance from its centre.* Let O be this centre, supposed to be a real point. The tangents from O touch the circle at I and J. By what we have seen (Chap. IX (17)), the

reduced symbol of any point of the circle is of the form $O + r(\theta I + \theta^{-1} J)$, where r is a constant number, and θ is a number varying from point to point of the circle. Thus every real point of the circle is at the distance r (or $-r$), from the centre, the number r being real because $r\theta$, $r\theta^{-1}$ are conjugate imaginaries for a real point. By changing the sign of θ, we may suppose r to be positive. It is then the number called the *radius* of the circle.

(6) *Let P be any real point of a conic, of which one focus is the real point O. Let PN, drawn to be perpendicular to the directrix which is the polar of O, meet this in N. We prove that the ratio of the distances $\overline{OP}$, $\overline{PN}$ is the same for all real points of the conic (supposed not a circle).*

Let the tangents OI, OJ of the conic touch this, respectively, in X and Z. The symbols of these will then be of the respective forms $O + pI$, $O + qJ$, where p, q are numbers. We assume that the directrix is a real line, and the mid point of X, Z, whose symbol is $O + \frac{1}{2}pI + \frac{1}{2}qJ$, is a real point; so that p, q are conjugate imaginaries. We can then, by choosing I, J suitably, suppose $X = O + I$, $Z = O + J$. Thus the directrix XZ meets IJ in the point, K, of symbol $X - Z$, or $I - J$; let H be the point on IJ whose symbol is $I + J$, so that $H = (I, J)/K$. Then, if P be any point of the conic, the line PH is perpendicular to PK, which is parallel to the directrix; so that PH is perpendicular to the directrix, meeting this in N.

The symbol of P, with a suitable constant number h, and a variable number θ, may be taken to be

$$P = \theta^2 X + 2h\theta O + Z, \quad \text{or} \quad \theta^2 I + J + (\theta^2 + 2h\theta + 1)\, O.$$

This gives $\qquad P + h\theta\,(I + J) = (\theta^2 + h\theta)\, X + (h\theta + 1)\, Z;$

wherefore, as $H = I + J$, the symbol of N, or (PH, XZ), is given by either of these forms. The reduced symbols of P and N are therefore, respectively,

$$P_1 = O + \theta^2 m^{-1} I + m^{-1} J, \quad N_1 = O + (\theta^2 + h\theta)\, m^{-1} I + (h\theta + 1)\, m^{-1} J,$$

where m stands for $\theta^2 + 2h\theta + 1$. Thus we have

$$\overline{OP} = (\theta^2 m^{-1} . m^{-1})^{\frac{1}{2}}, \quad \overline{PN} = (h\theta m^{-1} . h\theta m^{-1})^{\frac{1}{2}},$$

which give $\overline{OP}/\overline{PN} = \pm h^{-1}$. And this is independent of the position of P on the conic.

Ex. 1.   When the conic is a parabola, by expressing that the two points of the conic which lie on the Absolute line are coincident, we find the numerical value of h to be 1. Thus $\overline{OP} = \overline{PN}$.

Ex. 2.   The sum, or difference, of the distances of a point on the conic, from the two real foci, is independent of the point. For the sum, or difference, of the perpendicular distances from the point to the two (parallel) corresponding directrices, is independent of the point (by (2) above). Another proof of this, and a generalisation, occur below ((9), Ex. 14).

(7) We have already defined, and used, the symbol $\epsilon\,(m, n)$, if m, n be two lines, as meaning $(M, N/I, J)$, where M, N are the points in which the respective lines m, n meet the line IJ.

Now, let A, B, C be three real points of the plane; denote their joining lines BC, CA, AB respectively by l, m, n. We prove that *the ratios of the lengths $\overline{BC}, \overline{CA}, \overline{AB}$ depend only on $\epsilon\,(m, n)$, $\epsilon\,(n, l)$, and $\epsilon\,(l, m)$; and thus depend only on the points where the lines l, m, n meet the line IJ.*

If the lines BC, CA, AB meet IJ respectively in points of symbols $I+pJ$, $I+qJ$, $I+rJ$, we have $\epsilon\,(m, n) = q/r$, $\epsilon\,(n, l) = r/p$, and $\epsilon\,(l, m) = p/q$. So that $\epsilon\,(m, n)\,\epsilon\,(n, l)\,\epsilon\,(l, m) = 1$.

Taking pro**p**er fixed numbers, a, b, f, g, the symbols of A, B, C will be such that
$$fA = C + a\,(I+qJ), \quad gB = C + b\,(I+pJ);$$
hence, as $fA - gB$, or $(a-b)\,I + (aq-bp)\,J$, is the symbol of the point (AB, IJ), which we have represented by $I+rJ$, we have $r = (aq-bp)/(a-b)$; so that $a/b = -(r-p)/(q-r)$.

Also, from the definition of distance, neglecting for a moment the questions of sign, and reality, we have
$$\overline{BC} = \pm bp^{\frac{1}{2}}, \quad \overline{CA} = \pm aq^{\frac{1}{2}}, \quad \overline{AB} = [(a-b)\,(aq-bp)]^{\frac{1}{2}}, \; = \pm (a-b)\,r^{\frac{1}{2}};$$
and, since $b, a, a-b$ are in the ratios of $q-r$, $-(r-p)$, $p-q$, we thus see that $\overline{BC}, \overline{CA}, \overline{AB}$ are in the ratios of $\pm (q-r)\,p^{\frac{1}{2}}$, $\pm (r-p)\,q^{\frac{1}{2}}$, $\pm (p-q)\,r^{\frac{1}{2}}$. These depend only on the points where the lines BC, CA, AB meet the Absolute line; as was to be proved.

Now consider the question of the signs and reality of these expressions. By our convention for I, J, the multipliers ξ, η in the symbol of a real point, $O + \xi I + \eta J$, are such that η/ξ is of the form $e^{i\alpha}$, where α is real. Thus, in our case, we may suppose $p = e^{2i\theta}$, $q = e^{2i\phi}$, $r = e^{2i\psi}$, where θ, ϕ, ψ are real. Taking then $e^{i\theta}, e^{i\phi}, e^{i\psi}$ for $p^{\frac{1}{2}}, q^{\frac{1}{2}}, r^{\frac{1}{2}}$, respectively, the numbers
$$(q-r)\,p^{\frac{1}{2}}, \quad -(r-p)\,q^{\frac{1}{2}}, \quad (p-q)\,r^{\frac{1}{2}}$$
are in the ratios of
$$(e^{2i\phi} - e^{2i\psi})\,e^{i\theta}, \quad -(e^{2i\psi} - e^{2i\theta})\,e^{i\phi}, \quad (e^{2i\theta} - e^{2i\phi})\,e^{i\psi},$$
that is, of $\quad \sin\,(\phi - \psi), \quad -\sin\,(\psi - \theta), \quad \sin\,(\theta - \phi)$.

These are real; and, if $\pi > \theta > \phi > \psi > 0$, they are all positive.

Likewise $\epsilon\,(m, n) = e^{2i\,(\phi - \psi)}$, $\epsilon\,(n, l) = e^{2i\,(\psi - \theta)}$, $\epsilon\,(l, m) = e^{2i\,(\theta - \phi)}$. But we do not enter into the topological axioms necessary to put the theory into exact comparison with metrical geometry.

(8) We have defined, as the bisectors of the angles determined by two lines, which meet in a point O, not lying on the Absolute line, the two lines, through O, which are harmonic conjugates in regard to the two given lines, and also in regard to the lines OI, OJ. It was seen that these bisectors are real when I, J are conjugate imaginaries.

It is interesting to indicate the relation of this to the point of view taken in (7) preceding. Let two lines, l, m, through O, meet the Absolute line, respectively, in L, M. Let a line b be drawn through O such that $\epsilon\,(l, b) = \epsilon\,(b, m)$. Let the line b meet the Absolute line in B, and the perpendicular line b', through O, meet the Absolute line in B'. Thus $(B', B/I, J) = -1$. We can prove that also $(B', B/L, M) = -1$. Thus *the condition* $\epsilon(l, b) = \epsilon(b, m)$ *ensures that b is one of the bisectors of the angles determined by the lines l, m, as previously defined.*

To prove $(B', B/L, M) = -1$, we may use the symbols. Suppose the symbols of L, M, B, B', in terms of the symbols of I, J, to be, respectively, $L = I + pJ$, $M = I + qJ$, $B = I + hJ$, $B' = I - hJ$, where p, q, h are numbers. The relation $\epsilon(l, b) = \epsilon(b, m)$ is equivalent to $p/h = h/q$, or $pq = h^2$. In terms of the symbols of L, M, the symbols of B, B' are given by

$$(p-q)\,B = -(q-h)\,L + (p-h)\,M, \quad (p-q)\,B' = -(q+h)\,L + (p+h)\,M;$$

these give $(B', B/L, M) = [(p+h)/(q+h)] \div [(p-h)/(q-h)];$

in virtue of $pq = h^2$, this is -1.

But, if we interpret I, J, L, M, B, B' as points of a conic, the given relations $\epsilon(l, b) = \epsilon(b, m)$, $(B', B/I, J) = -1$ mean that B is one of the foci of the involution determined by (I, J) and (L, M), and that BB' passes through the pole of IJ. From this, BB' passes also through the pole of LM; so that $(B', B/L, M) = -1$.

Ex. Let four lines l, m, l', m' (which may be taken to pass through a point) meet the Absolute line, respectively, in L, M, L', M'. Denote $\epsilon(m, l)$, $\epsilon(m, l')$, $\epsilon(m', l)$, $\epsilon(m', l')$, respectively, by t, u, v, w, so that $tw = uv$. Prove that the cross ratio $(M, M'/L, L')$ is equal to P/Q, where, with proper signs for the square roots,

$$P = (t^{\frac{1}{2}} - t^{-\frac{1}{2}})/(u^{\frac{1}{2}} - u^{-\frac{1}{2}}), \quad Q = (v^{\frac{1}{2}} - v^{-\frac{1}{2}})/(w^{\frac{1}{2}} - w^{-\frac{1}{2}}).$$

(9) A simple result, related to (8) preceding, may be enunciated: let O, P, Q be three points, not in one line, no one on the Absolute line, such that the distances $\overline{OP}$, $\overline{OQ}$ are equal. Then PQ is parallel to one of the bisectors of the angles defined by the lines OP, OQ.

This is immediately proved by use of the symbols. Let OP, OQ meet the Absolute line in points whose respective symbols are $I + pJ$, $I + qJ$. Let the constant c be such that the symbol of P is $O + cq^{\frac{1}{2}}(I + pJ)$. Thus the condition $\overline{OP} = \overline{OQ}$ proves that the symbol of Q is

$$O \pm cp^{\frac{1}{2}}(I + qJ).$$

Hence the line PQ meets the Absolute line in the point whose symbol is $I \pm (pq)^{\frac{1}{2}}J$. The two points given by this are harmonic conjugates both in regard to I, J and in regard to the points whose symbols are $I + pJ$, $I + qJ$.

Ex. 1. Let a point O be joined to points P, Q, R. It follows of course that, if $\overline{OP} = \overline{OQ}$, and $\overline{OQ} = \overline{OR}$, then $\overline{OP} = \overline{OR}$. Prove, however, without using the formula for length, that, if PQ be parallel to one of the bisectors of the angles determined by OP, OQ; and also QR be parallel to one of the bisectors of the angles determined by OQ, OR; then PR is parallel to one of the bisectors of the angles determined by OP, OR.

Ex. 2. The two bisectors of the angles determined by the sides CA, CB, of a triangle, meet the remaining side AB in D and E. Prove, from the definition of length, that the ratios

$$\overline{AD}/\overline{DB}, \quad \overline{AE}/\overline{BE}, \quad \overline{AC}/\overline{BC}$$

are equal. Prove also that, if the line joining C to the mid point of A, B be perpendicular to AB, then $\overline{CA} = \overline{CB}$.

Ex. 3. It is a result of Euclid's geometry that, if OA, OB be perpendicular lines, and ON, perpendicular to AB, meet this in N, then $\overline{ON^2} = \overline{AN}.\overline{NB}$.

In fact, it can be seen that $\epsilon O\,(A, N)$, $\epsilon A\,(N, O)$, $\epsilon N\,(O, A)$ are respectively equal to $\epsilon B\,(O, N)$, $\epsilon O\,(N, B)$, $\epsilon N\,(B, O)$. So that the result follows from (7) above.

Prove also that $\qquad (\overline{ON})^{-2} = (\overline{OA})^{-2} + (\overline{OB})^{-2}.$

Ex. 4. Let X, Y be the foci of an involution on a line, supposed to be real points, O the mid point of X, Y, and P, Q any pair of the involution. Prove that

$$\overline{OX^2} = \overline{OY^2} = \overline{OP}.\overline{OQ}.$$

Ex. 5. Let PP', QQ' be two chords of a circle, meeting in O. Then $\overline{OP}.\overline{OP'} = \overline{OQ}.\overline{OQ'}$.

Let the Absolute line, meeting the circle in I, J, meet the chords PP', QQ', respectively, in L, M; and meet the other complementary pair of joins, PQ, $P'Q'$, respectively, in R and R'. By Desargues' theorem, the three pairs I, J; L, M; R, R' are in involution. Let C be one of the foci of this involution, and OC meet PQ, $P'Q'$, respectively, in U, U'. Because C is such a focus $(L, M/C, R) = (M, L/C, R')$; thus $(P, Q/U, R) = (Q', P'/U', R')$. The line OC is one of the bisectors of the angles determined by the lines OPP', OQQ'; thus, as IJ is the Absolute line, we have, by Ex. 1 of (2),

$$\overline{PU}/\overline{UQ} = \overline{Q'U'}/\overline{U'P'},$$

and hence, by Ex. 2 above, we have

$$\overline{OP}/\overline{OQ} = \overline{OQ'}/\overline{OP'}.$$

This is the required result. The comparison of this proof with that given by Euclid is interesting.

Ex. 6. Let PP', QQ' be chords of any conic, which meet in O. Prove (by Desargues' theorem, or otherwise) that the ratio $\overline{OP}.\overline{OP'}/\overline{OQ}.\overline{OQ'}$ is unaltered by taking any other position of O, with chords through this respectively parallel to PP' and QQ'.

Ex. 7. Let A, B, C, O be four points of a plane, none on the Absolute line, no three in a line. Denote the lines BC, CA, AB, OA, OB, OC, respectively, by l, m, n, l', m', n', and the corresponding lengths $\overline{BC}$, ..., respectively, by a, b, c, a', b', c'. Also denote $\epsilon\,(n, m)$, $\epsilon\,(m', n')$, $\epsilon\,(l, n)$, $\epsilon\,(n', l')$, $\epsilon\,(m, l)$, $\epsilon\,(l', m')$, respectively, by u, u_1, v, v_1, w, w_1. Then prove that $aa'/U = bb'/V = cc'/W$, where

$$U = (u_1 - u)/(u_1 u)^{\frac{1}{2}}, \quad V = (v_1 - v)/(v_1 v)^{\frac{1}{2}}, \quad W = (w_1 - w)/(w_1 w)^{\frac{1}{2}}.$$

If, in terms of the symbols of O, I, J, the symbols of A, B, C be, respectively, $O + \xi_1 I + \eta_1 J$, $O + \xi_2 I + \eta_2 J$, $O + \xi_3 I + \eta_3 J$, we have

$$u_1 = \xi_3 \eta_2 / \xi_2 \eta_3, \quad u = (\eta_1 - \eta_2)\,(\xi_3 - \xi_1)/(\eta_3 - \eta_1)\,(\xi_1 - \xi_2),$$
$$a = [(\xi_2 - \xi_3)\,(\eta_2 - \eta_3)]^{\frac{1}{2}}, \quad a' = (\xi_1 \eta_1)^{\frac{1}{2}}.$$

Thus we find that aa'/U is equal to $abc\,a'b'c'/\Delta$, where Δ is the determinant, of three rows, having, for elements of the three rows, respectively,

$$\xi_1\eta_1,\ \xi_2\eta_2,\ \xi_3\eta_3;\ \ \xi_1,\ \xi_2,\ \xi_3;\ \ \eta_1,\ \eta_2,\ \eta_3.$$

From this the result is immediate.

In ordinary elementary Trigonometry, the theorem is that denoted by

$$aa'/\sin(A'-A)=bb'/\sin(B'-B)=cc'/\sin(C'-C),$$

where A, B, C are the usual measures of the angles BAC, CBA, ACB, respectively, and A', B', C' the respective measures of the angles BOC, COA, AOB. We do not enter into the typological considerations which justify

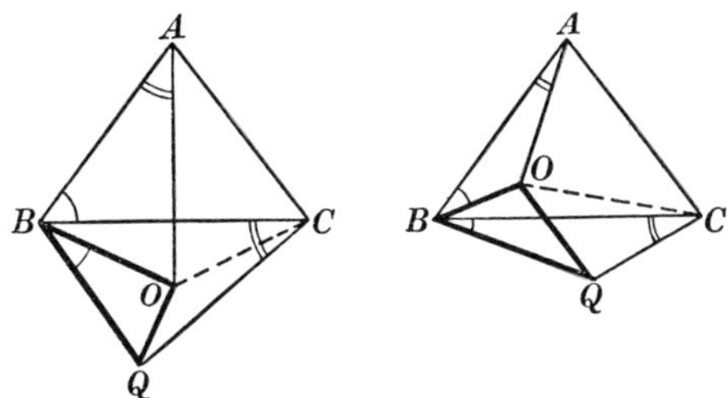

these measurements. But it seems proper to indicate the elementary proof which is possible when these are accepted: Take a point Q so that the lengths of the sides of the triangle OBQ are in the same ratios as the lengths of the sides of the triangle ABC, with also $\epsilon B\,(Q,\,O)=\epsilon B\,(C,\,A)$; the triangle OBQ is then said to be directly similar to the triangle ABC. It will follow that the triangle BQC is directly similar to the triangle BOA. Hence it can be proved that, for the triangle CQO, the lengths of the sides CQ, QO, OC are in the ratios of aa', bb', cc'; and the usual angular measurements of the angles QOC, OCQ, CQO are, respectively, $A'-A$, $B'-B$, $C'-C$. The theorem then follows from (7) above.

If the points A, B, C, O lie on a circle, it can be shewn that the determinant Δ vanishes. Then there holds the relation $aa'\pm bb'\pm cc'=0$, in one of its forms. Taking the case in which AC and the Absolute line meet the line BO in different segments BO, the points OCQ lie in a line, and we have $aa'+cc'=bb'$. This is the result usually stated by saying that, for a quadrangle inscribed in a circle, the sum of the products of the lengths of the two pairs of opposite sides is equal to the product of the lengths of the diagonals.

Ex. 8. The symbol of a variable point of a circle being $O+r(\theta I+\theta^{-1}J)$, where r is a fixed number, and θ a variable number, let the tangents at two points P, Q, with parameters θ, ϕ, meet in T. Prove that

$$\overline{TP}=\overline{TQ},\ =ir\,(\theta-\phi)/(\theta+\phi);$$

this may be supposed real and positive.

Deduce that, if A, B, C, D be four points of the plane such that the four joining lines AB, BC, CD, DA touch a circle, there exists a relation of one of the forms $\overline{AB}\pm\overline{CD}\pm\overline{AD}\pm\overline{BC}=0$; and draw diagrams illustrating the various cases.

Ex. 9. Prove, for a parabola, the following six metrical relations:

(a) The focus and vertex being respectively S and A, let the tangent at P meet the axis in T, and Y be the mid point of P, T. Then Y is on the tangent at A, and SY is perpendicular to PT.

(b) The normal at a point P of the curve, and the perpendicular from P to the axis, meet the axis, respectively, in G and N. Prove that

$$\overline{TA}=\overline{AN},\ \ \overline{PN}=2\overline{YA},\ \ \overline{NG}=2\overline{AS},\ \ (\overline{AY})^2=\overline{TA}.\overline{AS}.$$

(*c*) Two perpendicular lines AP, AQ from the vertex meet the curve again in P and Q. Then the chord PQ meets the axis in a point, F, which is fixed, such that $\overline{AF} = 4\overline{AS}$.

(*d*) Through points, P, Q, of the curve, lines are drawn both parallel to the tangent at a point V of the curve, and these meet the line through V which is parallel to the axis, respectively, in M and N, while PQ meets this line in R. Then $\overline{VM}.\overline{VN} = (\overline{VR})^2$.

(*e*) A variable tangent of the parabola meets three fixed tangents, respectively, in P, Q, R. Prove that $\overline{PQ}/\overline{QR}$ is constant.

(*f*) The tangents at two points P, Q of the curve meet in T, and the tangent at a point V meets TP, TQ, respectively, in L and M. Prove that $\overline{LT}/\overline{MQ} = \overline{PT}/\overline{TQ}$.

Ex. 10.   Let CP, CD be two conjugate diameters of an ellipse whose centre is C, and vertices are A', A. We think of P, D as being such that DP meets the axis $A'CA$ in a point not lying in the segment $A'A$ which contains C. Let the perpendicular from P, to the axis, meet the axis in N, and meet the auxiliary circle in Q and Q', the points $QPNQ'$ being in order. Let the

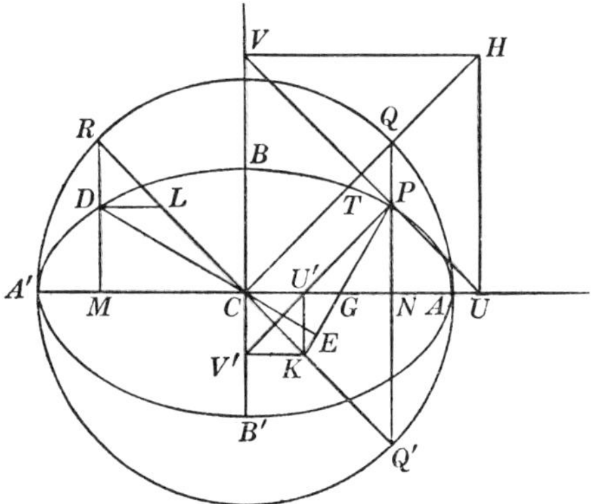

perpendicular from D to the axis meet the axis in M and meet the auxiliary circle in R, so that CR is perpendicular to CQ (the tangents of the ellipse and circle at P and Q, which are parallel, respectively, to CD and CR, meet on the axis). Let the normal of the ellipse at P meet the axis in G and meet CQ, CQ' respectively in H and K. Let the lines from H perpendicular to the axis $A'A$, and the conjugate axis $B'B$, meet these in U, V respectively; and the lines from K perpendicular to $A'A$ and $B'B$ meet these in U', V', respectively. Let the line DL, parallel to the axis $A'A$, meet CR in L. Prove that the triangles DLC, PQH are similar, and

$$\overline{LD} = \overline{QP}, \quad \overline{DC} = \overline{PH} = \overline{PK}, \quad \overline{QH} = \overline{CL} = \overline{CB}.$$

Also that the points U, P, V are in a line, as are P, U', V'; and that

$$\overline{PU} = \overline{PU'} = \overline{CB}, \text{ and } \overline{PV} = \overline{PV'} = \overline{CA};$$

while QG is parallel to the normal of the ellipse at the point T where CQ meets the ellipse; and PU, PU' are respectively parallel to CQ', CQ.

From these results we can give a construction for the positions (and lengths) of CA, CB when the conjugate diameters CP, CD are given in position, namely: Let PK, perpendicular to CD, be taken so that $\overline{PK} = \overline{CD}$; describe the circle having CK as diameter, and let the line joining P to the centre of this circle meet it in U', V'. Then the axes CA, CB are the lines

CU', CV', and, with appropriate notation, $\overline{CA} = \overline{PV'}$, and $\overline{CB} = \overline{PU'}$. The construction for K may give H, instead; but the results are similar.

Another result follows: If a line VPU, for which VP, PU are both of constant length, move with V, U respectively, always on two fixed lines CB, CA (which, in the case considered, are perpendicular), then the locus of P is an ellipse. It will be found (Ex. 13, below) that this is a particular case of a more general result.

Ex. 11.   We have already seen that the equation of a conic, in the co-ordinates we have used, is quadratic in these coordinates (Chap. ix (18)).   In the metrical theory, the equation of a conic is expressed in terms of coordinates which are distances. The reader may prove, that, if OA, OB are any lines, upon which are also the points A', B', respectively, for which O is the mid point of A', A and is also the mid point of B', B; and if, from any point P, there be drawn lines respectively parallel to OA, OB, to meet OB and OA in N and M, and we denote the distances PN, PM by x and y, respectively, then the cross ratio $P(B, B'/A, A')$ is a fractional quadratic (not homogeneous) function of x and y.   In fact, supposing, in terms of the symbols, that

$$A = O + aI + a_1J, \quad A' = O - aI - a_1J, \quad B = O + bI + b_1J, \quad B' = O - bI - b_1J,$$

while $P = O + \xi I + \eta J$, we may compute the value of the cross ratio in terms of ξ, η, a, a_1, b, b_1, and also the values of x, y; and thence find that the cross ratio is equal to

$$\{(Px + Qy)^2 - 1\}/\{(Px - Qy)^2 - 1\},$$

where P^{-1}, Q^{-1} are respectively the lengths of OA and OB.

Ex. 12.   Let A, B, C be three points, not in a line, and the lines CB, CA be denoted by l, m, respectively. Then denoting $\overline{BC}$, $\overline{CA}$, $\overline{AB}$, $\epsilon(l, m)$, respectively, by a, b, c, p, prove (see (7) above) that

$$c^2 = a^2 + b^2 - 2abq,$$

where q denotes $\frac{1}{2}(p^{\frac{1}{2}} + p^{-\frac{1}{2}})$, with a proper sign for the square root.

Ex. 13.   *If a triangle, whose sides are of constant lengths, move so that two angular points move on two fixed lines, the third angular point describes an ellipse.*

Let x, y be two fixed lines, meeting in O; let CAB be a triangle, of which A lies on x, and B lies on y. We assume that the lengths of CA, CB are constant, and also, denoting the lines CB, CA, respectively, by l and m, that $\epsilon\,(l, m)$ is constant. Let CM, CN, respectively parallel to y and x, meet x and y, respectively, in M and N, and denote the lengths of CN, CM by ξ and η, respectively. For the position of the triangle CAB in which C is on the line x, say at C_1, denote the length of OC_1 by $1/P$; for the position of the triangle CAB in which C is on the line y, say at C_2, denote the length of OC_2 by $1/Q$. Also denote $[\epsilon\,(x, y) . \epsilon\,(l, m)]^{\frac{1}{2}}$ by E, with an appropriate sign. Then prove that

$$(P\xi + Q\eta . E)(P\xi + Q\eta . E^{-1}) = 1.$$

The algebraic work is brief if we use the identity

$$(\lambda - \lambda^{-1})^2 + (\mu - \mu^{-1})^2 + (\lambda - \lambda^{-1})(\mu - \mu^{-1})(\lambda\mu + \lambda^{-1}\mu^{-1}) = (\lambda\mu - \lambda^{-1}\mu^{-1})^2,$$

for $\lambda = [\epsilon\,(x, m)]^{\frac{1}{2}}$, $\mu = [\epsilon\,(l, y)]^{\frac{1}{2}}$, applying (7) to the triangles AMC, BNC.

The relation connecting ξ and η proves, by Ex. 11, that the locus of C is a conic, with centre at O, which meets the line x in C_1 and C_1', where C_1' is such that O is the mid point of C_1' and C_1, and meets the line y in C_2 and C_2', where C_2' is such that O is the mid point of C_2' and C_2.

This proves the result stated. We can develop the matter further, in comparison with the figure of Ex. 10. Denote $\overline{BC}$, $\overline{CA}$, $\overline{AB}$ respectively by u, v, w; use E_0 for $[\epsilon\,(x, y)]^{\frac{1}{2}}$; also let the vertices of the ellipse (which are not generally on the lines x, y) be at distances α, β from the centre O. It may then.be proved that

$$Pu = Qv = (E_0 - E_0^{-1})/(E - E^{-1}), \quad \alpha\beta = uQ^{-1} = vP^{-1}, \quad \alpha \mp \beta = 2iw/(E_0 - E_0^{-1}).$$

Thus, conversely, consider an ellipse, of centre O, whose vertices are at distances α, β from the centre. Let two lines x, y be drawn through O to meet the ellipse respectively at points whose distances from the centre are P^{-1}, Q^{-1}. The *ellipse may be generated in general as the locus of the angular point C, of a triangle ABC, whose sides BC, CA, AB are of the respective lengths u, v, w, which are given by the formulae just put down, the points A, B being respectively on the lines x, y.* When, however, the lines x, y are the axes of the ellipse, so that $E_0 = -E_0^{-1} = i$, and $E + E^{-1} = 0$, then $\epsilon\,(l, m) = \epsilon\,(m, l) = 1$, and $E = E_0$, and the lines CA, CB coincide. In this case, the ellipse is the locus of a point C of a line, BA, of which B and A lie on the axes, the lengths CB, CA, BA being constant, namely, respectively $u = \alpha$, $v = \beta$, $w = \alpha - \beta$ (the line $PU'V'$ of Ex. 10).

In the general case the circle OAB has a diameter of constant length, $\alpha \mp \beta$. The whole theory is connected with that of the locus of a definite point in a circle of fixed radius, which rolls on another fixed circle.

Ex. 14. Prove that, if a circle touches a conic in two points, the line joining the points of contact is parallel to one of the axes of the conic. Consider the case of an ellipse. Let a tangent from a point P on the ellipse touch the circle in T, and the line from P perpendicular to the chord of contact meet this in N; prove that the ratio of the lengths of PT and PN is the same for every position of P on the conic, and for every circle taken with a parallel chord of contact. When the circle degenerates into the two tangents drawn to the conic from a focus, we obtain the result of (6).

As a corollary, taking two circles with their centres on the axis of the ellipse which contains the real foci, but their four points of contact imaginary, prove that the sum of the lengths of the tangents drawn to these circles, from a point P of the ellipse, is independent of P. When the four points of contact are all real, prove that the sum *or* difference of the lengths of the tangents is constant, according to the position of P on the ellipse.

The theorem is obtainable from the principles so far explained in this chapter. But, a simple metrical proof, in which the conic is regarded as a section of a *right circular* cone, is given below ((16), Ex. 1).

Ex. 15. The tangents at points P, Q, of an ellipse, of which the centre is C, and the real foci are S and H, meet in the point O. One of the intersections of CO with the ellipse is R. From R, the line RT is drawn, parallel to CS, to meet SO in T. Prove that the product of the lengths of SP and SQ is equal to the square of the length of ST.

Thus, in a sense referred to before, the triangles PST and TSQ are directly similar.

Ex. 16. In the preceding example, let the lines PQ, CO meet in V, and lines drawn from C parallel to OP, OQ meet the conic, respectively, in D and E. Prove that

$$\overline{OP}.\overline{OQ}/\overline{OV} = \overline{SO}.\overline{HO}/\overline{CO} = \overline{CD}.\overline{CE}/\overline{CV}.$$

Also, that each of these fractions is unaltered in value if the ellipse is replaced by a confocal ellipse.

Ex. 17. Let A, B, C, D be any four points in a plane, no three in a line; let the distances $\overline{BC}$, $\overline{CA}$, $\overline{AB}$, $\overline{AD}$, $\overline{BD}$, $\overline{CD}$ be denoted, respectively, by a, b, c, a', b', c'. Prove that

$$\begin{vmatrix} 0 & c^2 & b^2 & a'^2 & 1 \\ c^2 & 0 & a^2 & b'^2 & 1 \\ b^2 & a^2 & 0 & c'^2 & 1 \\ a'^2 & b'^2 & c'^2 & 0 & 1 \\ 1 & 1 & 1 & 1 & 0 \end{vmatrix} = 0.$$

If the symbols of the points be of the forms $O + \xi_i I + \eta_i J$, where ξ_i, η_i are four pairs of numbers, this result is obtained by multiplying, *row* with *row*, the two vanishing determinants, of five rows,

$$
\begin{vmatrix}
\xi_1\eta_1, & \xi_1, & \eta_1, & 1, & 0 \\
\cdots\cdots & & & & \\
\xi_4\eta_4, & \xi_4, & \eta_4, & 1, & 0 \\
1, & 0, & 0, & 0, & 0
\end{vmatrix},
\qquad
\begin{vmatrix}
1, & -\eta_1, & -\xi_1, & \xi_1\eta_1, & 0 \\
\cdots\cdots & & & & \\
1, & -\eta_4, & -\xi_4, & \xi_4\eta_4, & 0 \\
0, & 0, & 0, & 1, & 0
\end{vmatrix}.
$$

(10) When two points P, P' describe related ranges on a line, there are, in the general case, two points at which the related points P, P' coincide. The related ranges may be such that these two common points of the two ranges, coincide with one another. When this is so, and the point of coincidence lies on the Absolute line, then the distance $\overline{PP'}$, of two corresponding points of the two ranges, as here defined, is the same for every two corresponding points.

For, taking an arbitrary point O on the line, and denoting by U the point where the line meets the Absolute line, we may suppose corresponding points of the ranges to have symbols $O + \lambda U$, $O + \lambda' U$, where the variable number λ' is expressible by the variable number λ in the form $\lambda' = (a\lambda + b)/(c\lambda + d)$, in which a, b, c, d are fixed numbers. The common points of the two ranges correspond then to the values of λ given by the equation $\lambda (c\lambda + d) - (a\lambda + b) = 0$. If both roots of this be $\lambda = \infty$, so that the common points coincide at U, then $c = 0$ and $d = a$; and conversely. Then P, P' are given by symbols $P = O + \lambda U$, $P' = O + (\lambda + h) U$, where h is a fixed number, but λ is variable. By definition then, the distance $\overline{PP'}$ is given by $\pm hD$, where D is the distance of the points whose symbols are O and $O + U$.

Ex. 1.   Let three parallel lines be taken; on one of these let A and A' be two fixed points. On another of these lines take a variable point L; and let LA, LA' meet the third line in P and P'. Then the ranges (P), (P'), on this third line, are both related to the range (L), and are therefore related to one another. But the points P, P' can only coincide when L is at the point, U, on the Absolute line, which is common to the three parallel lines; the point of coincidence of P and P' is then also at U. Hence, as L varies, the length of PP' remains constant.

If instead of the line $AA'U$ we take another line through U, parallel therefore to the three lines taken, we can take, on this fourth line, two fixed points B, B', to take the places of A, A', respectively, such that the length PP' is the same as if found from A, A'. For, suppose that when L, moving on its line, takes the position M, the positions of P, P', lying on MA, MA', respectively, are Q and Q', so that the length of QQ' is equal to that of PP'. Let any line be drawn through U to meet LP, LP', respectively, in B and B'. Then, the point H in which QB, $Q'B'$ meet lies on the line LU, as we see by considering the two triangles AMA', BHB', which have $QPQ'P'U$ as axis of perspective, and have therefore L as centre of perspective. Thus QQ' has the same length as PP' when B, B' are used as fixed points, projection being from H and L, respectively; just as was the case when A, A' are the fixed points and projection is from M and L.

Ex. 2.   We may state Ex. 1 in more general form. Take a fixed conic; let U be one of the points in which the Absolute line meets the conic. Consider the tangent at U. Also let A, A' be any two fixed points of the conic.

If the lines which join a variable point L of the conic to A and A' meet this tangent at P and P', the ranges (P), (P'), on this tangent, are related to one another, and P, P' can only coincide when L, and therefore, also, P and P', are at U. Thus P, P' are at the same distance for all positions of L on the conic. In other words, the *chords of a hyperbola, which join a variable point L of the curve to two fixed points A, A' of the curve, meet an asymptote in two points P, P' which are at constant distance.*

Ex. 3.　Let A, B, C, E be fixed points of a conic, such that EB, CA are parallel, meeting in the point U on the Absolute line. Let the tangent at B meet the line CAU in T. Let L be a variable point of the conic, and CL meet the tangent BT in L'. Let EL, EL' meet CA in P and P'. Then the length of PP' is constant as L varies, and equal to the length of AT. For (P), (P') are related ranges, and P, P' can only coincide when L, L' coincide at B, in which case P, P' coincide at U.

Ex. 4.　Let the normal at a point P of a parabola, with focus S and vertex A, meet the axis AS in G; and the perpendicular to the axis, from P, meet the axis in N. Then the length of NG is constant as P varies, and equal to twice the length of AS.

For let the tangent at P meet the axis in T, and the axis AS meet the Absolute line in U. We have seen that A is the mid point of T, N, and S is the mid point of T, G, or say $N=(A, U)/T$, $G=(S, U)/T$. Wherefore N, G describe related ranges on the axis. This follows more directly by proving that either determines the other. But N, G can only coincide when both are at U. This shews that the length of NG is constant. When N is at S, the point T is on the directrix, and NG is of length equal to twice AS.

It may be remarked, in regard to Ex. 1, that we have reached the following results: (a) If A, A' be two points in a line, and C, C' two points in a parallel line, such that the lines AC, $A'C'$ are parallel, then CC' and AA' are of equal length; (b) If B, B' be two points in the line AA' such that BC, $B'C'$ are parallel, then BB' and AA' are of equal length; (c) The result in (b) is independent of the line CC', provided this remains parallel to AA'; (d) Two lines AP, AA', meeting in A, are of equal length, provided $A'P$ is parallel to one of the bisectors of the angles determined by the lines AP, AA'. These results, taken as postulates, would seem to be sufficient to build up a theory of length independent of a formula. But we do not enter further into this.

(11) Two points P, P', whose symbols, in terms of the symbols of three fixed points A, B, C, are $P=xA+yB+zC$ and $P'=x'A+y'B+z'C$, in which x, y, z, x', y', z' are numbers, are said to be obtained, either from the other, by a *linear transformation*, when x', y', z' are each expressible by a homogeneous linear function of x, y, z, with fixed numerical coefficients, say

$$x'=a_1 x+b_1 y+c_1 z,\quad y'=a_2 x+b_2 y+c_2 z,\quad z'=a_3 x+b_3 y+c_3 z.$$

It is understood, however, that the coefficients are such that the determinant, of three rows, (a_1, b_2, c_3), does not vanish; so that, conversely, each of x, y, z is a linear function of x', y', z'.

Such a transformation, which is also called a *homography*, determines either of the points P, P' when the other is given. There are certain obvious properties of such a relation: To all points P which lie on a line, correspond points P' also lying on a line, and conversely. Or, as we may say, the transformation changes a line into a line. Thus, the transformation changes all the lines passing through one point, P_0, into the lines passing through the corresponding point, P_0'; and, the points where the lines of the former pencil, of centre P_0, meet an arbitrary line, u, are changed into the points where the corresponding lines of the second pencil, of centre, P_0', meet the line, u', which corresponds to u. Evidently, the range of points, say P, on the line u, is related to the range of corresponding points, P', on the line u'; and, hence, the pencil of lines through P_0 is related to the pencil of corresponding lines through P_0'.

If ω, ω' be two given conics, such a linear transformation is obtainable by taking the polar line of P in regard to ω, and then the polar point, P', of this line, in regard to ω'. *It can, in fact, be shewn that every linear transformation, that is possible, is definable in this way, by taking the conics ω, ω' suitably.* Assuming this for the present, we see: (a), that, in general, when these conics ω, ω' meet in four distinct points, and so have a definite self-polar triangle, there are three points of the plane, the angular points of this triangle, each of which coincides with its transformed point; and, correspondingly, there are three lines, the sides of this triangle, each of which coincides with its transformed line. There are then, upon each of these three lines, two related ranges of corresponding points; and, upon this line, there are two self-corresponding points, which are two of the three self-corresponding points of the plane. Conversely, if we are given three arbitrary points A, B, C, forming a triangle, a linear transformation can at once be defined which leaves each of A, B, C unaltered, and, at the same time, changes an arbitrary point, D, into another arbitrary point, D'. This is the transformation changing a point P, of symbol $xA + yB + zC$, into the point P', of symbol $x'A + y'B + z'C$, wherein x'/x, y'/y, z'/z have the constant values which they take when P is at D and P' is at D'. It is supposed that D, D' are of such general positions that at most one of the three ratios x'/x, y'/y, z'/z, so found, has the value 1. But, (b), when the conics ω, ω' touch at one point, say A, there is specialisation in the form of the transformation. Let C be the definite point, on the common tangent at A, whose polar line is the same for the two conics. Take for B the point, other than A, where this polar meets one of the two conics. There are then *two* points which are unaltered by the transformation, namely C and A, of which A takes the place of two of the unaltered points which arise in case (a). Also there are *two* lines which are unaltered, namely the polar line AB, and the common tangent at A. By suitable choice of the symbols of the points A, B, C, it can be seen that the transformation changes the point P, of symbol $xA + yB + zC$, into the point P', of symbol $(x+y)A + yB + czC$, where c is a definite constant number, not equal to 1. Further, (c), when the conics ω, ω' have three coincident points in common, say at A, there is only *one* point of the plane unaltered by the transformation, namely A, and only one unaltered line, the common tangent at A. Let B be the remaining common point of the two conics and C the point where the tangent at A meets the tangent at B of one of the conics; then it can be seen that the transformation changes a point whose symbol is $xA + yB + zC$ into the point whose symbol is $(x+z)A + yB + (y+z)C$. When, (d), the conics ω, ω' have four coincident points in common, say at A, there is a point, C, on the tangent at A whose polar is the same for both conics; let this polar meet one of the conics in B. In this case, as in the case (b), there are *two* points unaltered by the transformation, namely A and C; and there are two lines unaltered, which are the tangent at A and the polar of the point C. The transformation now changes a point of symbol $xA + yB + zC$ into a point of symbol $(x+y)A + yB + zC$, which is the same result as in the case (b), except that the constant c is now 1. But the consequence is that not only A and C, but *every* point of the line AC, is unaltered; and *every* line through A is also unaltered. On any definite line through A there are pairs of corresponding points, describing related ranges whose two common corresponding points coincide in A. Finally, (e), let the conics ω, ω' touch in two points, say A and B; let C be the common pole of the chord AB. The transformation then changes a point $xA + yB + zC$ into a point $xA + yB + czC$, where c is a fixed number. In this case the point C, and every line through C, and every point of the line AB are unaltered by the transformation. If the line joining two corresponding points P, P' on any line through C, meet AB in U, the cross ratio $(P', P/U, C)$ is constant, being the number c. When $c = -1$, we have the transformation several times referred to as a *harmonic inversion*, with C as centre and AB as axis. And, as in that case, the line which arises by transformation of an arbitrary line meets this on AB. In the previous case, (d), two corresponding lines meet on AC, and the join of two corresponding points passes through A. (Cf. Note 6, below.)

The transformation (e) is called a *perspectivity*. It is the simplest of the five cases, perhaps, to describe geometrically. The previous case, (d), may be described as a *skew perspectivity*. In both cases there is a range of points each unaltered by the transformation, and also a pencil of lines each unaltered. But in the case of skew perspectivity the centre of the pencil of unaltered lines is one of the points of the range of unaltered points, and there is no unaltered point other than a point of this range.

(12) There is a particular transformation, having reference to the Absolute points of the plane, to which reference must be made.

This transformation, applicable to real points which do not lie on the Absolute line, is such that *if, thereby, two such points P, Q be changed into the points P', Q', respectively, then the length of $P'Q'$ is equal to the length of PQ.*

Let α, β, m be constant numbers; let P be any point whose symbol, in terms of the symbols of an arbitrarily taken point O, and those of the Absolute points I, J, is $O + \xi I + \eta J$, where ξ, η are numbers; then the transformation considered changes P into a point P' with the symbol $P' = O + (\alpha + m\xi)I + (\beta + m^{-1}\eta)J$. Making the convention we have adopted, that I, J are conjugate imaginaries, for P to be a real point the numbers ξ, η are conjugate imaginaries, and, for P' to be a real point, the numbers α, β are conjugate imaginaries, as are the numbers m and m^{-1}, so that m is of the form $e^{i\theta}$, where θ is a real number.

For clearness, distinguish the case where $m=1$. Then $P'=P+\alpha I+\beta J$. Thus P' is obtained from P by joining P to a particular point of the Absolute line (of symbol $\alpha I+\beta J$), which is the same for all positions of P, and taking P' on the joining line. If a point Q be changed to a point Q' by the same transformation, so that the lines PP', QQ' are parallel, the symbols are such that $Q'-P'=Q-P$, and the lengths $\overline{PQ}$ and $\overline{P'Q'}$ are equal. This transformation is often called a *translation*.

Suppose now that m is not equal to 1. Consider then the particular point C whose symbol is

$$C=O+\alpha\,(1-m)^{-1}I+\beta\,(1-m^{-1})^{-1}J;$$

this is transformed to the point whose symbol is

$$C'=O+[\alpha+m\alpha\,(1-m)^{-1}]\,I+[\beta+m^{-1}\beta\,(1-m^{-1})^{-1}]J,$$

so that $C'=C$, and the point C is unaltered. If we denote $\xi-\alpha\,(1-m)^{-1}$ by ξ_1 and $\eta-\beta\,(1-m^{-1})^{-1}$ by η_1, the transformation then changes the point of symbol $C+\xi_1 I+\eta_1 J$ into the point of symbol $C+m\xi_1 I+m^{-1}\eta_1 J$. Such a transformation is often called a *rotation*.

The transformation is formulated for points not lying on the Absolute line. But, for comparison with the general considerations already given, we may consider for a moment the transformation which changes a point of symbol $P=\zeta O+\xi I+\eta J$ into the point of symbol $P'=\zeta O+(\alpha\zeta+m\xi)\,I+(\beta\zeta+m^{-1}\eta)\,J$, so that the transformation of points on the Absolute line is given by putting $\zeta=0$. Then the translation is the case (d) considered above, called skew perspectivity, in which any point of the Absolute line remains unaltered, and every line through the point, on the Absolute line, whose symbol is $\alpha I+\beta J$, also remains unaltered. While the rotation belongs to the general case, (a), in which the three points C, I, J, and the joining lines of these, remain unaltered; but with the particularity that the distance of any point P from C is equal to the distance from C of the transformed point P'. And, further, the distance of two points P, Q of respective symbols $C+\xi_1 J+\eta_1 J$, $C+X_1 P+Y_1 J$, namely $[(X_1-\xi_1)\,(Y_1-\eta_1)]^{\frac{1}{2}}$, is equal to the distance of the two transformed points $P'=C+m\xi_1 I+m^{-1}\eta_1 J$, $Q'=C+mX_1 I+m^{-1}Y_1 J$; and also $\epsilon C\,(Q',\,P')=\epsilon C\,(Q,\,P)$.

In general, the transformation considered, depending on the constants α, β, m, is called a *movement*. And it is to be noticed that, if a point P is changed by one movement $(\alpha,\,\beta,\,m)$ to a point P', and then P' is changed to a point P'' by another movement $(\alpha_1,\,\beta_1,\,m_1)$, the result is as if P were changed to P'' by a single movement $(\alpha_1+m_1\alpha,\,\beta_1+m_1^{-1}\beta,\,mm_1)$. For, the equations

$$P'=O+(\alpha+m\xi)\,I+(\beta+m^{-1}\eta)\,J,$$

$$P''=O+[\alpha_1+m_1\,(\alpha+m\xi)]\,I+[\beta_1+m_1^{-1}\,(\beta+m^{-1}\eta)]\,J$$

lead to $\qquad P'' = O + (\alpha_2 + m_2\xi)\, I + (\beta_2 + m_2^{-1}\eta)\, J,$

where $\qquad \alpha_2 = \alpha_1 + m_1\alpha, \quad \beta_2 = \beta_1 + m_1^{-1}\beta, \quad m_2 = mm_1.$

For $m = 1$, $m_1 = 1$, the order in which the operations are taken is indifferent; namely, two successive translations are together equivalent to a translation, which is the same for both orders in which the translations may be made. Likewise two general movements are also *commutative* if $\alpha_1 + m_1\alpha = \alpha + m\alpha_1$, and $\beta_1 + m_1^{-1}\beta = \beta + m^{-1}\beta_1$, namely $\alpha\,(1-m)^{-1} = \alpha_1\,(1-m_1)^{-1}$ and $\beta\,(1-m^{-1})^{-1} = \beta_1\,(1-m_1^{-1})^{-1}$; that is, if the unaltered points, C and C', of the two rotations, coincide with one another.

Ex. 1. If, by a movement, the points P, Q, R become, respectively, P', Q', R', prove that, not only the distances QR, RP, PQ are respectively equal to the distances $Q'R', R'P', P'Q'$, but also $\epsilon P\,(Q, R) = \epsilon P'\,(Q', R')$, etc. In a usual phraseology, the triangles $PQR, P'Q'R'$ not only have sides of equal lengths, but are *directly similar* to one another.

Ex. 2. In Ex. 1, let lines be drawn, respectively through the mid points of P, P'; of Q, Q'; and of R, R', and respectively perpendicular to PP', QQ', RR'. Prove that in the general case the lines meet in a point, C, and the movement is a rotation with C as unaltered point. But in the case of a translation these lines are parallel, and meet on the Absolute line.

In fact, in the general case, two corresponding points P, P' are on a circle whose centre is the point C, of symbol $O + (1-m)^{-1}\,\alpha I + (1-m^{-1})^{-1}\,\beta J$; and, in the case of a translation, the lines PP' all pass through the point $\alpha I + \beta J$.

Ex. 3. Movement is not the only linear transformation in which the distances of points are unaltered. If l be a given line, and to any point P we make correspond the point P' which is the optical image of P in regard to the line l (so that l is perpendicular to PP' and contains the mid point of P, P'), then P' is obtained from P by a linear transformation, and the distance of two points P, Q is equal to the distance of their respective optical images P', Q'.

Let the line l be that joining a point O to a point $I + nJ$ of the Absolute line. Any point P has a symbol of the form $O + \xi I + \eta J$. Consider the point P' of symbol $O + n^{-1}\eta I + n\xi J$. Then PP' contains the point

$$(\xi - n^{-1}\eta)\, I + (\eta - n\xi)\, J,$$

or $I - nJ$, and is perpendicular to l; and the mid point of P, P' is the point $O + \tfrac{1}{2}(\xi + n^{-1}\eta)\, I + \tfrac{1}{2}(\eta + n\xi)\, J$, or $O + \tfrac{1}{2}(\xi + n^{-1}\eta)\,(I + nJ)$, which lies on l. The distance of two points $O + \xi I + \eta J$, $O + \xi' I + \eta' J$, namely $[(\xi' - \xi)(\eta' - \eta)]^{\frac{1}{2}}$, is then equal to the distance of the two points

$$O + n^{-1}\eta I + n\xi J, \quad O + n^{-1}\eta' I + n\xi' J.$$

But there is the difference, from a movement, that $\epsilon O\,(P, Q)$ is equal to $\epsilon O\,(Q', P')$, that is $[\epsilon O\,(P', Q')]^{-1}$; or, more generally, if P', Q', R' be the images of P, Q, R, respectively, then $\epsilon P\,(Q, R)$ is equal to $\epsilon P'\,(R', Q')$, etc. The triangles $P'Q'R'$, PQR may then be said to be *reversely similar* to one another, though their corresponding sides are of equal length.

The linear transformation here described is called *reflexion*, in the line l. Evidently it is a particular case of harmonic inversion, previously remarked, the axis being the line l, and the centre the point of the Absolute line common to all lines perpendicular to l.

(13) It is a familiar idea, arising from the metrical geometry of *rigid* plane figures, that the relations of the parts of such a figure are unaltered by a movement of the figure in its own plane; the linear

transformation we have called a perspectivity is also familiar, the transformed figure being obtained from the original by projection. It may seem that a general linear transformation is much more general than either of these transformations. In fact, it can be proved that any linear transformation of a plane figure can be obtained by a movement taken with a perspectivity. Precisely, if a linear transformation changes the figure Ω into the figure Ω', we prove that Ω' can be obtained by a perspectivity from a figure Ω_1 which is obtained from Ω by a movement. We notice that a linear transformation, containing, in the formulae, nine homogeneous coefficients, depends on eight constants, while a movement depends on three constants, and a perspectivity depends on five (two to determine the centre, two to determine the axis, and one to determine the relative cross ratio).

In proving this result, we use a certain abbreviated notation. We denote a linear transformation by a single functional symbol, say ϑ; and if, by this transformation, a point P becomes P', and a figure Ω becomes Ω', we write $P' = \vartheta P$ and $\Omega' = \vartheta\Omega$. The transformation by which we may pass back from the figure Ω' to the figure Ω, is denoted by ϑ^{-1}; and we write $\Omega = \vartheta^{-1}\Omega'$. If now the figure Ω' be further transformed by another linear transformation, ϕ, becoming Ω'', so that $\Omega'' = \phi\Omega'$, then Ω'' can be found from Ω by a single linear transformation. As $\Omega'' = \phi\Omega'$, and $\Omega' = \vartheta\Omega$, we write $\Omega'' = \phi\vartheta\Omega$, and Ω'' is found from Ω by the transformation denoted by $\phi\vartheta$. Algebraically, let ϑ change a point P, of symbol $xX + yY + z\mathbf{Z}$, to a point P' of symbol $x'X + y'Y + z'\mathbf{Z}$, where

$$x' = a_1x + b_1y + c_1z, \quad y' = a_2x + b_2y + c_2z, \quad z' = a_3x + b_3y + c_3z;$$

then we may write

$$(x', y', z') = \begin{pmatrix} a_1, & b_1, & c_1 \\ a_2, & b_2, & c_2 \\ a_3, & b_3, & c_3 \end{pmatrix} (x, y, z),$$

and the matrix, which we may denote by Θ, characterises the transformation, ϑ. If the transformation ϕ be similarly given by a matrix Φ, the single composite transformation $\phi\vartheta$ is given by the matrix $\Phi\Theta$, which is obtained (according to the ordinary rule for multiplying two matrices) by associating the *rows* of Φ with the *columns* of Θ.

Now assume a figure Ω to be changed to a figure Ω', by a linear transformation ϑ; we take the case when ϑ has three distinct unaltered points, not lying on the Absolute line, though the theorem stated is true also for more particular cases. Suppose Ω is changed into a figure Ω_1 by a movement, which we denote by μ, writing $\Omega_1 = \mu\Omega$, and Ω_1 is changed into Ω' by a perspectivity, which we denote by ϖ, writing $\Omega' = \varpi\Omega_1$. Then we have $\vartheta\Omega = \varpi\mu\Omega$, which is only true for a general figure Ω if $\vartheta = \varpi\mu$, or $\vartheta\mu^{-1} = \varpi$. We desire then to prove, ϑ being given, that there is a movement μ, whose inverse is μ^{-1}, such that the composite transformation $\vartheta\mu^{-1}$ is a perspectivity, ϖ, namely, is a linear transformation by which all the points of a certain line are unaltered, and all the lines through a certain point are also unaltered (the fixed point of intersection of these lines not being on the line of fixed points, in the general case).

Through the five points consisting of the supposed unaltered points A, B, C of the transformation ϑ, and the Absolute points I, J of the plane, there passes a conic, which we denote by ω; this is a circle. Applying the transformation ϑ to the points of this circle ω, we obtain a conic, which we denote by $\vartheta\omega$; this contains the points A, B, C of ω; let O' be the fourth common point of the two curves. As O' is a point of the conic $\vartheta\omega$, it must arise from a point O, of the circle ω, such that $O' = \vartheta O$. Draw the line through the mid point of the chord OO', of the circle ω, which is perpendicular to this chord; let K be one of the points where this line (which is a diameter of the circle ω) meets the circle ω. We may readily shew that there is a definite movement, a rotation, in which K is the unaltered point, or centre, which changes O into O'. This rotation we denote by μ, so that $O' = \mu O$. We are concerned then with two linear transformations ϑ, μ, of which the former has the unaltered points A, B, C; and the latter has the unaltered points K, I, J (by what was said above), and these six points all lie on the conic ω. We are to prove that the composite transformation $\vartheta\mu^{-1}$ is a perspectivity.

We prove in fact that every line through the point O', the common point, other than A, B, C, of the two conics ω and $\vartheta\omega$, is unaltered by the transformation $\vartheta\mu^{-1}$. Further, that the conics $\vartheta\omega$ and $\mu\omega$ touch one another at O'; these two conics have then two other common points, U, V; we prove that every point of the line UV is unaltered by the transformation $\vartheta\mu^{-1}$.

Consider the pencil of lines passing through O, the point $\vartheta^{-1}O'$, of the conic ω. By the transformation ϑ, this pencil is changed into a related pencil with O' as centre. The locus of the common point of corresponding rays of these two pencils is then a conic containing O and O'; as the points A, B, C are each unaltered by the transformation ϑ, the lines $OA, O'A$ of these two pencils correspond to one another, and the conic in question contains A; likewise it contains B and C. This conic is thus the conic ω. Thus, conversely, if P be any point of the conic ω, the transformation ϑ changes the line OP into the line $O'P$; and so changes P into a point of the line $O'P$, lying on the conic $\vartheta\omega$. A precisely similar argument holds for the transformation μ, of which the unaltered points K, I, J lie on ω, while $O' = \mu O$. From these two results, it follows that every line, say l', drawn through O', is changed into itself by the composite transformation $\vartheta\mu^{-1}$. For, by μ^{-1}, the line l' is changed into a line, l, through O, meeting the conic ω on the line l'; and then, by ϑ, this line l is changed again into the line l'.

Now consider the two conics $\vartheta\omega$ and $\mu\omega$. As O is on ω, while $O' = \vartheta O$ and $O' = \mu O$, both these conics pass through O'. We have seen that any line through O, meeting the conic ω again in P, is changed, both by ϑ and μ, into the line $O'P$. In particular the tangent line of ω, at O, is changed, both by ϑ and μ, into the line $O'O$. By a linear transformation of a conic a tangent at any point is changed into the tangent at the corresponding point of the new conic arising by the transformation. Thus $O'O$ is tangent at O' to both the conics $\vartheta\omega$ and $\mu\omega$. Let U, V be the other two points common to these conics. We have seen that, if P be any point of the conic ω, both the points Q, R, arising respectively as ϑP and μP, lie on the line $O'P$; so that these points Q, R are the second intersections of the line $O'P$ with $\vartheta\omega$ and $\mu\omega$, respectively; they are, moreover, such that $Q = \vartheta\mu^{-1}R$. We can then take a position of P such that $O'P$ meets the conics $\vartheta\omega$, $\mu\omega$ in the same point (other than O'); or, so that Q, R coincide; this can be done so that the point of coincidence is at U; or also so that it is at V. Thus both U and V are unaltered by the transformation $\vartheta\mu^{-1}$. Thus the line UV is unaltered. And as we have shewn that every line through O' is unaltered, therefore *every* point of the line UV is unaltered.

Thus the transformation $\vartheta\mu^{-1}$ is a perspectivity, with O' as centre, and the line UV as axis.

(14) We proceed now to some further detail in regard to the axis UV, in (13). As ABC, KIJ are two triangles inscribed in the conic ω, there exists a conic touching the sides of these triangles. This is in fact the parabola touching BC, CA, AB, having its focus at K; denote it by ω_0. This being triangularly inscribed in the circle ω, the two tangents drawn to ω_0, from any point of the circle ω, meet this circle again in two points whose join is a tangent of the parabola ω_0. We prove that *the axis UV, of the perspectivity $\vartheta\mu^{-1}$, is the tangent of the parabola obtained, in this way, from the point O of the circle.*

Denote the line UV by p. Let X, Y be any two points of this line, and denote the points $\mu^{-1}X$, $\mu^{-1}Y$, respectively, by X_0 and Y_0, and the line X_0Y_0 by p_0; so that $p = \mu p_0$. The point ϑX_0, or $\vartheta\mu^{-1}X$, is the point X, because $\vartheta\mu^{-1}$ leaves every point of the line UV, or p, unaltered; likewise for Y_0. Thus, besides $p = \mu p_0$, we have $p = \vartheta p_0$. As μ is a movement, the distance of X_0 and Y_0, for which $X = \mu X_0$, $Y = \mu Y_0$, is equal to the distance of X and Y. The line p_0 is therefore such that, when transformed into p by ϑ, any two transformed points of p have the same distance as the points of p_0 from which they arise. Further, as the range of points X, on p, which correspond to the points X_0, on p_0, either by μ or ϑ, is related to the range of points X_0, on p_0, the joining line X_0X, of two corresponding points on these lines, envelops a conic, touching both p_0 and p. In particular, the common point, of p_0 with the line BC, becomes, by the transformation ϑ, the point common to p and the line BC itself; thus BC is one of the tangents of the envelope; as are, in the same way, also CA and AB. And as $X = \mu X_0$, a similar argument shews that the envelope touches the sides of the triangle KIJ. Thus the conic enveloped by the lines X_0X is the parabola ω_0. And this shews, in particular, that the line p, or UV, is a tangent of the parabola. Also, from the fact that the lengths XY, X_0Y_0 are equal, we can infer that the lines p, p_0 meet on the axis of the parabola. Now let the line UV meet the circle ω in L and M. By what we have seen, both the points $\vartheta^{-1}L$, $\mu^{-1}L$ lie on the line OL; while also $L=\vartheta\mu^{-1}L$, so that the points $\vartheta^{-1}L$, $\mu^{-1}L$ are the same point; they are therefore the point common to OL and the line p_0. The line joining this point to L, which is the line OL, is thus tangent to the parabola ω_0. Similarly OM touches ω_0.

There is thus a duality in the figure. Just as every line through O' is unaltered by the transformation $\vartheta\mu^{-1}$, so every point of the line p is unaltered by this transformation; and, just as O' arises from O both by ϑ and μ, so the line p arises from p_0 both by ϑ and μ; and, as the circle ω is the locus of common points of corresponding rays drawn from O and O', so the parabola ω_0 is the envelope of joins of corresponding points of the lines p, p_0.

It seems right to see algebraically that this duality was to be expected. We have referred already to the *equation of a line*, $ux+vy+wz=0$, wherein u, v, w are constants, which connects the multipliers, or *coordinates*, x, y, z, of *every* point P, of symbol $x\mathbf{X}+y\mathbf{Y}+z\mathbf{Z}$, which lies on the line (Chap. IX (19)). If we take other constants u', v', w', such that

$$u=a_1u'+a_2v'+a_3w', \quad v=b_1u'+b_2v'+b_3w', \quad w=c_1u'+c_2v'+c_3w',$$

the condition $ux+vy+wz=0$ involves $u'x'+v'y'+w'z'=0$, provided x', y', z' are given by

$$x'=a_1x+b_1y+c_1z, \quad y'=a_2x+b_2y+c_2z, \quad z'=a_3x+b_3y+c_3z.$$

Thus the linear transformations we have considered, applied to points, can equally be applied to lines; just as a point is given by coordinates x, y, z, so a line is given by constants u, v, w. These are often called the *coordinates of the line*. When the coordinates of points are linearly transformed, the coordinates of lines are also linearly transformed. If we write $(x', y', z')=\Theta\,(x, y, z)$, the corresponding transformation for the coordinates of a line is given by $(u, v, w)=\overline{\Theta}\,(u', v', w')$, where $\overline{\Theta}$ means the matrix obtained from Θ by transposition of rows and columns; it was said by Sylvester that the transformation of u, v, w is *contragredient* to that of x, y, z.

Ex. 1.　Let the common points of the conic ω with the line p_0 be L_0, M_0. Prove that these lie on the lines $O'U$, $O'V$.

Ex. 2.　Let P, Q be any two points of the Absolute line IJ, and $P'=\vartheta P$, $Q'=\vartheta Q$. Prove that the lines PP', QQ' meet on the circle $P'Q'O'$.

Ex. 3.　Prove that the theorem, that a movement μ exists for which $\vartheta\mu^{-1}$ is a perspectivity, continues to hold when two, or when three, of the points unaltered in the transformation ϑ, coincide.

Further developments of the theory, in (13) and (14), may be found in H. J. S. Smith, *Proc. Lond. Math. Soc.* II, 1869 and III, 1871. Also in Chasles, *Géom. Supér.* 1880, p. 378.

(15) The following results, in regard to *envelopes*, whose statement involves the notion of distance, may be given as examples.

Ex. 1.   Let ellipses all having the same director circle be drawn through two points P, P', which are such that the common centre C is the mid point of P, P'. Prove that these ellipses envelop another ellipse, having P, P' as foci, which has double contact with the given director circle on the line PP'. The point of contact of the envelope, with the variable ellipse, is the point of this where the tangent is perpendicular to the tangent at P.

Ex. 2.   Recalling the definition of the subdirector circle of a conic, whose centre is at one focus, S, of the conic, and radius is of length $2a$, equal to that of the axis AA' of the conic (Chap. xiii (2)), let ellipses be described, with a common fixed point, P, a fixed focus, S, and a fixed subdirector circle, of centre S and radius equal to $2a$. Prove that these ellipses envelop another ellipse, with foci at S and P, whose axis is of length $4a - \overline{SP}$. The point of contact with the envelope is on the line joining P to the focus H, of the variable ellipse, which is complementary to S.

Ex. 3.   Let ellipses be described, with axes of given length $2a$, having one fixed focus S, the complementary focus H varying on a fixed line. Prove that all these ellipses touch two parabolas, both having S as focus, the directrices of these being two lines which are parallel to the given line and at perpendicular distances $2a$ from this. The points of contact of one of the variable ellipses with the two parabolas are on a line perpendicular to the given line.

Ex. 4.   Let H' be a variable point of that axis of a fixed ellipse which is perpendicular to the axis containing the real foci, S and H. Let a variable ellipse be described with foci at S and H', with axis of length $2a - \overline{SH'}$, where $2a$ is the length of the axis of the fixed ellipse. Prove that the variable ellipse touches the given ellipse, say at P, and also touches a fixed circle, of centre S and radius a, say at L, the line LH' containing the focus S, and the line PH' containing the complementary focus H.

(16) The reader may develop a theory of distance in space of three dimensions, analogous to the theory given here for a plane. For this, there is assumed an Absolute plane, and, lying therein, an Absolute conic. The Absolute points of any plane are then the two points of this plane which lie on the Absolute conic; and two lines, neither in the Absolute plane, are perpendicular if the points in which these lines meet the Absolute plane are conjugate in regard to the Absolute conic. If then X, Y, Z are the symbols of three points in the Absolute plane, which form a self-polar triangle in regard to the Absolute conic, and O be the symbol of a point not lying on the Absolute plane, the symbol of a point P, not on the Absolute plane, may be taken in the form $O + xX + yY + zZ$, where x, y, z are numbers. When O, X, Y, Z are real points, and x, y, z are real numbers, we speak of this point P as a real point. If the similar symbol of another real point P' be $O + x'X + y'Y + z'Z$, the distance of P and P' is the positive square root of $(x' - x)^2 + (y' - y)^2 + (z' - z)^2$. We assume that the reader is familiar with the more usual metrical results in space; and give here some examples which are of interest in connexion with what has preceded.

Ex. 1.   A *right circular cone* may be generated by drawing a line through the centre of a circle, perpendicular to the plane of the circle, and joining any point V of this perpendicular to all the points of the circle. The line is called the *axis of the cone*, the point V *the vertex*, and the lines joining V to the points of the circle are called *the generators*. We assume that any plane perpendicular to the axis meets the cone (that is, meets the generators) in points lying on a circle, whose centre is on the axis. We also assume that a sphere can be constructed, with centre at an arbitrary point of the axis, to touch the cone at every point of such a circle. In saying this, we assume familiarity with the notion of a tangent plane of a sphere, at any point; and also with that of a tangent plane of the cone, which touches the cone at all points of a single generator. We assume further that an arbitrary plane meets the cone in a conic, which may be a hyperbola, or a parabola, or an ellipse. We are concerned here with the last case. The section of a sphere by a plane is a circle; consider such a sphere as has been spoken of, which touches the cone at all points of a circle. The circle, in which this sphere is met by a plane section of the cone, touches the conic section of the cone by this plane, in two points, lying on the plane containing the points of contact of the sphere with the cone; the line joining the two points of contact is the line common to the plane of the conic section, and the plane of contact of the sphere with the cone. This we assume to be obvious. Also we assume that the axis, of the elliptic section of the cone which is under consideration, lies in a plane, through the axis of the cone, which is perpendicular to the plane of the section; the ellipse consists of pairs of points which are optical images of one another in regard to this plane; in particular, the two points of contact spoken of are such a pair. In the accompanying diagram this plane is the plane of the paper. Finally we assume that, if two lines, meeting in a point T, touch the sphere in P and Q, then the lengths of TP and TQ are equal.

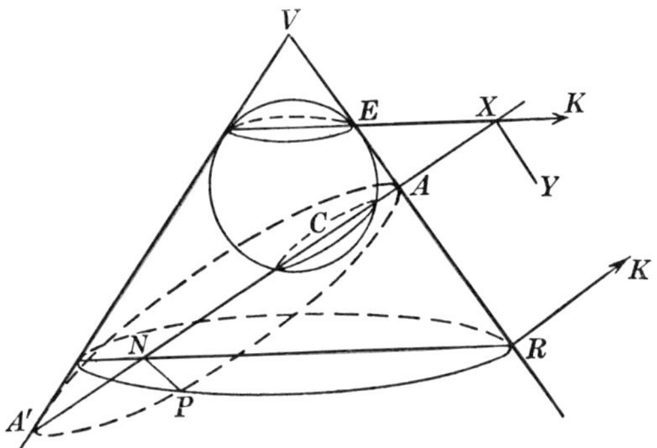

In the diagram, let VA, VA' be the two generators of the cone lying in the plane, through the axis of the conic, which is perpendicular to the plane section of the cone considered, the points A, A' being the vertices of this section. Let P be any point of this section, and PN the line drawn from P to the axis AA' of the section; let NR, in the plane of VA, VA', meet the generator VA in R. Let a sphere, which touches the cone in all points of a circle whose plane is perpendicular to the axis of the cone, meet the plane of the section in the circle C. Let the plane of contact of the cone and sphere meet VA in E, and meet the plane of the section in the line XY, which is then parallel to NP. We take X on the axis AA'. Then XY contains the two points of contact of the circle C with the elliptic section, as we may see most clearly by taking a diagram in which XY has real intersections with the elliptic section of the cone. Let RK, parallel to AX, meet EX in K.

Now consider the tangent from P, in the plane of the elliptic section, to the circle C. This touches the sphere at its point of contact with the circle. The length of this tangent is then equal to the length of the tangent to the sphere, drawn from P, which lies along the generator PV of the cone. This, however, is clearly of the same length as the line *RE*. The line drawn from *P* perpendicular to *XY* is clearly of the same length as *NX*, which is of the same length as the parallel line *RK*.

Wherefore, the length of the tangent from *P* to the circle *C* bears to the length of the perpendicular from *P* to the line of contact *XY*, the ratio of the length of *RE* to the length of *RK*. This ratio is, however, clearly independent of the position of *P* on the elliptic section, being the ratio of the lengths of *AE* and *AX*; and dependent only on the directions of these lines. It is thus also independent of the particular sphere, and so is independent of the circle *C* taken to have double contact with the elliptic section. In particular, if the sphere be taken to touch the plane of the elliptic section, say in *S*, the circle *C* degenerates into the lines joining *S* to the Absolute points of the plane of the section, and *S* is a focus of the section, with *XY* as the corresponding directrix. The ratio of lengths considered is then that of the line *SP* to that of the perpendicular from *P* to the directrix, commonly called the *eccentricity* of the ellipse. We thus have interesting proofs of the theorems given above in (6) and (9), Ex. 14.

Ex. 2. Let *ABC*, *DEF* be two triangles in a plane, which are in perspective, the lines *AD*, *BE*, *CF* meeting in a point, *S*, while the intersections of corresponding sides, (*BC*, *EF*); (*CA*, *FD*); (*AB*, *DE*), or say, respectively *P*, *Q*, *R*, lie in a line, *p*. Let another plane be drawn through this line *p*. We assume what we can at first describe by saying that we can regard the plane *DEF* as a rigid body, capable of rotation about the line *p*, in which the triangle *DEF* is rigidly fixed. That is, we assume, that, in the other plane through *p*, there exists a triangle *D'*, *E'*, *F'*, with sides *E'F'*, *F'D'*, *D'E'*, respectively, passing through the points *P*, *Q*, *R* of the line *p*, the distances *PE'*, *E'F'*, *QF'*, *F'D'*, *RD'*, *D'E'*, and also the quantities $\epsilon'(p, PE')$, $\epsilon'(p, QF')$, $\epsilon'(p, RD')$, measured in the new plane, being the same as for *D*, *E*, *F* in the original plane. As *E'F'*, *F'D'*, *D'E'* meet *EF*, *FD*, *DE* in the points *P*, *Q*, *R*, on the line *p*, which lie, respectively, on *BC*, *CA*, *AB*, it follows that the lines *AD'*, *BE'*, *CF'*, which join points of one plane to points of the other plane, meet in a point, say, *S'*.

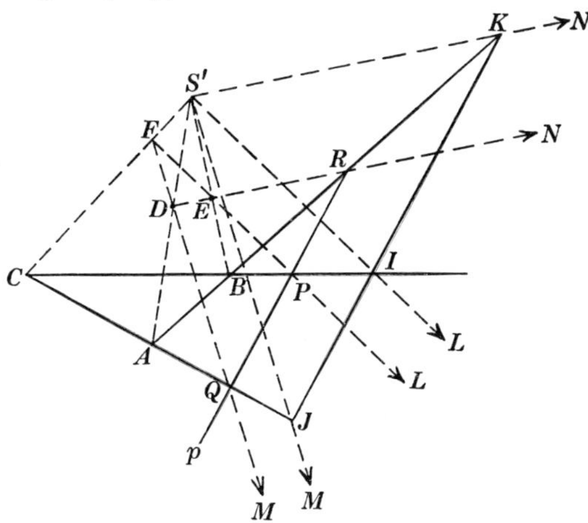

We prove that, *as the plane D'E'F' takes different positions through p, the point S' describes a circle, in a plane perpendicular to the line p, the centre of this circle being on a line IJK, parallel to p, lying in the original plane ABC.*

For brevity we denote the points *D'*, *E'*, *F'*, which are the positions of *D*, *E*, *F* after the plane has rotated about the line *p*, by *D*, *E*, *F*. In the plane

$S'EF$, which contains the line BC, draw the line, through S', parallel to EF, meeting BC in I; denote its intersection with EF, which lies on the Absolute plane, by L. Similarly, in the plane $S'FD$, draw from S' the line parallel to FD, meeting CA in J, and meeting the Absolute plane in M. And, in the plane $S'DE$, draw from S' the line parallel to DE, meeting AB in K, and meeting the Absolute plane in N. The lines $S'I$, $S'J$, $S'K$ thus lie in a plane parallel to the plane DEF, meeting the plane ABC in the line IJK, which is thus parallel to PQR, this latter being common to the planes DEF and ABC. The points A, B, R, K are in perspective, respectively, with D, E, R, N, from the centre S'; also A, B, R are fixed points, and DE, ER are, by hypothesis, of constant lengths as the plane DEF rotates about the line p. Thus K is the same point, in the plane ABC, whatever be the position of the plane DEF. Likewise I, J are fixed points. Also, as $S'K$ is parallel to ER, the line KS' is of constant length. So IS' and JS' are of constant lengths. Thus S' is a point common to three spheres, with centres at I, J, K, whose radii are definite constants. This shews that S' describes a circle, in a plane perpendicular to the line IJK (and therefore perpendicular to the line PQR, or p), the centre of this circle being on the line IJK.

The result is given by Chasles, *Géom. Supér.* 1880, pp. 253–255, who refers to Poncelet, *Prop. Proj.* pp. 55, 75. The curves described by D, E, F are circles, in planes perpendicular to the line p, with centres on this line; and the circle described by S' is in perspective with these circles, respectively from A, B and C.

Ex. 3. We have said that two lines in space are perpendicular if they meet the Absolute plane in points which are conjugate in regard to the Absolute conic; likewise, a line and a plane are perpendicular if they meet the Absolute plane in a point and a line of which the point is the pole of the line in regard to the Absolute conic; and two planes are perpendicular if the lines in which they meet the Absolute plane are conjugate in regard to the Absolute conic.

Let O, p, ϖ be respectively a point, a line, and a plane, in general positions. Prove that a definite plane can be drawn through O to be perpendicular to the line p, meeting this, say, in the point N; also, that a definite line can be drawn through O to meet the line p, and be perpendicular to this, this being the line ON. Prove also that a definite line, q, can be drawn through O to be perpendicular to the plane ϖ; and that, if m be any line lying in the plane ϖ, the plane through O, which is perpendicular to m, is perpendicular to the plane ϖ, and contains the perpendicular q.

Hence prove that, if A, B, C, D be four points in space, no three in a line, which form a *tetrahedron*, then the line drawn through the orthocentre of the triangle ABC, perpendicular to the plane ABC meets each of the lines, drawn from A, B, C, respectively perpendicular to the opposite planes DBC, DCA, DAB and these all meet in a line through D.

Ex. 4. Let A, B, C, D be four points in space, forming a tetrahedron. Prove that two definite planes can be drawn, one through the edge BC, the other through the opposite edge AD, to have their common line on the Absolute plane, with a similar statement for each of the other two pairs of opposite edges of the tetrahedron, CA, BD and AB, CD.

Hence, or otherwise, prove that, if the opposite edges CA, BD be perpendicular lines, and the opposite edges AB, CD be perpendicular lines, then also the edges BC, AD are perpendicular lines. This is in connexion with the theorem, for a quadrilateral in the Absolute plane, that if two pairs of complementary intersections, of the lines of the quadrilateral, consist each of points which are conjugate in regard to the Absolute conic, then so does the third pair. When the tetrahedron has this property, prove that the sum of the squares of the lengths of two opposite edges is the same for all three pairs of opposite edges; and, also, that the four lines, drawn from A, B, C, D, perpendicular respectively to the opposite planes $BCD, \ldots, ABC$, meet in a point.

CHAPTER XVI

THE EQUATION OF A CONIC, AND OF A LINE

When the symbol of a point P is given, in terms of the symbols A, B, C, of three fixed points of a plane, in the form $P = xA + yB + zC$, the ratios of the numbers x, y, z (or, sometimes, these numbers themselves) are called the *coordinates* of the point P. If the ratios of x, y, z are connected by a definite (homogeneous) polynomial relation, and vary subject to this, the point P describes what is called an *algebraic curve*; the numbers multiplying the monomial terms $x^p y^q z^r$, in this relation, are usually called the *coefficients* in this relation; their ratios are constant, and characterise the curve. The relation is called the *equation of the curve*.

We have, for instance, seen (Chap. IX (17)), for the case of a conic, that if A, C be two points of the conic, and B the pole of the line AC, then any point of the conic has a symbol of the form $\theta A + kB + \theta^{-1} C$, wherein k is a constant number, but θ is a number which varies from point to point of the curve. If this symbol be denoted by $xA + yB + zC$, we thus have $k^2 xz - y^2 = 0$; this equation, satisfied at every point of the conic, is the equation of the conic referred to A, B, C.

There are some properties which can be very easily proved by utilising the notion of the equation of a conic; and the algebraic method is often of value for testing the truth of a suspected theorem. Moreover, much labour has been devoted to the elaboration of the algebraic method. Thus we give some account of it in this chapter.

(1) **The equation of a line. The equation of a point.** Let $P_1 = x_1 A + y_1 B + z_1 C$, and $P_2 = x_2 A + y_2 B + z_2 C$, be the symbols of two points, referred to arbitrarily chosen fixed points A, B, C. Let u, v, w denote respectively $y_1 z_2 - y_2 z_1$, $z_1 x_2 - z_2 x_1$, $x_1 y_2 - x_2 y_1$. Then the equation $ux + vy + wz = 0$ is satisfied when x, y, z have the respective values $\theta x_1 + \phi x_2$, $\theta y_1 + \phi y_2$, $\theta z_1 + \phi z_2$, whatever numbers θ and ϕ may be. These, however, are the coordinates of any point, of symbol $\theta P_1 + \phi P_2$, lying on the line joining the points P_1 and P_2. Thus the equation $ux + vy + wz = 0$ is called *the equation of this line*. Conversely, if u, v, w, or their ratios, be arbitrarily given, any three points whose coordinates, (x, y, z), (x_1, y_1, z_1), (x_2, y_2, z_2) satisfy the equation $ux + vy + wz = 0$, are such that numbers, θ, ϕ, exist, for which $x = \theta x_1 + \phi x_2$, etc. So that these three points lie on a line. Thus it appears that a given line determines uniquely the ratios of the three numbers u, v, w; and, conversely, that, when these ratios are given, a definite line is determined. For this reason, the ratios of u, v, w are often called the *coordinates of the line*.

And then, this phraseology is extended: If a, b, c are numbers, of given ratios, and u, v, w are supposed variable, the equation $au + bv + cw = 0$ is called *the equation of the point whose co-ordinates are* (a, b, c); it is the equation connecting the coordinates of all the lines (u, v, w), which pass through the definite point (a, b, c); just as $px + qy + rz = 0$, in which p, q, r are numbers of fixed ratios, is the equation connecting the coordinates of every point (x, y, z) which lies on the definite line (p, q, r).

Ex. 1.   Let $u_1x + v_1y + w_1z = 0$ and $u_2x + v_2y + w_2z = 0$, or, say, $U_1 = 0$ and $U_2 = 0$, be the equations of any two lines. Prove that the equation of any line which passes through the common point of these two lines is of the form $U_1 + \lambda U_2 = 0$, where λ is a certain constant. Prove, also, that the equation of the line, through the common point of the two given lines, which is the harmonic conjugate of the line given by $U_1 + \lambda U_2 = 0$, in regard to the two given lines, is $U_1 - \lambda U_2 = 0$. More generally, let the four lines given by $U_1 = 0$, $U_2 = 0$, $U_1 + \lambda U_2 = 0$, $U_1 + \mu U_2 = 0$ cut an arbitrary line in the respective points P_1, P_2, L, M. Prove that the cross ratio $(L, M/P_1, P_2)$ is equal to λ/μ.

These facts may also be expressed by saying that the coordinates of any line which passes through the common point of the two given lines (u_1, v_1, w_1), (u_2, v_2, w_2) are $u_1 + \lambda u_2, v_1 + \lambda v_2, w_1 + \lambda w_2$, in which λ is a suitable number; and that the coordinates of the harmonically conjugate line spoken of are then $u_1 - \lambda u_2, v_1 - \lambda v_2, w_1 - \lambda w_2$. Also, that the cross ratio of the lines whose co-ordinates are $(u_1 + \lambda u_2, \ldots)$, $(u_1 + \mu u_2, \ldots)$, with respect to the lines $(u_1, \ldots,)$, $(u_2, \ldots)$, is λ/μ.

Further, the equation of any point on the line which joins the points whose respective equations are $a_1u + b_1v + c_1w = 0$, $a_2u + b_2v + c_2w = 0$, or, say, $X_1 = 0$, $X_2 = 0$, is capable of the form $X_1 + \lambda X_2 = 0$.

Evidently, just as, in building up the theory, we have used symbols for *points*, so we could, instead, use symbols for *lines*; a line of equation $u_1x + v_1y + w_1z = 0$ would be represented, say, by a symbol U_1; and so on.

Ex. 2.   The necessary and sufficient condition that three lines, with equations $U_1 = 0$, $U_2 = 0$, $U_3 = 0$, should meet in a point, is the vanishing of a certain determinant, Δ, formed from the coefficients in the equations $U_1 = 0$, $U_2 = 0$, $U_3 = 0$, its rows consisting respectively of the elements u_1, v_1, w_1; u_2, v_2, w_2; and u_3, v_3, w_3.

Suppose Δ not to be zero. Let X, Y, Z be the symbols for the common points of the respective pairs of lines $U_2 = 0$, $U_3 = 0$; $U_3 = 0$, $U_1 = 0$ and $U_1 = 0$, $U_2 = 0$; and, as before, let A, B, C be the symbols of the points in terms of which $U_1 = 0$, $U_2 = 0$, $U_3 = 0$ are defined. Prove that

$$U_1X + U_2Y + U_3Z = \Delta\,(xA + yB + zC),$$

the common point of which X is the symbol, having the coordinates $(u_2v_3 - u_3v_2, \ldots)$, etc.

This is true identically in regard to x, y, z. It shews that, *referred to the points X, Y, Z, the coordinates of the point (x, y, z) are (U_1, U_2, U_3)*.

Ex. 3.   Referring the four points of any quadrangle to the angular points of the vertex triangle, shew that the symbols of these angular points may be so chosen that the coordinates of the four points of the quadrangle are, respectively, those given by $(1, \pm 1, \pm 1)$. Referring the four lines of any quadrilateral to the angular points of the diagonal triangle, shew that the symbols of these angular points may be so chosen that the equations of the four lines of the quadrilateral are, respectively, those given by $x \pm y \pm z = 0$.

(2) **The point equation, and the tangential equation, of a conic.**
We have remarked above (Chap. IX (17)), that, referred to suitably

chosen points, A, B, C, the coordinates ξ, η, ζ of any point of a conic, whose symbol is $\xi A + \eta B + \zeta C$, satisfy an equation $k^2\zeta\xi = \eta^2$, where k is a constant number. Let X, Y, Z be any three points of the plane, not lying in a line, whose symbols, in terms of those of A, B, C, are, respectively,

$$X = a_1 A + b_1 B + c_1 C, \quad Y = a_2 A + b_2 B + c_2 C, \quad Z = a_3 A + b_3 B + c_3 C,$$

where a_1, ..., c_3 are constant numbers. Let x, y, z be the coordinates of a point in regard to X, Y, Z. Then the symbol of the point has both the forms $\xi A + \eta B + \zeta C$ and $xX + yY + zZ$; and the latter is

$$x\,(a_1 A + b_1 B + c_1 C) + y\,(a_2 A + b_2 B + c_2 C) + z\,(a_3 A + b_3 B + c_3 C);$$

thus we infer that

$$\xi = a_1 x + a_2 y + a_3 z, \quad \eta = b_1 x + b_2 y + b_3 z, \quad \zeta = c_1 x + c_2 y + c_3 z.$$

Thus the coordinates, relatively to X, Y, Z, of any point of the conic, satisfy the equation

$$k^2\,(c_1 x + c_2 y + c_3 z)\,(a_1 x + a_2 y + a_3 z) - (b_1 x + b_2 y + b_3 z)^2 = 0;$$

this is called the *point equation of the conic*, referred to X, Y, Z.

Conversely, it can be shewn that any homogeneous quadratic equation connecting x, y, z, say,

$$ax^2 + by^2 + cz^2 + 2fyz + 2gzx + 2hxy = 0,$$

wherein a, b, c, f, g, h, which are called the *coefficients* in the equation, are numbers of fixed ratios, involves that the point whose coordinates are x, y, z, lies on a conic. The nature of this conic depends on the ratios of the coefficients. In particular, if these coefficients are connected by a certain relation, the quadratic polynomial breaks into two factors, both linear in x, y, z; then the conic degenerates into two lines. We may prove this in several ways. Perhaps the most satisfying is to prove first that the quadratic polynomial can be written in the form

$$L\,(l_1 x + m_1 y + n_1 z)^2 + M\,(l_2 x + m_2 y + n_2 z)^2 + N\,(l_3 x + m_3 y + n_3 z)^2,$$

as a sum of three squares, in which L, M, N and the nine coefficients, l_1, ..., n_3, are real rational functions of the original coefficients a, b, c, f, g, h. The expression is variously possible. For this, remark first that a quadratic polynomial in two variables, x and y, say $Ux^2 + 2Vxy + Wy^2$, can be written similarly as a sum of *two* squares; if one at least of the coefficients U, W be other than zero, say W, this quadratic is

$$W^{-1}\,(UW - V^2)\,x^2 + W^{-1}\,(Vx + Wy)^2;$$

and, if both U, W vanish, it is $\tfrac{1}{2}V\,[(x+y)^2 - (x-y)^2]$. Now take the quadratic in x, y, z: if one at least of the coefficients a, b, c be not zero, say c is not zero, the quadratic is

$$c^{-1}\,(gx + fy + cz)^2 + c^{-1}\,[(ca - g^2)\,x^2 - 2\,(fg - ch)\,xy + (bc - f^2)\,y^2],$$

of which the last three terms are a quadratic polynomial in x and y

only. If all of a, b, c be zero, but none of f, g, h vanish, the quadratic in x, y, z is, for example,

$$\tfrac{1}{2}fg^{-1}h^{-1}\,(gz+hy+2f^{-1}ghx)^2 - 2f^{-1}ghx^2 - \tfrac{1}{2}fg^{-1}h^{-1}\,(gz-hy)^2.$$

While, if a, b, c be all zero, and one of f, g, h be zero, say $f=0$, the form is $2x\,(gz+hy)$, which is $\tfrac{1}{2}\,(x+gz+hy)^2 - \tfrac{1}{2}\,(x-gz-hy)^2$; and lastly, if a, b, c be all zero, and two of f, g, h be zero, say $f=0$, $g=0$, the form is $2hxy$, or $\tfrac{1}{2}h\,(x+y)^2 - \tfrac{1}{2}h\,(x-y)^2$. And we remark that, in the expression $L\xi^2 + M\eta^2 + N\zeta^2$ thus obtained for the general quadratic form, where ξ, η, ζ represent $l_1x+m_1y+n_1z$, $l_2x+m_2y+n_2z$, $l_3x+m_3y+n_3z$, the linear forms ξ, η, ζ are independent; that is, there is no equation $p\xi+q\eta+r\zeta=0$, in which p, q, r are constants, which is true for all values of x, y, z.

To see then what is represented by the quadratic equation connecting the coordinates x, y, z, we may suppose this equation to be $L\xi^2 + M\eta^2 + N\zeta^2 = 0$. If, herein, two of L, M, N (or the two linear functions whose squares are multiplied by these), vanish, the equation represents a single line taken twice over; if only one of L, M, N, say N, be zero, the equation represents the two lines whose equations are $L^{\frac12}\xi + iM^{\frac12}\eta = 0$, $L^{\frac12}\xi - iM^{\frac12}\eta = 0$; while, if none of L, M, N vanishes, the equation is
$$M\eta^2 + (L^{\frac12}\xi + iN^{\frac12}\zeta)\,(L^{\frac12}\xi - iN^{\frac12}\zeta) = 0;$$

this is obtainable from the two equations

$$L^{\frac12}\xi + iN^{\frac12}\zeta - \theta M^{\frac12}\eta = 0, \quad L^{\frac12}\xi - iN^{\frac12}\zeta + \theta^{-1}M^{\frac12}\eta = 0,$$

where θ is any number; these represent corresponding lines of two related pencils, as θ varies. The equation, therefore, represents a conic.

It appears that a single condition, connecting the original coefficients a, b, c, f, g, h, is necessary for the quadratic polynomial to break into two factors both linear in x, y, z. This condition is that the determinant Δ, of three rows, whose rows consist respectively of the elements a, h, g; h, b, f; g, f, c, should vanish. We may prove this by remarking that, if the quadratic polynomial be written as UP, where U denotes a linear form $ux+vy+wz$, and P denotes a linear form $lx+my+nz$, then we have, for all values of x, y, z,

$$2\,(ax+hy+gz) = uP + lU, \quad 2\,(hx+by+fz) = vP + mU,$$
$$2\,(gx+fy+cz) = wP + nU$$

(as may be proved by partial differentiation in regard to x, y, z). Thus the three linear forms $ax+hy+gz$, $hx+by+fz$, $gx+fy+cz$ all vanish for the values of x, y, z which are the common solution of $P=0$, $U=0$. This shews that the determinant vanishes. Conversely, it may be shewn that this condition is sufficient. More particularly, it may be shewn that the necessary and sufficient condition, for the quadratic equation to represent a single line taken twice over, is that all the first (quadratic) minors of the discriminantal determinant Δ should vanish, which follows if three, suitably chosen, of these minors all vanish. For these minors, which occur frequently in the algebraic work, it is usual to use the notation

$$A = bc - f^2,\ B = ca - g^2,\ C = ab - h^2,\ F = gh - af,\ G = hf - bg,\ H = fg - ch.$$

For instance, if a be not zero, these minors all vanish when $B = 0$, $C = 0$, $F = 0$. When all of a, b, c vanish, the quadratic polynomial can only be a perfect square by vanishing identically.

(3) *Ex.* 1.   It is very usual to suppose the coordinates used in the equation of a conic to be *orthogonal*. This means that the coordinates are referred to an arbitrary point Z, and two (real) points, X, Y, in which perpendicular lines through Z meet the Absolute line. The symbols of the Absolute points, in terms of the symbols of X, Y, are then $X \pm iY$. With such coordinates, x, y, z, we may put down the algebraic conditions for some of the commoner relations of the conic with the Absolute points; we use the preceding notation for the quadratic polynomial, and $xX + yY + zZ$ for the symbol of any point.

(*a*) The condition for the conic to be a parabola, touching the Absolute line, is $C = 0$; namely, the terms which are independent of the coordinate z, in the equation of the conic, are the square of a linear function of x, y only. This is true also when X, Y are *any* two points of the Absolute line.

(*b*) The conditions for the conic to be a circle, passing through the Absolute points, are two, namely $h = 0$, $a - b = 0$.

(*c*) The condition for the conic to be a rectangular hyperbola, having the Absolute points as conjugate points, is $a + b = 0$. In particular, when the conic breaks up into two lines (so that $abc + 2fgh - af^2 - bg^2 - ch^2 = 0$), this is the condition for these lines to be perpendicular. Thus, two lines, whose equations are $ux + vy + wz = 0$, $vx - uy + kz = 0$, are perpendicular.

(*d*) When the conic is not a parabola, its centre, the pole of the Absolute line, is the common point of the two lines represented by $ax + hy + gz = 0$, $hx + by + fz = 0$. Supposing the coefficients a, b, ..., h to have real ratios, the conic is a hyperbola or an ellipse according as the intersections with the Absolute line, given by $ax^2 + 2hxy + by^2 = 0$, are real or imaginary. This is so also if the points of reference X, Y be any two real points of the Absolute line.

(*e*) Whatever be the points of reference for the coordinates, the equation of the tangent line at the point (x', y', z') of the conic is $xU' + yV' + zW' = 0$, where $U' = ax' + hy' + gz'$, $V' = hx' + by' + fz'$, $W' = gx' + fy' + cz'$. This is also the equation of the polar line of an arbitrary point (x', y', z') in regard to the conic. This is most readily proved by finding the quadratic equation, in λ, which expresses that the point, whose coordinates are $x + \lambda x'$, $y + \lambda y'$, $z + \lambda z'$, lies on the conic; by the harmonic property of the polar line, the sum of the roots of this equation in λ must vanish.

(*f*) Also, for any coordinates, the condition, that the line whose equation is $ux + vy + wz = 0$ should touch the conic, is

$$Au^2 + Bv^2 + Cw^2 + 2Fvw + 2Gwu + 2Huv = 0,$$

where A, ..., H are the minors, in the discriminantal determinant Δ, which have been defined already. This equation is called the *tangential equation of the conic*. When this equation is given, the original point equation is immediately obtained, since there exist the identities

$$\Delta a = BC - F^2,\quad \Delta b = CA - G^2,\quad \Delta c = AB - H^2,$$
$$\Delta f = GH - AF,\quad \Delta g = HF - BG,\quad \Delta h = FG - CH.$$

(*g*) Also for any coordinates, if x, y, z be the coordinates of any point on either of the two tangents which can be drawn from the point (x_1, y_1, z_1) to the conic, the tangential equation is satisfied when we substitute therein, for u, v, w, the respective values $yz_1 - y_1 z$, $zx_1 - z_1 x$, $xy_1 - x_1 y$. The result of this substitution, if we regard x_1, y_1, z_1 as fixed, is the quadratic equation in x, y, z which belongs to the pair of tangents from (x_1, y_1, z_1) to the conic. This equation is also capable of the form $\Omega \Omega_1 - P^2 = 0$, where Ω is the quadratic polynomial $ax^2 + \ldots$, such that $\Omega = 0$ is the equation of the conic considered, Ω_1 is the result of substituting respectively x_1, y_1, z_1 for x, y, z in Ω, and P is $xU_1 + yV_1 + zW_1$, where $U_1 = ax_1 + hy_1 + gz_1$, etc., the equation $P = 0$ being that of the polar of (x_1, y_1, z_1) in regard to the conic.

(h) For orthogonal coordinates, the locus of a point (x_1, y_1, z_1) from which the tangents to the conic are perpendicular is found by applying, to the equation of these two tangents, given in (g), the condition, given in (c), for a quadratic equation to represent two perpendicular lines. Thus we find that (x_1, y_1, z_1) lies on the conic whose equation is

$$C(x^2 + y^2) - 2Gzx - 2Fyz + (A + B)z^2 = 0,$$

or
$$(Cx - Gz)^2 + (Cy - Fz)^2 + \Delta(a + b)z^2 = 0.$$

This then represents the director circle of the conic. When the given conic is a rectangular hyperbola, or $a + b = 0$, this degenerates into the two lines represented by $Cx - Gz \pm i(Cy - Fz) = 0$. The director circle also reduces to these two lines when the conic is itself degenerate, or $\Delta = 0$. And it degenerates when the conic is a parabola, or $C = 0$, into the two lines $z = 0$, $2Gx + 2Fy - (A + B)z = 0$; the latter is then the directrix, and the focus of the parabola is the pole of this line in regard to the conic.

(i) The rule which we have given for finding the equation of the polar of a point, in regard to a conic whose point equation is given, does in fact apply also for finding the equation of the pole of a given line in regard to a conic whose tangential equation is given. The equation of the pole of the line whose tangential coordinates are (u', v', w'), that is, of the line whose point equation is $u'x + v'y + w'z = 0$, is

$$u(Au' + Hv' + Gw') + v(Hu' + Bv' + Fw') + w(Gu' + Fv' + Cw') = 0;$$

namely $Au' + Hv' + Gw'$, etc., are the point coordinates of this pole.

As an exercise the reader may find the coordinates of the focus, when the conic is a parabola; and prove that the equation of the axis is

$$(a + b)(ax + hy) + (ag + hf)z = 0$$

(this being the line from the focus which is perpendicular to the directrix), which, in general, is the same as $(a + b)(hx + by) + (hg + bf)z = 0$.

(j) For orthogonal coordinates, the (four) foci of a conic are the common points of the pairs of tangents drawn from the Absolute points, whose coordinates are $(1, \pm i, 0)$. Using then (g) for the equations of these pairs of tangents, we find that the four foci are given by the two equations

$$(U^2 - V^2)/(a - b) = UV/h = \Omega,$$

where $\Omega = 0$ is the equation of the conic, and

$$U = ax + hy + gz, \quad V = hx + by + fz.$$

The equation $(U^2 - V^2)/(a - b) = UV/h$ represents the two axes of the conic, because this equation represents two lines through the centre of the conic which contain the foci.

(k) Particular orthogonal coordinates, for a conic whose centre is Z, are obtained by taking the points of reference X, Y, on the Absolute line, to be on the axes of the conic. The equation of the conic, with appropriate constants a, b, is then of the form

$$\frac{x^2}{a} + \frac{y^2}{b} - z^2 = 0.$$

In this case, the equation of a general tangent of the conic is of the form $ux + vy - (au^2 + bv^2)^{\frac{1}{2}} z = 0$. The equation of a confocal conic is

$$x^2/(a+\lambda) + y^2/(b+\lambda) - z^2 = 0,$$

for a suitable value of the constant λ. A tangent of this latter conic, which is perpendicular to the tangent of the original conic whose equation has been written down, has the equation

$$vx - uy - [(a+\lambda) v^2 + (b+\lambda) u^2]^{\frac{1}{2}} z = 0.$$

From the equations of the two tangents, forming $(ux + vy)^2 + (vx - uy)^2$, we verify that the locus of the common point of a tangent of the former conic, with a perpendicular tangent of the latter conic, is a circle whose equation is $x^2 + y^2 - (a+b+\lambda) z^2 = 0$. If this equation be written $\Psi = 0$, and the equations of the original conic, and the confocal, be respectively written $\Omega = 0$, $\Phi = 0$, we have $(a+\lambda)(b+\lambda) \Phi - ab\Omega = \lambda\Psi$; this shews that the common points of the conics lie on the circle.

Ex. 2. As before, let a conic be referred to two points X, Z of itself, and the pole, Y, of the chord which joins these points; so that, in terms of the symbols of these, the symbol of any point of the conic is $\theta X + Y + \theta^{-1} Z$, where θ is a parametric number; the symbol of any point of the plane may be written $xX + yY + zZ$. If (θ), (ϕ) be any two points of the conic, the equation $x - y(\theta + \phi) + z\theta\phi = 0$ represents a line; and this evidently contains both the points $(\theta, 1, \theta^{-1})$, $(\phi, 1, \phi^{-1})$. The line is thus the chord joining (θ), (ϕ). In particular, when $\phi = \theta$, the equation $x - 2y\theta + z\theta^2 = 0$ represents the tangent at the point (θ); this is also deducible from (e) of Ex. 1, the equation of the conic being $xz - y^2 = 0$. We can verify at once that

$$(x - 2y\theta + z\theta^2)(x - 2y\phi + z\phi^2) - [x - y(\theta + \phi) + z\theta\phi]^2 = (xz - y^2)(\theta - \phi)^2;$$

thus if $P = 0$, $R = 0$ be the equations of the tangents at any two points of the conic, and $Q = 0$ the equation of the chord joining the points of contact, the equation of the conic is $PR - Q^2 = 0$. Let $L = 0$, $M = 0$ be any two lines through the common point of $P = 0$, $R = 0$, which are harmonic conjugates in regard to these tangents. Then we have identities $P = pL + qM$, $R = pL - qM$, where p, q are constants. Thus the equation of the conic takes the form $p^2L^2 - q^2M^2 - Q^2 = 0$. This is the same as saying that, if coordinates ξ, η, ζ, be taken referred to three points which form a self-polar triangle in regard to the conic, then the equation of the conic takes the form

$$A\xi^2 + B\eta^2 + C\zeta^2 = 0,$$

in which A, B, C are constants. This is also deducible from the equation of a polar line in general, which is given in (e) of Ex. 1.

Ex. 3. Let $\Omega = 0$, $\Omega' = 0$ be the equations of any two conics; and λ a constant. The equation $\Omega - \lambda\Omega' = 0$, being quadratic in the coordinates, represents a conic; and clearly this contains the four points common to $\Omega = 0$, $\Omega' = 0$. A conic is determined by these four points and any fifth point; and λ can be so chosen that $\Omega - \lambda\Omega' = 0$ contains this fifth point. Thus, for a suitable value of λ, the equation $\Omega - \lambda\Omega' = 0$ represents any conic through the four common points of the two given conics. And this remains true when the four points are not all distinct; for instance, when $\Omega = 0$, $\Omega' = 0$ have a point of contact, $\Omega - \lambda\Omega' = 0$ represents the general conic which touches both the original conics at this point, and also contains their remaining two common points.

In particular, forming the discriminantal determinant Δ for the equation $\Omega - \lambda\Omega' = 0$, this vanishes for three values of λ. For these, the equation $\Omega - \lambda\Omega' = 0$ represents, respectively, the three pairs of complementary joins of the common points of the original conics.

Ex. 4. The two forms Ω, Ω', being respectively $ax^2 + ...$, $a'x^2 + ...$, shew that the tangential equation of the conic $\Omega - \lambda\Omega' = 0$ is $\Sigma - \lambda\Psi + \lambda^2\Sigma' = 0$, where $\Sigma=0$, $\Sigma'=0$ are the respective tangential equations, $\Sigma=Au^2+\dots$, $\Sigma'=A'u^2+\dots$, and

$$\Psi=(bc'+b'c-2ff')\,u^2+\dots+\dots+2\,(gh'+g'h-af'-a'f)\,vw+\dots+\dots=0.$$

Ex. 5. Let $P=0$, $Q=0$ represent two lines. A particular application of Ex. 3 is that the equation $\Omega+\lambda PQ=0$ represents any conic through the common points of these lines with the conic $\Omega=0$. In particular $\Omega+\lambda P^2=0$ represents any conic which touches $\Omega=0$ at the two points where $P=0$ meets $\Omega=0$.

Now let $\Omega=0$, $\Omega'=0$ be any two conics, and $P=0$, $Q=0$ be a pair of complementary common chords of these conics. There exists then an identity $p'\Omega'-p\Omega=PQ$, in which p', p are constants; for brevity we suppose these constants attached as multipliers to the coefficients in Ω' and Ω, and write $\Omega'-\Omega=PQ$. With any value of the constant λ we then have the identities

$$\lambda^2P^2+2\lambda\,(\Omega'+\Omega)+Q^2=(\lambda P+Q)^2+4\lambda\Omega=(\lambda P-Q)^2+4\lambda\Omega'.$$

This shews that the equation $\lambda^2P^2+2\lambda\,(\Omega'+\Omega)+Q^2=0$ represents a conic touching $\Omega=0$ at two points on the line $\lambda P+Q=0$, and also touching $\Omega'=0$ at two points on the line $\lambda P-Q=0$.

These two chords of contact pass through the common point of the two complementary common chords $P=0$, $Q=0$, of the two given conics; that is, through one angular point of the common self-polar triangle of the two conics; and these chords of contact are harmonic conjugates in regard to the two common chords of the conics which meet in this angular point. By choosing λ suitably, one of these chords of contact is an arbitrary line through the angular point; and this angular point is one of three. As λ occurs to the second order, there are two values of λ for which the conic

$$\lambda^2P^2+2\lambda\,(\Omega'+\Omega)+Q^2=0$$

passes through an arbitrary point. We may thus say: there are three families of conics, all having double contact with both of two given conics; of the conics of one family, the chord of contact with one of the two given conics may be taken arbitrarily; and, of the conics of one family, there are two which pass through an arbitrary point. The statement assumes that the common self-polar triangle of the two given conics has three definite angular points.

There are no conics, having double contact with both the two given conics, besides these put down. For, any conic with such contact, along respective chords $L=0$, $M=0$, has an equation capable of both the forms $a\Omega+L^2=0$, $a'\Omega'+M^2=0$, where a, a' are constants. The identity of these forms, involving $M^2-L^2=a\Omega-a'\Omega'$, shews that the lines $M+L=0$, $M-L=0$ are a complementary pair of common chords of $\Omega=0$, $\Omega'=0$; etc.

Ex. 6. Prove that the locus of the poles of a given line, in regard to all the conics of one family which have double contact with two given conics, is another conic. Also that the polar lines of a given point, in regard to these conics, envelope a conic. Indeed, it is easy to see, the relation between two conics, of having double contact with one another, is a self-dual relation.

Ex. 7. Let $P=0$, $P'=0$ be one pair of lines; also $Q=0$, $Q'=0$ another pair of lines; and $R=0$, $R'=0$ the other pair of complementary joins for the quadrangle formed by the four common points of $PP'=0$ and $QQ'=0$. By absorbing constant multipliers in the coefficients in the equations of, say, $P=0$ and $Q=0$, we may suppose that we have the identity $RR'=PP'-QQ'$. Prove then that one family of conics touching the four lines $P=0$, $P'=0$, $Q=0$, $Q'=0$ has the equation

$$\lambda^2R^2+2\lambda\,(PP'+QQ')+R'^2=0.$$

Find the other two families. And shew that the equation of a system of confocal conics is included in the given equation.

Ex. 8. Let P, Q be variable points which describe related ranges on a given conic. Prove that the line PQ envelopes a conic having double contact with the given conic.

In general, if points P, Q, lying on a given conic, be such that the line PQ is a variable tangent of another given conic, prove that the parameters, θ, ϕ, which determine P, Q, respectively, on the first conic, are connected by a polynomial equation, of the second order in both θ and ϕ, which is symmetrical in regard to θ and ϕ.

In particular, if the second conic have double contact with the first conic, this polynomial equation is of the form

$$(a\theta\phi + b\theta + c\phi + d)\,(a\theta\phi + b\phi + c\theta + d) = 0,$$

where a, b, c, d are constants. Thus, in this case, if Q_1, Q_2 be the two positions of Q which correspond to a position P, on the first conic, then the two points P and Q_1 describe related ranges on this conic; as do P and Q_2. Thus the line Q_1Q_2 touches a third conic. This has double contact with both the given conics, namely, at the two points where these touch.

Let $\Omega_1 = 0$, $\Omega_2 = 0$ be two conics which both have double contact with a conic $\Omega = 0$. Let P, Q_1 be points of $\Omega = 0$, such that PQ_1 touches $\Omega_1 = 0$; let Q_2 be another point of $\Omega = 0$ such that PQ_2 touches $\Omega_2 = 0$. Prove that the line Q_1Q_2 touches another conic which has double contact with $\Omega = 0$.

Ex. 9. The equation of a chord of an ellipse, whose equation is

$$x^2/a^2 + y^2/b^2 - z^2 = 0,$$

can be supposed to be

$$(1 - \lambda\mu)\,x/a + (\lambda + \mu)\,y/b - (1 + \lambda\mu)\,z = 0,$$

where λ, μ are constants. Let this chord be a tangent of a circle which has double contact with the ellipse on the line $x + hz = 0$, in which h is a constant. Prove (cf. Ex. 8) that

$$(1 - ha^{-1})\,\lambda\mu - (1 + ha^{-1}) - e^{-1}H\,(\lambda - \mu) = 0,$$

where $\qquad e^2 = 1 - b^2/a^2,\ H^2 = 1 - e^2h^2/a^2.$

Ex. 10. We may consider, in geometrical terms, and algebraically, the necessary condition for two conics $\Omega = 0$, $\Omega' = 0$ to have double contact. In geometrical terms, the point common to the polars of an arbitrary point P, taken in regard to the conics, must lie upon a line independent of P. Or, if P, Q, R be an arbitrary triangle, the polar triangles of this in regard to both the conics, which are separately in perspective with the triangle PQR, must be in perspective with one another. Algebraically, all the first (quadratic) minors of the discriminantal determinant, Δ, of the conic $\Omega - \lambda\Omega'$, must vanish for a proper value of λ. This will be so, if three properly chosen first minors vanish; the elimination of λ, from the pairs of these three conditions, leads to *two* functions, of the coefficients in Ω and Ω', whose vanishing constitutes the necessary and sufficient condition for the conics to have double contact.

Prove, for example, if Ω be $ax^2 + by^2 + cz^2 + 2fyz + 2gzx + 2hxy$, and Ω' be $xz - y^2$, that the conditions are, that two (and therefore the third) of the functions $af^2 - ch^2$, $Ch + 2Fa$, $Cf + 2Ha$ must vanish

$$(\text{where } \ C = ab - h^2,\ F = gh - af,\ H = fg - ch);$$

and that, then, the chord of contact of the conics has the equation

$$ax + hy + zaf/h = 0.$$

Ex. 11. We have found (Ex. 5) three families of conics having double contact with two given conics. Let one of these two conics degenerate into two pencils of lines. Then prove that there are *two* families of undegenerate conics, passing through two given points, which have double contact with a given conic. In particular there are two families of circles which have such contact, the chords of contact being parallel to the axes of the conic. Hence we can prove again the result, given in the preceding chapter (xv (9), Ex. 14), that, if PT be one of the tangents from a point P of an ellipse, drawn to a circle having double contact with an ellipse with chord of contact perpendicular to the axis SH, and PN be the perpendicular from P to the chord of contact, then the ratio of the lengths of PT, PN is independent of P, and independent of the chord of contact. The circle may be taken to have the equation

$$x^2/a^2 + y^2/b^2 - z^2 + (b^{-2} - a^{-2})(x + hz)^2 = 0,$$

where $x^2/a^2 + y^2/b^2 - z^2 = 0$ is the equation of the ellipse.

Of this theorem an anticipatory proof was also given by means of a right circular cone in space. It seems proper to add, to place the various results given here, for conics having double contact with a given one, in proper perspective, the following remark: If, from an arbitrary point O, a variable line be drawn to meet a *quadric surface* in points P, P', and thereon be taken the point R given by $R = (P, P')/O$, then the locus of R is a plane. This plane meets the surface in a conic, say Ω; and there is a tangent plane of the surface, at every point of the conic Ω, which passes through O. If now an arbitrary plane, not containing O, be taken, this plane meets the surface in another conic, Φ. If this conic be projected from O, on to the plane of the conic Ω, the result is a conic, Φ', having double contact with Ω, namely on the line in which the plane of Φ meets the plane of Ω. A consequence of this remark, for instance, is that a conic, in the plane in which the coordinates are x, y, z, with equation $x^2 + y^2 + z^2 - (lx + my + nz)^2 = 0$, having double contact with the conic $x^2 + y^2 + z^2 = 0$ at points on the line $lx + my + nz = 0$, may be regarded as the projection of the common section of the surface whose equation is $x^2 + y^2 + z^2 - t^2 = 0$ with either of the planes whose equations are $lx + my + nz \pm t = 0$. Here, t is the fourth coordinate which is necessary in space.

Ex. 12. Let A, B, C be three arbitrary points, not lying in a line. Through these points there can be drawn four conics having double contact with an arbitrary given conic. With coordinates referred to A, B, C, the equations of these conics are of the forms

$$ax^2 + by^2 + cz^2 + 2fyz + 2gzx + 2hxy - (xa^{\frac{1}{2}} \pm yb^{\frac{1}{2}} \pm zc^{\frac{1}{2}})^2 = 0,$$

these passing, for example, through A because this has coordinates $(1, 0, 0)$. Remark that the lines of contact of the four conics, with the given one, $ax^2 + \ldots + 2hxy = 0$, form a quadrilateral whose complementary pairs of intersection are conjugate pairs of points, in regard to the given conic; and that ABC is the diagonal triangle of this quadrilateral.

State the dual theorem in detail. And, by supposing the given conic to consist of the lines of two pencils, apply it to the case of the circles which touch the sides of a triangle.

Ex. 13. Let the common points of a conic Ω with the two sides AB, AC of a triangle be respectively F, F' and E, E'. Two conics can be drawn through these four points to touch the side BC. Prove that the tangents which can be drawn from A, to touch one of these two conics, are the tangents at A of two of the four conics which can be drawn through A, B, C to have double contact with the conic Ω.

Taking the dual of this, prove that a conic Φ, passing through the angular point A of a triangle ABC, to have B and C as complementary foci, meets the side BC in two points which are points of contact of BC with two of the four circles which touch the sides of the triangle. Prove also that the asymptotes of Φ are parallel to the axes of two of the parabolas which can be drawn through B and C to have A as focus.

Ex. 14. The equations, referred to A, B, C, of two conics through these three points, are of the forms $lyz + mzx + nxy = 0$, $pyz + qzx + rxy = 0$. Applying the reversible transformation $x = X^{-1}$, $y = Y^{-1}$, $z = Z^{-1}$, to these two conics, and their four common tangents, shew that we obtain the four conics which can be drawn through A, B, C to touch the two lines whose equations are $lX + mY + nZ = 0$, $pX + qY + rZ = 0$.

Ex. 15.   With coordinates x, y, z referred to the angular points of a triangle which is self-polar in regard to a given conic, the equation of the conic must be of the form $Ax^2 + By^2 + Cz^2 = 0$, where A, B, C are constants, not containing the three terms in yz, zx, xy. This is proved at once by re-marking that the polar of the point $(1, 0, 0)$ is the line $x = 0$, etc. The fact that the conic is its own harmonic image, in regard to one of the angular points of the triangle, and the opposite side, is the same as saying that, if (x, y, z) be a point of the conic, so is $(-x, y, z)$.

Consider now any two conics, with four distinct common points, so that they have a common self-polar triangle. Let the conics, referred to an arbitrary triangle, be $\Omega = 0$, $\Omega' = 0$, where

$$\Omega = ax^2 + \ldots + 2hxy, \quad \Omega' = a'x^2 + \ldots + 2h'xy.$$

After what we have just said, it must then be possible to replace x, y, z by such linear functions of new coordinates ξ, η, ζ, that the quadratic functions Ω, Ω' shall be simultaneously reduced to the forms

$$P\xi^2 + Q\eta^2 + R\zeta^2, \quad P'\xi^2 + Q'\eta^2 + R'\zeta^2,$$

involving only the square terms ξ^2, η^2, ζ^2. The algebra of this transformation is not uninteresting; and it has important generalisation to a pair of quadratic forms in more than three variables. We therefore give an account of it.

If the angular points A, B, C, of the common self-polar triangle, have symbols, in terms of the symbols of the general points X, Y, Z to which the conics are referred,

$$A = l_1 X + m_1 Y + n_1 Z, \quad B = l_2 X + m_2 Y + n_2 Z, \quad C = l_3 X + m_3 Y + n_3 Z,$$

the algebraic problem is to find the three sets of ratios $(l_1 : m_1 : n_1)$, $(l_2 : m_2 : n_2)$, $(l_3 : m_3 : n_3)$. For brevity, denote $al_1 + hm_1 + gn_1$, $hl_1 + bm_1 + fn_1$, $gl_1 + fm_1 + cn_1$ by U_1, V_1, W_1, respectively; and the three forms such as $a'l_1 + h'm_1 + g'n_1$ by U_1', V_1', W_1'. When l_2, m_2, n_2 are written for l_1, m_1, n_1, let the notation be U_2, U_2', etc. Further, let $l_2 U_1 + m_2 V_1 + n_2 W_1$, which is identically the same as $l_1 U_2 + m_1 V_2 + n_1 W_2$, be denoted by P_{12}; and so on.

The equation of the polar, of the point A, in regard to $\Omega = 0$, referred to X, Y, Z, is $xU_1 + yV_1 + zW_1 = 0$. Thus, in order that the point A should have the same polar in regard to both of $\Omega = 0$, $\Omega' = 0$, the ratios $l_1 : m_1 : n_1$ must satisfy the two equations $U_1/U_1' = V_1/V_1' = W_1/W_1'$. And we similarly require $U_2/U_2' = $ etc., and $U_3/U_3' = $ etc. Let λ_1 be the value of the three equal ratios U_1/U_1', V_1/V_1', W_1/W_1', and λ_2, λ_3, similarly, of the equal ratios U_2/U_2', etc., and U_3/U_3', etc., respectively. We see then that $\lambda_1, \lambda_2, \lambda_3$ are roots of the equation

$$\begin{vmatrix} a - \lambda a', & h - \lambda h', & g - \lambda g' \\ h - \lambda h', & b - \lambda b', & f - \lambda f' \\ g - \lambda g', & f - \lambda f', & c - \lambda c' \end{vmatrix} = 0,$$

of which the left side is the discriminantal determinant of the quadratic form $\Omega - \lambda\Omega'$. From the first three equal fractions, all equal to λ_1, we have $\lambda_1 = P_{12}/P_{12}'$, and $\lambda_1 = P_{11}/P_{11}'$, where P_{11} stands for $l_1 U_1 + m_1 V_1 + n_1 W_1$. From the second three fractions, all equal to λ_2, we similarly have

$$\lambda_2 = P_{21}/P_{21}' = P_{22}/P_{22}' = P_{23}/P_{23}'.$$

Also, the equations connecting A, B, C and X, Y, Z lead to

$$\xi A + \eta B + \zeta C = (l_1 \xi + l_2 \eta + l_3 \zeta) X + \text{etc.},$$

so that, if

$$x = l_1 \xi + l_2 \eta + l_3 \zeta, \quad y = m_1 \xi + m_2 \eta + m_3 \zeta, \quad z = n_1 \xi + n_2 \eta + n_3 \zeta.$$

then ξ, η, ζ are the coordinates referred to A, B, C. If these values of x, y, z be substituted in Ω and Ω', these take the forms

$$\Omega = P_{11}\xi^2 + P_{22}\eta^2 + P_{33}\zeta^2 + 2P_{23}\eta\zeta + 2P_{31}\zeta\xi + 2P_{12}\xi\eta,$$

with Ω' the same save for the substitution of P_{11}', P_{22}', etc. in place of P_{11}, P_{22}, etc.

Suppose, then, that λ_1, λ_2, λ_3 are found as the roots of the discriminantal cubic equation; *and suppose that no two of the roots of this cubic are equal.* Having λ_1, the values of the ratios of l_1, m_1, n_1 can be found from the three equations $U_1/U_1' = V_1/V_1' = W_1/W_1' = \lambda_1$. We shall suppose that the actual values of l_1, m_1, n_1 are chosen so that $P_{11}' = 1$ (this gives two sets l_1, m_1, n_1 and $-l_1$, $-m_1$, $-n_1$). Then from $\lambda_1 = P_{11}/P_{11}'$, we have $P_{11} = \lambda_1$. Similarly, having λ_2, λ_3, we can find l_2, m_2, n_2 and l_3, m_3, n_3, so that $P_{22}' = 1$, $P_{22} = \lambda_2$, $P_{33}' = 1$, $P_{33} = \lambda_3$. Then the two equations $\lambda_1 P_{12}' - P_{12} = 0$, $\lambda_2 P_{21}' - P_{21} = 0$, since $P_{12}' = P_{21}'$ and $P_{12} = P_{21}$, and λ_1 is not equal to λ_2, prove that $P_{12} = 0$ and $P_{12}' = 0$. These equations express that the points A, B, of respective symbols $l_1 X + m_1 Y + n_1 Z$, $l_2 X + m_2 Y + n_2 Z$, are conjugate to one another both in regard to $\Omega = 0$, and in regard to $\Omega' = 0$. The other similar equations, proved in the same way, shew that A, C and B, C are equally pairs of points conjugate to one another in regard to both the conics. Thus A, B, C are the angular points of the common self-polar triangle of the two conics; and the effect of the substitutions, with the values of l_1, m_1, etc., which have been found, is $\Omega = \lambda_1 \xi^2 + \lambda_2 \eta^2 + \lambda_3 \zeta^2$, $\Omega' = \xi^2 + \eta^2 + \zeta^2$.

The argument cannot be carried through when two of λ_1, λ_2, λ_3 are equal. It can be proved that, when this is so, the conics have contact with one another, at one, or possibly two, points; but we do not prove this here. When the conics touch at one point, they have no common self-polar triangle of distinct angular points. When the conics touch at two points, they have an infinite number of common self-polar triangles; in this case, referred to any one of these, the equations can still be reduced each to a sum of three squares, with two of the coefficients λ_1, λ_2, λ_3 equal. In the other case, another pair of reduced forms must be adopted. When all the roots of the discriminantal cubic equation are equal, the two conics have either three, or four, coincident common points; and there is no undegenerate common self-polar triangle; in these cases also other reduced forms must be adopted. (Note 6.)

Ex. 16. We consider now, briefly, the question of the reality of the roots λ_1, λ_2, λ_3, of the discriminantal cubic equation. In cases of quadratic forms of more than three coordinates, this has historical interest also, as having arisen in the gravitational theory of the motion of the planets.

Let a quadratic form, such as Ω' in the preceding example, have all its coefficients a', b', etc., *real*. It may then be such that it is *positive*, or zero, for all *real* values of the variables x, y, z; such, for instance, are $x^2 + y^2 + z^2$ and $(x+y)^2 + z^2$. But these examples shew that even then a distinction is possible; the form $x^2 + y^2 + z^2$ cannot vanish for any real values of x, y, z other than $x = 0$, $y = 0$, $z = 0$; but the form $(x+y)^2 + z^2$ vanishes for $z = 0$ and any real values of x and y for which $x + y = 0$. We say then that a form which is quadratic and homogeneous in the variables x, y, z, and has real coefficients, is a positive *definite* form when it is positive, or zero, for all real values of x, y, z, but can vanish only for $x = 0$, $y = 0$, $z = 0$.

Suppose, then, that both the forms Ω, Ω', in the preceding example, have real coefficients, and that one of these forms, say Ω', is a positive definite form. Then we prove that *the roots of the discriminantal cubic, λ_1, λ_2, λ_3, are necessarily real.* As the roots of an algebraic equation, with real coefficients, which are not real, occur in pairs of conjugate complex numbers, it will be sufficient to prove, assuming as before that λ_1, λ_2, λ_3 are different, that λ_1 and λ_2 cannot be conjugate imaginaries. For, if so, the ratios of l_1, m_1, n_1, determined from the equations $U_1/U_1' = \ldots = \lambda_1$, would be conjugate imaginaries of the ratios of l_2, m_2, n_2, determined from $U_2/U_2' = \ldots = \lambda_2$. We could then put $l_1 = p_1 + ip_1'$, $l_2 = p_1 - ip_1'$, with $m_1 = q_1 + iq_1'$, $m_2 = q_1 - iq_1'$ and $n_1 = r_1 + ir_1'$, $n_2 = r_1 - ir_1'$, where $p_1, ..., r_1'$ are all real. We have proved, because λ_1, λ_2 are different, that $P_{12}' = 0$. But, in P_{12}', the product $l_1 l_2$ is $p_1^2 + p_1'^2$ and the form $m_1 n_2 + m_2 n_1$ is $2(q_1 r_1 + q_1' r_1')$. Thus P_{12}' is

$$\Omega'(p_1, q_1, r_1) + \Omega'(p_1', q_1', r_1'),$$

where $\Omega'(p_1, q_1, r_1)$ is the value of Ω' when x, y, z are replaced by p_1, q_1, r_1, etc. Wherefore, as Ω' is a positive definite form, $P_{12}' = 0$ involves $\Omega'(p_1, q_1, r_1) = 0$ and $\Omega'(p_1', q_1', r_1') = 0$, and, hence, involves $p_1 = q_1 = r_1 = 0$ and $p_1' = q_1' = r_1' = 0$. This proves that λ_1, λ_2 cannot be conjugate imaginaries. Hence all of $\lambda_1, \lambda_2, \lambda_3$ are real. A proof, independent of the reduction of Ω, Ω' to sums of squares, may be found in *Proc. Lond. Math. Soc.* xxxv (1903), p. 380.

Ex. 17. We have seen that a quadratic form, Ω, whose coefficients, $a, b, ...$, are real, can be written in the form

$$A(p_1 x + q_1 y + r_1 z)^2 + B(p_2 x + q_2 y + r_2 z)^2 + C(p_3 x + q_3 y + r_3 z)^2,$$

with all of the numbers, $A, B, C, p_1, ..., r_3$, real; and this is possible in an infinite number of ways. Prove that, in whatever way this is done, the number of the three coefficients A, B, C, which are negative, is the same. A simple example of this is the distinction we have made between an ellipse and a hyperbola.

(4) Let $U = 0$, $U' = 0$ and $V = 0$, $V' = 0$ be two pairs of conics which are such that there is a conic through the four common points of the first pair which also passes through the four common points of the second pair of conics. The equation of this conic must then be capable of both the forms $U + \lambda U' = 0$, $V + \mu V' = 0$, for suitable values of the constants λ, μ. There must then exist an equation

$$aU + a'U' + bV + b'V' = 0,$$

wherein a, a', b, b' are numbers whose ratios are constant, which is identically true for all values of the coordinates x, y, z. We may then say that the four conics are *linearly connected*; or, also, that *any one of them belongs to the linear series founded on the other three*. This is not the case for any four conics, taken arbitrarily. Indeed, of conics taken arbitrarily, the least number, which are necessarily connected linearly, is *seven*; as appears from the fact that a quadratic polynomial contains six terms. When the relation spoken of holds, there is clearly a conic, through the common points of *any* two of the given conics, which also passes through the common points of the other two.

Ex. 1. Let two conics, $U = 0$, $U' = 0$, have double contact, say on a line whose equation is $p = 0$, and let $\Omega = 0$ be any other conic. Then, let $V = 0$ be a conic passing through the common points of $\Omega = 0$ and $U = 0$; say, $V = (\Omega, U)$. Likewise, let $V' = 0$ be a conic through the common points of $\Omega = 0$, $U' = 0$; say $V' = (\Omega, U')$.

Prove that if $V = 0$ be *any* such conic as described, the conic $V' = 0$ can be chosen to have double contact with $V = 0$, on the line $p = 0$. Or, if $V' = 0$ be *any* such conic as described, then $V = 0$ can be chosen to have double contact with $V' = 0$, on the line $p = 0$.

We can, in fact, choose the equations so that $U' - U = p^2$, and $V = \lambda \Omega + U$. Then V', given by $V' = \lambda \Omega + U'$, with the same value of the constant λ, is such that $V' - V = p^2$.

Ex. 2.   From Ex. 1, if $V=0$ represents a pair of complementary common chords of $\Omega=0$ and $U=0$, then a conic, $V'=0$, exists, through the common points of $\Omega=0$ and $U'=0$, which touches these chords on the line $p=0$.

Ex. 3.   Also, let $U'=0$ be a pair of tangents of a conic $U=0$, with points of contact on the line $p=0$. And, let $\Omega=0$ be a pair of chords, AB and CD, of the conic $U=0$, which meet the two tangents given by $U'=0$, respectively, in the points P, Q and R, T. Let $V=0$ be the complementary chords AC and BD, of the conic $U=0$. Then, a conic $V'=0$ exists, passing through the four points P, Q, R, T, which touches both the lines AC, BD (or $V=0$), on the line $p=0$.

Ex. 4.   Let $U=0$, $\Omega=0$, $p=0$ be as in Ex. 3; let $V'=0$ be the lines PT, QR. There is then a conic, $V=0$, through the four points A, B, C, D of the conic U, which has double contact with the lines $V'=0$, on the line $p=0$.

Ex. 5.   The tangents at two points P, Q, of a conic, meet in the point Y; take points, S, H, not lying on the conic. Let the lines SP, HQ meet in X, and the lines SQ, HP meet in Z. Also let I, J be a complementary pair, other than S, H, of the four tangents drawn to the conic from S and H. Prove

(*a*) There exists a conic, touching the four lines SP, HP, SQ, HQ, which passes through I and J, whose tangents at I and J meet in Y.

(*b*) There exists a conic, touching the four tangents from S, H to the given conic, which passes through X and Z, whose tangents at X and Z also meet in Y.

These results are the duals, respectively, of those given in Ex. 3 and Ex. 4. For, consider the duals of these results in the present example: Let t, u be two tangents of a conic, with points of contact on a line y. Let s, h be two lines, meeting the conic in four points; and i, j be a pair, other than s, h, of complementary joins of these four points. Then the duals of (*a*), (*b*) are

(*a'*) There exists a conic, through the four intersections of the tangents t, u with the lines s, h, which touches the lines i, j, on the line y. This is Ex. 3.

(*b'*) There exists a conic, through the four common points of the given conic with the lines s, h, which touches the line joining the points (s, t), (h, u), and also touches the line joining the points (s, u), (h, t), the points of contact being on the line y. This is Ex. 4.

Thus the statements (*a*), (*b*) are proved. When I, J are the Absolute points, and S, H are a pair of complementary foci of the given conic, the statement (*a*) is, that Y is the centre of a circle which touches the four lines XP, PZ, ZQ, QX; and the statement (*b*) is, that there is a conic confocal with the given one, through the points X, Z, with Y as pole of the chord XZ. This is the confocal which may be defined as one of the two which pass through Z. The proof we have given shews that the theorems are not metrical. But, if we allow the use of length, explained in the preceding chapter, the existence of the circle, of centre Y, follows from the facts that PY is one of the bisectors of the angles determined by PX, PZ; while QY is one of the bisectors of the angles determined by QX, QZ; and YS is one of the bisectors of the angles determined by SP, SQ. To prove, similarly, the existence of the confocal through X, Z, touched by XY, YZ, suppose, for definiteness, that the original conic is an ellipse, and utilise the topological considerations which are suggested by drawing a diagram. Then, because PZ, ZQ, QX, XP are tangents of a circle, the difference of the lengths of PZ, QZ is equal to the difference of the lengths of PX and QX; while the sum of the lengths of SP, PH is equal to the sum of the lengths of SQ, QH. Hence we infer that the difference of the lengths of SZ, HZ is equal to the difference of the lengths of SX, HX. This proves that Z, X lie on a hyperbola having S, H as foci, whose tangents at Z, X are bisectors of the angles determined, respectively, by ZS, ZH and by XS, XH.

The comparison of the two methods of proof is interesting; but the descriptive proof, founded only on the equations, seems much to be preferred.

(5) We consider now the algebraical condition, that the conic, whose point equation is $ax^2 + \ldots + 2hxy = 0$, should be outpolar to the conic whose tangential equation is $A'u^2 + \ldots + 2H'uv = 0$. We prove that this is

$$aA' + bB' + cC' + 2fF' + 2gG' + 2hH' = 0.$$

We assume that the condition for two points (x, y, z), (x', y', z') to be conjugate in regard to the former conic is $axx' + \ldots + h(xy' + x'y) = 0$; this follows by considering the quadratic equation in λ/μ which expresses that the point $(\lambda x + \mu x', \lambda y + \mu y', \lambda z + \mu z')$ lies on the conic. Similarly, the condition for the two lines $u_1 x + v_1 y + w_1 z = 0$, $u_2 x + v_2 y + w_2 z = 0$ to be conjugate in regard to the second conic is

$$A'u_1 u_2 + \ldots + H'(u_1 v_2 + u_2 v_1) = 0,$$

the pole of the line $ux + vy + wz = 0$ being (ξ, η, ζ), where

$$\xi = A'u + H'v + G'w, \quad \eta = H'u + B'v + F'w, \quad \zeta = G'u + F'v + C'w.$$

Now, suppose that a triangle can be inscribed in the first conic which is self-polar in regard to the second conic. Let two sides of this triangle, meeting in the point (ξ, η, ζ), have the equations $u_1 x + v_1 y + w_1 z = 0$, $u_2 x + v_2 y + w_2 z = 0$, the third side being $ux + vy + wz = 0$. The tangent of the first conic, at (ξ, η, ζ), being $px + qy + rz = 0$, we have $p\xi + q\eta + r\zeta = 0$; and thence, as (ξ, η, ζ) is the pole of the third side in regard to the second conic, we have, substituting for ξ, η, ζ their values in terms of u, v, w, the relation $p(A'u + H'v + G'w) + \ldots + = 0$, which is

$$A'pu + \ldots + H'(pv + qu) = 0.$$

The first conic contains the four intersections of the two degenerate conics

$$(u_1 x + v_1 y + w_1 z)(u_2 x + v_2 y + w_2 z) = 0,$$

and

$$(ux + vy + wz)(px + qy + rz) = 0,$$

say $U = 0$ and $P = 0$. Whence, for suitable constant numbers λ, μ, there is an identity $ax^2 + \ldots + 2hxy = \lambda U + \mu P$. This gives three pairs of equations such as $a = \lambda u_1 u_2 + \mu up$, $2h = \lambda(u_1 v_2 + u_2 v_1) + \mu(pv + qu)$. Also because the sides $u_1 x + \ldots = 0$, $u_2 x + \ldots = 0$ are conjugate in regard to the second conic, we have $A'u_1 u_2 + \ldots + H'(u_1 v_2 + u_2 v_1) = 0$. Thus the three pairs of equations lead, by means of $A'pu + \ldots + H'(pv + qu) = 0$, to the result $aA' + \ldots + 2hH' = 0$. This is then a *necessary* condition for the first conic to be outpolar to the second conic. Conversely, suppose this condition to be satisfied. Take any point (ξ, η, ζ) on the first conic; let the polar of this point, in regard to the second conic, be $ux + vy + wz = 0$; and let the lines joining (ξ, η, ζ) to the two points in which this line meets the first conic be $u_1 x + \ldots = 0$, $u_2 x + \ldots = 0$. Then, by reversing the algebra we have used, we can infer that

$$A'u_1 u_2 + \ldots + H'(u_1 v_2 + u_2 v_1) = 0.$$

Thus the triangle, whose sides are the three lines, is self-polar in regard to the second conic; and it is inscribed in the first conic. The condition found is therefore not only necessary but also sufficient for the outpolar relation.

By using the dual equations at every step of the argument, we can similarly prove that the condition found, $A'a + \ldots + 2H'h = 0$, is the necessary and sufficient condition for the existence of triangles which are circumscribed to the second conic $A'u^2 + \ldots + 2H'uv = 0$ and self-polar in regard to the first conic. Thus when one conic is outpolar to another, this other is inpolar to the former.

A particular application of the formula is when the first conic consists of two lines which are conjugate in regard to the second; or consists of a single tangent of this latter, taken twice.

(6) An immediate consequence of (5) preceding gives the equation of the harmonic locus of two conics. Let these conics have the equations $ax^2 + \ldots + 2hxy = 0$ and $a'x^2 + \ldots + 2h'xy = 0$, their respective tangential equations being $Au^2 + \ldots + 2Huv = 0$ and $A'u^2 + \ldots + 2H'uv = 0$. We have remarked that the equation of the pair of tangents drawn to the former conic from the point (x_1, y_1, z_1) is

$$A (yz_1 - y_1z)^2 + \ldots + 2H (yz_1 - y_1z)(zx_1 - z_1x) = 0.$$

If we express that this pair of lines forms a conic outpolar to the second conic $A'u^2 + \ldots + 2H'uv = 0$, then this will express that the pair of lines from (x_1, y_1, z_1), tangent to the first conic, are harmonic conjugates in regard to the pair of tangents from (x_1, y_1, z_1) to the second conic; so that (x_1, y_1, z_1) will lie on the harmonic locus of the two conics. The result is that (x_1, y_1, z_1) lies on the conic given by

$$(BC' + B'C - 2FF') x^2 + \ldots + 2 (FG' + F'G - CH' - C'H) xy = 0.$$

Dually, the tangential equation of the harmonic envelope of the two conics is

$$(bc' + b'c - 2ff') u^2 + \ldots + 2 (fg' + f'g - ch' - c'h) uv = 0.$$

We may notice that when the two conics coincide the former equation becomes the point equation of either, and the latter becomes the tangential equation of either (neither being degenerate).

Ex. 1. Prove that the condition, that the harmonic locus of the two conics should be outpolar to the conic $ax^2 + \ldots + 2hxy = 0$, is the same as the condition $(aA' + \ldots = 0)$ that this conic should be outpolar to the other (the former conic not being degenerate).

Ex. 2. Denote the left side of the equation of the harmonic locus by ϕ, and the quadratic $ax^2 + \ldots + 2hxy$ by ω; also denote the form $aA' + \ldots + 2hH'$ by Θ'. Assume that the two given conics are in general positions. Consider the pair of points in which a tangent of the second conic $a'x^2 + \ldots + 2h'xy = 0$ meets the first conic $\omega = 0$. Prove that these two points are conjugate to one another in regard to the conic whose equation is $2\phi - \Theta'\omega = 0$.

Prove also that the condition, that the conic $2\phi - \Theta'\omega = 0$ should be outpolar to the second conic $A'u^2 + \ldots + 2H'uv = 0$, is

$$\tfrac{1}{4} (aA' + \ldots + 2hH')^2 = (bc - f^2)(B'C' - F'^2) + \ldots + 2 (fg - ch)(F'G' - C'H').$$

This form is symmetrical in regard to the coefficients in the point equation of the first conic ω, and the coefficients in the tangential equation of the second. It is the same as $\tfrac{1}{4}\Theta'^2 = \Delta'\Theta$, where Δ' is the discriminantal determinant of $a'x^2 + \ldots + 2h'xy = 0$, and $\Theta = a'A + \ldots + 2h'H$.

Ex. 3. Let the conic $\omega = 0$ be triangularly circumscribed to the conic $a'x^2 + \ldots + 2h'xy = 0$. It follows then, from Ex. 2, that the triangles inscribed in the former conic which are circumscribed about the latter are all self-polar in regard to the conic $2\phi - \Theta'\omega = 0$.

Thus, the conic $a'x^2 + \ldots + 2h'xy = 0$ is inpolar to the conic $2\phi - \Theta'\omega = 0$. This is then outpolar to the former conic. Whence, by Ex. 2, we have $\frac{1}{4}\Theta'^2 = \Delta'\Theta$. This is then a consequence of the fact that $\omega = 0$ is triangularly circumscribed to $a'x^2 + \ldots + 2h'xy = 0$.

Conversely, if this relation is satisfied, the conic $a'x^2 + \ldots + 2h'xy = 0$ is inpolar to $2\phi - \Theta'\omega = 0$. Thus triangles can be circumscribed about $a'x^2 + \ldots + 2h'xy = 0$ which are self-polar in regard to $2\phi - \Theta'\omega = 0$. Of such a triangle, if one angular point be taken on the conic $\omega = 0$, then, because of the relation of this conic to $2\phi - \Theta'\omega = 0$, the other two angular points lie on $\omega = 0$. This conic is therefore triangularly circumscribed to $a'x^2 + \ldots + 2h'xy = 0$.

When this is so, we see that $\omega = 0$ is also outpolar to $2\phi - \Theta'\omega = 0$.

Ex. 4. From Exx. 2, 3 it follows that when $\omega = 0$ is triangularly circumscribed to $a'x^2 + \ldots + 2h'xy = 0$, then these conics are polar reciprocals of one another in regard to $2\phi - \Theta'\omega = 0$. It can be proved, conversely, that when the two conics are polar reciprocals in regard to $2\phi - \Theta'\omega = 0$, then ω is triangularly circumscribed to $a'x^2 + \ldots + 2h'xy = 0$.

For, suppose the coordinates to be chosen so that the two conics have the respective equations $ax^2 + by^2 + cz^2 = 0$, $a'x^2 + b'y^2 + c'z^2 = 0$; we have proved this to be possible when the conics are in general positions, so as to have a common self-polar triangle. We can then compute that ϕ and $2\phi - \Theta'\omega$ are, respectively, $aa'(bc' + b'c)\,x^2 +$ two similar terms, and $a\,(-ab'c' + bc'a' + ca'b')\,x^2 +$ two similar terms. A conic in regard to which the given conics are polar reciprocals is, however, $Ax^2 + By^2 + Cz^2 = 0$, where $A = (aa')^{\frac{1}{2}}$, etc., as we readily prove. By expressing that $2\phi - \Theta'\omega = 0$ is the same as this, we deduce that $(a/a')^{\frac{1}{2}} + (b/b')^{\frac{1}{2}} + (c/c')^{\frac{1}{2}} = 0$. This leads to $\frac{1}{4}(ab'c' + \ldots)^2 = a'b'c'(a'bc + \ldots)$; which is the form taken by $\frac{1}{4}\Theta'^2 = \Delta'\Theta$.

Ex. 5. From the definitions we have given for the forms Δ, Θ, Θ', Δ', for the two conics $ax^2 + \ldots + 2hxy = 0$, $a'x^2 + \ldots + 2h'xy = 0$, or say $\omega = 0$, $\omega' = 0$, it follows that the discriminantal determinant of $\omega - \lambda\omega'$ is

$$\Delta - \lambda\Theta + \lambda^2\Theta' - \lambda^3\Delta'.$$

The vanishing of the discriminant is, however, the condition for the conic $\omega - \lambda\omega' = 0$ to break up into two lines. Suppose $\omega = 0$, $\omega' = 0$ to have a common self-polar triangle, and, referred to this, let $\omega = \alpha\xi^2 + \beta\eta^2 + \gamma\zeta^2$, and $\omega' = \alpha'\xi^2 + \beta'\eta^2 + \gamma'\zeta^2$. Then, as ξ, η, ζ are linear functions of x, y, z, the new form of $\omega - \lambda\omega'$, namely $(\alpha - \lambda\alpha')\,\xi^2 + \ldots$, will break into linear factors for the same values of λ. Thus the values of Δ, Θ, Θ', Δ', computed for the new forms, will be in the same ratios as when computed for the original forms; also the roots of the cubic determinantal equation are α/α', β/β', γ/γ', as was independently found above. If any general linear transformation of the variables, from x, y, z to x_1, y_1, z_1, change the quadratics ω, ω', respectively, to $a_1x_1^2 + \ldots + 2h_1x_1y_1$, $a_1'x_1^2 + \ldots + 2h_1'x_1y_1$, the functions Δ_1, Θ_1, Θ_1', Δ_1', computed from these new forms, will equally be in the same ratios as Δ, Θ, Θ', Δ'. For this reason, the four functions Δ, Θ, Θ', Δ' are called *invariants* of the two quadratics ω, ω'. In general, any homogeneous polynomial in Δ, Θ, Θ', Δ', which is also homogeneous in the coefficients of each of ω and ω', is called an invariant of ω and ω'. The vanishing of such an invariant involves a relation between the roots λ_1, λ_2, λ_3, of the cubic equation, and a geometrical condition for the conics. For instance, the condition, $\Theta' = 0$, that $\omega = 0$ should be outpolar to $\omega' = 0$, is expressed by $\lambda_1 + \lambda_2 + \lambda_3 = 0$. And the condition that the former conic should be triangularly circumscribed to the latter, or $(\Theta'/\Delta')^2 - 4\,(\Theta/\Delta') = 0$, is expressed by $(\lambda_1 + \lambda_2 + \lambda_3)^2 - 4\,(\lambda_2\lambda_3 + \lambda_3\lambda_1 + \lambda_1\lambda_2) = 0$, that is, by one of the four equations $\lambda_1^{\frac{1}{2}} + \lambda_2^{\frac{1}{2}} + \lambda_3^{\frac{1}{2}} = 0$.

There are also functions involving both the coefficients of the conics and the coordinates, whose vanishing expresses a geometrical property. Thus, if any general linear transformation of the coordinates be made, these functions may be computed from the derived forms of the conics. Such functions, as involving the coordinates, are called covariants. The function ϕ, and the function $2\phi - \Theta'\omega$, are examples.

Ex. 6.   Let the functions ω, ω' be expressed by coordinates which are referred to a triangle that is inscribed to the conic $\omega = 0$, and circumscribed to the conic $\omega' = 0$, it being supposed that such a triangle exists. In the point equation of $\omega = 0$ the terms involving x^2, y^2, z^2 can then be shewn to be absent; and, in the tangential equation of $\omega' = 0$, the terms involving u^2, v^2, w^2 can be shewn to be absent. Compute the functions Θ, Θ', Δ' for these new forms of the equations, and shew that these satisfy the equation $\Theta'^2 - 4\Delta'\Theta = 0$. The left side of this equation is homogeneous in Θ, Θ', Δ', and is separately homogeneous in the coefficients of ω and ω' (respectively of degrees 2 and 4). Thus, by what we have stated, the equation $\Theta'^2 - 4\Delta'\Theta = 0$ continues to hold, whatever triangle of reference is employed, when there exists a triangle which is both inscribed in $\omega = 0$ and circumscribed to $\omega' = 0$. We have proved this directly; but, in general, invariantive relations may be obtained, as in this example, by suitable choice of a triangle of reference.

Ex. 7.   For an example of the method suggested in Ex. 6, we take this theorem: Let the two tangents be drawn to the conic $\omega' = 0$ from any point of the conic ω. The line that joins the other two points in which these tangents meet the conic $\omega = 0$, envelopes a conic, passing through the common points of the conics $\omega = 0$, $\omega' = 0$, whose equation is

$$(\Theta'^2 - 4\Delta'\Theta)\,\omega + 4\Delta'\Delta\omega' = 0.$$

It is assumed that $\omega = 0$, $\omega' = 0$ are undegenerate, and in general positions.

Let the conics $\omega = 0$, $\omega' = 0$, referred to their common self-polar triangle, be such that $\omega = \alpha\xi^2 + \beta\eta^2 + \gamma\zeta^2$ and $\omega' = \alpha_0\xi^2 + \beta_0\eta^2 + \gamma_0\zeta^2$; for brevity denote the roots $\lambda_1, \lambda_2, \lambda_3$ of the discriminantal equation of $\omega - \lambda\omega'$, which are α/α_0, β/β_0 and γ/γ_0, respectively by p, q, r. And denote $-p+q+r$, $p-q+r$, $p+q-r$ respectively by P, Q, R. The points of $\omega = 0$ can be given, in terms of a variable parameter θ, by

$$\xi\alpha^{\frac{1}{2}} = 1 - \theta^2, \quad \eta\beta^{\frac{1}{2}} = 2\theta, \quad \zeta\gamma^{\frac{1}{2}} = i\,(1 + \theta^2);$$

then the equation of the chord joining the points θ, θ_1 of this conic is

$$\xi\alpha^{\frac{1}{2}}\,(1 - \theta\theta_1) + \eta\beta^{\frac{1}{2}}\,(\theta + \theta_1) + i\zeta\gamma^{\frac{1}{2}}\,(1 + \theta\theta_1) = 0,$$

for this, being linear in ξ, η, ζ, represents a line, and is easily verified to contain both the points θ, θ_1. The tangential equation of $\omega' = 0$ is

$$\alpha_0^{-1}u^2 + \beta_0^{-1}v^2 + \gamma_0^{-1}w^2 = 0.$$

Wherefore, if the two tangents to $\omega' = 0$, drawn from the point θ, meet the conic ω again in the points θ_1, θ_2, we see, substituting $u = \alpha^{\frac{1}{2}}\,(1 - \theta\theta_1)$, etc., in the tangential equation of $\omega' = 0$, that θ_1, θ_2 are the roots of the quadratic equation in ϕ,

$$p\,(1 - \phi\theta)^2 + q\,(\phi + \theta)^2 - r\,(1 + \phi\theta)^2 = 0.$$

Whence we find $\theta_1 + \theta_2$ and $\theta_1\theta_2$, and hence the ratios of $1 - \theta_1\theta_2$, $\theta_1 + \theta_2$, $1 + \theta_1\theta_2$; and so find, for the equation of the line joining the points θ_1, θ_2 of $\omega = 0$,

$$\xi\alpha^{\frac{1}{2}}P\,(1 - \theta^2) + 2\eta\beta^{\frac{1}{2}}Q\theta + i\zeta\gamma^{\frac{1}{2}}R\,(1 + \theta^2) = 0.$$

But, if $U = 0$, $V = 0$, $W = 0$ be the equations of any lines, a line whose equation is $\theta^2 U + 2\theta V + W = 0$ envelopes a conic whose equation is $V^2 - WU = 0$, this being the condition for the two lines of the family, which pass through an arbitrary point, to coincide. Wherefore, the line joining the points θ_1, θ_2 of the conic $\omega = 0$ envelops the conic whose equation is

$$\alpha P^2\xi^2 + \beta Q^2\eta^2 + \gamma R^2\zeta^2 = 0.$$

This is, however, the same as $\sigma\omega+\omega'=0$, with

$$\sigma=(p^2+q^2+r^2-2qr-2rp-2pq)/4pqr.$$

And, as the invariants for ω, ω' are

$$\Delta=\alpha\beta\gamma,\quad \Theta=\Delta'\,(qr+rp+pq),\quad \Theta'=\Delta'\,(p+q+r),\quad \Delta'=\alpha_0\beta_0\gamma_0,$$

this is the same as the equation already put down,

$$(\Theta'^2-4\Delta'\Theta)\,\omega+4\Delta'\Delta\omega'=0.$$

Herein, the left side is homogeneous in Δ, Θ, Θ', Δ', and in ξ, η, ζ, of degree 3 in α, β, γ, and of degree 4 in α_0, β_0, γ_0. Hence, this equation is valid also for any general forms of ω and ω'.

When $\Theta'^2-4\Delta'\Theta=0$, both conics being undegenerate, so that neither Δ nor Δ' is zero, this equation reduces to $\omega'=0$. In this case, then, the conic $\omega=0$ is triangularly circumscribed to $\omega'=0$.

Ex. 8.   Let the conic found in Ex. 7 be called $\Omega=0$. From the argument there given, we see that if, from the point of contact, with $\omega=0$, of a common tangent of $\omega=0$, $\omega'=0$, we draw the other tangent, say t, to the conic ω', then t touches also $\Omega=0$. Prove that, if the point of contact of t with $\omega'=0$ lie on $\omega=0$, then $\Theta'^2-4\Delta'\Theta=0$.

Thus, as was previously seen (Chap. XI (12), Ex. 10), the condition for $\omega=0$ to be triangularly circumscribed to $\omega'=0$ is, that the tangent to $\omega'=0$, at one of its intersections with $\omega=0$, should contain the point of contact, with $\omega=0$, of one of the common tangents of $\omega=0$, $\omega'=0$.

Ex. 9.   Let θ, ϕ, ψ be three variable numbers; we have, in Ex. 7, seen that there can exist cases in which every two of these, say ϕ, ψ, are connected by an equation $p\,(1-\phi\psi)^2+q\,(\phi+\psi)^2-r\,(1+\phi\psi)^2=0$ in which the numbers p, q, r are the same for each of the three pairs ϕ, ψ; ψ, θ; θ, ϕ. Algebraically, this possibility is very interesting.

Assume the three equations, for ϕ, ψ; for ψ, θ; and for θ, ϕ, with the same values of p, q, r, to be satisfied by values of θ, ϕ, ψ of which no two are equal. Prove, by elimination of p, q, r, that then the values of θ, ϕ, ψ must be such that $\theta\phi\psi\,(\phi\psi+\psi\theta+\theta\phi)-(\theta+\phi+\psi)=0$. Hence, from two of the equations, used to determine the ratios of p, q, r, prove that

$$p^2+q^2+r^2-2qr-2rp-2pq=0.$$

In fact, if $A=q/(p-r)$, $B=-2\,(p-q+r)/(p-r)$, the three equations, from which we start, are of such forms as $A\,(\phi^2+\psi^2)+B\phi\psi+1+\phi^2\psi^2=0$; from the last two of these we find

$$A=(\theta^2\phi\psi-1)/(\theta^2-\phi\psi),\quad B=-(\theta^4-1)\,(\phi+\psi)/\theta\,(\theta^2-\phi\psi),$$

which lead to

$$(A^2-1-AB)\,\theta\,(\theta^2-\phi\psi)^2=(\theta^4-1)\,[\theta\phi\psi\,(\phi\psi+\psi\theta+\theta\phi)-(\theta+\phi+\psi)];$$

while, from the definitions of A, B, we have

$$(A^2-1-AB)\,(p-r)^2=-(p^2+q^2+r^2-2qr-2rp-2pq).$$

Thus, the three given equations cannot all be satisfied by values of θ, ϕ, ψ, of which no two are equal, unless p, q, r are connected by

$$p^2+q^2+r^2-2qr-2rp-2pq.$$

But, conversely, when this relation connects p, q, r, the three equations have an infinite number of solutions, in which, for instance, θ may be taken arbitrarily and ϕ, ψ then determined from the second and third equations. The values so found will be such that

$$\theta\phi\psi\,(\phi\psi+\psi\theta+\theta\phi)-(\theta+\phi+\psi)=0.$$

A more general case of such a set of quations, only consistent under a proper condition for the coefficients which enter therein, but then capable of an infinite number of solutions, arises in the Appendix on Poncelet's

theorem, at the end of this chapter. Such a set of equations is often called
porismatic.

Ex. 10.   Let A, B, C be constants, and let the four roots of the quartic equation in x,
$$A\,(x^2-1)^{-1}+Bx^{-1}+C\,(x^2+1)^{-1}=0,$$
be denoted by θ, ϕ, ψ, t. Prove that
$$\theta\phi\psi\,(\phi\psi+\psi\theta+\theta\phi)-(\theta+\phi+\psi)=0,$$
and $t=-1/\theta\phi\psi$. Hence prove that if a triangle can be inscribed in the conic
$\omega=0$ whose sides touch the conic $\omega'=0$, then the three angular points of
this triangle, and the three angular points of the common self-polar triangle
of $\omega=0$, $\omega'=0$, lie upon a conic.

Ex. 11.   The conic $\Omega=0$, of Ex. 7, given by $(\Theta'^2-4\Delta'\Theta)\,\omega+4\Delta'\Delta\omega'=0$,
is the polar reciprocal of $\omega=0$, taken in regard to the conic $2\phi-\Theta'\omega=0$. This
is clear geometrically, or also, from the forms referred to the common self-
polar triangle of $\omega=0$, $\omega'=0$, which are
$$\omega=\alpha\xi^2+\beta\eta^2+\gamma\zeta^2,\quad 2\phi-\Theta'\omega=\alpha P\xi^2+\beta Q\eta^2+\gamma R\zeta^2,$$
$$\Omega=\alpha P^2\xi^2+\beta Q^2\eta^2+\gamma R^2\zeta^2.$$
Also, when $\omega=0$ is triangularly circumscribed to $\omega'=0$, the conic $\omega=0$ is
outpolar to $2\phi-\Theta'\omega=0$.

(7) We consider now, briefly, the case when $\omega=0$ is a conjugate line
locus of $\omega'=0$; or, as is sometimes said, when $\omega=0$ is *quadrilaterally,*
or *quadrangularly,* circumscribed to $\omega'=0$ (cf. Chap. xiv (21)).

The condition for this, as we have seen (Chap. ix (23), Ex. 18), is that
the tangents of $\omega'=0$, at two common points of $\omega=0$, $\omega'=0$, should
meet on $\omega=0$. Suppose the conics to have a common self-polar triangle,
and, referred thereto, let $\omega=\alpha\xi^2+\beta\eta^2+\gamma\zeta^2$, $\omega'=\alpha_0\xi^2+\beta_0\eta^2+\gamma_0\zeta^2$. A
common chord of the conics has an equation $\eta/\mu+\zeta/\nu=0$, where
$\mu=(\gamma\alpha_0-\alpha\gamma_0)^{\frac12}$, $\nu=(\alpha\beta_0-\beta\alpha_0)^{\frac12}$. The pole of this chord, in regard to
$\omega'=0$, is $(0,\eta_1,\zeta_1)$, where $\eta_1\beta_0\mu=\zeta_1\gamma_0\nu$. The condition that this pole
should lie on $\omega=0$ is $\beta\gamma_0^2\nu^2+\gamma\beta_0^2\mu^2=0$. With the notation, $p=\alpha/\alpha_0$,
$P=-p+q+r$, etc., used in (6), this condition is $P\,(q-r)=0$. We
suppose the conics $\omega=0$, $\omega'=0$ not to have double contact on the line
$\xi=0$; so that $q-r$ is not zero. Thus the condition for $\omega=0$ to be a
conjugate line locus of $\omega'=0$, with choice of a certain pair of complementary chords of these conics, is $P=0$. The complete condition is
thus $PQR=0$.

But we have, for all values of λ, the equation
$$\Delta-\lambda\Theta+\lambda^2\Theta'-\lambda^3\Delta'=\Delta'\,(p-\lambda)\,(q-\lambda)\,(r-\lambda).$$
If herein we take $\lambda=\tfrac12\,(p+q+r)$, $=\tfrac12\Theta'/\Delta'$, we thence have
$$-\Delta'^3PQR=\Theta'\,(\Theta'^2-4\Delta'\Theta)+8\Delta'^2\Delta.$$

Here, on the right, is a form homogeneous, of degree 3, in Δ, Θ, Θ', Δ',
which is homogeneous, of degree 3, in the coefficients of the point
equation of $\omega=0$, as also in the coefficients of the tangential equation
of $\omega'=0$. Thus, Δ' not being zero, the vanishing of this form is the
condition for ω to be a conjugate line locus of $\omega'=0$.

The condition is the same as that the conic $\Omega=0$, or
$$(\Theta'^2-4\Delta'\Theta)\,\omega+4\Delta'\Delta\omega'=0,$$

should degenerate (as we see from $PQR=0$). It is also the same as that the conic $2\Delta'\omega-\Theta'\omega'=0$ should degenerate; which is a convenient remark for computing the condition in any given case. And, it is also the condition that the conics $\Omega=0$, $2\Delta'\omega-\Theta'\omega'=0$ should be the same conic. These are readily verified by considering the simplified forms of $\omega=0$, $\omega'=0$.

Ex. When $\omega=0$ is a conjugate line locus of $\omega'=0$, with points A, B as centres, the conic $2\phi-\Theta'\omega=0$ reduces to the two lines which touch $\omega'=0$ at the two points where the line AB meets $\omega'=0$. A particular case is that a tangent of a conic meets the director circle in two points which are harmonic conjugates in regard to the two points where the tangent meets the asymptotes of the conic.

(8) The condition that the harmonic envelope of the conics

$$\alpha\xi^2+\beta\eta^2+\gamma\zeta^2=0, \quad \alpha_0\xi^2+\beta_0\eta^2+\gamma_0\zeta^2=0,$$

whose equation is $(\beta\gamma_0+\gamma\beta_0)\,u^2+\ldots+=0$, should degenerate is that the product of three factors, $\beta\gamma_0+\gamma\beta_0$, $\gamma\alpha_0+\alpha\gamma_0$, $\alpha\beta_0+\beta\alpha_0$, should vanish. Assuming $\omega'=0$ not to degenerate, this, in terms of a previous notation, is $(q+r)\,(r+p)\,(p+q)=0$. If, in the identity

$$\Delta-\lambda\Theta+\lambda^2\Theta'-\lambda^3\Delta'=\Delta'\,(p-\lambda)\,(q-\lambda)\,(r-\lambda),$$

we put $\lambda=p+q+r$, or Θ'/Δ', we thus find the general form of the condition that the harmonic envelope should degenerate to be $\Delta\Delta'-\Theta\Theta'=0$. Assuming also that $\omega=0$ is not degenerate, this is the same as the condition that the harmonic locus of $\omega=0$, $\omega'=0$ should degenerate.

Ex. 1. Let $t=0$, $t'=0$ be the equations of a pair of common tangents of the conics $\omega=0$, $\omega'=0$. Let the polars of the common point of these tangents, taken in regard to the conics respectively, be $p=0$, $p'=0$. Then the harmonic locus of the two conics, which, as we have remarked, contains the four points of contact of the two common tangents, must have an equation of the form $Att'+Bpp'=0$, wherein A, B are constants.

Referring the conics to their common self-polar triangle, and putting

$$h=\beta\gamma_0-\beta_0\gamma, \quad l=(\alpha\alpha_0 h)^{\frac{1}{2}}, \text{ etc.}, \quad \phi=\alpha\alpha_0\,(\beta\gamma_0+\beta_0\gamma)\,\xi^2+\ldots,$$

prove, in fact, that

$$h\phi=(\beta\gamma_0+\beta_0\gamma)\,[l^2\xi^2-(m\eta+n\zeta)^2]-2\,(\beta n\eta-\gamma m\zeta)\,(\beta_0 n\eta-\gamma_0 m\zeta);$$

thus, if $\beta\gamma_0+\beta_0\gamma=0$, but $\beta\gamma_0-\beta_0\gamma$ is not zero, the harmonic locus, $\phi=0$, degenerates into the two polars $p=0$, $p'=0$.

When $\beta\gamma_0-\beta_0\gamma=0$, the two conics have double contact; and the harmonic locus touches both the conics at their points of contact.

Ex. 2. For two circles, the Absolute points being given in tangential form by $u^2+v^2=0$, the equations being

$$a\,(x^2+y^2)+cz^2+2fyz+2gzx=0, \quad a'\,(x^2+y^2)+\ldots+2g'zx=0,$$

prove that the harmonic envelope, in tangential form, is given by

$$(ca'+c'a-2ff'-2gg')\,(u^2+v^2)+2\,(gu+fv-aw)\,(g'u+f'v-a'w)=0.$$

When $ca'+c'a-2ff'-2gg'=0$, this degenerates into the two pencils of lines whose centres are at the centres of the circles. In this case the circles are perpendicular.

(9) The tangential equation of the four common points of the two conics $\alpha\xi^2 + \beta\eta^2 + \gamma\zeta^2 = 0$, $\alpha_0\xi^2 + \ldots = 0$, formed as the product of the equations of the points, whose point coordinates are $(\beta\gamma_0 - \gamma\beta_0)^{\frac{1}{2}}$, $(\gamma\alpha_0 - \alpha\gamma_0)^{\frac{1}{2}}$, $(\alpha\beta_0 - \beta\alpha_0)^{\frac{1}{2}}$, or say $h^{\frac{1}{2}}$, $k^{\frac{1}{2}}$, $l^{\frac{1}{2}}$, is

$$u^4h^2 + v^4k^2 + w^4l^2 - 2v^2w^2kl - 2w^2u^2lh - 2u^2v^2hk = 0.$$

This is at once verified to be $\Phi^2 - 4\Sigma\Sigma' = 0$, where $\Phi = 0$ is the tangential equation, $(\beta\gamma_0 + \gamma\beta_0)\,u^2 + \ldots = 0$, of the harmonic envelope of the two conics, and $\Sigma = 0$, $\Sigma' = 0$ are the tangential equations $\beta\gamma u^2 + \ldots = 0$, $\beta_0\gamma_0 u^2 + \ldots = 0$, of the two conics.

This equation expresses the geometrically obvious fact that the harmonic envelope touches the eight tangents of the two conics whose points of contact are their common points. The equation may be formed directly by remarking that the tangential equation of the conic $\omega + \lambda\omega' = 0$, which, for varying λ, is any conic through the four common points, is $\Sigma + \lambda\Phi + \lambda^2\Sigma' = 0$. For there are two conics through the four points which touch an arbitrary line; and the corresponding values of λ are found from this equation, when u, v, w are the coordinates of the line. When the line passes through one of the four common points, the two values of λ are equal.

Similarly, the eight common tangents of $\omega = 0$, $\omega' = 0$ are given, in point coordinates, by $\phi^2 - 4\Delta\Delta'\omega\omega' = 0$, ($\phi = 0$ is the harmonic locus).

(10) By means of the equations more definiteness may perhaps be given to the ideas explained in (19) and (20) of Chap. xiv. Let ϕ_0, ϕ_1, $\ldots$, ϕ_r be $(r+1)$ polynomials, all homogeneously of the same order in the coordinates x, y, z, which are linearly independent; that is, not connected by any identical equation $a_0\phi_0 + \ldots + a_r\phi_r = 0$, in which a_0, $\ldots$, a_r are constants. Then the equation $\lambda_0\phi_0 + \ldots + \lambda_r\phi_r = 0$, in which λ_0, $\ldots$, λ_r are independent of x, y, z, represents, as different values are given to λ_0, $\ldots$, λ_r, a series of curves. This is called a *linear series of curves, of freedom r*. Evidently, there is a curve of the series which passes through r given points, of general positions. Also, if ψ_0, ψ_1, $\ldots$, ψ_r be any $(r+1)$ curves of the system which are linearly independent, then the general curve of the series may equally be given by $\mu_0\psi_0 + \ldots + \mu_r\psi_r = 0$, where μ_0, $\ldots$, μ_r are constants varying from curve to curve of the series. This we may express by saying that the curves of the series may be *based* upon any $(r+1)$ curves of the series which are linearly independent.

For instance, the aggregate of all the conics in a plane form a linear series of freedom 5. The conics which are outpolar to a given conic, however, form a linear series of freedom 4. For the condition that a conic $ax^2 + \ldots + 2hxy = 0$ should be outpolar to a given conic whose tangential equation is $A_0u^2 + \ldots + 2H_0uv = 0$, is $aA_0 + \ldots + 2hH_0 = 0$, which is one linear condition among the six (homogeneous) coefficients in $ax^2 + \ldots + 2hxy = 0$. So, if this conic be outpolar to two given conics, there are two such linear conditions for its coefficients; there are then only four (homogeneous) arbitrary coefficients left in $ax^2 + \ldots + 2hxy = 0$.

In general, the conics outpolar to m given conics ($m \leqslant 5$) form a linear series of freedom $5-m$. Dually, the conics inpolar to m given conics form, when expressed by their tangential equations, a linear *system* of freedom $5-m$.

As an illustration, let four arbitrary conics be given, in point coordinates x, y, z. We shew that it is possible to take four linear functions, of the quadratic polynomials which form the left sides of the equations of these four conics, which are all squares of linear functions of the coordinates x, y, z. To see this, consider the conics, expressed in tangential coordinates, which are inpolar to the four given conics. By what we have said, these inpolar conics are a system expressed by a tangential equation $\lambda\Sigma + \lambda'\Sigma' = 0$, in which Σ, Σ' are definite quadratic polynomials in the tangential coordinates u, v, w, and λ, λ' are independent of the coordinates. This system consists then of the conics which touch the four definite lines that are the common tangents of the conics $\Sigma = 0$, $\Sigma' = 0$; and the original conics are those outpolar to all the conics of this system. If $U = 0$, $V = 0$, $W = 0$, $Z = 0$ be the four common tangents of the conics of the system, expressed in point coordinates, we have seen that $U^2 = 0$ is a particular conic outpolar to all the conics of the system (Chap. XIV (16)); thus $U^2 = 0$, $V^2 = 0$, $W^2 = 0$, $Z^2 = 0$ are four conics of the series outpolar to the system. In general the four lines $U = 0$, etc. are in general positions, and the four polynomials U^2, etc. are linearly independent. Wherefore, any conic outpolar to the conics of the system can be expressed in the form $pU^2 + qV^2 + rW^2 + sZ^2 = 0$, where p, q, r, s are independent of the coordinates. In particular, every one of the original four conics can be so expressed. From this, by solving the four equations so arising, there is a linear function of the four given conics which reduces to U^2, another linear function which reduces to V^2, and others reducing to W^2 and Z^2, respectively. This is the statement made above. We have already remarked (in (1)) that, with proper coordinates ξ, η, ζ, the equations of four lines can be taken in the forms $\xi \pm \eta \pm \zeta = 0$. Thus there are linear functions of the quadratic polynomials which are the left sides of the equations of the four given conics, which can be expressed in each of the forms $(\xi \pm \eta \pm \zeta)^2$.

Another illustration of the ideas arises from the circles which are perpendicular to a given circle. For a conic to pass through two given points, in particular to be a circle, two linear homogeneous conditions are required, among the coefficients in the equation of the conic; for the circle to be perpendicular to a given circle, a single further such condition is required ($ac_0 + a_0c - 2ff_0 - 2gg_0 = 0$). Thus the circles, which are perpendicular to a given circle, form a linear series of freedom 2. This series can be based on any three circles, linearly independent of one another, which are perpendicular to the given circle. Now, the pair of lines, joining any point of the given circle to the Absolute points, satisfy the necessary condition. Thus, the circles which are perpendicular to a given circle form a linear series which may be based on the three "point circles" whose centres are any three points of the given circle.

Next consider the circles which are outpolar to a given conic. These, likewise, satisfying the two conditions for being a circle, and the one condition for being outpolar, form a linear series of freedom 2. This series can be based on any three linearly independent circles which are outpolar to the given conic. The lines which join the common point of two perpendicular tangents, of the given conic, to the Absolute points, being harmonic in regard to these tangents, are however such an outpolar circle. The common point of two perpendicular tangents is also on the director circle of the conic. Thus, the series of circles outpolar to a given conic may be based on the "point circles" whose centres are any three points of the director circle of the conic.

Comparing this series with that just considered, we see that *the series of circles outpolar to a conic is identical with the series of circles which are perpendicular to the director circle of the conic*. The argument is capable of extension to space of any number of dimensions.

Ex. 1.   The equation of a "point circle", of centre (a, b, c), referred to a point O, not lying on the Absolute line, and two points X, Y, on the Absolute line, which are such that the symbols of the Absolute points are $X \pm iY$ (in terms of the symbols of these two points), is

$$(xc - za)^2 + (yc - zb)^2 = 0.$$

The condition that one conic, in point coordinates, should be outpolar to another conic, given in tangential coordinates, is an equation which is of the first degree in the coefficients of both the conics. Thus the statement made, that the director circle of a conic is the locus of the centres of point circles which are outpolar to the given conic, gives the point equation of the director circle. This equation is thus linear in the coefficients which enter in the tangential equation of the conic.

Therefore, if $\Sigma = 0$, $\Sigma' = 0$ be the tangential equations of two conics, the director circle of any conic, $\Sigma + \lambda\Sigma' = 0$, which touches the four common tangents of $\Sigma = 0$, $\Sigma' = 0$, has an equation in which the parameter λ enters to the first degree. The director circles of the conics which touch four given lines thus form a linear series, of freedom 1; that is, they are a coaxal series of circles; as was previously proved.

Ex. 2.   Consider the harmonic envelope of one given conic, $\Omega = 0$, and any conic $\omega + \lambda\omega' = 0$, which is one of a series of conics through four points. These, for all values of λ, form a linear system, of freedom 1, given by an equation $\Sigma + \mu\Sigma' = 0$, where $\Sigma = 0$, $\Sigma' = 0$ are the tangential equations of two definite conics. This has been seen geometrically (Chap. xiv (18)), and is clear from the fact that the equation of the harmonic envelope, of two conics, is linear in the coefficients in the point equations of the two conics.

Theorems analogous to those proved here can be considered for equations which are homogeneous, but not linear, in the coefficients of a conic—as has been illustrated in Chap. xiv.

(11) We have considered linear series of curves in a plane. We may also consider linear series of sets of points lying on a conic; a few simple ideas in regard to this matter serve to place some well-known theorems in a clearer light.

We have proved that an involution, of pairs of points on a conic, is determined by the intersections of the conic with lines passing through a point. The equation of such a line is of the form $P + \lambda Q = 0$, where $P = 0$, $Q = 0$ are two lines through this point, and λ is a constant varying from line to line. We have seen that, with general points of reference, the coordinates of a point of a conic are proportional to three quadratic polynomials, in a parameter, θ, which varies from point to point of the conic; say these are $a\theta^2 + a_1\theta + a_2$, $b\theta^2 + b_1\theta + b_2$, $c\theta^2 + c_1\theta + c_2$, where $a, \ldots, c_2$ are nine numbers whose ratios are constant. Thus the points where the line $P = 0$ meets the conic are given by the two values of θ for which a certain quadratic polynomial in θ vanishes, say, by $\Phi = 0$; and the points where $Q = 0$ meets the conic are given by the two roots of another quadratic polynomial, say, by $\Psi = 0$. Thence a pair of points, of the involution on the conic, are given by an equation $\Phi + \lambda \Psi = 0$, where λ varies with the pair of the involution which is considered.

More generally, let Φ_0, Φ_1, $\ldots$, Φ_r be $(r+1)$ definite polynomials in a parameter, θ, all of the same order, say m, these being linearly independent; that is to say, not satisfying any equation, of the form $c_0\Phi_0 + \ldots + c_r\Phi_r = 0$, identically in regard to θ, where $c_0, c_1, \ldots, c_r$ are independent of θ. As a polynomial in θ, of order m, contains $m+1$ powers of θ, values of $c_0, c_1, \ldots, c_r$ can be found, independent of θ, to satisfy this identity, if $r + 1 > m + 1$. For instance, any four quadratic polynomials in θ are connected by such an identity. Thus we suppose $r \leqslant m$. Corresponding to each of $\Phi_0 = 0$, $\ldots$, $\Phi_r = 0$ we have a set of m given points on the conic; and, if $\lambda_0, \ldots, \lambda_r$ be any $r+1$ parameters, whose ratios are independent of θ, the equation $\lambda_0\Phi_0 + \ldots + \lambda_r\Phi_r = 0$ determines also a set of m points on the conic. The *various sets so obtainable, by taking different values of $\lambda_0, \ldots, \lambda_r$, are said to form a generalised involution, of sets of m points, on the conic, based on the $r+1$ arbitrary sets separately determined by $\Phi_0 = 0$, $\ldots$, $\Phi_r = 0$.* Evidently, $\lambda_0, \ldots, \lambda_r$ can be so chosen as to give a set of m points, belonging to the involution, of which r of the points are arbitrarily chosen. For this reason the involution may be said to be *of freedom r*. The $m - r$ remaining points of the set are then determined by the r assigned points. In particular, if $r = m$, any set of m points of the conic is a set of the involution; thus we suppose $r < m$. The case of an involution of pairs of points, based on two arbitrary pairs, is that for which $m = 2$ and $r = 1$.

Consider the slightly more general case, for which $m = 3$, but r is still $= 1$. On a conic, $\Omega = 0$, of which the points are given by a parameter θ, let D, E, F and P, Q, R be any two sets of three points; let the parameters for D, E, F be the roots of a cubic polynomial in θ, say Φ; and the parameters for P, Q, R the roots of a cubic polynomial Ψ. We consider the set of three points given, for a particular value of the constant λ, by the equation $\Phi + \lambda \Psi = 0$. A geometrical construction may be given: Let H be an arbitrary point of the conic, $\Omega = 0$; let $h\Omega + \Omega_1 = 0$ represent, for a proper value of h, any conic through the four points H, D, E, F; so, let $k\Omega + \Omega_2 = 0$ represent a conic through H, P, Q, R. Let these two conics meet in the points A, B, C (besides H), these not lying on $\Omega = 0$. Then, the general conic through H, A, B, C has an equation $h\Omega + \Omega_1 + \lambda(k\Omega + \Omega_2) = 0$, for a proper value of λ; and the intersections of this with $\Omega = 0$ lie on the conic $\Omega_1 + \lambda\Omega_2 = 0$. The common points of $\Omega = 0$, $\Omega_1 = 0$ have, for parameters, the parameter of H, and the parameters arising for $\Phi = 0$; likewise, the common points of $\Omega = 0$, $\Omega_2 = 0$ have the parameter of H and the parameters arising for $\Psi = 0$. Thus, the general set given by $\Phi + \lambda\Psi = 0$ is determined by variable conics through the fixed points H, A, B, C.

When Φ, Ψ are of order m in θ, any set given by $\Phi + \lambda\Psi = 0$ is determined when one point of the set is assigned; if this point be of parameter θ_0, and Φ_0, Ψ_0 denote the values of Φ, Ψ when $\theta = \theta_0$, the other $m - 1$ points of the set are determined by $\Phi\Psi_0 - \Psi\Phi_0 = 0$, after this has been divided by $\theta - \theta_0$. In the reduced polynomial, of order $m - 1$, the coefficients will be rational polynomials in θ_0. And a converse theorem, which is important, is true: *If we have any aggregate of sets of m points on a conic, with the property that, when the parameter, θ_0, of any one point of a set is given, the parameters for the remaining $m - 1$ points are the roots of a polynomial equation whose coefficients are rational in θ_0, then all the sets of the series are given by an equation $\Phi + \lambda\Psi = 0$; so that the series is based on two sets of the series, which may be chosen arbitrarily from the series.* (For a proof of this, we refer to *Principles of Geometry*, Vol. II, p. 136.) As an example, let DEF and PQR be two triangles inscribed in the given conic $\Omega = 0$. The sides of these triangles then touch another conic; and there exists a triangle, inscribed in $\Omega = 0$, having one angular point at any arbitrary point, U, of the conic $\Omega = 0$, of which the sides also touch the second conic. Thus, by the converse theorem we have quoted, the angular points, say U, V, W, of such a triangle, have parameters which are roots of a cubic equation $\Phi + \lambda\Psi = 0$, in which $\Phi = 0$, $\Psi = 0$, respectively, give D, E, F and P, Q, R; the value of λ depends on the position of the chosen angular point U. Or again, there exists a (third) conic in regard to which the inscribed triangles DEF, PQR, of $\Omega = 0$, are both self-polar; and every triangle, inscribed in $\Omega = 0$, which is self-polar in regard to this third conic, is equally given by the equation $\Phi + \lambda\Psi = 0$. This gives the result, already proved ((6), **Ex. 3**), that the triangles inscribed in one conic and circumscribed about another conic, when such triangles are possible, are all self-polar in regard to a third conic.

We may also consider, from this point of view, the theorem already proved (Chap. XIV (21), **Ex. 22**), that, if three triangles be inscribed in a conic, so that the six sides of any two of these triangles touch another conic, then the three conics so arising, each touching the sides of two of the inscribed triangles, have a common tangent. On the given conic, let the parameters, which are the roots of the cubic polynomials Φ_1, Φ_2, Φ_3, give, respectively, the angular points of the three inscribed triangles. The general triangle whose sides touch the conic which touches the sides of the two triangles given by $\Phi_2 = 0$, $\Phi_3 = 0$ is given by $\Phi_2 + \lambda_1 \Phi_3 = 0$, for varying values of λ_1. Similarly, two other sets of triangles are given by equations $\Phi_3 + \lambda_2 \Phi_1 = 0$ and $\Phi_1 + \lambda_3 \Phi_2 = 0$. For the truth of the theorem enunciated, we require to shew that values of λ_1, λ_2, λ_3 exist, such that the cubic polynomials in θ, namely, $\Phi_2 + \lambda_1 \Phi_3$, $\Phi_3 + \lambda_2 \Phi_1$, $\Phi_1 + \lambda_3 \Phi_2$, have two roots in common; the line joining the points with these parameters being then the common tangent of the three conics spoken of. The polynomials Φ_1, Φ_2, Φ_3 are three quite arbitrary cubic polynomials. The theorem is thus reduced to an interesting algebraic problem. One solution is that indicated in Ex. 1 following.

Ex. 1. Let Φ_1, Φ_2, Φ_3 be any three cubic polynomials in a parameter θ. Prove that six constants, l_1, m_1, l_2, m_2, l_3, m_3, can be found so that the equation
$$(l_1\theta + m_1)\, \Phi_1 + (l_2\theta + m_2)\, \Phi_2 + (l_3\theta + m_3)\, \Phi_3 = 0$$
is true identically in regard to θ. Then, denoting $l_2 m_3 - l_3 m_2$, $l_3 m_1 - l_1 m_3$, $l_1 m_2 - l_2 m_1$, respectively, by h_1, h_2, h_3, prove that the three cubic polynomials in θ, namely, $h_3 \Phi_2 - h_2 \Phi_3$, $h_1 \Phi_3 - h_3 \Phi_1$, $h_2 \Phi_1 - h_1 \Phi_2$, are in the ratios of $l_1\theta + m_1$, $l_2\theta + m_2$, $l_3\theta + m_3$. These three cubic polynomials thus have a common quadratic factor.

Ex. 2. Prove that a point in space, whose four coordinates are in the ratios of θ^3, θ^2, θ, 1, describes a curve, when θ varies, which has three points lying upon an arbitrary plane. Prove further that, from an arbitrary point, not lying on this curve, a single line can be drawn to meet this curve in two points. A cubic polynomial in θ determines the plane joining the three points of the curve whose parameters are the three roots of the polynomial. Thus three cubic polynomials Φ_1, Φ_2, Φ_3 determine three planes, and the common point of these planes. Shew that the parameters of the two points of the curve which lie on the line, drawn from this common point to meet the curve in two points, are the roots of the quadratic factor common to $\Phi_2 + \lambda_1 \Phi_3$, $\Phi_3 + \lambda_2 \Phi_1$, $\Phi_1 + \lambda_3 \Phi_2$ for proper values of λ_1, λ_2, λ_3.

APPENDIX TO CHAPTER XVI

PONCELET'S THEOREM FOR VARIABLE TRIANGLES

In this section we give a geometrical proof of Poncelet's theorem for variable triangles, with indications of how this may be extended to the case of polygons. We also give analytical results whose interest arises from their connexion with the theory of elliptic integrals, developed since Poncelet's time.

(1) Consider a linear series of conics, denoted by (ω, σ), defined as passing through the four common points of two conics, ω and σ, in general positions. Let P, Q, R be points lying on the conic ω, and σ_1 a conic of the series (ω, σ). Let PQ touch σ, and QR touch σ_1. Let P', Q', R' be other points of the conic ω, such that $P'Q'$ touches σ, and $Q'R'$ is a proper one of the two tangents drawn from Q' to touch σ_1. We prove that $R'P'$ touches one of the two conics of the series (ω, σ) which can be drawn to touch RP, say σ_2. Also, that the three lines PP', QQ', RR', joining corresponding angular points of the two triangles, all touch a conic of the series (ω, σ), say ϕ.

The dual of a series (ω, σ) of conics is a system (Ω, Σ) of conics touching the four common tangents of two conics Ω, Σ. Let p, q, r be tangents of Ω, and Σ_1 a conic of the system (Ω, Σ). Let the point (p, q) lie on the conic Σ, and the point (q, r) lie on Σ_1. Let p', q', r' be other tangents of Ω, such that the point (p', q') lies on Σ, and the point (q', r') is a proper one of the two points in which the line q' meets Σ_1. Then, as the dual of what is stated above, the point (r', p') lies on one of the two conics of the system (Ω, Σ) which pass through the point (r, p), say Σ_2. Also, the three points (p, p'), (q, q'), (r, r'), intersections of corresponding sides of the two triangles (p, q, r), (p', q', r'), lie on a conic also belonging to the system (Ω, Σ), say Φ.

We prove the former statement; from which the dual follows. For this, we first remark, as a lemma, that, if a quadrangle of four points be taken on the conic ω (not supposed to be also on σ) such that a pair of complementary joins of these four points are tangents of σ, then, either of the other two pairs of complementary joins, of these four points, consists of two tangents of a properly chosen conic of the series (ω, σ), the chord of contact being the same as for the first pair of complementary joins. We prove this lemma, using coordinates: Let the two chords of ω which touch σ have equations $\xi = 0$, $\zeta = 0$, and $\eta = 0$ represent the chord of contact. We can then suppose the equation of σ to be $\zeta\xi - \eta^2 = 0$, the polynomial σ being $\zeta\xi - \eta^2$. Another pair of complementary joins of the quadrangle given by the intersections of ω with $\xi = 0$, $\zeta = 0$, will have an equation $\lambda\omega + \zeta\xi = 0$, where λ is a constant. This, however, is $\lambda\omega + \sigma + \eta^2 = 0$. Thus the two new chords have double contact with the conic $\lambda\omega + \sigma = 0$, with $\eta = 0$ as chord of contact. The proof shews that *any* conic, through the four points $\omega = 0$, $\zeta\xi = 0$, has double contact, on $\eta = 0$, with some conic of the series (ω, σ). For the lemma cf. Hart, *Quart. J. Math.* 1857.

Now, let PQ, QR be two chords of the conic ω, such that PQ touches σ, and QR touches another conic, σ_1, of the series (ω, σ). Let P', Q' be two other points of the conic ω, such that $P'Q'$ also touches σ. As the joins PQ, $P'Q'$, of four points of the conic ω, touch the conic σ, it follows, by the lemma, that there is a conic, ϕ, in the series (ω, σ), which touches the lines PP', QQ'; and that the line, say LM, joining the points L, M where ϕ touches PP', QQ', is the line, say $\eta = 0$, which joins the points of contact of PQ and $P'Q'$ with σ. From Q', two tangents can be drawn to σ_1, meeting ω again, say, in R_1' and R_2'. As QR, $Q'R_1'$ both touch σ_1, and Q, R, Q', R_1' are points of ω, it follows, also from the lemma, that there is a conic, of the series (ω, σ), which touches both QQ' and RR_1'. Likewise, there is a conic, of the series (ω, σ), which touches both QQ' and RR_2'. But there are only two conics of the series (ω, σ) which touch the line QQ'. Thus the conic ϕ, already considered, of the series (ω, σ), which touches PP' and QQ', coincides with the conic of this series touching QQ' and RR_1', or coincides with that touching QQ' and RR_2'. We choose, of the points R_1', R_2', that one, which we denote by R', such that ϕ touches QQ' and RR'. Then, as P, P', R, R' are points of ω such that PP', RR' both touch ϕ, it follows, again from the lemma, that there is a conic of the series (ω, σ) which touches RP and $R'P'$. Whence, $R'P'$ touches one of the two conics, of the series (ω, σ), which touch RP, say σ_2.

It is thus possible to have an infinite number of triangles $P'Q'R'$, inscribed in ω, whose sides, in order, touch the conics σ, σ_1, σ_2 of the series (ω, σ). And a similar argument shews that we may have a polygon, of any number of sides, whose angular points lie on ω, whose sides, in order, touch, respectively, conics σ, σ_1, σ_2, ..., all belonging to the series (ω, σ). This is Poncelet's theorem.

(2) The examination of the relation between the conics σ, σ_1, σ_2, in the case of triangles, leads to algebra which has great interest in connexion with the theory of elliptic integrals. We give some account of this, dealing first, in a descriptive way, with the theory of (1); but we do not here obtain complete clearness in regard to the ambiguous signs which enter; this requires more details of the theory of elliptic functions than we assume. In (4) below, using the language and topology of metrical geometry, more definiteness is obtained.

Let the fundamental conic ω have the equation $x^2 + y^2 + z^2 = 0$, an any other conic of the series (ω, σ) have the equation

$$(a + \theta) x^2 + (b + \theta) y^2 + (c + \theta) z^2 = 0,$$

where a, b, c are definite numbers, and θ a number varying from conic to conic of the series. We speak of θ as the *parameter* of the conic of the series which is considered. A particular point of the conic ω has coordinates x_1, y_1, z_1 given by

$$(b-c)^{\frac{1}{2}}(a+p)^{\frac{1}{2}}, \quad \epsilon(c-a)^{\frac{1}{2}}(b+p)^{\frac{1}{2}}, \quad \epsilon'(a-b)^{\frac{1}{2}}(c+p)^{\frac{1}{2}},$$

where we suppose the square roots to be definitely chosen, and each of ϵ, ϵ' to be 1 or -1. The value of p varies from point to point of ω. The value of p is a definite function of the ratios of x_1^2, y_1^2, z_1^2, so that the same value of p occurs four times over on ω, the four points being distinguished by the values of ϵ and ϵ'. The value of p may be defined geometrically as the parameter of the second conic of the series (ω, σ) which touches the tangent of ω at the point considered. For the equation of this tangent is $xx_1 + yy_1 + zz_1 = 0$; and the tangential equation of the conic, of the series (ω, σ), whose parameter is p, is $u^2(a+p)^{-1} + v^2(b+p)^{-1} + w^2(c+p)^{-1} = 0$. This equation is satisfied by replacing u, v, w by x_1, y_1, z_1, which are given by the values above. The four points of ω for which the values of p are the same arise for the four common tangents of ω and this conic of parameter p.

We seek now the condition that a common tangent of two conics of the series (ω, σ), with repective parameters ϕ, ψ, should pass through a point of ω whose parameter is θ; that is, should contain a point of ω whereat the tangent touches the conic, of the series (ω, σ), whose parameter is θ. We may denote the conics of parameters θ, ϕ, ψ, respectively, by $\sigma_1, \sigma_2, \sigma_3$. A common tangent of the conics $(\phi), (\psi)$ is given by the equation

$$\begin{aligned}
x(b-c)^{\frac{1}{2}}(a+\phi)^{\frac{1}{2}}(a+\psi)^{\frac{1}{2}} &+ y\zeta(c-a)^{\frac{1}{2}}(b+\phi)^{\frac{1}{2}}(b+\psi)^{\frac{1}{2}} \\
&+ z\zeta'(a-b)^{\frac{1}{2}}(c+\phi)^{\frac{1}{2}}(c+\psi)^{\frac{1}{2}} = 0,
\end{aligned}$$

where the square roots are supposed to be chosen definitely, but each of ζ, ζ' is $+1$ or -1. This is clear because this equation satisfies the tangential equations of both σ_2 and σ_3. The condition for this common tangent to contain the point of the conic ω, where the parameter is θ, is

$$(b-c)(a, \theta, \phi, \psi) + \epsilon\zeta(c-a)(b, \theta, \phi, \psi) + \epsilon'\zeta'(a-b)(c, \theta, \phi, \psi) = 0,$$
$$\dots\dots(A)$$

where (a, θ, ϕ, ψ) means $(a+\theta)^{\frac{1}{2}}(a+\phi)^{\frac{1}{2}}(a+\psi)^{\frac{1}{2}}$, etc. If this relation is satisfied with definite values of $\epsilon\zeta$ and $\epsilon'\zeta'$, then every one of the four common tangents of σ_2 and σ_3 contains an appropriately chosen one of the four points θ which lie on ω; for, when $\epsilon\zeta, \epsilon'\zeta'$ are given, and ζ, ζ' are assigned, proper values of ϵ, ϵ' can be found. The relation is symmetrical in regard to θ, ϕ, ψ. Thus, when it is satisfied, a common tangent of σ_3 and σ_1 contains a point ϕ on ω; and a common tangent of σ_1 and σ_2 contains a point ψ on ω.

But in fact, if no two of a, b, c be equal, the rationalised form of this relation is symmetrical in regard to the two sets a, b, c and θ, ϕ, ψ. For, let A_1, A_2, A_3 denote $a+b+c$, $bc+ca+ab$, abc, respectively, and P_1, P_2, P_3 denote $\theta+\phi+\psi$, $\phi\psi+\psi\theta+\theta\phi$, $\theta\phi\psi$, respectively. It can be proved that the product of the four functions on the left of the equation, for the various values of $\epsilon\zeta$ and $\epsilon'\zeta'$, is $(b-c)^2\,(c-a)^2\,(a-b)^2\Delta$, where $\Delta=(A_2-P_2)^2-4\,(A_1+P_1)\,(A_3+P_3)$; and Δ is symmetrical in the two sets a, b, c and θ, ϕ, ψ.

We may therefore replace the four equations (A) by the four equations
$$(\phi-\psi)\,\Theta\pm(\psi-\theta)\,\Phi\pm(\theta-\phi)\,\Psi=0, \qquad \ldots\ldots\text{(B)}$$

wherein Θ denotes $(\theta+a)^{\frac12}\,(\theta+b)^{\frac12}\,(\theta+c)^{\frac12}$, etc.; it being supposed that no two of θ, ϕ, ψ are equal.

It is clear that Δ is symmetrical in regard to θ, ϕ, ψ, and is a polynomial of order 2 in each of these. It can be verified that the two roots of the quadratic equation in θ, given by $\Delta=0$, are
$$\theta=[(\phi, \psi)\pm 2\Phi\Psi]/(\phi-\psi)^2, \qquad \ldots\ldots\text{(C)}$$
where $\qquad (\phi, \psi)=\phi\psi\,(\phi+\psi)+2A_1\phi\psi+A_2\,(\phi+\psi)+2A_3,$

so that (ϕ, ϕ) is $2\Phi^2$. By the properties of the linear series of conics (ω, σ), a common tangent of σ_2, σ_3 meets ω in two points which are harmonic conjugates in regard to the two points of contact.

We shall prove in (3) that the condition $\Delta=0$, of which we have here given the geometrical interpretation, by means of a common tangent of two of the conics $\sigma_1, \sigma_2, \sigma_3$, is the sufficient condition for the existence of a triangle, inscribed in ω, whose sides touch $\sigma_1, \sigma_2, \sigma_3$.

But first, we give greater clearness to the meaning of the square roots which have entered. In a plane in which the coordinates of a point are $(\xi, \eta, 1)$, the equation
$$\eta^2=(\xi+a)\,(\xi+b)\,(\xi+c)$$

represents a cubic curve, meeting an arbitrary line in three points. This curve contains the six points for which ξ, η have, respectively, any of the pairs of values $(\theta, \pm\Theta), (\phi, \pm\Phi), (\psi, \pm\Psi)$. Thus the vanishing of the expression on the left, in equation (B) above, given by

$$\begin{vmatrix} \Theta, & \pm\Phi, & \pm\Psi \\ \theta, & \phi, & \psi \\ 1, & 1, & 1 \end{vmatrix}=0,$$

means that, for some choice of the ambiguous signs, the three points of the cubic curve, for which (ξ, η) have the values $(\theta, \Theta), (\phi, \pm\Phi),$ $(\psi, \pm\Psi)$, *lie in a line*. The equation $\Delta=0$ therefore expresses this geometrical fact for the cubic curve. This curve contains, for every point $(\xi, \eta, 1)$, also the point $(\xi, -\eta, 1)$; and we see, by considering the curve, why it is that, when ϕ and ψ are given, there are two associated values of θ arising when $\Delta=0$; since we may join the point for which (ξ, η) are (ϕ, Φ) to either of the points for which (ξ, η) are (ψ, Ψ) or $(\psi, -\Psi)$.

There is another interpretation of the fact that three points of this cubic curve lie on an arbitrary line. The curve differs essentially from a conic, in that the coordinates of a point of it are not expressible as rational functions of a parameter. An alternative way of stating this property of a conic is to say that, the coordinates of a point on a conic are expressible as *trigonometric* functions of a variable argument; these are single-valued functions of the argument, with a single *period*, so that the increase of the argument by this period corresponds to a complete circuit of the conic. The coordinates of a point of the cubic curve under consideration are likewise expressible as single-valued functions of a variable argument; but these functions are generalisations of the trigonometric functions, and have two periods. If we denote the variable argument by u, the general value of this which belongs to any point of the curve is of the form $u + m\varpi + m'\varpi'$, where ϖ, ϖ' are the periods in question, and m, m' are arbitrary positive or negative integers. When this argument is given, the values of both the coordinates, ξ, η, of the point are definite, with no ambiguity of sign.

Now let the arguments of the three points in which the cubic curve is met by an arbitrary line be u_1, u_2, u_3, each subject to an additive ambiguity in terms of the two periods, as explained. Then it can be proved, these arguments being properly defined, *that* $u_1 + u_2 + u_3$ *vanishes.* Or, what is the same so far as the curve is concerned, this sum is of the form $m\varpi + m'\varpi'$, where m, m' are integers. This significant fact was proved by Abel, and is the simplest case of what is commonly called Abel's theorem.

Therefore we can say, *another form of the relation* $\Delta = 0$, *above, connecting* θ, ϕ, ψ, *is* $u_1 \pm u_2 \pm u_3 = 0$, *where* u_1, u_2, u_3 *are the arguments respectively associated with the points of the cubic curve for which* (ξ, η) *are* (θ, Θ), (ϕ, Φ), (ψ, Ψ).

The argument u, here employed, associated with a point $(\xi, \eta, 1)$ of the cubic curve, may be defined by the equation

$$ u = \int_{\infty}^{(\xi,\,\eta)} \frac{d\xi}{\eta}, $$

the path of integration being through values of ξ, η, including complex values, which satisfy the equation of the curve. The lower limit of the integral, denoted by ∞, is the point where the cubic curve touches the Absolute line.

(3) We proceed next to find the condition for the existence, as in (1), of a triangle PQR, inscribed in the conic ω, whose sides, respectively, touch conics of the series (ω, σ) which have the parameters θ, ϕ, ψ. We do not, however, enter into sufficient detail to justify fully all the signs we choose, when these are ambiguous with our present specifications.

Let P, Q, R be points on the conic ω, with respective parameters p, q, r. So that, for instance, the tangent of ω at P touches the conic whose equation is $(a+p)\,x^2 + (b+p)\,y^2 + (c+p)\,z^2 = 0$. Let the two conics of the series (ω, σ) which touch the side QR be those of parameters θ and θ'; likewise, let RP touch the conics (ϕ), (ϕ'), and PQ

touch the conics (ψ), (ψ'), all belonging to the series (ω, σ). Denote the argument

$$\int_{\infty}^{(\theta,\,\Theta)} \frac{d\xi}{\eta},$$

integrated on the cubic curve, up to the point for which ξ, η are θ and Θ, for brevity, by $[\theta]$. We may then regard the points Q, R as determined from the parameters θ, θ', just as θ was determined from ϕ, ψ, in (2) preceding (see equation (C)). Therefore, by what has been said, attaching proper signs to Θ, Θ', we have the relations, connecting the integrals, which are given by

$$[q]=[\theta']-[\theta], \quad [r]=[\theta']+[\theta].$$

For, in (2), the two roots θ, of $\Delta=0$, when ϕ, ψ were given, changed into one another by changes in the signs of Φ, Ψ; and, a change in the sign of Θ, in the definition here given of the integral $[\theta]$, gives rise to the change of $[\theta]$ into $-[\theta]$. Similarly, adopting proper signs for Φ, Φ', Ψ, Ψ', we have

$$[r]=[\phi']-[\phi], \quad [p]=[\phi']+[\phi]; \quad [p]=[\psi']-[\psi], \quad [q]=[\psi']+[\psi].$$

These equations give

$$-[q]+[r]=2\,[\theta], \quad -[r]+[p]=2\,[\phi], \quad -[p]+[q]=2\,[\psi],$$

and lead to
$$2\,[\theta]+2\,[\phi]+2\,[\psi]=0.$$

In (2), we have shewn, in regard to three conics of the series (ω, σ), with parameters θ, ϕ, ψ, that, in order that a common tangent of two of these, say (ϕ) and (ψ), should contain a point of ω whereat the tangent touches the third (θ), the condition is $\Delta=0$. Further that, with a proper sign for Θ, this is equivalent to the equation $[\theta]+[\phi]+[\psi]\equiv 0$, whatever signs be given to Φ and Ψ. It appears then, that the geometrical relation between the conics (θ), (ϕ), (ψ), investigated in (2), is sufficient to secure that a triangle PQR, inscribed in ω, should have sides touching the conics (θ), (ϕ), (ψ). Conversely, if triangles $P'Q'R'$ be inscribed in ω, for which the sides $P'Q'$, $Q'R'$, $R'P'$, respectively, touch the conics σ, σ_1, σ_2, then, when R' coincides with P', the line $P'Q'$ is a common tangent of σ and σ_1, and $R'P'$ becomes the tangent of ω at P', and touches σ_2. The *condition considered in* (2) *is thus both a necessary and sufficient condition for the three conics to be touched by the sides of a triangle inscribed in* ω.

We remark that we have seen before (Chap. XI (12), Ex. 10), that a necessary and sufficient condition for a conic ω to be triangularly circumscribed to a single conic σ, is that the tangent of σ, at one of the common points of ω and σ, should meet ω again in a point whereat the tangent is also a tangent of σ.

(4) We can obtain greater definiteness, and with less reference to the ideas of the theory of elliptic functions, if we assume the topological relations implicit in elementary metrical geometry, and an acquaintance with the trigonometric functions.

We take, not the case so far considered, of conics of the series (ω, σ), but the dual case stated in (1), for the system of conics (Ω, Σ); these we describe as confocals.

We consider an ellipse, Ω, of equation $x^2/a + y^2/b = 1$, in which the coordinates are $(x, y, 1)$, and a, b are real positive numbers, with $a > b$. From a point T, not on this ellipse, tangents TP_1, TP_2 are drawn to this ellipse. Through T is drawn a confocal ellipse, of equation $x^2/(a+\psi) + y^2/(b+\psi) = 1$, and also a confocal hyperbola, of equation $x^2/(a+\chi) + y^2/(b+\chi) = 1$. Thus ψ is a real positive number, and χ is a real negative number, lying between $-b$ and $-a$. When $\psi = 0$, the confocal ellipse coincides with Ω. When $\chi = -a$, the confocal hyperbola coincides with the axis of Ω which is given by $x = 0$ (taken twice over). This we call the minor axis of Ω. When $\chi = -b$, the confocal hyperbola coincides with a segment, on $y = 0$, of what we call the major axis of Ω.

Near to T, on the confocal ellipse (ψ), a point T' is taken, from which the two tangents to Ω are $T'P_1'$ and $T'P_2'$. We denote the lengths of the diameters of Ω which are parallel to the tangents TP_1 and TP_2, respectively, by $2r_1$ and $2r_2$; also the measure of the (small) angle (of contingence) between the tangents TP_1 and $T'P_1'$ by $d\epsilon_1$, and that between TP_2 and $T'P_2'$ by $d\epsilon_2$. As the tangents TP_1, TP_2 have, as bisectors, the tangents of the confocals through T, it follows that the ratio $d\epsilon_1/d\epsilon_2$ has, for limit, the ratio of the lengths of the tangents TP_2, TP_1; which is r_2/r_1 (Chap. xv (9), Ex. 6). If ds_1, ds_2 be the lengths of the chords P_1P_1', P_2P_2', the limit of $ds_1/d\epsilon_1$ is the length of the radius of curvature of the conic Ω at P_1, as we see by considering the circle of curvature. And we have essentially proved (Chap. xiii (13)) that the radius of curvature has a length $r_1^3/(ab)^{\frac{1}{2}}$. Similarly at P_2.

Thus, ultimately, ds_1/ds_2, being $(r_1^3/r_2^3)\,(d\epsilon_1/d\epsilon_2)$, is r_1^2/r_2^2. These are elementary results of metrical geometry which can be justified from preceding results. Using trigonometrical functions, we can suppose any point of the ellipse Ω to be given by $x = a^{\frac{1}{2}} \sin\theta$, $y = b^{\frac{1}{2}} \cos\theta$. Then, supposing P_1 and P_1' to be given by θ_1 and $\theta_1 + d\theta_1$, we can prove that $ds_1/d\theta_1$ is ultimately $a^{\frac{1}{2}}(1 - k^2 \sin^2\theta_1)^{\frac{1}{2}}$, where $k^2 = 1 - b/a$, and that $r_1^2 = a(1 - k^2 \sin^2\theta_1)$; with similar results for P_2, given by θ_2. Thus the equation connecting the variations of P_2 and P_1, on Ω, as T varies on the confocal ellipse (ψ), is $(1 - k^2 \sin^2\theta_1)^{-\frac{1}{2}}\,d\theta_1 - (1 - k^2 \sin^2\theta_2)^{-\frac{1}{2}}\,d\theta_2 = 0$. When T varies on the confocal hyperbola (χ), we similarly find $(1 - k^2 \sin^2\theta_1)^{-\frac{1}{2}}\,d\theta_1 + (1 - k^2 \sin^2\theta_2)^{-\frac{1}{2}}\,d\theta_2 = 0$.

With the usual phraseology, in which $\pi/2$ is the measure of the angle between two perpendicular lines $(= 1{\cdot}57\ldots)$, and $-\pi < \theta < \pi$, we shall put

$$u = \int_0^\theta (1 - k^2 \sin^2\theta)^{-\frac{1}{2}}\,d\theta,$$

the square root being taken positively; it is then clear that any value of u which arises defines a definite point of the ellipse Ω, for the specified range of θ. We assume that the points P_1, P_2 are thus included.

If these correspond to values u_1, u_2, we hence conclude that $u_1 - u_2$ remains constant when T moves, within a certain range, on the ellipse (ψ); and $u_1 + u_2$ similarly remains constant as T moves on the hyperbola (χ).

These respective values of $u_1 - u_2$ and $u_1 + u_2$ depend, respectively, on ψ and χ. We proceed to evaluate them. For this purpose we introduce two trigonometrical arguments μ, ν to replace ψ and χ, respectively. For μ, limited in the first instance by $0 \leqslant \mu \leqslant \pi/2$, we take $\tan \mu = (\psi/b)^{\frac{1}{2}}$, so that

$$\psi = b \tan^2 \mu, \quad a + \psi = a \sec^2 \mu \, (1 - k^2 \sin^2 \mu), \quad b + \psi = b \sec^2 \mu;$$

for ν, similarly limited by $0 \leqslant \nu \leqslant \pi/2$, we take $\sin \nu = [(a + \chi)/(a - b)]^{\frac{1}{2}}$, so that, with $c = a - b$, $= ak^2$,

$$\chi = -a \, (1 - k^2 \sin^2 \nu), \quad a + \chi = c \sin^2 \nu, \quad b + \chi = -c \cos^2 \nu,$$

which are in accord with $-a \leqslant \chi \leqslant -b$. Then we prove that, when T is on the confocal ellipse (ψ),

$$u_1 - u_2 = 2 \int_0^\mu (1 - k^2 \sin^2 \theta)^{-\frac{1}{2}} \, d\theta,$$

the standard case being that for which T is a point of the ellipse (ψ) for which $y > 0$, and $\theta_1 > \theta_2$. And, when T is on the confocal hyperbola (χ),

$$u_1 + u_2 = 2 \int_0^\nu (1 - k^2 \sin^2 \theta)^{-\frac{1}{2}} \, d\theta,$$

provided both coordinates x, y, of T, are positive; while, when $y > 0$ but $x < 0$, the value of $u_1 + u_2$ is the negative of the integral written.

To prove the former equation, let the tangent of the ellipse Ω, at the point B for which $x = 0$, $y = b^{\frac{1}{2}}$, meet the ellipse (ψ) in the point T_0 for which $x > 0$. One of the tangents, from T_0 to Ω, touches Ω at B for which $\theta_2 = 0$, and $u_2 = 0$. Let θ_1 belong to the point of contact of the other tangent from T_0. To find θ_1, we may express that the point of contact lies on the polar of T_0 in regard to Ω. The point T_0 is an intersection of the line $y = b^{\frac{1}{2}}$ with the ellipse (ψ), whose equation, in terms of μ, is

$$x^2/a \, (1 - k^2 \sin^2 \mu) + y^2/b = \sec^2 \mu.$$

The result is that

$$\tan \tfrac{1}{2}\theta_1 = \tan \mu \, (1 - k^2 \sin^2 \mu)^{\frac{1}{2}}.$$

The constant value of $u_1 - u_2$, which we seek, is the value of u_1 when θ_1 has the value given by this equation for $\tan \tfrac{1}{2}\theta_1$. This equation leads to

$$\sec^2 \tfrac{1}{2}\theta_1 = \sec^2 \mu \, (1 - k^2 \sin^4 \mu),$$

$$(1 - k^2 \sin^2 \theta_1)^{\frac{1}{2}} = [1 - k^2 \sin^2 \mu \, (1 + \cos^2 \mu)]/(1 - k^2 \sin^4 \mu).$$

By substitution we thence find that

$$\int_0^{\theta_1} (1 - k^2 \sin^2 \theta)^{-\frac{1}{2}} \, d\theta = 2 \int_0^\mu (1 - k^2 \sin^2 \theta)^{-\frac{1}{2}} \, d\theta.$$

To find the constant value of $u_1 + u_2$ when T moves on the confocal hyperbola (χ), we distinguish the cases according to which of the real branches of the hyperbola it is, on which T lies. Consider the two intersections of this hyperbola (χ), with Ω, for which respectively $y > 0$, $x > 0$ and $y > 0$, $x < 0$. The equation of the hyperbola is

$$x^2/\sin^2 v - y^2/\cos^2 v = c,$$

and the four points, $(a^{\frac{1}{2}} \sin \theta, b^{\frac{1}{2}} \cos \theta)$, of Ω, which lie thereon, are given by $\theta = v$, $\theta = -v$, $\theta = \pi - v$, $\theta = \pi + v$. When T, moving on (χ), is at either of the first two of these points of Ω, we have $u_1 = u_2$, and hence

$$u_1 + u_2 = \pm 2 \int_0^v (1 - k^2 \sin^2 \theta)^{-\frac{1}{2}} \, d\theta,$$

where the upper or lower sign is to be taken according as, on the branch of the hyperbola on which T lies, we have $x > 0$ or $x < 0$. This proves the statement we have made.

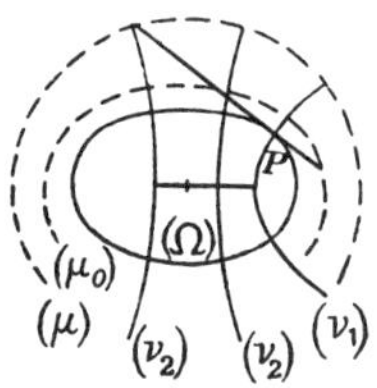
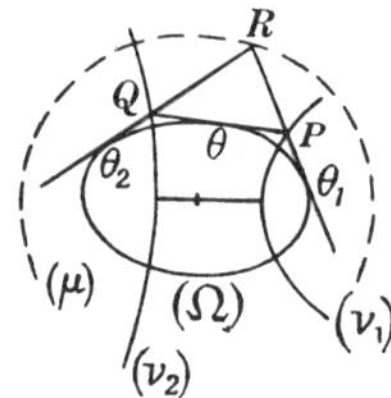
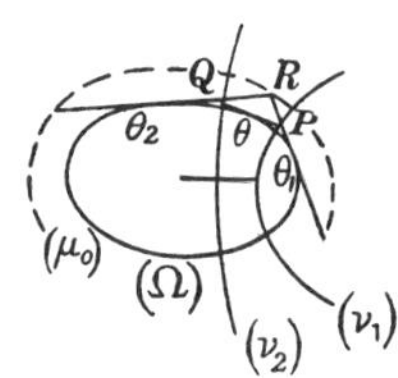

We now apply these results to the Poncelet figure. Suppose a tangent, at the point θ of the ellipse Ω, to meet the confocal hyperbola (v_1) at a point P, at which x and y are both positive, and to meet the confocal hyperbola (v_2) at a point Q. It may happen that, at Q, both x and y are positive, or that, at Q, while y is positive, x is negative. We distinguish these cases. Let the other tangents drawn to Ω, from P and Q, touch this at θ_1 and θ_2, respectively, and meet one another in R. Let the confocal ellipse through R be called either (μ_0), or (μ), according to the cases. If we denote the integral $\int_0^{} (1 - k^2 \sin^2 \theta)^{-\frac{1}{2}} \, d\theta$ with the upper limits v_1, v_2, μ_0, μ, respectively, by N_1, N_2, M_0, M, we then have, by what has been proved,

$$u + u_1 = 2N_1, \quad u + u_2 = 2N_2, \quad u_1 - u_2 = 2M_0,$$
or
$$u + u_1 = 2N_1, \quad u + u_2 = -2N_2, \quad u_1 - u_2 = 2M.$$

Thus, respectively, in the two cases,

$$u_1 - u_2 = 2N_1 - 2N_2 = 2M_0, \quad \text{and} \quad u_1 - u_2 = 2N_1 + 2N_2 = 2M;$$

so that, the ranges being limited as described, after division by 2, we have

$$\int_0^{\mu_0} (1 - k^2 \sin^2 \theta)^{-\frac{1}{2}} \, d\theta = \int_0^{v_1} (1 - k^2 \sin^2 \theta)^{-\frac{1}{2}} \, d\theta - \int_0^{v_2} (1 - k^2 \sin^2 \theta)^{-\frac{1}{2}} \, d\theta,$$

$$\int_0^{\mu} (1 - k^2 \sin^2 \theta)^{-\frac{1}{2}} \, d\theta = \int_0^{v_1} (1 - k^2 \sin^2 \theta)^{-\frac{1}{2}} \, d\theta + \int_0^{v_2} (1 - k^2 \sin^2 \theta)^{-\frac{1}{2}} \, d\theta,$$

each of ν_1, ν_2, μ_0, μ being positive. Thus, as θ varies, since $u_1 - u_2$ is constant, the point R moves on a confocal ellipse, respectively (μ_0) and (μ). The last two equations written, which are independent of θ, give the relations connecting the parameters of the three confocals, on which P, Q, R move so that QR, RP, PQ touch Ω, when such a figure is possible.

A particular case is when the point θ is taken on Ω, *in coincidence with* P, at the point $(x > 0,\ y > 0)$ where (ν_1) meets Ω. Then Q, R coincide on the tangent of Ω at P. Thus, the tangent of Ω, at this point P, common to Ω and (ν_1), meets the two branches of (ν_2), respectively, on the confocal ellipses (μ_0) and (μ). And, between the parameters ν_1, ν_2, μ_0, μ, there exist the equations $M_0 = N_1 - N_2$, $M = N_1 + N_2$.

It is of interest, for the relation between the parameters, to give an independent verification of this particular case: A point common to Ω, whose equation is $x^2/a + y^2/b = 1$, and the hyperbola whose equation is $x^2/(a + \chi_1) + y^2/(b + \chi_1) = 1$, has coordinates x, y respectively equal to $[a\,(a + \chi_1)/c]^{\frac{1}{2}}$, $[-b\,(b + \chi_1)/c]^{\frac{1}{2}}$, both here taken positive, these satisfying the equations of both curves. The tangent of Ω, at this point, has the equation

$$x\,[(a + \chi_1)/ac]^{\frac{1}{2}} + y\,[-(b + \chi_1)/bc]^{\frac{1}{2}} = 1.$$

Also, a point where the ellipse $x^2/(a + \psi) + y^2/(b + \psi) = 1$, meets the hyperbola $x^2/(a + \chi_2) + y^2/(b + \chi_2) = 1$, has coordinates

$$\pm\,[(a + \psi)\,(a + \chi_2)/c]^{\frac{1}{2}},\ \ [-(b + \psi)\,(b + \chi_2)/c]^{\frac{1}{2}},$$

the upper or lower signs being taken according to the branch of (χ_2) which is considered.

In order, now, that this common point of (ψ) and (χ_2) should be on the tangent of Ω at the intersection of Ω with (χ_1), considered, we require

$$\pm\,[(a + \psi)\,(a + \chi_1)\,(a + \chi_2)/a]^{\frac{1}{2}} + [(b + \psi)\,(b + \chi_1)\,(b + \chi_2)/b]^{\frac{1}{2}} = c;$$

and if, as before,

$$a + \psi = a\,\sec^2\mu\,(1 - k^2\sin^2\mu),\ \ b + \psi = b\,\sec^2\mu;$$

$$a + \chi_1 = c\,\sin^2\nu_1,\ \ b + \chi_1 = -c\,\cos^2\nu_1;\ \ a + \chi_2 = c\,\sin^2\nu_2,\ \ b + \chi_2 = -c\,\cos^2\nu,$$

this equation is

$$\pm\,\sin\nu_1\sin\nu_2\,(1 - k^2\sin^2\mu)^{\frac{1}{2}} = \cos\mu - \cos\nu_1\cos\nu_2. \quad\ldots\ldots(a)$$

Taking the square of this, and then solving for $\cos\mu$, we find

$$(1 - k^2\sin^2\nu_1\sin^2\nu_2)\cos\mu$$
$$= \cos\nu_1\cos\nu_2 + \epsilon\,\sin\nu_1\sin\nu_2\,(1 - k^2\sin^2\nu_1)^{\frac{1}{2}}\,(1 - k^2\sin^2\nu_2)^{\frac{1}{2}},$$

where $\epsilon = \pm 1$, which leads to

$$\frac{\cos\mu - \cos\nu_1\cos\nu_2}{\sin\nu_1\sin\nu_2}$$
$$= \epsilon\,\frac{(1 - k^2\sin^2\nu_1)^{\frac{1}{2}}\,(1 - k^2\sin^2\nu_2)^{\frac{1}{2}} + \epsilon k^2\sin\nu_1\cos\nu_1\sin\nu_2\cos\nu_2}{1 - k^2\sin^2\nu_1\sin^2\nu_2}.$$

Thus, by equation (a) preceding, the right side here is the value, in terms of ν_1 and ν_2, of $\pm (1 - k^2 \sin^2 \mu)^{\frac{1}{2}}$. By considering the case $k = 0$, we infer that, when k is small enough, we should take $\epsilon = 1$ when the intersection with the hyperbola (ν_2) of the tangent at the point P of Ω at which the tangent is drawn to Ω has the coordinate x positive, but $\epsilon = -1$ when x is negative; and obtain the value of $(1 - k^2 \sin^2 \mu)^{\frac{1}{2}}$ in both cases. To the reader to whom the addition formulae for elliptic functions are known, the formulae we have obtained for $\cos \mu$, $(1 - k^2 \sin^2 \mu)^{\frac{1}{2}}$ are familiar, as being those for $\operatorname{cn}(N_1 - N_2)$ and $\operatorname{dn}(N_1 - N_2)$, when $\epsilon = 1$; and for $\operatorname{cn}(N_1 + N_2)$ and $\operatorname{dn}(N_1 + N_2)$, when $\epsilon = -1$. Comparing these results obtained for the particular case with the equations $N_1 - N_2 = M_0$, $N_1 + N_2 = M$ found in general, we thus prove again, as in (3) above, that *a necessary and sufficient condition for triangles to exist, circumscribing the fundamental ellipse* Ω, *with their angular points on the confocal hyperbolas* (ν_1), (ν_2) *and the confocal ellipse* (μ), *is, that the tangent of the fundamental ellipse, at a point where this is met by* (ν_1), *should pass through a common point of* (ν_2) *and* (μ). The ellipse (μ) is given by $M = N_1 - N_2$, or by $M = N_1 + N_2$, according as the common point of (μ) and (ν_2), which is considered, lies on one or other branch of (ν_2). The integrals N_1, N_2, M characterise the conics in question.

The symmetry of the figure shews that the tangent of Ω, at every one of the common points of Ω and (ν_1), contains an appropriately taken intersection of (ν_2) and (μ), when $M = N_1 + N_2$; and the same tangent contains an intersection of (ν_2) and (μ_0) if $M_0 = N_1 - N_2$. The ambiguities of sign in the general case are thus resolved by the topology inherent in the metrical figure.

Ex. 1.   Taking $\nu_1 = \nu_2$, a variable tangent of the fundamental ellipse Ω meets the two branches of the confocal hyperbola (ν), or (N), at points, wherefrom the other tangents to Ω meet on the confocal ellipse (μ), or (M), which is given by $M = 2N$. Also, a tangent of Ω, at a point lying on the confocal hyperbola (ν), contains a common point of (ν) with the ellipse (μ).

This is the dual of a theorem given above (in (6), Ex. 7), that two tangents to a conic ω', drawn from a point of another conic ω, meet this latter conic again in two points whose joining line touches a third conic, which has the equation

$$(\Theta'^2 - 4\Delta'\Theta) \, \omega + 4\Delta'\Delta\omega' = 0.$$

For the case considered here, $\tan \frac{1}{2}\mu = \tan \nu \, (1 - k^2 \sin^2 \nu)^{\frac{1}{2}}$.

Ex. 2.   Prove that the formulae which have been given lead to

$$\int_0^\psi a^{\frac{1}{2}} \, [\psi \, (a + \psi) \, (b + \dot{\psi})]^{-\frac{1}{2}} \, d\psi = 2 \int_0^\mu (1 - k^2 \sin^2 \mu)^{-\frac{1}{2}} \, d\mu;$$

$$\int_{-a}^\chi a^{\frac{1}{2}} \, [\chi \, (a + \chi) \, (b + \chi)]^{-\frac{1}{2}} \, d\chi = 2 \int_0^\nu (1 - k^2 \sin^2 \nu)^{-\frac{1}{2}} \, d\nu.$$

Ex. 3.   We have used a parametric representation for a confocal ellipse, in terms of μ, and a representation for a confocal hyperbola, in terms of ν. By use of the formulae of elliptic functions, these may be connected. Taking $k'^2 = 1 - k^2 = b/a$, and using K, K' for

$$K = \int_0^{\pi/2} (1 - k^2 \sin^2 \theta)^{-\frac{1}{2}} \, d\theta, \quad K' = \int_0^{\pi/2} (1 - k'^2 \sin^2 \theta)^{-\frac{1}{2}} \, d\theta,$$

the functions $\operatorname{sn}(u+K+iK')$, $\operatorname{cn}(u+K+iK')$, $\operatorname{dn}(u+K+iK')$ are respectively equal to
$$\operatorname{dn} u/k\operatorname{cn} u, \quad -ik'/k\operatorname{cn} u, \quad ik'\operatorname{sn} u/\operatorname{cn} u.$$

Thus, the integrals N, M, whose upper limits are ν and μ, are such that $N=M+K+iK'$.

Ex. 4. The confocal ellipse (μ), for which μ has the value ρ, meets the confocal hyperbola (ν), for which ν has the same value, $\nu=\rho$, at a point for which $y=b^{\frac{1}{2}}$, $x=a^{\frac{1}{2}}\tan\rho\,(1-k^2\sin^2\rho)^{\frac{1}{2}}$ (lying on a tangent of Ω); as well as in three other points.

Ex. 5. We have considered the case of a variable triangle, with sides touching the ellipse Ω, of which two angular points lie on confocal hyperbolas, and one lies on a confocal ellipse. In fact, from equations of the forms $\pm 2N_1=u_2\pm u_3$, $\pm 2N_2=u_3\pm u_1$, $\pm 2N_3=u_1\pm u_2$, the arguments u_1, u_2, u_3 can be eliminated only, (1), if two of the ambiguous signs on the right, in these equations, be positive, and one negative, which is the case we have considered, or (2), if all the ambiguous signs on the right be negative. In this latter case, the angular points of the varying triangle all lie on confocal ellipses. The reader may consider what are the possible cases when the fundamental conic is a hyperbola.

Ex. 6. Through a point T there passes an ellipse (μ), and a hyperbola (ν), both confocal to the fundamental ellipse Ω, the notation being that we have used. From T, tangents TP_1, TP_2 are drawn to Ω, and the hyperbola (ν) meets Ω in the point U. We think of the points T, U, P_1 as having both coordinates x, y positive. Let t_1, t_2 be the lengths of the tangents TP_1, TP_2, respectively, and s_1, s_2 be the lengths of the arcs P_1U and UP_2 of the ellipse Ω.

Prove, geometrically, from the fact that the tangents of (μ), (ν), at T, are the bisectors of the angles determined by TP_1 and TP_2, that $(t_1-s_1)-(t_2-s_2)=0$; also that $(t_1-s_1)+(t_2-s_2)$ is unaltered by variation of T on the ellipse (μ).

If (ν_1) be the parameter of the confocal hyperbola through P_1, whose coordinates are then $a^{\frac{1}{2}}\sin\nu_1$, $b^{\frac{1}{2}}\cos\nu_1$, prove that

$$t_1=a^{\frac{1}{2}}\tan\mu\,(1-k^2\sin^2\nu)^{\frac{1}{2}}\,(1-k^2\sin^2\nu_1)^{\frac{1}{2}},$$

$$(t_1-s_1)/a^{\frac{1}{2}}=\tan\mu\,(1-k^2\sin^2\mu)^{\frac{1}{2}}-\int_0^\mu (1-k^2\sin^2\theta)^{\frac{1}{2}}\,d\theta,$$

of which the latter depends only on μ. Prove also, by differentiation, assuming the result proved above which was expressed by $N_1=N+M$, that

$$\int_0^\nu (1-k^2\sin^2\theta)^{\frac{1}{2}}\,d\theta+\int_0^\mu (1-k^2\sin^2\theta)^{\frac{1}{2}}\,d\theta-\int_0^{\nu_1}(1-k^2\sin^2\theta)^{\frac{1}{2}}\,d\theta$$
$$-k^2\sin\nu\sin\mu\sin\nu_1=0.$$

If we take for the point T the point common to the tangents at the ends A, B of the two axes of the fundamental ellipse Ω,

$$(x=a^{\frac{1}{2}},\ y=0 \ \text{and} \ x=0,\ y=b^{\frac{1}{2}}),$$

shew that the point U, where the confocal hyperbola drawn through T meets the ellipse Ω (with x, y both positive), is such that, for arcs of the ellipse, arc $BU-$arc $UA=a^{\frac{1}{2}}-b^{\frac{1}{2}}$. This particular point U is often called Fagnano's point. The hyperbola through T is characterised by a value of ν such that

$$\int_0^{\pi/2}(1-k^2\sin^2\theta)^{-\frac{1}{2}}\,d\theta=2\int_0^\nu (1-k^2\sin^2\theta)^{-\frac{1}{2}}\,d\theta.$$

The points of contact, P_1 and P_2, of the tangents to the fundamental ellipse Ω, drawn from any point of this hyperbola, are such that $\tan\nu_1\tan\nu_2=(a/b)^{\frac{1}{2}}$. Prove that a circle can be drawn touching the tangents $x=a^{\frac{1}{2}}$, $y=b^{\frac{1}{2}}$, of the ellipse Ω, which also touches Ω at the Fagnano point. Examine the application of the theorem of this example to the case when tangents are drawn from a varying point of an ellipse to touch a confocal hyperbola. For the matters here dealt with, the reader may consult Fagnano, *Produzioni Matematiche*, 1750; and Chasles, *Compt. rendus*, 1843, 1844.

Ex. 7. We have found, when tangents at points, P_1, P_2, of the fundamental ellipse Ω, meet on a confocal ellipse, the equation $u_1 - u_2 = e$, where e depends on the parameter of this ellipse; and when these tangents meet on a confocal hyperbola, the equation $u_1 + u_2 = h$, where h depends on the parameter of the hyperbola. These formulae are necessary and sufficient for the figure. Hence it is clear that, if any number, n, of tangents of Ω form a polygon, of which all the angular points, except one, move on conics confocal with Ω, then the remaining angular point equally moves on a confocal. For, by addition, or subtraction, of any number of successive equations of these forms, we can eliminate all the arguments u_1, u_2, ... except the first and last, and so obtain an equation $u_1 \mp u_n = k$, where k is a constant. This equation proves that the tangents at the points P_1, P_n, of Ω, meet on a confocal.

This is the dual of the result remarked in (1). And this result may be stated in metrical terms: If any polygon P_1, P_2, ..., with angular points lying on a circle, be such that all the sides P_1P_2, P_2P_3, ..., of the polygon, except one, touch each a circle of a series coaxal with the given circle, then so also does the remaining side. For this form, a proof may be constructed by shewing that, if p_1, p_2, ... be the lengths of the perpendiculars drawn from P_1, P_2, ..., respectively, to the radical axis of the series of circles, and p_1', p_2', ... be the lengths of these perpendiculars drawn from the points P_1', P_2', ... of another position of the polygon, then the lengths of P_1P_1', P_2P_2', ... are in the ratios of $p_1^{\frac{1}{2}} + p_1'^{\frac{1}{2}}$, $p_2^{\frac{1}{2}} + p_2'^{\frac{1}{2}}$, ...

(5) We may add the following examples, of which Exx. 1–5 have already occurred.

Ex. 1. Two conics, Ω, Ω', have double contact. Another conic, ω, meets each of these in four points, respectively (P) and (P'). Any conic, Φ, is drawn through the common points (P) of Ω and ω. Prove that a conic, Φ', can be drawn through the common points (P') of Ω' and ω to have double contact with Φ; the line of contact of Φ, Φ' is the same as of Ω and Ω'.

Ex. 2. In particular, if Ω' consist of any pair of tangents of Ω, and Φ consist of any pair of complementary common chords of Ω and ω, prove that: Through the common points (P'), of a conic ω, with a pair of tangents of another conic Ω, a conic Φ' can be drawn, to touch both of a complementary pair of common chords of Ω and ω.

Ex. 3. Dually, if, from any two points of a conic Ω, be drawn the pairs of tangents to another conic ω, these four tangents touch a conic Φ' containing any complementary pair of intersections of the common tangents of Ω and ω.

Ex. 4. If Ω, ω be confocal conics, and, from both of two points T, T' of Ω, be drawn the pair of tangents to ω, respectively p, q and p', q', then any conic touching these four lines has double contact with a certain conic confocal with Ω and ω, the pole of the chord of contact being also the pole of TT' in regard to Ω.

In particular, p, q, p', q' touch a circle, with centre at this pole. And there is a confocal conic, passing through the two points (p, p') and (q, q'), whose tangents at these points pass through the pole of TT' in regard to Ω.

Ex. 5. Let the two tangents from a point H, to a conic ω, meet a confocal Ω respectively in P, R and in P', R'. Prove that two of the diagonals of the quadrilateral which is formed by the tangents of Ω, at the points P, R, P', R', are the tangents, at H, of the two confocals which pass through this point.

Ex. 6.   Let ξ, ζ, η denote respectively

$$(x-a)\cos\theta + y\sin\theta - r, \quad (x-a)\cos\theta' + y\sin\theta' - r,$$

$$(x-a)\cos\tfrac12(\theta+\theta') + y\sin\tfrac12(\theta+\theta') - r\cos\tfrac12(\theta'-\theta),$$

where r is $(a^2-h^2)^{\frac12}$. Verify the two identities

$$(a-h)(x^2+y^2-2bx+h^2)+(b-a)[(x-h)^2+y^2]$$
$$=(b-h)\operatorname{cosec}^2\tfrac12(\theta'-\theta)(\eta^2-\xi\zeta)=(b-h)(x^2+y^2-2ax+h^2),$$

and interpret these geometrically in connexion with the foregoing theory.

Ex. 7.   Prove that the equation $\theta=[(\phi,\ \psi)\pm 2\Phi\Psi](\phi-\psi)^{-2}$, of (2), leads to

$$\pm(\phi-\psi)(a+\theta)^{\frac12}=(a+\phi)^{\frac12}(b+\psi)^{\frac12}(c+\psi)^{\frac12}\pm(a+\psi)^{\frac12}(b+\phi)^{\frac12}(c+\phi)^{\frac12}.$$

Ex. 8.   If (x, y, z) and $(x+dx,\ y+dy,\ z+dz)$ be near points of the conic $x^2+y^2+z^2=0$, such that

$$x=(b-c)^{\frac12}(a+p)^{\frac12}, \quad y=(c-a)^{\frac12}(b+p)^{\frac12}, \quad z=(a-b)^{\frac12}(c+p)^{\frac12},$$

while h_1, h_2, h_3 and θ are arbitrary, prove that

$$\frac{(dx, y, h_3)}{XU^{\frac12}}=\frac{i\,dp}{2(a+p)^{\frac12}(b+p)^{\frac12}(c+p)^{\frac12}},$$

where X denotes $h_1x+h_2y+h_3z$, U denotes $(a+\theta)x^2+(b+\theta)y^2+(c+\theta)z^2$, and (dx, y, h_3) denotes the determinant whose three rows consist, respectively, of the elements (dx, dy, dz), (x, y, z), (h_1, h_2, h_3).

Interpreting metrically, shew that the differential on the left arises from

$$\frac{ds}{R}\Big/\frac{t}{r},$$

where ds is the arc of the conic $x^2+y^2+z^2=0$ from P to P_1 and $2R$ is the length of the diameter of this conic parallel to the tangent at (x, y, z), while t is the length of a tangent drawn from (x, y, z) to the conic $(a+\theta)x^2+(b+\theta)y^2+(c+\theta)z^2=0$, and $2r$ the length of a diameter of this conic parallel to this tangent.

For Poncelet's brief indication of the proof of his theorem, see *Prop. Proj.* (1865), I, pp. 312 ff. For the theory of (2) and (3) above, see Cayley, *Papers*, IV, p. 292 (1861), and other papers. A geometrical proof of Poncelet's theorem for triangles is given in *Principles of Geometry*, Vol. III, p. 116, as a deduction from the theory of confocal quadric surfaces. An account, with use of Elliptic Functions, is in Halphen, *Fonct. Ellipt*, 2nd Partie (1888), Chap. X. This, and Cayley's investigation, lead to the case of the porism for a polygon of any number of sides, whose angular points lie on one conic and whose sides touch another conic. For this problem a complete investigation is sketched in *Proc. Camb. Phil. Soc.* (On chains of two-two relations), XXIII (1926) p. 92.

MISCELLANEOUS EXAMPLES

[If A, B, C be three points of a line, we frequently denote the point A', which is the harmonic conjugate of A in regard to B and C, by writing $A' = (B, C)/A$. And we sometimes express the fact that A, A' are harmonic conjugates, in regard to B and C, by writing A, $A' \sim B$, C. But it must be borne in mind that A, $A' \sim B$, C and D, $D' \sim B$, C do not together involve A, $A' \sim D$, D'. For the notation $\epsilon\,(l, m)$, $\epsilon P\,(Q, R)$, see Chap. x (4).]   For co-director circle see Chap. XIII (12).

1.   If A, A' and B, B' be two pairs of points defining an involution, the two points $(A, B)/A'$ and $(A', B')/A$ form a pair of this involution.

2.   In any involution, to any pair A, A' there corresponds another pair B, B', such that B, $B' \sim A$, A'. If P, P' be any other pair of this involution, the points $(A, A')/P$ and $(B, B')/P'$ are the same point.

3.   If D, D'; E, E'; F, F' be three pairs of an involution, and $P = (E, F)/D'$, $Q = (F, D)/E'$, $R = (D, E)/F'$, then the three pairs P, D'; Q, E'; R, F' are in involution. Interpreting this on a conic, we have the result: If the sides BC, CA, AB of a triangle ABC are touched by a conic respectively in D, E, F, and the lines joining D, E, F to an arbitrary point meet the conic again respectively in D', E', F', then the lines AD', BE', CF' meet in a point.

4.   Let A, B, C, D be four points on a line. There are three involutions determined respectively by the three pairs B, C and A, D; C, A and B, D; A, B and C, D. Prove that the two foci of any one of these involutions are harmonic conjugates in regard to the two foci of each of the other two involutions. Thus the foci of any one of the involutions are the two points which are the common pair of the other two involutions.

5.   Let A, A'; B, B'; C, C'; I, J be four pairs of an involution on a line, and the points P, Q, R be given by $P = (I, J)/A'$, $Q = (I, J)/B'$, $R = (I, J)/C'$. Prove that A, P; B, Q; C, R; I, J are also four pairs belonging to an involution.

6.   Let I, J be the symbols of two points of a line, and a, b, c be arbitrary numbers. Prove that the three pairs of points, whose symbols are

$$(I + a^2 J,\ I - bcJ),\quad (I + b^2 J,\ I - caJ),\quad (I + c^2 J,\ I - abJ),$$

belong to an involution.

7.   Let A, B, C be the symbols of three points of a line, and λ, μ, ν be numbers. Prove that the condition that the three pairs of points, whose symbols are

$$(A,\ B - \lambda C),\quad (B,\ C - \mu A),\quad (C,\ A - \nu B),$$

should belong to an involution, is $\lambda\mu\nu = 1$.

8.   Let A, B, C be the symbols of three points of a line, and λ, μ, ν be numbers. Prove that the condition that the three pairs of points, whose symbols are

$$(B + \lambda C,\ B - \lambda C),\quad (C + \mu A,\ C - \mu A),\quad (A + \nu B,\ A - \nu B),$$

should belong to an involution, is $\lambda^2 \mu^2 \nu^2 = 1$.

9.   Let A, B, C be points of a line, and $A' = (B, C)/A$, $B' = (C, A)/B$, $C' = (A, B)/C$. Prove that A, A'; B, B'; C, C' are three pairs of an involution. If I, J be the foci of this involution, prove that the six cross ratios

$$(B, C/A, I),\ (C, A/B, I),\ (A, B/C, I),\ (C, B/A, J),\ (A, C/B, J),\ (B, A/C, J)$$

are equal, the values being $-\omega$ or $-\omega^2$, where ω is an imaginary cube root of unity. Also that the three cross ratios $(B, C/I, J), (C, A/I, J), (A, B/I, J)$ are equal, the values being, correspondingly, ω or ω^2.

Prove that $A=(B', C')/A', B=(C', A')/B', C=(A', B')/C'$; and find the values of $(B', C'/A', I)$ and $(B', C'/I, J)$.

Prove also that $B, C'; B', C; A, A'$ are pairs of an involution; as are $B, C; B', C'; I, J$. Also that the ranges $A, B', C, A', B, C', I, J$ and $B', C, A', B, C', A, I, J$ are related.

The points I, J constitute what are called *the Hessian pair for the three points A, B, C*, a relation clearly unaltered by projection. The interpretation of these points on a conic; and the interpretation of what are called the (four) Hessian points of four given points, occur in subsequent examples (Exx. 108, 109, 250).

10. If A, B, C, D be four points of a line, prove that the product, of three cross ratios,

$$(B, C/A, D) \cdot (C, A/B, D) \cdot (A, B/C, D) = -1.$$

11. Let A, B, C, P, Q, P', Q' be seven points of a line. Express the cross ratio $(P', Q'/P, Q)$ in terms of the four cross ratios

$$(P, C/A, B), \quad (Q, C/A, B), \quad (P', C/A, B), \quad (Q', C/A, B).$$

12. Let A, B, C, O, I, J be six arbitrary points of a plane. Let lines AP, BQ, CR, be drawn respectively through A, B, C to satisfy the three conditions that $(AP, AO); (AB, AC); (AI, AJ)$ are pairs of lines of a pencil in involution, also $(BQ, BO); (BC, BA); (BI, BJ)$ are pairs of lines of a pencil in involution, and $(CR, CO); (CA, CB); (CI, CJ)$ are pairs of lines of a pencil in involution. Prove that the lines AP, BQ, CR meet in a point.

13. Let A, B, C, D be four points in a plane, of which no three are in line, and E, F be the respective points of intersection $(AB, CD), (AD, BC)$; so that EF, BD, CA are the three diagonals of a quadrilateral. Let an arbitrary line meet these diagonals, respectively, in L, M, N. Prove that the points $P=(E, F)/L, Q=(B, D)/M, R=(C, A)/N$ are in a line (the conjugate of the given line in regard to the quadrilateral).

We may also prove this as follows: Let the arbitrary line meet the lines ABE, ADF, respectively, in X and Y. By Pappus' theorem, considering the cross-joins of B, E, X and D, F, Y, the two intersections $U=(EY, XF)$ and $V=(XD, BY)$ are in a line with C. Let this line meet XY in O. Let AU, AV, AC meet XY, respectively, in P', Q', R'. Then the points P, Q, R are, respectively, $(A, U)/P', (A, V)/Q'$ and $(A, C)/R'$. Thus the points O, P, Q, R lie on the line which is the harmonic conjugate of the line XOY, in regard to the two lines $OA, OUVC$. (The points R', N are identical.)

By Brianchon's theorem, the point O is the point of contact with XY of the conic which touches AB, BC, CD, DA, XY; and $OPQR$ is the tangent at O of the second of the two conics touching AB, BC, CD, DA which passes through O.

14. An arbitrary line meets the respective joins AB, CD, BC, AD, of four general points in a plane, in E, F, H, K. The intersections (AC, BD), $(BF, CE), (DH, CK)$ are, respectively, O, P, Q. Prove that OP, OQ are harmonic conjugates in regard to OC, OD.

Also, if U, V be the respective intersections $(AB, CD), (BC, AD)$, the three pairs of lines $OP, OU; OQ, OV; OC, OD$ are in involution.

15. Lines a, a', passing through the angular point A of a triangle ABC, meet BC respectively in D and D'; likewise lines b, b' through B meet CA respectively in E, E'; and lines c, c' through C meet AB in F, F' respectively. And it is supposed that the three cross ratios $(D, D'/B, C)$, $(E, E'/C, A), (F, F'/A, B)$ have equal values. Prove that the join of the two points $(b, c'), (b', c)$ passes through A, with a similar result for $(c, a'), (c', a)$, and for $(a, b'), (a', b)$.

16. Let the lines to the orthocentre drawn from the angular points A, B, C of a triangle meet the opposite sides BC, CA, AB, respectively, in L, M, N. Prove that the lines from the same points A, B, C, drawn perpendicular respectively to the sides MN, NL, LM of the triangle LMN, meet in the circumcentre of the triangle ABC. (See Ex. 150 below.)

17. Shew that the diagonal triangle of a quadrilateral, of four lines, is both circumscribed to, and in perspective with, the triangle formed by any three of the four given lines, the axis of perspective of the two triangles being the fourth of the given lines. The centres of perspective, in the four cases, form a quadrangle whose vertices are the angular points of the diagonal triangle of the quadrilateral.

18. The lines AO, BO, CO, joining the angular points of a triangle to a point O, meet the respective opposite sides BC, CA, AB respectively in D, E, F. Prove that the three intersections $(BC, EF), (CA, FD), (AB, CD)$ lie on a line, say l; called the polar of O in regard to the triangle. Let l' be the polar of another point O', in regard to the triangle. Let l, l' meet in X and $EF, E'F'$ meet in P. Prove that A, P, X are in a line, while the point $(P, X)/A$ lies on BC. Prove also that the three points $(BC, E'F), (CA, F'D'), (AB, DE)$ lie in a line.

19. Let ABC be a triangle, and N, D two arbitrary points, the lines AN, BN, CN meeting BC, CA, AB, respectively, in P, Q, R. Prove that three points U, V, W, lying respectively on AD, BD, CD, can be found such that the lines BW, CV, DP meet in a point, say K; while the lines CU, AW, DQ meet in a point, say L; and the lines AV, BU, DR meet in a point, say M. Then prove that the seven lines $UP, VQ, WR, AK, BL, CM, DN$ meet in a point.

20. Points D, E, F, respectively on the lines BC, CA, AB, are such that AD, BE, CF meet in a point. Points U, V, W, respectively on the lines EF, FD, DE, are such that AU, BV, CW meet in a point. Prove that the lines DU, EV, FW meet in a point.

21. The *antipoints* of two given points are defined as the two intersections of the lines joining these to the Absolute points of the plane. If four points are such that two of them are the antipoints of the other two, prove that any one of the four points is the orthocentre of the other three.

22. Points D, E, F, on the lines BC, CA, AB, respectively, are such that AD, BE, CF meet in a point. An arbitrary line meets BC, CA, AB respectively in L, M, N. Let D' be taken on BC so that L is one of the foci of the involution determined by the pairs B, C and D, D'; and a similar construction be made for E' on CA and F' on AB. Prove that AD', BE', CF' meet in a point.

23. The lines joining the angular points A, B, C of a triangle to a point O meet the circumcircle of ABC again, respectively in D, E, F. Two arbitrary circles are drawn, respectively through B, E and through C, F. Prove that the common points of these two circles lie on a circle through A, D; and that O is the radical centre of the three circles $(A, D), (B, E), (C, F)$.

24. The common chord HK, of two circles, meets the line of centres in O; the pole of HK in regard to one of the circles is X. A common tangent of the two circles meets the line of centres in T, and the polar of T in regard to the other circle meets the line of centres in M. Prove that $\epsilon H(O, X) = \epsilon H(M, T)$. The symbol $\epsilon H(O, X)$ means the cross ratio $(HO, HX/HI, HJ)$, where I, J are the Absolute points of the plane.

25. Let A, B, C, D be four points lying on a circle. Any pair of complementary joins of the points meet in an angular point of the vertex triangle of the quadrangle formed by the four points. Prove that the bisectors of the angles determined by any pair of complementary joins are parallel to the bisectors for each of the other two pairs.

26.   If A, B, C, D be four general points, and AB, DC meet in E, while AD, BC meet in F, the mid points of A, C and of B, D and of E, F lie in a line, called the *median* line of the quadrilateral formed by the four lines AB, BC, CD, DA. If a quadrilateral be formed by four tangents of a circle, prove that the median line contains the orthocentre of the diagonal triangle of the quadrilateral.

27.   The points S, H, P, Q are four points in a plane. One of the bisectors of the angles determined by the lines SP, SQ, meets one of the bisectors of the angles determined by HP, HQ, in the point T. Prove that, for all the four points T so obtained, $[\epsilon T\,(S,\,H)]^2$ has the same value, being equal to the product $\epsilon P\,(S,\,H)\,.\,\epsilon Q\,(S,\,H)$.

28.   If the centre of a conic be at the orthocentre of a self-polar triangle of the conic, the conic is a circle.

29.   Let P be a variable point of a fixed circle, and C any fixed point. Prove that the line through the mid point of C, P, perpendicular to CP, meets the tangent at P in a point lying on a fixed line.

30.   The centres of similitude of two circles are the foci of the involution, on the line of centres, determined by the two pairs of points which consist of the centres of the circles, and the limiting points of the circles.

31.   Points P, Q, lying respectively on two circles, are such that the mid point of P, Q lies on the radical axis. Shew that the line perpendicular to PQ, drawn through this mid point, contains the mid point of the centres of the two circles.

32.   The orthocentre of a triangle ABC is conjugate to one angular poin A in regard to the circle having BC as diameter.

33.   Let a chord, LM, of one circle, meet another circle in points X, Y, which are conjugate to one another in regard to the first circle, so that X, $Y \sim L$, M. Prove that the point in which the line LM meets the radical axis is conjugate, in regard to the first circle, to the mid point of X, Y.

34.   From a variable point P, of one side AB of a triangle ABC, lines PX, PY are drawn respectively perpendicular to the sides CA, CB to meet CB and CA, respectively, in Y and X. Prove that XY envelopes a parabola, which touches CA and CB.

35.   If two tangents of a parabola be given, prove that a third tangent can be found so that the orthocentre of the triangle formed by the three tangents is any given point of the directrix.

36.   Given four lines of which three form the triangle ABC, and the fourth meets BC, CA, AB respectively in L, M, N, we know that the circles AMN, BNL, CLM, ABC meet in a point, say O. Let OL, OM, ON meet the circle ABC again, respectively, in A', B', C'. Prove that the lines AA', BB', CC' are all parallel to the line LMN.

37.   Let A, B be two fixed points, and l, m be two fixed lines meeting the line AB, respectively, in L and M; let H, K be the foci of the involution determined by the two pairs A, B and L, M. If conics be drawn through A, B to touch the lines l, m, the line joining the points of contact passes through H or K.

   If a focus, and two points, of a conic be given, the directrix corresponding to the focus passes through one of two fixed points.

   Of circles touching two given lines, the chord of contact has one of two definite directions.

38.   For a conic, of which S is one focus, the tangents at two points P, Q meet in T. Prove that ST is one of the bisectors of the angles determined by SP and SQ.

39. Let A, B be two points which are conjugate in regard to a given conic. Let an arbitrary line through A meet the conic in P and Q, and BP, BQ meet the conic again, respectively, in P' and Q'. Prove that A, P', Q' lie in a line. And, if PQ', $P'Q$ meet in C, then C is the pole of AB.

40. The perpendicular from a focus S of a conic, to the corresponding directrix, meets this in X. Also P, Q are two points of the conic such that PQ contains S. Prove that XS is one of the bisectors of the angles determined by XP, XQ.

41. Let A, B be two points of a conic, and P the pole of AB. Let T, T' be two arbitrary points. Let TP, $T'P$ meet AB in U, U', respectively; let the tangents from T to the conic meet AB in X and Y, and the tangents from T' meet AB in X' and Y'. Prove that

$$(X, X'/A, B).(Y, Y'/A, B)=(X, Y'/A, B).(Y, X'/A, B)=[(U, U'/A, B)]^2.$$

[This is the general form of the result ordinarily expressed, in metrical terms, by saying that the angle between two tangents of a conic is equal, or supplementary, to the half-sum, or half-difference, of the angles which the two points of contact subtend at the two real foci.]

42. The axes of a conic are parallel to the bisectors of the angles determined by any pair of complementary chords common to the conic and any circle.

43. The tangents to a conic, drawn from any point of a directrix, meet the line drawn through the corresponding focus S parallel to the directrix, in P and Q. Prove that S is the mid point of P, Q.

44. A chord PQ of a conic meets a directrix in U, the tangents at P, Q meet in T, and S is the focus corresponding to the directrix. Prove that ST, SU are the bisectors of the angles determined by SP and SQ.

45. Prove that two conics which have the same centre have a pair of common chords passing through the centre. Prove that a third conic can be drawn having double contact with one of the given conics on an arbitrary line through the centre, and having also double contact with the other given conic on a certain line through the centre; and that these chords of contact are harmonic conjugates in regard to the common central chords of the two given conics.

46. If, for two confocal conics, a tangent of one be perpendicular to a tangent of the other, the lines joining the common centre to the respective points of contact are harmonic conjugates in regard to the common central chords of the two conics.

47. The lines joining an arbitrary point of a conic to the ends of a chord, which contains a focus, meet the corresponding directrix in two points, which form, with the focus, a self-polar triangle of the conic. The sides of this triangle which meet in the focus are perpendicular.

48. Given a focus, and the corresponding directrix, of a conic, and one point of the conic, construct the points in which the conic meets an arbitrary line drawn through the focus.

49. Let a circle drawn through the (real) foci of a hyperbola meet one asymptote in T, T', and the other asymptote in U, U'. Prove, with suitable notation, that TU, $T'U'$ are tangents of the curve. Also, if S be one of the foci, then $_\epsilon S\,(T, U)$ is the same for all such circles.

50. The foci of a conic, which lie on an axis whose end points are A and A', are S and S'. The tangent at any point P of the curve meets the tangents at A and A', respectively, in T and T'. Prove that TS, $T'S'$ meet on the normal of the curve at P; as do TS', $T'S$.

51.   The tangent and normal at a point P, of a conic, meet an axis in T and G, respectively. Prove that the circle TPG contains the foci which lie on the other axis. If S be one of these foci, the lines ST, SG are conjugate in regard to the conic.

52.   A pair of conjugate diameters of a conic, whose centre is C, meet the directrix corresponding to a focus S in H and K. Prove that S is the orthocentre of the triangle CHK.

53.   Let A, A' be fixed points. Let a point P vary so that the bisectors of the angles determined by PA, PA' are in fixed directions (that is, pass through two definite points on the Absolute line). Prove that the locus of P is a rectangular hyperbola, having its centre at the mid point of A, A' and its asymptotes parallel to the fixed directions.

54.   Prove that two rectangular hyperbolas can be drawn to touch four given lines; and that the centres of these are at the points where the median line of the quadrilateral, formed by the four lines (see Ex. 26), meets the circumcircle of the diagonal triangle of the quadrilateral.

55.   A chord, QR, of a rectangular hyperbola is drawn perpendicular to the tangent at a point P of the curve. Prove that the chords PQ, PR are perpendicular.

56.   Let R, R' be two points of a rectangular hyperbola such that RR' contains the centre of the curve. Let PQ be any chord of the curve. Prove that $\epsilon R\,(P,\,Q)=\epsilon R'\,(P,\,Q)$.
Let P, Q, R, R' be points of a rectangular hyperbola such that RP, RQ are perpendicular, and $R'P$, $R'Q$ are perpendicular. Prove that RR' passes through the centre of the curve, and is perpendicular to the line through the centre which is conjugate to PQ.

57.   Let O be the centre of curvature at a point P of a parabola; prove that the line through the focus S which is perpendicular to SP contains the mid point of P, O.

58.   Let O be the centre of curvature at a point P of a rectangular hyperbola, whose centre is C. Let O_1 be taken on OP such that P is the mid point of O, O_1. Prove that CP, CO_1 are perpendicular.

59.   Let the real complementary foci of any conic be S, H; and P be any point of the conic. Let the lines drawn through S, H respectively perpendicular to SP and HP meet the normal of the curve, drawn at P, in U and V. Prove that the centre of curvature at P is $(U,\,V)/P$.

60.   Let the polar of a point P of a conic, taken in regard to the director circle, meet the normal of the curve at P in the point O_1. Prove that the centre of curvature at P is the point O, on the normal at P, such that P is the mid point of O, O_1. The polar of P, in regard to the director circle, meets the tangent at P, of the conic, in the point which is the pole, in regard to the conic, of the normal at P.

61.   The circle of curvature, at a point P of a conic, meets the conic again in Q. Prove that the Apollonius rectangular hyperbola, of the centre of this circle, touches the conic at P, and contains the mid point of the chord P, Q. If this mid point be V, the centre of the hyperbola is at the mid point of P, V.

62.   A conic, Ω, contains the angular points, C, X, Y, of a triangle which is self-polar in regard to another conic, ω; and the conics meet in P, Q, R, S. Let T be the pole, in regard to the conic ω, of the chord PQ. Prove that T is the pole of the complementary common chord RS, *in regard to the triangle CXY*. Thus, in particular, CT and RS meet the side XY in two points which are harmonic conjugates in regard to X and Y.

When Ω is the Apollonius rectangular hyperbola of a point in regard to the conic ω, we can thence derive a construction for the two other normals which can be drawn to a conic from the common point of two given normals.

If the normals at the points P, Q, R, S, of a conic whose centre is C, meet in a point, and S' is the point of the conic for which C is the mid point of S and S'', prove that P, Q, R, S' lie on a circle.

63. The points, in which the axis of an ellipse, that contains the imaginary foci, is met by a circle containing the real foci, are the poles, in regard to the ellipse, of the common chords of the ellipse and circle which are parallel to the other axis. If UP, UQ be tangents to the ellipse, drawn from any point U of the first axis, prove that the circle UPQ contains the real foci.

64. If A, A' and P, P' be two pairs of points on a line, the points $Q=(A, A')/P$, $Q'=(A, A')/P'$ are a pair of the involution determined by A, A' and P, P'.

The involution on a line, which is determined by the pairs of complementary joins of four points of a conic, is the same involution as that determined by the pairs of lines which join the pole, of the line, to the pairs of complementary intersections of the tangents of the conic at the four points.

The tangents to a conic, from an arbitrary point, have the same bisectors as have the lines joining this point to a pair of complementary foci of the conic; and these are also the bisectors for the two lines which join the point to the two points, in which the polar of the point meets the directrices corresponding to the foci.

65. The lines joining a variable point P to two fixed points H, K, meet the fixed line, perpendicular to HK, through the mid point of H, K, in the points X, X'. If these are a pair of a definite involution on this line, prove that the locus of P is a conic, having the line XX' as an axis. And that this conic contains the foci of the involution.

66. A line which meets two given circles in two harmonically conjugate pairs of points, envelopes a conic having one pair of foci at the centres of the circles; and the auxiliary circle of this conic is coaxal with the given circles.

67. A line meets the sides BC, CA, AB, of a triangle inscribed in a conic, respectively, in D, E, F; and meets the conic in O and Q. Prove that the range D, E, F, O is related to the pencil of lines which join A, B, C, Q to any point of the conic. Deduce that the sides of two triangles, inscribed in a conic, touch another conic.

68. The two conics which can be drawn touching four definite tangents of a given conic, to pass through a particular focus of this conic, have their tangents at this focus perpendicular to one another.

69. Two conics have three of their common points coincident, at a point X. Prove that three of their common tangents coincide with the tangent touching both conics at X.

70. The lines joining a variable point, P', of a conic, respectively to the centre C, and a focus S, meet the conic again in P and Q. Prove that PQ envelopes a conic, which has double contact with the given conic on the axis CS. Thus the locus of the pole Z, of the chord PQ, in regard to the original conic, is also a conic. Prove that SZ is perpendicular to the tangent PZ. Infer that the locus of Z is a circle (the auxiliary circle of the conic).

71. The locus of the centres of the series of conics drawn through four given points is the same as the locus of centres of conics drawn through the four points where any fixed conic of the original series meets the asymptotes of any other conic of this series.

72.   The four circles which touch the sides of a triangle ABC have, for their points of contact with BC, CA, AB, respectively, P, Q, R; P_1, Q_1, R_1; P_2, Q_2, R_2; P_3, Q_3, R_3. It is a consequence of Brianchon's theorem that each of the four triads of lines AP, BQ, CR; AP_1, BQ_1, CR_1; AP_2, BQ_2, CR_2; AP_3, BQ_3, CR_3 have a common point. Prove that, also, AP_1, BQ_2, CR_3 meet in a point, as do AP, BQ_3, CR_1. If the sides AB, AC remain the same, but the side BC varies, find the locus of the common point of AP_1, BQ_2, CR_3.

73.   The tangent at a point P of one circle meets another circle in Q and R. If S be one of the limiting points of the two circles, prove that the line SP is one of the bisectors of the angles determined by SQ and SR.

74.   The centroid of four points of a circle is the mid point of the centre of the circle and the centre of the rectangular hyperbola which passes through the four points.

75.   From a variable point lying on a fixed line through the centre of a conic (a diameter of the conic) the perpendicular is drawn to the polar of the point. Shew that the locus of the point where this perpendicular meets the polar is a rectangular hyperbola containing the four foci of the conic, having one asymptote **conjugate to the given** diameter. Prove also that the polar, and the perpendicular thereon from its pole, meet an axis of the conic in points harmonically conjugate in respect of the foci on this axis.

76.   Let ABC, $A'B'C'$ be two triangles inscribed in a conic, and the four intersections $(BC, A'B')$, $(BC, A'C')$, $(B'C', AB)$, $(B'C', AC)$ be respectively denoted by Q, R, Q', R'. Prove that the cross ratios $(Q', R'/B', C')$, $(B, C/Q, R)$ are equal; and, hence, that the sides of the two triangles touch another conic.

77.   Let the line which joins a fixed point O to a variable point P, of a given conic, meet the polar of the point O, in the point T. Let a point Q be taken on OP such that the cross ratio $(Q, P/O, T)$ is constant. Prove that the locus of Q is a conic which touches the given conic at the points where the polar meets the original conic.

78.   Let A, A' be fixed points of a conic, whose centre C is the mid point of A, A'. Lines are drawn from A, A' to a variable point P of the conic. Through a fixed point S on AA' lines are drawn parallel to AP and $A'P$ to meet CP in Q and Q'. Prove that the locus of Q (and of Q') is a conic with the same centre and asymptotes as the given conic, which passes through S.

79.   The points X, Z, T and the lines SX, SZ are given. The point L varies on XT. The lines SL, ZT meet in M, and the lines XM, ZL meet in P. Prove that P describes the conic through Z, X, T which has SX, SZ as tangents at X and Z. Hence also, given a self-polar triangle of a conic, and a point, on each of two sides of this, which lies on the conic, construct the conic. Further, from this, given a self-polar triangle of a conic, and also the polar line of a given point, construct the conic.

80.   Let X, Z be fixed points of a given conic, and let P, Q be variable points of the conic, such that the pencil of lines joining X, Z, P, Q to any point of the conic has a constant cross ratio. Prove that the locus of the common point of the tangent at P, with the line joining Q to the fixed pole of XZ, is a conic touching the given conic at X and Z.

If P, Q be variable points of a given conic, of which S is one focus, such that $_\epsilon S\,(P, Q)$ is constant, prove that the locus of the pole of PQ is another conic, having the same focus S and the same corresponding directrix, as the given conic.

81.   Two conics touch at the points X, Z; the tangent at a variable point P of one of the conics meets the other conic in Q and Q', and meets the common tangents at X and Z respectively in L and M. Prove that the cross ratio $(Q, P/L, M)$ is independent of P.

82. The four arbitrary lines ABC, $B'C'A$, $C'A'B$, $A'B'C$ have B, B' and C, C' as two pairs of complementary intersections; and a conic is such that B, B' are conjugate in respect thereto, as are C, C'. We can prove that the other pair A, A' are conjugate in regard to this conic as follows: let the polar of B, in regard to this conic, which contains B', meet BC in Q; also the polar of C, which passes through C', meet BC in R; let these polars meet in P, and the line PA' meet BC in H. From the involution determined on BC by the pairs of complementary joins of P, B', A', C', it can then be proved that PA' is the polar of A.

83. One conic meets the sides BC, CA, AB of a triangle, respectively, in D, D'; in E, E'; and in F, F'. Another conic meets these sides, respectively, in L, L'; in M, M'; and in N, N'. Prove that the product of the six cross ratios

$$(L, D/B, C), (L', D'/B, C), (M, E/C, A), (M', E'/C, A),$$
$$(N, F/A, B), (N', F'/A, B)$$

is the number 1.

84. Three conics, ω_1, ω_2, ω_3, have two points in common. It is then the case that the complementary common chords, of the three pairs of these conics, meet in a point. We may prove this by considering, on the complementary chord common to ω_2, ω_3, the two involutions determined thereon respectively by the pencil of conics (ω_1, ω_2), and the pencil of conics (ω_1, ω_3).

85. Two conics have three given points in common; D, E are two other points of the first, and D_1, E_1 of the second conic. Construct the remaining common point of the two conics.

86. Two conics have double contact; the other tangents, one to each, are drawn to these conics from a point of one of the common tangents. Prove that the line joining the points of contact of these two tangents passes through the point of contact of the other common tangent.

87. The tangent at a point V of a hyperbola, whose centre is C, meets the asymptotes in P and P'; the line is drawn through V parallel to the asymptote CP, and T is any point of this line. Prove that the line through T perpendicular to TP' is the polar of P in regard to a certain conic, varying as T varies, which is confocal with the hyperbola.

88. Two conics touch at O, and meet again in U and V; a variable line through O meets the conics again in P, Q, respectively, and meets UV in T. Prove that $(P, Q/O, T)$ is constant.

89. Let A, B, C, D, I, J be points of a conic. As the pencils $I(A, B, C, D)$, $J(B, A, D, C)$ are related, the intersections of corresponding rays of the two pencils lie on another conic containing I and J.

If the variable chord OPP' of a fixed circle be drawn through a fixed point O to meet the circle in P and P', prove that the antipoints of P and P' lie on the circle, with centre at O, which is perpendicular to the given circle.

90. The tangents at two fixed points I, J of a conic meet in Y, and O, S are fixed points on a line which passes through Y. Let P be any point of the conic, and the line YP meet the conic again in P', and meet IJ in H. If OP, OP' meet the line SH in Q and Q', prove that Q and Q' describe another conic which touches SI, SJ at I, J, respectively. The tangent at Q of this conic meets the tangent at P of the original conic on the line IJ.

91. Of two conics which have a common focus there is one pair of common chords which meet at the common point of the directrices of the two conics corresponding to the focus.

92. Shew how to construct the four conics, having a given focus, which pass through three given points.

93.   If a line perpendicular to an axis of a conic be called an ordinate, prove that the lines drawn through the vertices A, A' of a hyperbola parallel to one asymptote meet an ordinate, perpendicular to AA', in points K, K' which are conjugate in regard to the hyperbola. Also that the line drawn, through the mid point of K, K', perpendicular to the asymptote, meets the axis AA' on the normals of the curve at the two points where the ordinate meets the curve.

94.   If P, Q be variable corresponding points of two related ranges on a conic, prove that the line PQ envelopes a conic touching the given conic at the two common corresponding points of the related ranges.

95.   A conic touches the lines BC, CA, AB, respectively, at D, E, F; and P is any point of this conic. Prove that the conic through A, which touches PB, PC, respectively, at B and C, touches the line EF at the point where DP meets EF.

96.   The auxiliary circle of a conic is the circle through the points common to the asymptotes and two parallel directrices. The director circle is the circle through the common points of the conic and its directrices.

97.   The poles of the sides BC, CA, AB of a triangle, in regard to a conic, are respectively P, Q, R; and QR meets BC in O. Prove that the pairs of lines AB, PR; PQ, AC; PA, PO meet QR in pairs of points of an involution. Infer that AP, BQ, CR meet in a point.

98.   Points D, E, F, respectively on the sides BC, CA, AB of a triangle, are such that AD, BE, CF meet in a point. A line EU, through E, meets FD in Q, and a line FU, through F, meets ED in R, the points A, Q, R being on a line. Prove that U describes a conic touching BC, CA, AB, respectively, at D, E, F; and that the tangent of this conic at U meets BC on the line QR.

99.   Two conics meet in the four points A, P, Q, R; a line through A meets the conics again in B and C, respectively. Lines BE, CF, through B and C, meet one another in D, and meet the conics again respectively in E and F. Prove that the six points P, Q, R, D, E, F lie on a conic.

100.   The tangents at two points X, Z of a conic meet in Y. A second conic is drawn through Y and Z to touch XZ at Z, and this meets the original conic again in P, Q, R. A third conic is drawn through P, Q, R, X to touch XZ at X. Prove that the polar of Z, in regard to this third conic, is the line XY.

101.   Prove that the two lines drawn from any point of a rectangular hyperbola parallel to the asymptotes are conjugate lines in regard to any conic which has the ends of any diameter of the rectangular hyperbola for foci.

102.   Consider the locus, ϖ, of the poles of a given line, p, in regard to the series of conics drawn through four given points. Any point, C, of the conic ϖ, thus corresponds to that particular conic of the series for which C is the pole of p. Let P be the pole of p in regard to the locus conic ϖ, and an arbitrary line through P meet ϖ in A and A'. Prove that the conics of the series which correspond to the points A and A' meet p in two pairs of points which are harmonic conjugates of one another.

103.   Let a rectangular hyperbola through three points A, B, C meet the circumcircle of ABC again in the point D. If H be the orthocentre of ABC, the points H, D are ends of a diameter of the hyperbola, and the lines AH, AD are therefore parallel to conjugate diameters of the hyperbola. Thus the asymptotes are parallel to the bisectors of the angles determined by AH, AD. Considering two such hyperbolas, meeting the circumcircle in points D, D' which are the ends of a diameter of the circle, prove that the asymptotes of either hyperbola are parallel to the axes of the other.

104. Let A, B, C, D be arbitrary points, and p the polar line of D in regard to the triangle ABC. Prove that the locus of the poles, of the line p, in regard to the series of conics through A, B, C, D, is the conic which touches BC, CA, AB, respectively, at the points where these are met by AD, BD, CD.

105. The locus of the centres of the conics which pass through four points A, B, C, D is a conic which, (a) is a circle when D is the orthocentre of ABC; (b) is a rectangular hyperbola when D is on the circumcircle of ABC; (c) is the conic touching BC, CA, AB at their mid points when D is the centroid of ABC.

106. Two conics Ω, ω meet in A, B, C, D. A line p meets Ω in M, N, and meets ω in I, J, such that $M, N \sim I, J$. The pole of p in regard to Ω is P, and PD meets the conic Ω again in H. Prove that the lines BC, AH meet p in two points Q, R such that $Q, R \sim I, J$.

107. A conic touches the sides BC, CA, AB, of a triangle, respectively, in D, E, F. The lines AO, BO, CO, joining A, B, C to a point O, meet the conic respectively in X, X'; in Y, Y'; and in Z, Z'. Prove that, with properly chosen notation, the lines DX, EY, FZ meet in a point, as do DX, EY', FZ'; and DX', EY, FZ'; and DX', EY', FZ.

108. The tangents of a conic at the angular points, A, B, C, of an inscribed triangle, meet the sides BC, CA, AB, respectively, in points L, M, N which lie on a line. The points in which this line meets the conic are called the Hessian points of A, B, C, regarded as three points of the range on the conic. (See Ex. 9.)

Let A, B, C and P, Q, R be any two sets of three points on a conic. We can determine two related ranges on the conic by the condition that the points A, B, C, of one of these, correspond, respectively, to the points P, Q, R of the other. Each of the three pairs of cross-joins of A, B, C and P, Q, R, such as AQ, BP, determines a point (of intersection), and, by Pascal's theorem, these points lie on a line, which meets the conic in the two common corresponding points of the two related ranges.

Thus the Hessian points of A, B, C are the common corresponding points of the two related ranges, on the conic, whose correspondence is determined by the fact that A, B, C correspond respectively to B, C, A. It follows, if these points be denoted by I and J, that

$$(B, C/A, I) = (C, A/B, I) = (A, B/C, I).$$

As the product of these three cross ratios is -1, each of them is equal to the negative of an imaginary cube root of unity, say equal to $-\omega$. Wherefore each of $(B, C/A, J)$, $(C, A/B, J)$, $(A, B/C, J)$ is equal to $-\omega^2$. Hence $(B, C/I, J)$, which is equal to the product of $(B, C/I, A)$ and $(B, C/A, J)$, is equal to ω; as, likewise, are $(C, A/I, J)$ and $(A, B/I, J)$. From this we see that, if the parameter of the points of the conic be chosen so that, at I, J, the values are 0 and ∞, respectively, then the values of the parameter at A, B, C will be $\lambda, \lambda\omega^2, \lambda\omega$, respectively, where λ is an arbitrary number; there is in fact an infinite series of sets of three points on the conic for which the Hessian points are the same.

From the Hessian pair of three points of a conic, we may define the Hessian pair of three points, P, Q, R, on a line. And we may construct these: Draw three arbitrary lines, PBC, QCA, RAB, through P, Q, R, which meet in pairs in the angular points A, B, C of a triangle. Let O be the pole of the line PQR in regard to this triangle, and AO, BO, CO meet BC, CA, AB, respectively, in D, E, F. A conic touches BC, CA, AB in D, E, F, respectively. The points in which this conic meets the line PQR are the Hessian points of P, Q, R.

109.   If four arbitrary points A, B, C, D be given on a conic ω, we can determine geometrically another set of four points on ω, forming a set called the *Hessian set* of A, B, C, D on ω. (Cf. also Ex. 250.)

For, considering the vertex triangle of the quadrangle A, B, C, D, each vertex, such as P, is the common point of a pair of complementary joins of A, B, C, D; and these joins meet the opposite side, QR, of this triangle, in two points, say Q', R'. Let X, X' be the foci of the involution determined by the two pairs Q, R and Q', R'; let Y, Y' and Z, Z' be the pairs of foci correspondingly defined on the other two sides RP, PQ of the vertex triangle. The six points X, X', Y, Y', Z, Z' lie upon a conic, say ω_0, as may be readily proved. This depends only on A, B, C, D. The common points of ω_0 and ω are the four points, A', B', C', D', which are the Hessian points of A, B, C, D on ω.

It may be verified that, if the parameter θ by which the points of the conic ω are specified is so chosen that A, B, C, D are defined by the four roots of the equation $\theta^4 + 2\mu\theta^2 + 1 = 0$, where μ is a number, then the four points A', B', C', D' are given by the equation $\theta^4 + (3\mu^{-1} - \mu)\,\theta^2 + 1 = 0$.

Another derivation of A', B', C', D' may be given. There is a unique conic through A, B, C, D which is outpolar to ω; say Ω. The points A', B', C', D' are the points of contact with ω of the four common tangents of ω and Ω. It can be shewn that ω_0 is inpolar to ω and to Ω, and therefore inpolar to all conics through A, B, C, D, and has the vertex triangle of these points as self-polar triangle. This determines ω_0 in another way.

110.   The point P', conjugate to P in regard to a quadrangle A, B, C, D, is defined as the point conjugate to P in regard to all conics through A, B, C, D. Let p be an arbitrary line. As P moves on p, the point P' describes a conic. This is the locus of poles of the line p in regard to all conics through A, B, C, D.

111.   Take four definite tangents of a given conic, and a pair of complementary intersections of these four lines. We may consider the conjugate line locus of the given conic when these two intersections are taken as centres; this conic contains the four points of contact of the four selected tangents with the given conic. There are thus three such conics, all containing these four points of contact. The polars of an arbitrary point P, in regard to these three conics, thus meet in a point, P', which is also on the polar of P in regard to the original conic.

Now consider the system of conics touching the four selected tangents of the original conic. The envelope of the polars of P, in regard to this system of conics, is another conic. Prove that the point of contact of this conic, with the polar of P taken in regard to a particular conic of the system, is the point P', derived from P, as above, by means of this particular conic.

112.   A case of Ex. 111. The envelope of the polars of a point P, in regard to the conics of a confocal system, is a parabola. This parabola touches the polar of P, taken in regard to a particular confocal, at a point which lies on the polar of P in regard to the director circle of this particular confocal.

113.   Let QVQ' be one of a series of parallel chords of a conic, whose centre is C, and let the diameter CV, which contains the mid point, V, of Q, Q' meet this conic in S and S'. Prove that the circle having QQ' as diameter envelopes a conic, with foci at S and S', whose auxiliary circle is the director circle of the original conic. Consider the case when the original conic is a rectangular hyperbola.

114.   Let an arbitrary line meet the sides BC, CA, AB, of a triangle inscribed in a conic, respectively in L, M, N. On this line, take the points L', M', N', respectively conjugate to L, M, N, in regard to the conic. Prove that the lines joining L', M', N' to any point of the conic meet BC, CA, AB, respectively, in points lying on another line.

115.   Let S be a focus of a conic which touches the sides of a triangle ABC, and lines be drawn through S respectively perpendicular to SA, SB, SC. Let these lines meet any tangent of the conic in P, Q, R, respectively. Prove that the lines AP, BQ, CR meet in a point.

116.   Let a circle have double contact with a hyperbola, with chord of contact parallel to the axis joining the real foci. Prove that the line joining one of these foci to the centre of the circle meets the circle in points lying on the tangents of the hyperbola at its vertices.

117.   As in Ex. 116, let a circle have double contact with a hyperbola, with chord of contact parallel to the axis joining the real foci S, H; and A, A' be the vertices of the hyperbola, lying on this axis SH. Let the circle meet one of the asymptotes in P and P'. Prove that the tangents of the circle at P and P' each contain one of the foci S, H; and that the mid point of P, P' is on the chord of contact. Prove also that the length of PP' is equal to the distance AA'.

If a series of such circles be taken, prove that the pole of PP', in regard to the circle, describes a rectangular hyperbola, containing the four foci of the original hyperbola, and having one asymptote coincident with the asymptote PP' of the original hyperbola. Prove also that this rectangular hyperbola is the conjugate line locus of the original hyperbola, when the centres are at the points where the rectangular hyperbola meets the Absolute line.

118.   Let A, B, O, O' be four points such that the lines OA, OB are perpendicular, and the lines $O'A$, $O'B$ are also perpendicular. Find the focus of the parabola which touches these four lines, and the points of contact therewith.

119.   Let A, B, C, D be four points on a conic; consider the four lines AB, BC, CD, DA. Take the two points where the tangent of the conic at A meets the lines CB, CD which do not contain A; and determine two points similarly from each of B, C, D. Prove that the eight points so obtained lie on a conic.

[Six of the eight points are obtainable from two of them by harmonic inversion in regard to the angular points of the vertex triangle of the quadrangle $ABCD$, each taken with the opposite side; and a conic can be described through two given points to have a given self-polar triangle. The theorem is given by Moebius (*Baryc. Cal.* § 281), as an application of Pascal's theorem, which he quotes as due to Robert Simson, *Conic Sections*, v, 47.]

120.   Let PQ be a chord of one particular conic, ω, of a confocal system; the locus of the poles of PQ in regard to the conics of the system is a line. Prove that this line meets the normal at P, of the conic ω, in the point which is the pole of PQ in regard to the other confocal which passes through P (and similarly for Q).

121.   If two conics touch the sides of a triangle, their common self-polar triangle is circumscribed to this triangle, and also in perspective with it, the axis of perspective being the fourth common tangent of the two conics.

122.   Let ABC be a self-polar triangle of a conic, and O an arbitrary point; and let AO, BO, CO meet BC, CA, AB, respectively, in D, E, F. Let the conic meet EF, FD, DE, respectively, in L, L'; in M, M'; and in N, N'. Prove, with proper notation, that DL, EM, FN meet in a point, and DL', EM', FN' meet in a point.

123.   Let I, J be the common corresponding points of two related ranges of points on a conic. If I, J be taken as Absolute points, so that the conic is a circle, with centre, say, at O, prove that the relation between a point, P, of one range, and the corresponding point, P', of the other range, is that

$\epsilon O\,(P',\,P)$, which means $O\,(P',\,P/I,\,J)$, is independent of P. Or, in metrical terms, the relation between the two ranges is, that one is obtained from the other, by a rotation about O, through a certain angle.

Let P, Q, R be three points of one range, and P', Q', R' the respectively corresponding points of the other range; let A, B, C be the poles, in regard to the circle, respectively of the chords PP', QQ', RR'. Prove that the three points of intersection $(QR, Q'R')$, $(RP, R'P')$, $(PQ, P'Q')$ are the respective mid points of B, C; of C, A; and of A, B, in regard to the Absolute line IJ. This Absolute line is the Pascal line of the hexagon P, Q', R, P', Q, R'.

124. Points D, E, F, lying respectively on the sides BC, CA, AB of a triangle, are such that AD, BE, CF meet in a point; the lines EF, FD, DE meet BC, CA, AB, respectively, in L, M, N, which lie in a line. Let O be any point; on OA, OB, OC respectively, let points X, Y, Z be taken so that YZ, ZX, XY, respectively, contain L, M, N. Prove that DX, EY, FZ meet in a point.

125. Let P, Q, R, T and P', Q', R', T' be two sets of four points lying on a conic. Denote the respective poles of PP', QQ', RR', TT' by A, B, C, O. Also, denote the six intersections $(QR, Q'R')$, $(RP, R'P')$, $(PQ, P'Q')$, $(TP, T'P')$, $(TQ, T'Q')$, $(TR, T'R')$, respectively, by D, E, F, X, Y, Z. Prove that these six points lie, respectively, on BC, CA, AB, OA, OB, OC. Prove also that YZ, EF, BC meet in a point, say L; while ZX, FD, CA meet in a point, say M; and XY, DE, AB meet in a point, say N; and that the points L, M, N lie in a line. Hence it follows, by Ex. 124, that the lines DX, EY, FZ meet in a point. The line LMN is the Pascal line of the hexagon P, Q', R, P', Q, R'.

A particular corollary is that, if A, B, C be the respective poles of PP', QQ', RR', then the lines, joining A, B, C, respectively, to the intersections $(QR, Q'R')$, $(RP, R'P')$, $(PQ, P'Q')$, meet in a point. When the range P', Q', R', T' is related to the range P, Q, R, T, the points D, E, F, X, Y, Z lie on a conic.

126. Another statement of the result of Ex. 125 is as follows: let P, Q, R, T and P', Q', R', T' be two sets of four points on a conic. Associate any pair of complementary joins, of P, Q, R, T, with the corresponding pair of complementary joins of P', Q', R', T', in this way—consider the involution on the conic determined by the two pairs of points Q, R and Q', R'; and also the involution determined by the two pairs P, T and P', T'; and consider the two points of the conic forming the common pair of these two involutions (these are the points of the conic lying on the line joining the two points $(QR, Q'R')$, $(PT, P'T')$). Adopt the similar procedure in the two other possible ways. Then, the three common pairs so found, on the conic, belong to an involution (so that the three joining lines, such as the one described, meet in a point).

127. Let P, P'; Q, Q'; R, R'; T, T' be four pairs of points on a conic, and A, B, C, O the respective poles of the lines PP', QQ', RR', TT'. Consider the Pascal lines of the four hexagons, P, Q', R, P', Q, R'; T, Q', R, T', Q, R'; T, R', P, T', R, P'; T, P', Q, T', P, Q'. Prove that the common point of any two of these four lines lies on the join of two of the four points A, B, C, O.

When the two ranges P, Q, R, T and P', Q', R', T', on the conic, are related, the four Pascal lines coincide.

128. It has been proved that the four circles, each circumscribed to a triangle formed by three of four arbitrary lines, meet in a point. We may prove in more general terms that, if four lines and two points be given, the four conics through these two points, each circumscribing a triangle formed by three of the four lines, have all one point in common. (Chap. xiv (9).)

Denote the two given points by I, J, and the given lines by a, b, c, d. Let E, F be the two respective intersections (b, c) and (a, d); and A, B, C, D be the other intersections of pairs of the lines, so chosen that the lines CB, BA, CD, DA are, respectively, a, b, c, d. The conics (A, D, E, I, J) and (B, C, E, I, J) have, besides E, I, J, a further common point, say K; and IJK is a triangle inscribed in both these conics. Denote the sides JK, KI, IJ of this triangle respectively by l, m, n. The triangles ADE, IJK are both inscribed to the conic (A, D, E, I, J); thus the sides of these triangles, which are the lines b, c, d, l, m, n, touch a conic. In a similar way the lines b, c, a, l, m, n touch a conic. This proves that the conic defined as touching b, c, l, m, n touches also d and a; it is thus a conic touching a, b, c, d, l, m, n. But then, since a conic touches the lines a, d, c, l, m, n, which form two triangles CDF and IJK, it follows that K lies on the conic (C, D, F, I, J). Similarly, because a conic touches a, d, b, l, m, n, we can infer that K lies on the conic (A, B, F, I, J). We have thus proved that K is common to the four conics through I, J which circumscribe the triangles formed by a, b, c, d. This highly ingenious proof is due to Steiner; and it is interesting to remember that he would probably regard the theorem as more general than that for the four circles.

129. Let ϑ, ϕ, ψ be three arbitrary conics, and P an arbitrary point. Let the conic through P and the four common points of ϕ, ψ be denoted by (ϕ, ψ, P); and so on. Prove that the three conics (ϕ, ψ, P), (ψ, ϑ, P), (ϑ, ϕ, P) have four points in common.

130. Let A, B, C, D be four points, and X, Y, Z denote, respectively, the points of intersection (AB, CD), (AD, BC), (AC, BD). Let H be any point on YZ, and HC, HD meet XZ, respectively, in I and J. Prove that IB, JA meet on YZ, say in S. Prove also that a conic touches IH, HJ, JS, SI, respectively, at C, D, A, B. Further, P being an arbitrary point, let the points (PS, AB) and (PH, CD) be respectively L and M. Prove that the six points X, Y, Z, P, L, M lie on a conic. Prove also that the points (SI, JH), (IH, JS) lie on XY.

131. If one focus of a conic inscribed in a triangle be at the circumcentre of the triangle, the conjugate focus is at the orthocentre.

132. Let A, B, C, O, O' be arbitrary points; let AO, BO, CO meet BC, CA, AB, respectively, in L, M, N, and AO', BO', CO' meet BC, CA, AB, respectively, in L', M', N'. Then the six points L, L', M, M', N, N' lie on a conic, say ω. Prove that the line which is the polar of O' in regard to the triangle ABC is met by any conic through A, B, C, O in two points which are conjugate in regard to the conic ω.

133. Prove that the polar reciprocal of a parabola in regard to a circle whose centre, O, is on the directrix (the locus of the pole, in regard to the circle, of a variable tangent of the parabola) is a rectangular hyperbola, shewing that this hyperbola touches the directrix at O, has asymptotes parallel to the tangents of the parabola drawn from O, and has, for centre, the pole, in regard to the circle, of the polar line of O in regard to the parabola.

134. Any two parabolas whose axes are perpendicular meet in four points lying on a circle.

135. Given four lines, and two points, there are two conics touching the lines for each of which the points are conjugate to one another.

136. With a given centre, a single conic exists which either, (*a*) touches three given lines; or (*b*) has a given self-polar triangle; or (*c*) contains three given points.

137.   The centres of rectangular hyperbolas which touch the sides of a triangle lie on the circle in regard to which the triangle is self-polar. The centres of rectangular hyperbolas for which a given triangle is self-polar lie on the circumcircle of the triangle. And the centres of rectangular hyperbolas which circumscribe a given triangle lie on the nine-point circle of the triangle. These three circles, for the same triangle, have two points in common.

138.   The tangent being drawn at a given point of a given conic, which is not a parabola, and a point H being given, not lying on the conic, find two other tangents of the conic such that H is the orthocentre of the triangle formed by the three tangents. Of three tangents of a parabola, which form a triangle whose orthocentre is a given point of the directrix, two may be taken arbitrarily (Ex. 35).

139.   A rectangular hyperbola, ρ, whose asymptotes are parallel to the axes of a given conic ω, meets ω in A, B, C, D. If K be the centre of ω, prove that the circle ABC contains the point D' on ω which is such that K is the mid point of D, D'. If the hyperbola ρ contains the centre K of ω, then the normals at A, B, C, D, of the conic ω, meet in a point.

140.   Two conics have parallel asymptotes; and we consider the common chord of the conics which is complementary to the Absolute line. Shew that two circles are possible, meeting on this chord, each of which has double contact with one of the conics, the chords of contact being the same line for both circles. Shew that this is possible with the common chord of contact passing through a given point.

141.   Prove that the so-called self-polar circles, of the four triangles formed by the threes of four arbitrary lines, have two points in common; and that these circles are perpendicular to the director circles of the conics which touch the four lines.

142.   On a given conic ω, passing through four points A, B, C, O, there is given a definite involution of pairs of points, of which P, P' is an arbitrary pair. Prove that the conic touching BC, CA, AB, OP, OP' has a fixed tangent, which meets ω in a pair of the involution. If A', B', C', on the conic, be such that A, A'; B, B'; C, C' are pairs of the involution, and OA', OB', OC' meet BC, CA, AB, respectively, in L, M, N, prove that L, M, N lie on the fixed tangent. Also, that every conic through O, A, M, N meets ω in a pair of the involution; as does every conic through O, B, N, L and every conic through O, C, L, M.

143.   The conics having double contact with two given conics consist of three families. The locus of the poles of an arbitrary line, in regard to the conics of one family, is a conic. Hence, two families of conics exist, passing through two given points, having double contact with a given conic. In particular, considering one of the two families of conics, passing through two points I, J, touching two given lines OA, OB, prove that the chord of contact with OA, OB passes through a fixed point H, of the line IJ, which is such that OH is the locus of poles of the line IJ in regard to the other family. Interchanging the two families, let K be the other point found on IJ; prove that $H, K \sim I, J$.

144.   Prove that, among the conics drawn through three (real) points to touch a line, there are four parabolas, and two rectangular hyperbolas. Also, that the envelope of the polars of a given point, in regard to all the conics, is a conic.

145.   If a conic, ω, meet the sides BC, CA, AB of a triangle, respectively, in D, D'; E, E'; F, F', the six points formed by the foci of the three involutions $(B, C; D, D')$, $(C, A; E, E')$, $(A, B; F, F')$ lie in threes upon four

lines. If another conic be taken, passing through A, B, C, a complementary pair of common chords, of this conic with the conic ω, meets BC in a pair of the involution $(B, C; D, D')$.

146. From Ex. 145, if L, M be two points of the conic ω, the conic A, B, C, L, M meets ω in two other points whose joining line meets BC in a point determinable by the involution $(B, C; D, D')$. Hence prove that four conics can be drawn through A, B, C to have double contact with the conic ω, the chords of contact being the four lines each containing a focus of all the involutions $(B, C; D, D')$, $(C, A; E, E')$, $(A, B; F, F')$.

Dually, four conics can be drawn to touch BC, CA, AB and have double contact with a given conic. If this given conic degenerates into two pencils, it follows that four conics can be drawn to touch BC, CA, AB and pass through two arbitrary points. In particular, construct the radical axis of the circumcircle ABC and any other given circle.

147. Let a conic ω meet the lines CA, AB, respectively, in E, E' and F, F'; prove that the tangents drawn from A, to one of the two conics through E, E', F, F' which touch BC, are the tangents at A of two of the four conics which can be drawn through A, B, C, to have double contact with ω.

Deduce that the points of contact, with BC, of the four circles which touch BC, CA, AB, lie in pairs upon the two conics through A which have B, C as a complementary pair of foci. Also, that the asymptotes of these two conics are parallel to the axes of the parabolas, having A as focus, which pass through B and C.

Construct the parabolas through two given points which have a given focus.

148. Shew that, by means of an algebraic transformation of the form $x' = x^{-1}$, $y' = y^{-1}$, $z' = z^{-1}$, the conics through three given points are changed into lines. Shew that the theorem, that four conics can be drawn through three points to touch two lines, is deducible, by such a transformation, from the fact that two conics have four common tangents.

149. A conic, ω, meets the sides BC, CA, AB of a triangle in D, D'; E, E'; F, F', respectively. Prove that another conic, ω', exists meeting BC, CA, AB, respectively, in $D_1 = (B, C)/D$, $D_1' = (B, C)/D'$; $E_1 = (C, A)/E$, $E_1' = (C, A)/E'$; $F_1 = (A, B)/F$, $F_1' = (A, B)/F'$. We may call ω, ω' harmonically conjugate in regard to the triangle ABC.

Prove that a conic can be drawn through the common points of ω and ω' to contain A, B, C; and another conic through these common points to have ABC as a self-polar triangle.

Now suppose four points A, B, C, D, and a line l, to be given. Let l' be the polar of the point D in regard to the *triangle ABC*. Prove that the harmonic conjugate, ω', of the degenerate conic $\omega = (l, l')$, is the locus of the poles of l in regard to all conics through A, B, C, D. Thus ω' meets the lines l, l' in four points which lie on a conic through A, B, C; and lie on a conic for which ABC is a self-polar triangle. Prove also that ω' meets l in the foci of the involution determined thereon by all conics through A, B, C, D.

150. If the lines drawn from three points A, B, C, respectively perpendicular to the joins, $B'C'$, $C'A'$, $A'B'$, of three other points, meet in a point, then the lines drawn from A', B', C', respectively perpendicular to BC, CA, AB, also meet in a point (Steiner, *Werke*, I, p. 157).

More generally, if the lines drawn from A, B, C, respectively conjugate to $B'C'$, $C'A'$, $A'B'$, in regard to any conic, meet in a point, then the lines drawn from A', B', C', respectively conjugate to BC, CA, AB, in regard to this conic, also meet in a point.

151. A line meets the sides BC, CA, AB, $B'C'$, $C'A'$, $A'B'$ of two triangles respectively in L, M, N, L', M', N'. Prove that, if $A'L$, $B'M$, $C'N$ meet in a point, then also AL', BM', CN' meet in a point.

152. A triangle $A'B'C'$ has its angular points A', B', C', respectively, on the sides BC, CA, AB of another triangle, and the lines AA', BB', CC' meet in a point. The triangle $A''B''C''$ has its angular points respectively on the sides $B'C'$, $C'A'$, $A'B'$, and $A'A''$, $B'B''$, $C'C''$ meet in a point. Prove that AA'', BB'', CC'' meet in a point.

153. Consider the series of conics through four points A, B, C, D. Shew that these conics determine, on a line drawn through the point D, a range which is related to the pencil formed by the tangents of the conics at the point D. If a line be drawn through C, shew that the range determined thereon by the conics is in perspective with the range which the conics determine on the line through D. Thus, if two lines be drawn through D, the series of conics determines related ranges on these two lines; the joins of corresponding points of these ranges thus envelope a conic, which touches the two lines drawn through D; and also touches the lines BC, CA, AB. For example, if two lines be drawn through one of the common points of a series of coaxal circles, and these be met by a variable circle of the series respectively in P and P', then PP' envelopes a parabola whose focus is at the other common point of the circles.

154. If three circles be respectively denoted by 1, 2, 3, prove that the six points constituted by the two centres of similitude of the circles 1, 2, the two centres of similitude of the circles 1, 3, and the limiting points of the circles 2, 3, lie upon a conic.

155. Let A, B, C, O be four points of a conic; let a line meet BC, CA, AB respectively in L, M, N; and OL, OM, ON meet the conic again in P, Q, R, respectively. Shew that AP, BQ, CR meet in a point, lying on the line LMN.

156. Let two circles be such that the line drawn through one limiting point, L, of the circles, perpendicular to the line of centres, meets the circles, respectively, in two harmonically conjugate pairs of points. Prove that L is then the mid point of the centres of the two circles. Shew that, also, on the line drawn through L, there are two points such that any line through either meets the circles, respectively, in two harmonically conjugate pairs of points. Further, the points of contact of the four common tangents of the two circles lie (in fours) on two lines. Determine how many of the points referred to can be real.

157. Lines through a point P of a hyperbola whose centre is C, respectively parallel to the asymptotes x, y, meet a diameter through C which meets the curve in A and B, respectively, in H and K. Prove that the tangent at P meets this diameter in the point T given by $T=(H, K)/C$. And, if BM, parallel to x, meets PK in M, while AN, parallel to y, meets PH in N, prove that M, T, N lie in a line.

158. Let ABC, $A'B'C'$ be two triangles inscribed in a conic; let $A'C'$, $A'B'$, respectively, meet BC in X and Y; let AX, AY meet the conic again, respectively, in L and M. Prove that the lines $B'L$, $C'M$ meet on BC.

159. A radius, HL, through the centre H of a subdirector circle of an ellipse, meets the ellipse in two points. Prove that one of these, say P, is such that the tangents of the circle and ellipse, at L and P, respectively, meet on the directrix of the ellipse corresponding to the focus S which is complementary to H, say in T; and the polars of T, in regard to the two curves, meet on this directrix. [For the definition of the subdirector circle, see Chap. XIII (2).]

160. If a parabola contains the mid points of the sides BC, CA, AB, of a triangle, and meets these sides, respectively, also in D, E, F, prove that AD, BE, CF are parallel to the axis of the parabola. State the theorem in general terms.

161. Three conics α, β, γ have each double contact with a conic ω, along chords a, b, c, respectively. Through the common point of the chords b, c there pass two complementary common chords of the conics β, γ. Taking similarly the three pairs β, γ; γ, α; α, β, prove that the six chords so obtained meet in threes in four points. Deduce, as a dual theorem, that the centres of similitude, of the pairs of three circles, lie in threes on four lines. Consider also the particular case when γ degenerates into a pair of lines passing through the common point of a pair of common tangents of the conics α and β.

162. Two triangles ABC, $A'B'C'$ are inscribed in a conic. Let BC, $B'C'$ meet the line AA', respectively, in L and L', and D, D' be the foci of the involution determined by the two pairs A, L and A', L'. Make a similar construction for E, E' on the line BB', and for F, F' on the line CC'. Prove that D, D', E, E', F, F' lie on a conic.

163. If four arbitrary points be taken, prove that the nine-point circles, of the four triangles formed by threes of the points, have a point in common.

164. Through two points P, Q, of a parabola, chords PL, QM, of the curve, are drawn, respectively parallel to the tangents at Q and P. Prove that the chord LM is parallel to PQ.

165. Let A, B, C, D be given points, and l a given line. Let l meet AB in P, and Q be taken on AB such that $Q=(A, B)/P$. Construct a conic through A, B, C, D, such that Q is the pole of l, in regard to this conic.

166. Let conics Ω, ω, which have double contact, be met by a line respectively in I, J and in U, V. Let the tangents from I, to ω, touch this in A, A'; and the tangents from J, to ω, touch this in B, B'. Let S, H be the respective intersections (JB, IA'), (JB', IA). Let the tangents to ω at U, V meet Ω, respectively, in P, P' and in Q, Q'. Prove, with proper choice of notations, that the tangents of Ω at P', Q meet in S, and the tangents at P, Q' meet in H.

167. Let ABC and SIJ be two circumscribing triangles of a conic. Let IJ be met by AC, AB, SB, SC, respectively, in U, V, G, H. Prove that I, J; U, G and V, H are three pairs in involution. Infer that

$$S\,(I, J/B, C) = A\,(I, J/B, C),$$

and hence that A, B, C, S, I, J lie on a conic.

168. If ABC, SIJ be two triangles inscribed in a conic, prove that the general tangent of the conic which touches the sides of these triangles may be constructed thus: let BC, CA, AB meet IJ respectively in L, M, N; and L', M', N' be the points on IJ given by $L'=(I, J)/L$, $M'=(I, J)/M$, $N'=(I, J)/N$. Let SL', SM', SN' meet BC, CA, AB, respectively, in D, E, F. Prove that these points D, E, F lie in a line. Let K be an arbitrary point of this line, and SK meet IJ in K'; on IJ take the point R given by $R=(I, J)/K'$. Then KR is the general tangent of the inscribed conic.

169. The tangents at the points D, U of one conic Ω touch another conic, Ω', at the points D', U', respectively, and the lines DD', UU' meet in O. The line DU' meets the conic Ω' again in K', and the line $D'U$ meets the conic Ω again in K. Prove that the tangents of the two conics, at K, K' respectively, meet on the common tangent DD', in the point $T=(D, D')/O$.

170. The inscribed circle of a triangle ABC touches BC, CA, AB, respectively, in D, E, F. The mid points of B, C; of C, A; and of A, B are respectively D', E', F'. The lines joining A, B, C, respectively, to the centre of the inscribed circle, meet BC, CA, AB, respectively, in L, M, N. Prove that the points $L'=(D, D')/L$, $M'=(E, E')/M$, $N'=(F, F')/N$ are in line.

And that this line is the tangent of the inscribed circle at the point where it is touched by the nine-point circle of the triangle. (Cf. Ex. 169.) Consider the cases of all the four circles which touch the sides of the triangle.

171.   The Apollonius rectangular hyperbola, with respect to any conic, of a point P which lies on the conic, meets the tangent at P not only at P, where the normal of the conic touches the hyperbola, but also in the pole, in regard to the conic, of the normal at P. Thus also, the points where the tangent of the conic at P meets a conjugate pair of directrices (which, as may be proved, are harmonic conjugates in regard to P and the pole of the normal at P) are conjugate points in regard to the Apollonius hyperbola.

172.   Prove that the harmonic loci of the pairs of conics formed by any conic and a confocal conic, all confocals being taken in turn, constitute a series of conics with four common points. Prove also that the foci of the involution, which the conics of this series determine on the tangent at a point P of the conic, lie on the Apollonius hyperbola of P in regard to the conic.

173.   Prove that the parabola which is the envelope of the polars, of a point P, in regard to the conics of a confocal system, touches the normals at P, of the two confocals which pass through P, at the centres of curvature, at this point, of these two confocals. The directrix of this parabola is the line joining P to the centre of the confocals; its focus is the point P' which is conjugate to P in regard to all conics through the four foci. The polar reciprocal of the parabola, in regard to any one of the confocals, is the Apollonius hyperbola of P for this confocal.

If the four normals be drawn from P to any one of the confocals, the parabola is that which touches the tangents of the confocal at the feet of these normals. Also, as P, P' are in symmetrical relation, P is the focus of the parabola which arises from P'.

174.   Points C, A, B, P are taken in order upon a circle; the intersections (BC, AP), (CA, BP), (AB, CP) are respectively D, E, F; the lengths of BC, CA, AB, AP, BP, CP are respectively a, b, c, p, q, r. Prove that the lengths of EC, EA are in the ratio of ar to pc, and the lengths of AD, DP are in the ratio of bc to qr. Also that the lengths of EC, EA are respectively

$$r\,(br + cq)/(c^2 - r^2), \quad c\,(bc + qr)/(c^2 - r^2),$$

and the square of the length of AD is

$$b^2 c^2\,(bq + cr)/(br + cq)\,(bc + qr).$$

Find the distances of every two points of the figure in terms of b, c, q, r. And prove that the square of the length of EF is equal to the sum of the squares of the lengths of the tangents drawn from E and F to the circle, with similar results for DE and DF. (Cf. Ex. 5, Note 4, below.)

175.   Let ABC be any triangle, and G, K be any two points. Let the polar line of G, in regard to the triangle, meet BC, CA, AB respectively in L, M, N. Let KL meet AC, AB respectively in E and F'; let KM meet BA, BC respectively in F and D'; and let KN meet CB, CA respectively in D and E'. Then, by Pascal's theorem, the points E, F', F, D', D, E' lie on a conic, say ω, the line LMN being the Pascal line of this hexagon. Let the conic ω meet the line LMN in U and V. Denote by Ω the conic through A, B, C, U, V. Prove that the lines joining A, B, C respectively to the poles of BC, CA, AB, taken in regard to the conic Ω, pass through the point K. Prove also that the lines AU, AV are a pair of the pencil in involution determined by the two pairs (AB, AC) and (AG, AK); with a like result for the pairs BU, BV and CU, CV. Further, prove that the common chord of the conics ω, Ω, which is complementary to UV, is the polar line of K in regard to the conic ω, and is the Pascal line of the inscribed hexagon

E, E', F, F', D, D'. Also, that this chord meets LMN, which is the polar line of G in regard to the triangle ABC, on the line which is the polar line of K in regard to this triangle.

When U, V are the Absolute points, Ω is the circumcircle of the triangle ABC, and ω is the so-called *Lemoine circle*.

176. The points A, B, C, O lie on a conic. The tangent of the conic at O meets BC, CA, AB respectively in L, M, N. The lines AL, BM, CN meet the conic again respectively in P, Q, R. Let U, V, W denote the respective intersections (BM, CN); (CN, AL); (AL, BM). Prove that O, U, P lie in a line; as do O, V, Q, and also O, W, R.

177. Given an arbitrary conic, and three arbitrary points S, S', R, denote the polar lines of S, S', R, in regard to the conic, respectively by s, s', r; let the tangents, p, q, drawn from R, touch the conic in P, Q, lying on r. Let the tangent p meet the polars s, s' respectively in T and T', and the tangent q meet s, s' respectively in V and V'. Let the line RS meet the polars r, s respectively in L and X; and the line RS' meet the polars r, s' respectively in L' and X'. On the lines p, q, respectively, take the points P', Q', given by $P' = (T, T')/P$ and $Q' = (V, V')/Q$. Prove that the four lines XL', $X'L$, PQ', $P'Q$ meet in a point.

178. For any conic, consider a focus S, and the corresponding directrix; let P be any point of the conic. Let the normal of the conic at P meet the directrix in N, and meet the curve again in Q, and let R be the pole of PQ. Let PR meet the directrix in T, and N' be taken on the directrix so that T is the mid point of N, N'. Prove that the common point, W, of NS and $N'R$, lies on the normal of the conic at the point Q.

If the conic is a parabola, RN' and PQ are parallel, and RN, PN' are parallel to WQ.

179. Prove that the Apollonius rectangular hyperbola, of a point O, in regard to a conic, which contains the feet of the normals drawn from O to the conic, meets the polar line, p, of O, in regard to the conic, in two points P, Q, which are such that OP, OQ are perpendicular, and OPQ is a self-polar triangle of the conic (so that the Apollonius hyperbola, which passes through O, is outpolar to the conic). Also that the tangent of the hyperbola at O is the perpendicular from O to the polar line p.

Prove also that the Apollonius hyperbola is the locus of the poles of the line p in regard to all conics through the four points in which the given conic meets its directrices. Thus we can specify eleven points which lie on the hyperbola: (a) three points consisting of the centre of the conic and the points where the axes of the conic meet the Absolute line; (b) two points consisting of the poles, in regard to the given conic, of the tangents at O to the two confocals which pass through O; (c) four points consisting of the points where the lines joining O to the four foci of the given conic each meets the corresponding directrix; and, finally, (d), if O' be the point conjugate to O in regard to all conics through the four foci, two points which are the poles, in regard to the original conic, of the lines joining O' to the Absolute points.

The polar reciprocal of the Apollonius hyperbola, in regard to the original conic, is the parabola, which is the envelope of the polars of O in regard to all conics confocal with the original conic. This parabola is also the harmonic envelope for the two conics which consist of the original conic, and the Apollonius hyperbola considered for this conic.

180. Let four chords be drawn from an arbitrary point O to meet a conic, in A, A'; B, B'; C, C' and D, D', respectively. Prove that the two conics O, A, B, C, D and O, A', B', C', D' touch one another at O (however the notation be chosen).

181. Prove that conics can be described to touch two given lines, l, m, for which both the given points, A, A', are conjugate, and also the given points, B, B', are conjugate. Prove that the line joining the points of contact of such a conic with the lines l, m envelopes another conic. This may be defined as the conic which touches the four cross joins, AB, $A'B'$, AB', $A'B$, of the given points, and has l, m for a pair of conjugate lines.

Deduce that four conics exist with two given tangents which have three given pairs of mutually conjugate points. The given tangents are conjugate lines in regard to all the conics which touch the chords of contact of these four conics with these tangents.

182. In general, four conics exist passing through three given points and touching two given lines. Prove, however, that, if O, A, B be three given points on a given conic, Ω, and D be a point not lying on Ω, there is a unique conic, ω, touching DA, DB, which has three coincident intersections with Ω at the point O. Let D' be the point, on the line joining D to the pole of AB, taken in regard to Ω, which is conjugate to D. Prove that the conic ω', touching $D'A$, $D'B$, which has three coincident intersections with Ω, also at the point O, has, for its chord of contact with $D'A$, $D'B$, the same line as the chord of contact of ω with DA, DB.

183. The line joining two fixed points, U, V, meets the sides AB, AC, BC, of a fixed triangle, respectively, in B', C', O. A variable point P, on BC, is joined to U by a line which meets AB, AC in M and M', respectively; and is joined to V by a line which meets AB, AC in N and N', respectively. On the line UV there is a definite involution, which is determined by the two pairs of points U, V and B', C'; we choose a definite pair of this involution, say X, Y. Prove that a conic can be drawn through M, M', N, N', X, Y. This conic will vary as P varies on BC. Prove that this variable conic has double contact with a certain fixed conic Ω, which touches AB, AC at B and C, respectively. On the line UV, take the fixed point Z which is given by $Z = (U, V)/O$. Shew that the chord of contact with Ω, of the variable conic, is the line ZP. Shew also that the polar of Z, in regard to Ω, meets BC in a point, T, such that AZ, AT are harmonic conjugates with respect to AB, AC.

184. Let $P_1, Q_1, R_1, P_2, Q_2, R_2, O$ be seven arbitrary points of a given conic, and A the intersection OP_1, R_2P_2, not lying on the conic. Let B, C be the respective intersections (OQ_2, Q_1R_1), (Q_1R_1, R_2P_2). The two degenerate conics, consisting of the line-pairs (OA, BC), (OB, CA), which both contain O, determine, on the given conic, respectively, the two sets of three points P_1, Q_1, R_1 and P_2, Q_2, R_2 (besides O). Thus we infer that the so-called involution, of sets of three points, determined on the given conic by the two given sets $P_1Q_1R_1$ and $P_2Q_2R_2$, are those determined, besides O, by the conics through the four points A, B, C, O. Of such four points two, A and O, could be taken arbitrarily, the point O being on the conic, and B, C as the other common points of two general conics O, A, P_1, Q_1, R_1 and O, A, P_2, Q_2, R_2. Cf. Chap. xvi (11).

185. Prove that the axes of the conics which touch four given tangents of a circle are, in aggregate, the polars of the centre of the circle in regard to the conics, and touch a parabola.

186. Let ω_0 be a particular conic of a confocal system, A and B two points of this, and P the pole of AB. Let the pairs of tangents be drawn from A and B to another conic of the confocal system. Prove that P is the orthocentre of the diagonal triangle of the quadrilateral formed by the four tangents, and is the centre of a circle which touches these tangents. Prove also that the parabola which is the envelope of the polars of P in regard to all the conics of the confocal system is also the conic which is the envelope of the polars of P taken in regard to all conics which touch the four par-

ticular tangents chosen. The directrix of this parabola is the line joining P to the centre of the confocals; and is the median line of the quadrilateral formed by these particular tangents. The focus is the point, P', conjugate to P in regard to all conics through the four foci of the confocals. The parabola touches the line AB, as well as the normals at A, B, of the conic ω_0. The point common to these normals has for polar, in regard to the parabola, the line which is the locus of poles of AB in regard to all the confocals. The parabola also touches the axes of the confocals.

187.   Prove that the director circles of all conics touching the sides of a triangle are the circles which are perpendicular to the circle in regard to which the triangle is self-polar.

188.   Prove that all conics touching the sides of a triangle, whose director circles have a given common point, have another common tangent. When the given common point is the orthocentre of the triangle, the conics in question are parabolas, whose director circles degenerate.

189.   Let l, m and l', m' be two given pairs of lines, and another given line be taken as the Absolute line. Prove that two Absolute points can be chosen on this last line so that both of $\epsilon(l,m)$ and $\epsilon(l',m')$ have given values; and that this is possible in two ways unless $\epsilon(l,m) = \epsilon(l',m') = -1$. [Compare the theorem generally expressed by saying that two pairs of lines can be projected into two pairs of lines of which each pair meet at a given angle.]

190.   Let A, B, C, D be four arbitrary points in a plane, with AB, DC meeting in E and BC, AD meeting in F, while AC and BD meet EF in L and M, respectively. Take the two points I, J on EF for which both $I, J \sim E, F$ and $I, J \sim L, M$. If I, J be taken as the Absolute points, the four points A, B, C, D are the angular points of a *parallelogram*, of which one pair of parallel sides are perpendicular to the other pair, so that the figure is a *rectangle*; but, as the diagonals AC, BD are perpendicular, this rectangle is a *square*.

191.   Let the complementary joins AC, BD, of a quadrangle A, B, C, D, meet in O. Determine points H, K on BD such that the cross ratios $(D, O/H, K)$, $(O, B/H, K)$ have for values the respective arbitrary numbers m, n, which is possible in two ways. Also, on the same line BD, determine two points H', K', such that $(D, O/H', K') = m'$ and $(O, B/H', K') = n'$, where m', n' are given numbers. Let I, J be the respective intersections (AH, CH') and (AK, CK'). Regarding I, J as the Absolute points, we then have $\epsilon A (D, C) = m,\ \ \epsilon A (C, B) = n,\ \ \epsilon C (D, A) = m',\ \ \epsilon C (A, B) = n'$.

Thus there are four positions possible for the pair I, J, such that, if these be taken as Absolute points, the quadrangle has, in the metrical phraseology, the lengths of its sides, and also the measurements of its angles, given.

192.   Prove that two points can be found on an arbitrary line, such that, if these be taken as Absolute points, we have, for any other three given lines a, b, c, the equalities $\epsilon(b, c) = \epsilon(c, a) = \epsilon(a, b)$. For any two triangles which are in perspective, a pair of Absolute points can be found such that these relations hold for the sides of both triangles.

193.   Let four general pairs of lines be given; these determine four pairs of points upon any line. Shew that there are three lines for which these four pairs are in involution. The three lines are the diagonals of a certain quadrilateral, the three pairs of foci of the respective involutions thereon being the pairs of complementary intersections of the four lines of this quadrilateral.

194.   Let two conics ϑ, ϑ' be polar reciprocals of one another in regard to a conic ω. Let ϕ be the polar reciprocal of ϑ in regard to ϑ', and ϕ' the polar reciprocal of ϑ' in regard to ϑ. Prove that ϕ, ϕ' are polar reciprocals of one another in regard to ω.

195. Let the tangents be drawn from a point P of one conic ω, to touch a confocal conic Ω, these tangents meeting ω again in T and U. We know (Chap. XIII (11)) that the line TU contains the pole, N, in regard to ω, of the normal of ω at P. Let the same construction, with the same conics ω, Ω, be made, starting from another point P' of ω. Prove that, as Ω is varied in the confocal system, the common point of TU and $T'U'$ describes a rectangular hyperbola, which contains the five points given by the common centre of the confocals; the points of the Absolute line lying on their common axes; and the points N, N'.

196. If five arbitrary lines be given, a parabola touches every four of these. Prove that the foci of the five parabolas lie on a circle. (Proofs are given below, Note 1, and Ex. 264.)

197. Prove that if, of the six conics touching the fives of six arbitrary lines, three be triangularly inscribed to the same conic, Ω, then all are triangularly inscribed to Ω.

198. A conic, ω, is triangularly inscribed to a conic Ω; five tangents of ω are taken. It can be shewn that there are two conics triangularly inscribed to Ω which touch four of these tangents, one of these being ω. For the various sets of four, of the five tangents of ω, let the five other conics, triangularly inscribed to Ω, which touch four of the tangents, be taken. Shew that these have a common tangent. Thus these, taken with ω, are six conics, triangularly circumscribed to Ω, which touch the fives of six lines.

199. Let an arbitrary line meet the sides BC, CA, AB, of a triangle inscribed in a conic, respectively, in U, V, W. Let the tangents to the conic, from U, touch the conic in L and L'; similarly, let VM, VM' be the tangents from V, and WN, WN' the tangents from W. Prove that the three pairs of lines AL, AL', BM, BM', CN, CN' meet in threes in four points, being the complementary joins of the points of a quadrangle, of which ABC is the vertex triangle.

Shew that this result is a particular case of the theorem that, if a conic meet the sides BC, CA, AB, of a triangle, respectively, in U, D; in V, E; and in W, F, and X, X' be the foci of the involution determined by the two pairs (B, C), (U, D), with Y, Y' and Z, Z' similarly defined on CA and AB, respectively, then the six lines AX, AX', BY, BY', CZ, CZ' meet in threes in four points. The points D, E, F are to be taken, in the particular case, to be the points where the tangents of the conic at A, B, C meet BC, CA, AB, respectively.

Conversely, if the lines which join the angular points A, B, C, of an inscribed triangle of a conic, to a point O, meet the conic again, respectively, in L, M, N; then the tangents at L, M, N, respectively, meet BC, CA, AB in points which are in a line.

200. Let two triangles ABC, $A'B'C'$ be in perspective, the lines AA', BB', CC' meeting in a point. Prove that a conic, ω, exists, in regard to which A', B', C' are the respective poles of BC, CA, AB. Let $A'B'$, $A'C'$ meet BC in L, L'; and D, D' be the foci of the involution, determined by (B, C) and (L, L'). Similarly, determine M, M' and E, E' on CA; and N, N' and F, F' on AB. Prove that a conic can be drawn through the four points E, E', F, F' to meet both BC and $B'C'$ in points lying on the conic ω. Also, that D, D'; E, E' and F, F' are three pairs of conjugate points in regard to the conic ω.

201. The line t is a tangent of a conic touching the sides BC, CA, AB of a triangle; and O is an arbitrary point. The lines through O, respectively conjugate to OA, OB, OC, with regard to the conic, meet the tangent t in

A', B', C', respectively. Prove that AA', BB', CC' meet in a point. And that this point lies on a conic which contains A, B, C, O and the two points where the tangents from O, to the original conic, meet the tangent t.

202.　Let ABC, $A_1B_1C_1$ and $A_2B_2C_2$ be three triangles, of which the angular points, A, A_1, A_2, lie on one line, the angular points, B, B_1, B_2, lie on another line, and the angular points, C, C_1, C_2, lie on a third line, these three lines meeting in a point. Consider the three triangles of which the sides are, respectively, the lines BC, B_1C_1, B_2C_2; the lines CA, C_1A_1, C_2A_2; and the lines AB, A_1B_1, A_2B_2. Prove that the angular points of these new triangles, one from each triangle, equally lie, in threes, on three lines which meet in a point.

203.　Two fixed lines, which meet in O, contain, respectively, the fixed points A, B and X, Y. The point C on one line is given by $C=(A, B)/O$; on this line are also the variable points H, K such that $K=(A, B)/H$. On the fixed lines XB, YA take, respectively, the intersections $M=(XB, YK)$, $N=(YA, XH)$. Prove that the intersection (AB, MN) is the point $T=(H, K)/C$. Also, that the intersection $P=(XH, YK)$ describes a conic through A, B, X, Y, of which the tangent at P contains T.

204.　From a variable point, O, on the radical axis of two fixed circles, whose centres are A and B, tangents OP, OP' are drawn to the former circle, and tangents OQ, OQ' to the latter. Prove that the locus of the common point of AP, BQ is one of two hyperbolas, having A, B for conjugate foci.

205.　Let AB and CD be complementary common chords of two conics. Let the tangents of the conics, at two points P, Q, one on each conic, meet in a point T which lies on the chord CD. Let AP, BQ meet in X, and AQ, BP meet in Y. Prove that T, X, Y lie in a line.

206.　Let A, B, C, D be four points of a conic. Deduce from the cross-ratio definition of the conic that the product of the lengths of the perpendiculars drawn to the chords AB, CD from any point P of the conic is in a constant ratio to the product of the lengths of the perpendiculars drawn from P to the chords BC, AD.

207.　If the lines joining a variable point P, to two fixed points, meet a fixed line in two points whose distance is constant, prove that the locus of P is a hyperbola, of which the fixed line is one asymptote.

208.　An axis of a conic, containing the real focus S, meets the directrix corresponding to S in the point X. On one of the tangents of the conic, drawn from X, there is taken the variable point Q, and the line from Q, perpendicular to the axis, meets this axis in M, and meets the conic in P, P'. Prove that the lengths of SP and QM are equal.

209.　From a fixed point X, on a fixed diameter of a given circle, is drawn a fixed line. From a variable point Q, of this line, the line is drawn perpendicular to the fixed diameter, meeting this in M. On QM is taken a point P such that the tangents from P to the circle have (both) a length equal to the length of QM. Prove that the locus of P is a conic, having double contact with the circle, the line of contact being the line through X which is perpendicular to the fixed diameter.

210.　The tangents at points C, X of a conic meet in L. The tangents at two other points A, B meet the tangent CL in U and V, respectively. If the intersections (AB, CX) and (AB, XL) be E and M, respectively, prove that $(U, V/C, L)=(A, B/E, M)$.

When XL is the Absolute line, this becomes a theorem, for a parabola, given by Apollonius, which states that the ratio of the lengths of CU, CV is equal to the ratio of the lengths of EA, EB.

211. Let A, V, V' be three points of a conic. Through the pole, O, of the chord VV', draw an arbitrary line, meeting AV, AV' in M, M', respectively, meeting the conic in P and Q, and meeting the tangent at A in K. Prove that $(P, M/O, K) = (M, Q/O, K)$, and $(P, M'/O, K) = (M', Q/O, K)$.

Hence, for a parabola, let a chord PQ, which contains the pole, O, of a chord VV', be met by the lines drawn through V, V', which are parallel to the axis, in M and M'. Prove that the product of the lengths of OP and OQ is equal to the square of the length of OM, or of OM', these being equal. Also, if the line through O, parallel to the axis, meet the curve in R, and H be the point of the curve where the tangent is parallel to PQ, prove that the product of the lengths of OP and OQ is equal to four times the product of the lengths of SH and OR, where S is the focus. From the ends of a chord of a parabola which passes through the focus, two chords are drawn both perpendicular to this focal chord. Prove that the product of the lengths of these two parallel chords is equal to the square of the length of the focal chord which is parallel to them.

212. The tangents at points P, Q of a conic, of which C is the centre, and S is a (real) focus, meet in O; and OC meets the conic in R and R'. The line through R, parallel to the axis SC, meets SO in T. Prove that the product of the lengths of SP and SQ is equal to the square of the length of ST. Thus, in particular, when the chord PQ contains the focus S, the circle having PQ as diameter contains the point T. This circle contains also the point T', where the line through R', parallel to CS, meets OS.

213. Let R, Q be two points on a conic, and C, S be any two points not on the conic. Let the tangents SU, SV, from S, to the conic, meet the polar of C in I and J, respectively, and SC meet this polar in H. Let RH meet the conic again in K, also RC meet the conic again in R', and meet the tangent at Q in O; let the intersection (OS, RK) be T, and N be the pole of the chord KQ. If UV meet the polar of C in Y, the triangle CHY is self-polar in regard to the conic, so that KY passes through R'. Let the second tangent to the conic from O touch the conic in P. Prove that the cross ratios of the pencils $T(P, O/I, J)$, $Q(T, S/I, J)$, $S(C, N/I, J)$ are equal to one another, and equal to the cross ratio $(R', Q/U, V)$ on the conic. We know also that $S(P, T/I, J) = S(T, Q/I, J)$.

If we take I, J for Absolute points, the result of Ex. 212 is obtained.

214. Let A, B, C and I, J be fixed points, of which I, J lie on AB. Consider a variable quadrilateral, of which one pair of complementary intersections are the fixed points I, J, while another complementary pair of intersections D, E are such that D varies on CB and E varies on CA. Let the third pair of complementary intersections consist of the points P, P', lying respectively on JE and IE. Shew how to construct P' when P is given. Using the symbols, prove that if $P = \xi I + \eta J + zC$, where ξ, η, z are numbers, then $P' = n\eta I + m\xi J + zC$, where m, n are fixed numbers such that $I = A - mB$ and $J = B - nA$. Thus P' is obtained from P by a *linear transformation*, which has three unaltered points, of which one is at C, and the other two are the foci of the involution on AB of which (A, B) and (I, J) are two pairs. Prove that CP, CP' meet AB in a pair of this involution.

By this transformation, any line, through J, is changed to a line, through I, which meets the former line on CA. Thus, if ω be any conic which is changed by the transformation to the conic ω', any tangent of ω, passing through J, is changed to a tangent of ω', passing through I, and these tangents meet on CA. In particular, if CA be taken as Absolute line, and J be a focus of ω, then I is a focus of ω'. Also, the polar of any point, in regard to ω, is changed into the polar of the transformed point, taken in regard to ω'; so that, if ω be a circle of centre J, and CA be the Absolute line, then ω' has I for focus, the corresponding directrix being CB, this being the line arising from CA by the transformation. In this case, the tangents from I, to the circle ω, meet the tangents from J to ω' at points lying on the line CB. This enables us to draw the tangents from J to the conic ω', when we have drawn the tangents from I to the circle ω; it is Boscovich's construction for drawing the tangents to a given conic, of given focus I and given directrix, from an arbitrary point J taken as centre of a certain circle (1757); it is assumed that we know how to draw the tangents from I to the circle.

With CA as Absolute line, the transformation may be stated thus: let I, J, B be three fixed points of one line, and BC be also a fixed line. To any point P, find D, where IP meets CB; then draw from I a line parallel to JP; the point P' is the point where this line meets DJ. As P describes a circle with centre at J, the point P' describes a conic with I as focus and BC as directrix.

215. Prove that, in general, six conics can be drawn through four arbitrary points to touch a given conic; determine in particular what these conics are when the given conic consists of two lines. But, when a pair of complementary joins of the four given points are two tangents of the given conic, prove that there are four conics through the given points which touch the given conic. Thus, dually, of conics of a definite confocal system, there are four which touch a given circle. The number 6 may be verified by a process of degeneration, in which the given conic consists of two lines, having a general relation with the four given points. There are two conics through the four points which touch each of the lines of the degenerate conic; and we must assume that the conic, through the four points and the common point of these two lines, arises from two of the tangent conics in the general case (the tangential equation of a conic which consists of two lines is the square of the tangential equation of their common point). Thus we obtain $2+2+2$ tangent conics.

The number of conics which touch five given conics in general position is 3264. This may likewise be verified by a process of degeneration, assuming each of the five given conics to break up into two lines. For let the points common to these five pairs of lines be $O_1, O_2, ..., O_5$. Pick out five lines, one from each of the five pairs; and let each of these lines be regarded as containing one of the five points $O_1, ..., O_5$. Consider the conics which can be drawn through r of these five points ($r=0, 1, ..., 5$), to touch $5-r$ lines, taken one from each of the pairs of lines which meet in the other $5-r$ points. For $r=1$, there are two conics passing through a point which touch 4 lines; and, for $r=4$, there are two conics through four points which touch a line. For $r=2$, there are four conics through two points which touch three lines; and, for $r=3$, there are four conics through three points which touch two lines. There is a single conic through five points, and a single conic touching five lines. As there are 2^5 sets of five lines (one through each of $O_1, ..., O_5$), each with its associated point, we obtain thus, in all,

$$2^5 \left[1 + (5, 1).2 + (5, 2).4 + (5, 3).4 + (5, 4).2 + 1 \right]$$

conics, wherein $(5, s)$ denotes the binomial coefficient $5!/s!(5-s)!$. This number is $32.2.51$, which is $2^6(2^4+1)(2+1)$, or 3264.

The original authority for the number 3264 is Schubert, *Abzählenden Geometrie*, 1879, p. 97. The method of degeneration here employed is put forward only as tentative.

216. In general two polygons, of a specified number of sides, can be inscribed in a given conic, so that the sides of the polygon pass each through a given point, not lying on the conic.

Consider first the case of a triangle, whose sides are to pass through given points, D, E, F, which do not lie on the conic. When DEF form a triangle which is self-polar for the conic, there is an infinite series of such triangles, one angular point thereof being an arbitrary point of the conic. In general, let the tangents to the conic from D touch the conic in I and J. Recall the fundamental theorem that the locus of a point, when the pencil of four lines joining this to four given points has a constant cross ratio, is a conic passing through the four points. If a triangle, ABC, can be inscribed in the conic with BC passing through D, the rays AB, AC are harmonic conjugates in regard to AI, AJ; and we suppose that AB, AC contain, respectively, the fixed points F and E. Consider then the locus of a point, P, such that PE, PF are harmonic conjugates in regard to PI and PJ. This locus is a conic, meeting the given conic in I and J. The other two points in which this conic meets the original conic are two possible positions of the point A; and when A is chosen, the associated possible positions of B and C can be constructed. The construction remains valid when D, E, F are in a line. And, as appears below, provided no two of the points D, E, F are conjugate to one another in regard to the conic, the same two inscribed triangles are obtained when, instead of the tangents DI, DJ, we take the tangents from E, or from F. Otherwise the point D must be appropriately chosen. When every two of D, E, F are conjugate in regard to the conic, any one of the points A, B, C may be taken arbitrarily on the conic. If we regard I, J as the Absolute points of the plane, so that the given conic is a circle, with centre at D, the positions of A are the points of this circle (other than I, J) which lie on the circle having EF as diameter. This latter circle degenerates when D is conjugate to E, or to F, in regard to the given circle. When D, E, F have general positions, let D', E', F' be the respective poles of EF, FD, DE in regard to the given conic, and DD', EE', FF' meet in O; let OD, OE, OF meet $E'F'$, $F'D'$, $D'E'$, respectively, in L, M, N. Then the line MN meets the given conic in the two possible positions of the angular point A; and NL, LM similarly contain, respectively, the two possible positions of B and C.

Now suppose that, instead of the three points D, E, F, we have n given points, not on the given conic, say O_1, O_2, ..., O_n. Let P_0 be an arbitrary point on the conic, and P_0O_1 meet the conic again in P_1; let P_1O_2 meet the conic again in P_2; and so on. Finally, let $P_{n-1}O_n$ meet the conic again in P_n. As P_0P_1 passes through a fixed point, the pairs P_0, P_1, on the conic, as P_0 varies, are the pairs of an involution; and so on. Thus, when P_0 varies, P_n describes a range on the conic which is related to that described by P_0. There are therefore two positions for P_0 for which P_n coincides with P_0, unless P_n coincides with P_0 for every position of this. Each of these gives such a polygon as is desired. The same argument is applicable when $n = 3$. The line P_0P_n touches a conic having double contact with the given conic.

217. Let three circles be given. A common tangent of two of the circles passes through one of the centres of similitude of these two circles; and may, for a moment, be said to belong to this centre of similitude. Suppose that the circles are such as to have three tangents, each touching two of the circles, which meet in a point, the centres of similitude to which these tangents respectively belong not being in a line. It can then be proved that there is a second set of three tangents, each touching two of the circles, which meet in a point; these belonging respectively to the three centres of similitude to which the first set belongs.

A proof may be given, using the symbols, which makes no reference to circles. Let O be an arbitrary point, and l, m, n be arbitrary lines passing through O. Three other lines, a, b, c, are drawn through O, of which a is either one of the bisectors of the angles determined by m and n, and similarly b is one bisector for n and l, and c is one bisector for l and m. On the lines a, b, c, respectively, three arbitrary points, A, B, C, not lying in a line, are taken; and the intersections (l, BC), (m, CA), (n, AB) are denoted by D, E, F, respectively. It can then be shewn that either, AD, BE, CF meet in a point, or, D, E, F lie in a line.

For this, let I, J be the symbols of the Absolute points, and $I + \lambda J$, $I + \mu J$, $I + \nu J$ be the symbols of the points in which l, m, n, respectively, meet the Absolute line, while $I + pJ$, $I + qJ$, $I + rJ$ are the symbols of the points in which a, b, c, respectively, meet the Absolute line. The definition of a, b, c requires that $p = \epsilon_1 (\mu\nu)^{\frac{1}{2}}$, $q = \epsilon_2 (\nu\lambda)^{\frac{1}{2}}$, $r = \epsilon_3 (\lambda\mu)^{\frac{1}{2}}$, where $\epsilon_1{}^2 = \epsilon_2{}^2 = \epsilon_3{}^2 = 1$. These lead to $\lambda = \epsilon qr/p$, $\mu = \epsilon rp/q$, $\nu = \epsilon pq/r$, where $\epsilon = \epsilon_1 \epsilon_2 \epsilon_3$. It can be proved that AD, BE, CF meet in a point when $\epsilon = 1$, but D, E, F lie in a line when $\epsilon = -1$.

We suppose $\epsilon = 1$. Let l', m', n' be the lines, through D, E, F, respectively, defined uniquely by the fact that BC is a bisector of the angles determined by l, l', while CA is a bisector for m, m', and AB is a bisector for n, n'. The lines l', m', n' then meet in a point, which we denote by X. If the symbols of A, B, C be taken, respectively, to be $xO + I + pJ$, $yO + I + qJ$, $zO + I + rJ$, so that the symbol of O is $(q - r) A + (r - p) B + (p - q) C$, it may be verified that l', m', n' all pass through the point X whose symbol is $\xi A + \eta B + \zeta C$, where, using u, v, w, respectively, for $p (q - r)$, $q (r - p)$, $r (p - q)$, the numbers ξ, η, ζ are, respectively, $u^{-1} (y - z) (yq^{-1} - zr^{-1})$, $v^{-1} (z - x) (zr^{-1} - xp^{-1})$ and $w^{-1}(x - y) (xp^{-1} - yq^{-1})$. These shew that AO, AX are a pair of the involution pencil determined by (AB, AC) and (AI, AJ), etc. Thus O, X are isogonal in regard to ABC. It is supposed here that $x(q - r) + y(r - p) + z(p - q)$ is not zero, that is, that A, B, C are not in a line. The symbols of D, E, F are, respectively, $v^{-1}B + w^{-1}C$, $w^{-1}C + u^{-1}A$, $u^{-1}A + v^{-1}B$.

Now we can describe circles, with A, B, C as centres, and D, E, F as centres of similitude of the pairs of circles (B, C), (C, A), (A, B), having l, m, n as concurrent tangents, each touching two of the circles; then l', m', n' are the complementary tangents of the same pairs of the circles. And conversely, the circles may be taken arbitrarily, with the common tangents l, m, n meeting in a point; then the complementary tangents l', m', n' are proved to meet in a point.

218. A line l meets the sides BC, CA, AB, of a triangle, respectively, in L, M, N. A point O is taken outside the plane of the triangle, and P, Q, R are points on the lines OL, OM, ON, respectively. The lines QR, RP, PQ meet the line l, respectively, in D, E, F. Prove that AD, BE, CF meet in a point.

219. Four planes, α, β, γ, δ, pass through a point O in space. On the six lines $(\beta, \gamma), (\gamma, \alpha), (\alpha, \beta), (\alpha, \delta), (\beta, \delta), (\gamma, \delta)$, common to the pairs of the planes, are taken, respectively, the points A, B, C, P, Q, R. Prove that the planes ABC, AQR, BRP, CPQ meet in a point. Or, if the planes of two triangles ABC, $A'B'C'$ meet in a line which intersects BC, CA, AB, $B'C'$, $C'A'$, $A'B'$, respectively, in L, M, N, L', M', N', and AL', BM', CN' meet in a point, then the pairs L, L'; M M'; N, N' belong to an involution, and $A'L$, $B'M$, $C'N$ also meet in a point. This is essentially a theorem stated by Moebius (1828). (See Chap. IV (6), Ex. 3.)

Shew also that it follows immediately from this that the circumcircles of the four triangles formed by four lines in a plane, meet in a point. [A plane section of a sphere projects, from a point of the sphere, into a circle lying in the tangent plane touching the sphere at the point diametrically opposite to the centre of projection.]

220. Four points A, B, C, D lie in a plane, no three of these lying in a line. Through the joining lines BC, CA, AB, AD, BD, CD, of pairs of the points, there pass, respectively, the planes α, β, γ, λ, μ, ν. Shew that the points of intersection of the four sets of three planes, (α, β, γ), (α, μ, ν), (β, ν, λ), (γ, λ, μ), lie in a plane. This is the dual of Ex. 219.

221. If in Ex. 219, we take five planes through a point, each four give rise to a point, as explained. These five points can be shewn to lie in a plane. If we take six planes through the point, each five of these gives rise to a

plane, as explained here. The six planes so arising meet in a point. And so on indefinitely. This result was explicitly stated by Homersham Cox (*Quart. Journ. of Math.* xxv, 1891).

222.   Let A, B, C, D be four general points in a plane ϖ, and A', B', C', D' be four general points in a plane ϖ', which may coincide with ϖ. Let P be any point of the first plane. We can then determine a point P' of the second plane to satisfy the two conditions that the pencils $D\,(A, B, C, P)$, $D'\,(A', B', C', P')$ are related, and the pencils $A\,(B, C, D, P)$, $A'\,(B', C', D', P')$ are related. Prove that, then, the pencils $B\,(C, A, D, P)$, $B'\,(C', A', D', P')$ are related, and the pencils $C\,(A, B, D, P)$, $C'\,(A', B', D', P')$ are related.

Prove that when P describes a line, then P' describes a related range on another line. Also, when the planes ϖ, ϖ' coincide, prove that, in the most general case, there are three positions of P which coincide with their, respectively, corresponding points P'; in that case, to any point of the line which joins two of these three positions of P, there corresponds another point of the same line.

223.   Through a fixed point O, a line l is drawn meeting in P the variable tangent t, of a fixed conic, so that $\epsilon\,(l, t)$ remains constant. Prove that the locus of P is a curve which cuts an arbitrary line in four points (a quartic curve); and that this curve has three double points, one at O and the others at the Absolute points of the plane. Consider the particular cases when (*a*) the point O is a focus of the conic, and (*b*) the conic is a parabola.

224.   If $\Sigma_1 = 0$, $\Sigma_2 = 0$, $\Sigma_3 = 0$ be the tangential equations of three conics, prove, for the system of conics given by $\mu_1\Sigma_1 + \mu_2\Sigma_2 + \mu_3\Sigma_3 = 0$, wherein μ_1, μ_2, μ_3 are parametric numbers, which are such that a focus describes a given line, that the complementary focus describes a conic. Dually, and more generally, prove that the pairs of complementary chords common to a conic $\Omega = 0$ and the conics of the series $\lambda U + \mu V + \nu W = 0$, wherein $U = 0$, $V = 0$, $W = 0$ are given conics, and λ, μ, ν are parameters, are the line-pairs which are conjugate to one another in regard to all the conics touching four appropriate lines.

225.   Consider two quadratic forms

$$\omega = ax^2 + by^2 + cz^2 + 2fyz + 2gzx + 2hxy, \quad \omega' = a'x^2 + \ldots + 2h'xy\,;$$

let $A = bc - f^2$, $F = gh - af$, etc., being minors in the discriminantal determinant, Δ, of ω; and let A', F', etc., Δ' be the similar functions for ω'; we suppose neither Δ nor Δ' to be zero. Also let

$$\Omega = \Delta^{-1}\,(Au^2 + Bv^2 + Cw^2 + 2Fvw + 2Gwu + 2Huv),$$

$$\Omega' = \Delta'^{-1}\,(A'u^2 + \ldots + 2H'uv).$$

And let the discriminantal determinants of the respective quadratics, $\omega + \lambda\omega'$, $\Omega' + \lambda\Omega$, be denoted by d and D. Prove that $\Delta\Delta'D = d$.

Shew that the cross ratio of the pencil joining the four points of the conics $\omega = 0$, $\omega' = 0$ to any point of $\omega = 0$ is equal to that of the range which the four common tangents of these conics determine on any tangent of $\omega' = 0$.

226.   Consider the seven points consisting of the four points of a quadrangle and the three vertices of the quadrangle. Prove that a cubic curve can be drawn through these seven points and two other arbitrary points. Also, that the tangents of any one such cubic curve, at the four points of the quadrangle, meet in a point lying on the curve, say P_0; and, that any line through P_0 meets the cubic curve in two further points which are conjugate to one another in regard to the quadrangle; in particular, the tangent of the cubic curve at P_0 meets the curve again in the point P_0' which is conjugate to P_0 in regard to the quadrangle. With suitable coordinates, referred to the vertex triangle of the quadrangle, the equation of the cubic

curve is $x_0 x (y^2 - z^2) + y_0 y (z^2 - x^2) + z_0 z (x^2 - y^2) = 0$, where P_0 has the co-ordinates (x_0, y_0, z_0). Points which are conjugate to one another in regard to the quadrangle have coordinates (x, y, z), (x^{-1}, y^{-1}, z^{-1}). When one of these points describes a line, the other describes a conic. When P_0 is the pole of the line in regard to the conic, the conic meets the line in the Hessian points of the points where the cubic meets the line (see Ex. 265).

227.   Let A, B, C, D be four points of a conic, and A', B', C' the re-spective poles of BC, CA, AB. Take the tangent of the conic at D, and the point D', which is the pole of this tangent in regard to the triangle $A'B'C'$. Prove that the polar line of D', in regard to the conic, coincides with the polar line of D in regard to the triangle ABC; also that D' lies on the Hessian line of A, B, C, in regard to the conic (this is the line which contains the points where the tangents at A, B, C, respectively, meet BC, CA, AB).

Also, shew that, if H_1, H_2 are the so-called Hessian points of A, B, C on the conic, where the Hessian line meets the conic, then, when D is at H_1, the point D' is at H_2; and, in this case, that D' coincides with the pole of the tangent at D taken in regard to the triangle ABC. Conversely, the only two points D, on the conic, for which D' coincides with the pole of the tangent at D, taken in regard to the triangle ABC, are H_1 and H_2.

Let A_1, B_1, C_1; A_2, B_2, C_2; A_3, B_3, C_3 be three sets of an involution of sets of three points, on the conic, the involution being founded on two arbitrary sets of three points (Chap. xvi (11)). Any point D of the conic forms, with the points of a set, a quadrangle, such as A_1, B_1, C_1, D. Prove that the points D', derived, as above, from the three quadrangles so formed, are in a line; also, that the polar lines of D, in regard to the three triangles $A_1 B_1 C_1$, $A_2 B_2 C_2$, $A_3 B_3 C_3$, meet in a point.

228.   It is also true, in Ex. 227, when the conic is a circle, that the pedal lines of D, in regard to the three triangles $A_1 B_1 C_1$, $A_2 B_2 C_2$, $A_3 B_3 C_3$, meet in a point. This requires a separate proof.

229.   With rectangular coordinates $(x, y, 1)$, the equation of a circle is of the form $x^2 + y^2 - 2ax - 2by = r^2 - a^2 - b^2$; prove that, for three circles, the equation of one line containing three centres of similitude, of pairs of the circles, is

$$\begin{vmatrix} x, & y, & 0, & 1 \\ a_1, & b_1, & r_1, & 1 \\ a_2, & b_2, & r_2, & 1 \\ a_3, & b_3, & r_3, & 1 \end{vmatrix} = 0.$$

230.   Prove that the circle of curvature of the ellipse whose equation is $x^2/a^2 + y^2/b^2 = 1$, at the point whose coordinates are $(a \cos \theta, b \sin \theta, 1)$, has the equation

$$x^2 + y^2 - 2xa^{-1}c^2 \cos^3 \theta + 2yb^{-1}c^2 \sin^3 \theta + a^2 + b^2 - 3(a^2 \sin^2 \theta + b^2 \cos^2 \theta) = 0,$$

where $c^2 = a^2 - b^2$. Find the equation of the common tangent of this circle and the ellipse other than the tangent at $(a \cos \theta, b \sin \theta, 1)$.

231.   If, in Ex. 230, the tangents from a point P, or $(\xi, \eta, 1)$, touch the ellipse in U, U', prove that the circle PUU' contains the point $(\xi', \eta', 1)$, where $\xi' = c^2 \xi / (\xi^2 + \eta^2)$, $\eta' = -c^2 \eta / (\xi^2 + \eta^2)$. This is the same for all conics confocal with the ellipse. If S, H be the real foci of the ellipse, the conic P, U, U', S, H contains the point $(\xi_1, \eta_1, 1)$, where $\xi_1 = -\xi$, $\eta_1 = (\xi^2 - c^2)/\eta$.

232.   The envelope of the polars of $(\xi, \eta, 1)$, in regard to conics confocal with the ellipse in Ex. 230, has the equation obtained by rationalising $(x\xi)^{\frac{1}{2}} + (-y\eta)^{\frac{1}{2}} + c = 0$.

233.   In Ex. 230, the equation of the Apollonius rectangular hyperbola of the point $(\xi, \eta, 1)$, in regard to the ellipse, is $c^2 xy + b^2 x\eta - a^2 y\xi = 0$.

234. In Ex. 231, the equation of the circle PUU' is

$$(x^2+y^2)\left(\frac{\xi^2}{a^2}+\frac{\eta^2}{b^2}\right)-(\xi^2+\eta^2)\left(\frac{x\xi}{a^2}+\frac{y\eta}{b^2}\right)+c^2\left(\frac{\xi^2}{a^2}-\frac{\eta^2}{b^2}-\frac{x\xi}{a^2}+\frac{y\eta}{b^2}\right)=0.$$

235. The conics, whose equations are $x^2+yz=0$, $x^2+h^2yz=0$, are polar reciprocals of one another in regard to $x^2+hyz=0$. But are also polar reciprocals in regard to the conic $2x^2-h\,(my^2+m^{-1}z^2)=0$, for arbitrary values of m.

236. Find, in explicit form, the point equation of the general conic which touches the four lines whose equations are $a_ix+b_iy+c_iz=0$ $(i=1, 2, 3, 4)$.

237. If the squares of the lengths of the sides of a triangle be given, the Absolute points of the plane can be found.

Let ABC be the triangle, and the lengths of BC, CA, AB be respectively a, b, c. In terms of the symbols of A and of the Absolute points I, J, let the symbols of B, C be respectively $A+p_3I+q_3J$ and $A+p_2I+q_2J$. Thus, by the definition of length,

$$b^2=p_2q_2,\ \ c^2=p_3q_3,\ \ a^2=(p_2-p_3)\,(q_2-q_3)=b^2+c^2-(p_2q_3+p_3q_2).$$

If P be any point of symbol $xA+yB+zC$, or

$$(x+y+z)\,A+(yp_3+zp_2)\,I+(yq_3+zq_2)\,J,$$

the distance AP, or say d_1, is given by $d_1{}^2\,(x+y+z)^2=(yp_3+zp_2)\,(yq_3+zq_2)$; or, if we denote the ratios $x/(x+y+z)$, $y/(x+y+z)$, $z/(x+y+z)$ by ξ, η, ζ, respectively, $d_1{}^2=\eta^2c^2+\zeta^2b^2+\eta\zeta\,(b^2+c^2-a^2)$. We see also that, for points on the Absolute line, we have $x+y+z=0$, and, for points not on the Absolute line, we have $\xi+\eta+\zeta=1$. More generally, if P, P' be any two points of respective symbols $\xi A+\eta B+\zeta C$, $\xi'A+\eta'B+\zeta'C$, the distance PP', or say d, is given by

$$d^2=(\eta p_3+\zeta p_2-\eta'p_3-\zeta'p_2)\,(\eta q_3+\zeta q_2-\eta'q_3-\zeta'q_2),$$

or

$$Y^2c^2+Z^2b^2+YZ\,(b^2+c^2-a^2),$$

where Y, Z denote $\eta-\eta'$, $\zeta-\zeta'$, respectively; if also $X=\xi-\xi'$, so that $X+Y+Z=0$, this is the same as $d^2=-a^2YZ-b^2ZX-c^2XY$.

The symbol $A+(\eta p_3+\zeta p_2)\,I+(\eta q_3+\zeta q_2)\,J$, of any point, shews that, for a point lying on one of the lines AI, AJ, we must have

$$(\eta p_3+\zeta p_2)\,(\eta q_3+\zeta q_2)=0,$$

that is,

$$\eta^2c^2+\zeta^2b^2+\eta\zeta\,(b^2+c^2-a^2)=0.$$

The Absolute points are thus obtainable by combining two equations such as this, say

$$\zeta^2a^2+\xi^2c^2+\zeta\xi\,(c^2+a^2-b^2)=0,\ \ \xi^2b^2+\eta^2a^2+\xi\eta\,(a^2+b^2-c^2)=0.$$

With proper interpretation of the ambiguous signs it may thence be proved that, at one of the Absolute points, the numbers $-a\eta/b\xi$, $-a\zeta/c\xi$ are, respectively,

$$[\epsilon C\,(A,\,B)]^{\frac{1}{2}},\ \ [\epsilon B\,(A,\,C)]^{\frac{1}{2}},\ \ \text{or}\ \ [\epsilon C\,(B,\,A)]^{\frac{1}{2}},\ \ [\epsilon B\,(C,\,A)]^{\frac{1}{2}}.$$

238. From Ex. 237, prove that the condition that a conic, whose point equation is $a_1\xi^2+\ldots+2h_1\xi\eta=0$, should be a rectangular hyperbola, is

$$a_1a^2+b_1b^2+c_1c^2-f_1\,(b^2+c^2-a^2)-g_1\,(c^2+a^2-b^2)-h_1\,(a^2+b^2-c^2)=0,$$

where a, b, c are the lengths of the sides of the triangle to which the co-ordinates ξ, η, ζ are referred, these being such that $\xi+\eta+\zeta=1$ (or $\xi+\eta+\zeta=0$ on the Absolute line).

We may prove this by substituting $-\eta-\zeta$ for ξ in the equation of the conic, and thence expressing that the conic meets the Absolute line in two points harmonically conjugate in regard to the Absolute points given by $\eta^2c^2+\zeta^2b^2+\eta\zeta\,(b^2+c^2-a^2)=0$.

239. Prove that the tangential equation of a conic, touching the sides of the triangle to which the coordinates are referred, is of the form $fu^{-1}+gv^{-1}+hw^{-1}=0$; and that the point equation of this conic is obtained by rationalising the equation $(f\xi)^{\frac{1}{2}}+(g\eta)^{\frac{1}{2}}+(h\zeta)^{\frac{1}{2}}=0$. Also that the pole of the line $\xi+\eta+\zeta=0$ in regard to this conic has coordinates $(g+h, h+f, f+g)$.

240. With coordinates, such as those used, for which the Absolute line has equation $\xi+\eta+\zeta=0$, prove the equation of a circle touching the sides of the triangle of reference is as in Ex. 239, with $f=b+c-a$, $g=c+a-b$, $h=a+b-c$, where a, b, c are, as before, the lengths of the sides of the triangle. Also that the equation of the circumcircle of the triangle is

$$C=a^2\eta\zeta+b^2\zeta\xi+c^2\xi\eta=0.$$

Also that the equation of any circle whatever is

$$a^2\eta\zeta+b^2\zeta\xi+c^2\xi\eta+(p\xi+q\eta+r\zeta)(\xi+\eta+\zeta)=0,$$

where p, q, r are suitable constants. Also that the equation of the so-called self-polar circle of the triangle is

$$P=(b^2+c^2-a^2)\,\xi^2+(c^2+a^2-b^2)\,\eta^2+(a^2+b^2-c^2)\,\zeta^2=0,$$

and the equation of the nine-point circle is $2C-P=0$, where C, P are the quadratics so named.

241. Prove that the lengths of the perpendiculars, p, q, r, drawn to an arbitrary line from the angular points A, B, C of a triangle whose sides are of lengths a, b, c, are connected by the relation

$$\begin{vmatrix} 0, & c^2, & b^2, & p, & 1 \\ c^2, & 0, & a^2, & q, & 1 \\ b^2, & a^2, & 0, & r, & 1 \\ p, & q, & r, & 1, & 0 \\ 1, & 1, & 1, & 0, & 0 \end{vmatrix}=0.$$

242. Two circles have a common centre O, and A, B are fixed points on a line through O. A variable line through O meets the two circles respectively in P and Q. If AP, BQ meet in R, and RC, drawn parallel to OPQ, meet the line AB in C, prove that C is a fixed point, and R varies on a circle whose centre is C. (J. H. Grace.)

243. Three conics ϕ_1, ϕ_2, ϕ_3, all passing through two points I, J, meet another conic ω respectively in $A_1, B_1, C_1, D_1; A_2, B_2, C_2, D_2; A_3, B_3, C_3, D_3$. The four lines A_1A_2, B_1B_2, C_1C_2, D_1D_2 meet the line IJ respectively in P, Q, R, S; and the lines PA_3, QB_3, RC_3, SD_3 meet the conic ω again respectively in A_4, B_4, C_4, D_4. Prove that these points lie on a conic containing I and J.

244. Let O, D, I, J be any points of a conic σ, and X, Y any two points of the chord IJ. If a variable conic ϕ, drawn through X and Y, meet the conic σ in P, Q, R, S, and $\epsilon O(D, P)$ be written for the cross ratio $O(D, P\,/\,I, J)$, etc., prove that the product

$$\epsilon O(D, P).\epsilon O(D, Q).\epsilon O(D, R).\epsilon O(D, S)$$

is the same for all such conics ϕ.

245. Let ABC be a triangle, and X, Y, Z be three points on a line. Denote the intersections (BZ, CY), (CX, AZ), (AY, BX) respectively by A', B', C'. Prove that A, B, C, A', B', C' lie on a conic. Let the lines joining an arbitrary point of this conic respectively to X, Y, Z meet this conic again in L, M, N. Shew that AL, BM, CN meet in a point, as do $A'L$, $B'M$, $C'N$.

246. Prove that the locus of the centre of a conic which touches the sides of a given triangle and passes through an arbitrary fixed point, is a conic which touches the three lines joining the mid points of the sides of the given triangle.

247. Prove that the director circle of a conic which touches the sides of a given triangle and passes through the circumcentre of the triangle, touches both the circumcircle and the nine-point circle of the triangle.

248. If the Absolute points of the plane be given by the two equations $x+y+z=0$, $fyz+gzx+hxy=0$, and p, q, r respectively denote $\frac{1}{2}(g+h-f)$, $\frac{1}{2}(h+f-g)$, $\frac{1}{2}(f+g-h)$, prove that the equation of the director circle of a conic, touching the sides of the triangle of reference, with tangential equation $\lambda vw + \mu wu + \nu uv = 0$, in which λ, μ, ν are constants, is

$$\Omega = (\lambda+\mu+\nu)(fyz+gzx+hxy) - (x+y+z)L = 0,$$

where L is $\lambda px + \mu qy + \nu rz$. Also that the co-director circle (v. Chap. XIII (12)) of the conic and the confocal thereto which touches the line k, or $lx+my+nz=0$, has the equation

$$\Delta\Omega + d(\lambda mn + \mu nl + \nu lm)(x+y+z)^2 = 0,$$

where $\qquad d = qr+rp+pq, \quad \Delta = p(m-n)^2 + q(n-l)^2 + r(l-m)^2.$

This equation is of the form $\lambda\Omega_1 + \mu\Omega_2 + \nu\Omega_3 = 0$.

249. With the same Absolute points as in the preceding example, prove that the pedal circle of the point (ξ, η, ζ), in regard to the triangle of reference, has the equation

$$\sigma(fyz+gzx+hxy) - (x+y+z)(Px+Qy+Rz) = 0,$$

where $\sigma = (\xi+\eta+\zeta)(f\eta\zeta+g\zeta\xi+h\xi\eta)$, and

$$P = \xi(p\eta+g\zeta)(p\zeta+h\eta), \quad Q = \eta(q\zeta+h\xi)(q\xi+f\zeta), \quad R = \zeta(r\xi+f\eta)(r\eta+g\xi).$$

Prove, further, that the circles, with centres at the angular points of the triangle, which are perpendicular to this pedal circle, have equations $U=0$, $V=0$, $W=0$, where

$$U = \sigma(hy^2+gz^2+2pyz) - P(x+y+z)^2, \text{ etc.}$$

The pedal circle is thus perpendicular to any circle whose equation is $aU+bV+cW=0$, in which a, b, c are constants; also, for all points (ξ, η, ζ) lying on the cubic curve whose equation is $aP+bQ+cR+\sigma=0$, the pedal circle is perpendicular to the circle ω, of centre (a, b, c), whose equation is

$$a(hy^2+gz^2+2pyz) + \ldots + \ldots + (x+y+z)^2 = 0.$$

The cubic curve in ξ, η, ζ referred to is interesting. It contains the angular points of the triangle of reference, and consists of pairs of points, (ξ, η, ζ), $(f\xi^{-1}, g\eta^{-1}, h\zeta^{-1})$, which are isogonal in regard to this triangle; its other intersections with the sides lie on a line. Values of a, b, c can be chosen so that the curve breaks up into an arbitrary line and the isogonally corresponding conic; also values, $(-p^{-1}, -q^{-1}, -r^{-1})$, such that the curve reduces to the sides of the triangle. For general values of a, b, c it has the property, belonging to a triangle, that the pedal circles of three points of it which lie on an arbitrary line are coaxal.

It is possible, however, to approach the matter in another way (cf. *Math. Gazette*, xv, 1931, pp. 258, 259, a reference supplied by Mr J. H. Grace), using the result in Ex. 248. For the equations $\Omega_1=0$, $\Omega_2=0$, $\Omega_3=0$ obviously represent the auxiliary circles of the three conics, touching the given line k, which have for a pair of conjugate foci, respectively, the pairs B, C; C, A; A, B of the points of reference. The common perpendicular circle, say ω_0, of these three auxiliary circles, is thus perpendicular to the general co-director circle of Ex. 248, for all values of λ, μ, ν. One set of values for these, however, gives the conic, touching the sides, which has one focus, S, at an arbitrary point of the line k. The confocal to this which touches k is the pencil pair which has its centres at S and the point isogonal to S in regard to the triangle. The co-director circle of these two confocals is the pedal circle of S. Thus the pedal circles of all points of the line k are perpendicular to the

circle ω_0. The centre of this circle ω_0, as we see by considering the auxiliary circles spoken of, is the point of concurrence of the lines, perpendicular to BC, CA, AB, drawn respectively from the feet of the perpendiculars to the line k which pass through A, B, C; this centre lies on the radical axis of the circles having for diameters the (limited) diagonals of the quadrilateral formed by the line k and the sides of the triangle (the radical axis is the perpendicular, to the median line of the quadrilateral, drawn from the orthocentre of the triangle of reference. Its equation is $xpl\,(m-n)+\ldots=0$). It can be shewn that the circle ω_0 is the circle ω for such values, a_0, b_0, c_0, of a, b, c, that ω is perpendicular to the three auxiliary circles; and that these are the values also for which the cubic curve $aP+\ldots+\sigma=0$ breaks up into the line k and the isogonally corresponding conic. The pedal circles of the points in which the line k meets the general cubic curve are, we have seen, all perpendicular to the circle ω, of centre (a, b, c); these pedal circles must therefore be coaxal, with radical axis passing through (a_0, b_0, c_0). It can be shewn that a_0, b_0, c_0 are given by
$$l\,(a_0 p+1)=m\,(b_0 q+1)=n\,(c_0 r+1)=d^2 lmn E^{-1},$$
where
$$E=d\,(mnqr+\ldots)-\Delta pqr.$$

250. Four points A, B, C, D of a line (or conic) are often said to be *equianharmonic* when their cross ratios have the two values $-\omega$, $-\omega^2$, in which ω is an imaginary cube root of unity. Shew that if three points A, B, C be given, either point, D, of the Hessian points, D, D', of A, B, C, forms with these an equianharmonic set; and that the Hessian set of A, B, C, D consists of D' with three other points A', B', C', whose Hessian set also consists of D and D', the set A', B', C', D' being also equianharmonic. Prove that if the points be taken on the conic $x^2+y^2+z^2=0$, the two sets may be taken to be the intersections of this with the two conics $x^2+\omega y^2+\omega^2 z^2=0$, $x^2+\omega^2 y^2+\omega z^2=0$. (Cf. Ex. 109, and Sylvester, *Papers*, i, p. 296.)

251. If $u=0$, $v=0$, $w=0$ are the equations of three conics, prove that all the conics with equation $\theta^2 u+\theta v+w=0$, for different values of θ, are inpolar to another determinate conic. (Form the tangential equation of the general conic of this system.) There are six pairs of lines constituting degenerate conics of this system; the six points of intersection of the lines of such a pair thus lie on a conic.

Let $x=0$, $y=0$, $z=0$, $x'=0$, $y'=0$, $z'=0$ be the equations of six arbitrary lines, and $u=xx'$, $v=yy'-xx'-zz'$, $w=zz'$. The six pairs of lines contained in the system of conics given by $\theta^2 u+\theta v+w=0$ are $xx'=0$, $yy'=0$, $zz'=0$, arising respectively for $\theta=\infty$, $\theta=1$, $\theta=0$, together with, say, $\xi\xi'=0$, $\eta\eta'=0$, $\zeta\zeta'=0$, arising respectively for $\theta=\theta_1$, $\theta=\theta_2$, $\theta=\theta_3$, suppose. Denote by σ the conic which touches the lines x, y, z, η, ζ. The condition for the conic $\theta^2 u+\theta v+w=0$ to be triangularly circumscribed to the conic σ, in general of the fourth order in θ, reduces then to a form $\theta(\theta-1)\,(U+V\theta)=0$, in which U, V are independent of θ. As the conic σ touches $\eta=0$, $\zeta=0$, we thus have $U+V\theta_2=0$ and $U+V\theta_3=0$, and hence, assuming $\theta_2 \neq \theta_3$, we have $U=0=V$, and *all* the conics $\theta^2 u+\theta v+w=0$ are triangularly circumscribed to σ; thus, in particular, one of $\xi=0$, $\xi'=0$ touches σ. In this way, with an appropriate notation, we infer that the six lines consisting of five (or three) of x, y, z, ξ, η, ζ, taken with the one (or three) complementary lines of x', y', z', ξ', η', ζ', touch a conic. (The conics of the θ-system all touch the general quartic curve whose equation is $(xx')^{\frac{1}{2}}+(yy')^{\frac{1}{2}}+(zz')^{\frac{1}{2}}=0$. The reader may compare *Proc. Lond. Math. Soc.* ix, 1911, pp. 151, 167, and *ibid.* xi, 1913, p. 301. The proof here given of the sets of six lines which touch a conic was suggested by Mr J. H. Grace.)

252. Consider the (2, 2) relation connecting the parameters θ, ϕ which is expressed by
$$\phi^2(a_1\theta^2+b_1\theta+c_1)+\phi(a_2\theta^2+b_2\theta+c_2)+a_3\theta^2+b_3\theta+c_3=0,$$

where the nine constants a_1, b_1, ..., c_3 have such values that the relation is not a (2, 1), or (1, 2) relation, is not satisfied by a fixed value of θ, or a fixed value of ϕ, and is not (the square of) a (1, 1) relation. Prove that, by two fractional linear substitutions for θ, ϕ, properly chosen, the relation can be reduced to one of the three forms

$$(\phi_1 - \theta_1)^2 + \phi_1 + \theta_1 = 0, \quad \phi_1{}^2 + \theta_1{}^2 + B\phi_1\theta_1 + C = 0,$$
$$\phi_1{}^2(a\theta_1{}^2 + c) + b\phi_1\theta_1 + c\theta_1{}^2 + a = 0,$$

where B, C, a, b, c are constants.

253. Let six tangents of a conic be taken, and two fixed points on one of these. Prove that the ten conics through these two fixed points, each containing the angular points of a triangle formed by three of the other five lines, have a point in common. For a particular case, take two perpendicular lines through the orthocentre of a triangle, and consider the ten circles circumscribing triangles formed by threes, from these two and the sides of the triangle.

254. Deduce from the preceding example (using a transformation of coordinates of the form $x' = x^{-1}$, $y' = y^{-1}$, $z' = z^{-1}$) that, if A, B, C, D, E be five points of a cubic curve which has a double point at O, with the lines t, u as tangents thereat, and a conic be drawn, through O and four of the points A, B, C, D, E, to meet the tangents t, u again in P and Q, then the five lines PQ meet in a point of the conic $ABCDE$. Dually, if a, b, c, d, e be tangents of a curve, of class three, which has a double tangent touching the curve at I and J (and has therefore three cusps with concurrent cuspidal tangents), and a conic be drawn, touching the double tangent, and also touching four of the lines a, b, c, d, e, the tangents from I, J to this conic meeting in S, then the five points S lie on a line which touches the conic touching a, b, c, d, e. The six conics are thus triangularly inscribed to the degenerate conic consisting of the double tangent and the line of the points S. (The transformation $x' = x^{-1}$, etc., spoken of, is taken with respect to the triangle formed by the two fixed points in one of the six lines, in Ex. 253, and a point common to two of the other five lines. For the theorem proved here, cf. (9) in Note 1, below.)

255. A triangle ABC has its angular points on one conic Ω, and its sides touching another conic ω. Arbitrary points I, J are taken on the side BC, and the other tangents to ω from I and J meet in S, while AS meets the conic Ω again in D. Prove that a variable conic through I, J, S, D meets Ω again in three points A', B', C', such that $B'C'$, $C'A'$, $A'B'$ touch ω. (Let the tangent SI, of ω, meet Ω in P, R; and SJ meet Ω in P_1, R_1. Let DI, DJ respectively meet Ω again in Q_1 and Q. Then, three particular conics through I, J, S, D are the line pairs BC, DA; SPR, DQ; SP_1R_1, DQ_1. Thus the nine sides of the triangles ABC, PQR, $P_1Q_1R_1$ touch a conic, which is ω because this touches BC, CA, AB, PR, P_1R_1. Thus $B'C'$, $C'A'$, $A'B'$ touch ω.)

256. If triangles ABC can be inscribed in a conic Ω whose sides touch a conic ω, then the angular points of the common self-polar triangle of Ω and ω, lie, with A, B, C, upon a conic. In metrical terms, if the two conics have common axes, the normals of Ω, at A, B, C, meet in a point.

257. Let a conic Ω be triangularly circumscribed about a conic ω, and also outpolar to this. If ABC be a triangle inscribed in Ω and circumscribed about ω, the tangents of Ω at A, B, C form a triangle which is self polar in regard to ω.

258. Let A, B, C, A_1, A_2, A_3 be any six points, the coordinates of A_1, in regard to ABC, being (a_1, b_1, c_1), etc. Let P be any other point, with coordinates (ξ, η, ζ) in regard to ABC. Take three other points P_1, P_2, P_3, of which the coordinates of P_1 are $(a_2a_3\xi^{-1}, b_2b_3\eta^{-1}, c_2c_3\zeta^{-1})$, and those of

P_2, P_3 are derived by cyclical change of the suffixes 1, 2, 3. If Δ denote the determinant (a_1, b_2, c_3), and the minors in this be α_1, β_2, γ_3, etc., the equation of the line $P_2 P_3$, or p_1, is then $a_1^{-1}\alpha_1\xi x + b_1^{-1}\beta_1\eta y + c_1^{-1}\gamma_1\zeta z = 0$, etc. Let Σ_1 denote the conic touching BC, CA, AB, p_2, p_3, so that

$$\Sigma_1 = \alpha_2\alpha_3\,\xi vw + \beta_2\beta_3\,\eta wu + \gamma_2\gamma_3\,\zeta uv = 0,$$

in tangential coordinates u, v, w. A general conic which touches the tangents drawn to Σ_1 from A_1 and P_1 has then the tangential equation

$$(a_1 u + b_1 v + c_1 w)\,(a_2 a_3\,\eta\zeta u + b_2 b_3\,\zeta\xi v + c_2 c_3\,\xi\eta w) + \lambda_1\Sigma_1 = 0.$$

It can be shewn that, if we take $\lambda_1 = (\alpha_1\xi + \beta_1\eta + \gamma_1\zeta)/\Delta$, this conic is unaltered by cyclical change of the suffixes 1, 2, 3; it is therefore similarly derivable from A_2, P_2, Σ_2 or A_3, P_3, Σ_3. Let it then be denoted by ω. It touches the nine lines consisting of p_1, p_2, p_3, the pair of tangents from A_1 to Σ_1, the pair of tangents from A_2 to Σ_2, and the pair of tangents from A_3 to Σ_3. It can be shewn that ω is triangularly inscribed to the six conics containing the fives of the six points A, B, C, A_1, A_2, A_3.

Let S_1 denote the conic containing A, B, C, A_2, A_3; and so on. We first prove that ω is triangularly inscribed to S_1. For this, we draw the tangents to ω from A_3, which lies on S_1, and we find the other two points of S_1 lying on these tangents, and shew that the join of these touches ω. The tangents to ω from A_3 are the tangents from A_3 to Σ_3, and Σ_3 (like Σ_1, Σ_2) is triangularly inscribed to S_1. The two tangents to ω from A_3 thus meet S_1 in two points whose join touches Σ_3. This join is easily seen to be p_2. As p_2 touches ω it thus follows that ω is triangularly inscribed to S_1. By a similar proof, it is triangularly inscribed to S_2 and S_3.

Now it can be shewn that if the points A_1, A_2, A_3 be taken for triangle of reference, putting $u' = a_1 u + b_1 v + c_1 w$, etc., and $\xi' = \alpha_1\xi + \beta_1\eta + \gamma_1\zeta$, etc., then the equation of the conic ω assumes a form bearing the same relation to the determinant formed from α_1, β_2, γ_3, etc., as does the original to the determinant Δ. Whence it follows that ω is triangularly inscribed to the three conics each containing A_1, A_2, A_3 and a pair of A, B, C. We have shewn that it is triangularly inscribed to the other three conics arising from the six points.

The tangential equation of ω involves the arbitrary ξ, η, ζ homogeneously to the second order. Thus, if two lines be given, in general position, there are four sets of values of the ratios of ξ, η, ζ for which the corresponding conics ω touch these lines. The case in which the two given tangents meet in one of the six points A, B, C, A_1, A_2, A_3 is exceptional. Whatever ξ, η, ζ may be, there are two tangents of ω through every one of these six points; and conversely it can be shewn that ξ, η, ζ can be chosen so that these two tangents are arbitrary for *one* of the six points. The other three sets of ratios of ξ, η, ζ securing that ω should touch these two lines, through such a point, say A, are not in fact distinct, and the other three conics ω corresponding thereto are merged into an infinite series of degenerate conics, each consisting of the pencil of the lines whose centre is A taken with a pencil whose centre is any point of the conic containing B, C, A_1, A_2, A_3. When the six points satisfy the single condition of lying on a cubic curve which has a double point at A, whose tangents thereat are the two given lines through A, the conic ω that should touch these lines coincides with a particular one of these degenerate conics. (Cf. Exx. 197, 198 above.)

259. We may consider, in connexion with the six conics containing the fives of six points, another set of six conics. In explaining this, for the sake of brevity, we use the notation of matrices. For a conic whose point equation is $ax^2 + by^2 + cz^2 + 2fyz + 2gzx + 2hxy = 0$, we consider the matrix

$$\begin{pmatrix} a, & h, & g \\ h, & b, & f \\ g, & f, & c \end{pmatrix}$$

which we may denote by m. The elements of the inverse matrix m^{-1} are then the corresponding minors in the determinant Δ, of the matrix m, each divided by Δ, namely $A\Delta^{-1}$, $B\Delta^{-1}$, etc. The matrix of the coefficients in the tangential equation of the conic is m^{-1}; we denote Δ by $|\,m\,|$. Now consider two conics, with respective matrices m, m_0, of which the former is triangularly circumscribed to the latter. It can be shewn that the condition for this is that the sum of the square roots of the roots of the cubic equation in λ which is the discriminantal determinant of $m - \lambda m_0$, or of $mm_0^{-1} - \lambda$, should be zero. According to the signs given to the square roots this can arise in one of four ways. When one way is chosen, there is another conic, of matrix m' say, in regard to which all the possible triangles inscribed in m and circumscribed about m_0, are self-polar triangles, the conics m, m_0 being polar reciprocals of one another in regard to the conic m'. It can easily be proved that the condition for this last relation is that the matrices satisfy the equation $m = m'm_0^{-1}m'$, which is equivalent to $m(m')^{-1} = m'(m_0)^{-1}$. In the figure taken, the conic m is outpolar to m', and m_0 is inpolar to m', so that m' is outpolar to m_0. The condition, that a conic of matrix M should be outpolar to a conic of matrix M', is that the sum of the roots of the cubic equation in λ, which is the discriminantal determinant, $|\,M - \lambda M'\,|$, of the conic whose matrix is $M - \lambda M'$, or of the determinant $|\,M(M')^{-1} - \lambda\,|$, should be zero. Thus, in our case, the sum of the roots for both the cubic equations in λ expressed by $|\,m(m')^{-1} - \lambda\,| = 0$, $|\,m'(m_0)^{-1} - \lambda\,| = 0$ is zero. Also, because the two tangents to the conic m_0, drawn from any point of the conic m, are conjugate lines in regard to the conic m', it follows that the conic m is the harmonic locus of the two conics m_0, m'. Further, from the equation $mm_0^{-1} = m'm_0^{-1}m'm_0^{-1}$, or $mm_0^{-1} = (m'm_0^{-1})^2$, we have, for any value of λ,

$$mm_0^{-1} - \lambda = (m'm_0^{-1} - \lambda^{\frac{1}{2}})\,(m'm_0^{-1} + \lambda^{\frac{1}{2}}),$$

by which is expressed the consistence of the fact that the sum of the roots of the cubic equation in θ, $|\,m'm_0^{-1} - \theta\,| = 0$, is zero, with the fact that the sum of the square roots of the roots of the equation $|\,mm_0^{-1} - \theta\,| = 0$ is zero.

Suppose now we are given six points, and that m_1, m_2, ..., m_6 are the matrices of the six conics each containing five of these points. Let m_0 be the matrix of a conic, investigated in the preceding example, which is triangularly inscribed to all of m_1, m_2, ..., m_6. There will then be six other conics m_1', m_2', ..., m_6', such that m_i is outpolar to m_i', and m_i' is outpolar to m_0 (for $i = 1$, ..., 6). These six conics m_i', being all outpolar to the conic m_0, will be linearly connected. Further, the pair of tangents to m_0, drawn from the common point of the conics m_1, m_2, ..., m_5 are a degenerate conic outpolar to all of m_1', m_2', ..., m_5'. In other words, these five conics are such that the conic generally existing which is outpolar to all of them, breaks up into two lines; so that, if m_1', ..., m_4' were assigned, m_5' is such as to be inpolar to one of the three pairs of lines which exist forming a conic outpolar to m_1', ..., m_4'.

260. If A, A', B, B', C, C' be any six points of a plane, the locus of a point P such that the three pairs of lines PA, PA'; PB, PB'; PC, PC' belong to a pencil in involution, is a cubic curve, containing the six points A, A', etc., and also the six points of intersection $(BC, B'C')$, $(CA, C'A')$, $(AB, A'B')$, $(BC', B'C)$, $(CA', C'A)$, $(AB', A'B)$. In particular when the point A' lies on the line BC, the cubic curve degenerates, consisting then of the line BC and a conic.

261. Four tangents of a conic have six points of intersection. Prove that these six points, together with the six intersections of any other four tangents of the conic, lie on a cubic curve. If any cubic curve be drawn through the six intersections of four tangents of a conic, and, from the three points of the curve which lie on any other tangent of the conic, there be drawn the three other tangents of the conic, prove that the angular points of the triangle formed by these lie on the cubic curve. (Cf. p. 372.)

262. A cubic curve passing through the six intersections of any four lines is drawn to touch the two lines which can be drawn from an arbitrary point T to be conjugate in regard to all the conics touching the four given lines. The points of contact of the cubic curve, with these two lines from T, are denoted by Q and R; and the conic which touches the four lines and touches QR is denoted by Σ. The other two tangents to Σ drawn from Q and R meet in P. Then the conic touching TQ, TR at Q and R, respectively, and passing through P is drawn, say ω. Shew that the cubic curve passes through P and touches the conic ω at P. Prove that the common tangent of the two curves at P is the polar of T in regard to the conic Σ, and is a tangent of the conic which is the envelope of the polars of T in regard to all conics touching the four given lines.

A curve of class three touching the six joins of four points which are such that any one is the orthocentre of the other three, is drawn to contain the Absolute points of the plane. Prove that this curve touches the nine-point circle of the four points. (F. Morley, H. W. Richmond.)

263. [See Note 1, below.] Prove Clifford's theorem (*Collected Papers*, 1882, p. 52) that the Miquel circle of five general lines in a plane, namely the circle containing the foci of the five parabolas which touch the fours of these lines, may be defined as the locus of a point, P, which has the property that the feet of the perpendiculars drawn from P to these lines determine a conic containing P.

Shew that the condition that the Clifford conic should be a rectangular hyperbola is that the point P should be the vanishing point of the determinants in the matrix M_4 arising in (12) in Note 1, below. This vanishing point is found as the point of concurrence of four circles, each containing centres of circles associated with four of the five lines (Kantor and de Longchamps); and is proved to lie on the Miquel circle. The remark that the Clifford conic for this point is a rectangular hyperbola was made by Mr J. H. Grace.

264. [See Note 1, below.] The geometrical proof in (14), Note 1, below, of the theorem of de Longchamps, with Dr Bath's addition, takes a very simple form if we use algebra. Consider a rational quartic curve, in space of four dimensions, of which the coordinates x, y, z, t, u of a point are θ^4, θ^3, θ^2, θ, 1. Let a plane, ϖ, with equations $y/b = z/c = t/d$, meet this curve in the points I, J, for which, respectively, $\theta = \infty$, $\theta = 0$. For coordinates in this plane we can take x, z, u; so that J is $x = 0 = z$ and I is $u = 0 = z$. The chords of the curve which meet the plane ϖ describe thereon a conic, μ, whose equation is

$$(c^2 x - b^2 z)\,(c^2 u - d^2 z) = (bd - c^2)^2 z^2;$$

this conic contains I, J, and we speak of it as the Miquel circle. On this circle there is a point which lies on a chord of the quartic curve whose end points are (on the curve) harmonic conjugates in regard to I and J. The coordinates of this point, M, are $(bcd^{-1}, b, c, d, cdb^{-1})$.

If we project the quartic curve, on to the plane ϖ, from the chord joining the two points O, R of the curve, whose parameters are respectively θ, ϕ, we obtain a circle whose equation is

$$\{x - zc^{-1}[b(\theta + \phi) - c\theta\phi]\}\,\{u\theta\phi - zc^{-1}[d(\theta + \phi) - c]\} = z^2 c^{-2}\{b - c(\theta + \phi) + d\theta\phi\}^2.$$

The circles so obtained when R (and ϕ) varies, all pass through the point of the Miquel circle lying on the chord from O which meets this circle. The centre of such a circle, for general θ, ϕ, is the point

$$x = b(\theta + \phi) - c\theta\phi, \quad z = c, \quad u = d(\theta^{-1} + \phi^{-1}) - c\theta^{-1}\phi^{-1}.$$

The locus of such centres, as R varies, O remaining fixed, is a circle, say Ω, whose equation is

$$(x - b\theta z c^{-1})\,(u - d\theta^{-1} z c^{-1}) = (b - c\theta)\,(d - c\theta^{-1}) c^{-2} z^2.$$

Whatever O may be, all these circles pass through the point M (which is

Dr Bath's result). Again the centre of the circle Ω has coordinates $x = b\theta$, $z = c$, $u = d\theta^{-1}$, and, as O varies, all these centres lie on the circle whose equation is $xu = bdc^{-2}z^2$. Mr J. H. Grace remarks that the point M is the inverse of the centre of the Miquel circle in regard to the circle $xu = bdc^{-2}z^2$.

The justification for describing μ as the Miquel circle is explained below, in Note 1 (14). The general theorem of de Longchamps for $n+1$ lines in a plane may likewise be obtained algebraically as here, by considering a rational curve of order n in space of n dimensions; and it appears that Dr Bath's result has a generalisation to this case.

265.   Prove that three parabolas can be drawn to touch the sides of a triangle, with axes containing the circumcentre. If S be the focus of any one of these, and ST the diameter of the circumcircle, lying on the axis of this parabola, prove that the pedal line of T, in regard to the triangle, is parallel to this axis, and is one of the cuspidal tangents of the Steiner curve which is the envelope of the pedal lines (see Note 1 (11), below). Thus the axes of the three parabolas are parallel to the three cuspidal tangents (J. A. Todd).

266.   Let the circle containing the common points of two confocal conics be called the co-director circle for these confocals. Take a fixed ellipse, and a fixed point P, external thereto. Consider the concentric circles which are the co-director circles of the ellipse taken in turn with the two confocal conics which pass through P; one of these, of radius r_1, meets the ellipse in four real points, the other, of radius r_2, contains P in its interior. Let two circles be described, with centre at P, of respective radii $R_1 = r_2 - r_1$, and $R_2 = r_2 + r_1$. Prove that both these circles are triangularly circumscribed to the ellipse. Consider the infinite series of triangles which can be inscribed in one of these circles, say (R_1), all circumscribed to the ellipse. Prove that the nine-point circles of the triangles of this series envelope the two co-director circles, all having the diameter R_1. Likewise the nine-point circles of the triangles circumscribing the ellipse which are inscribed in the circle (R_2) envelope the two co-director circles. When P is on the ellipse we obtain a theorem already given (Ex. 247).

267.   Consider an involution of ∞^1 sets of three variable points A, B, C, upon a circle Ω. As remarked in Ex. 228, the pedal lines of a fixed point D, of the circle, in regard to the various triangles ABC, pass through a point D_0. Taking another point E, of the circle, the pedal lines of E pass through a point E_0. Prove that the lines of the pencils (D_0), (E_0), for the same triangle ABC, meet in a point which describes a circle, ω, as ABC varies. In particular, when DE is a diameter of the circle Ω, then $D_0 E_0$ is a diameter of ω. Prove that, in this case, ω is the co-director circle of that conic, Σ, which is touched by the sides of all the triangles ABC, and the confocal to this which touches the line DE. In general there is another diameter $D'E'$, of the circle Ω, which touches the latter confocal. But, when this latter contains the centre of Ω, the circle ω touches the nine-point circles of all the triangles ABC (cf. Ex. 266).

268.   On an arbitrary conic Ω, take sets of three points A, B, C forming an involution founded on two such sets, and consider, for the triangle ABC, any one of the three circles—the circumcircle, the nine-point circle, or the self-conjugate circle. Then, as ABC varies, the circle considered remains perpendicular to a fixed circle, and its centre describes a conic. The circle in fact envelops a quartic curve having double points at the Absolute points of the plane (cf. *Principles of Geometry*, vol. IV, chap. III). Determine geometrically the circle to which the variable circle is perpendicular, in the three cases; and the condition for the quartic curve to break up into two circles. [We may take Ω to have the equation $zx - y^2 = 0$, the Absolute points being given by $y = 0$, $ax^2 + 2gxz + cz^2 = 0$, and the sets ABC by an equation $\theta(\theta^2 + \theta m + n) + \lambda(\theta^2 + \theta u + v) = 0$, in which λ varies but m, n, u, v are constants. (Cf. also Roberts, *Examples for Conics*, 1884, Section II.)]

NOTE 1

Steiner's hypocycloid. Systems of lines in a plane

(1) It was proved by Steiner (*Ges. Werke*, ii, p. 641, 1856) that the pedal line, of a variable point on the circumcircle of a fixed triangle, envelops a curve (of the third class, and of order four) having three cusps. These cusps are symmetrically placed upon a circle, forming an equilateral triangle; this circle has its centre at the centre of the nine-point circle of the triangle, its radius being three times that of the nine-point circle. It is interesting to state Steiner's construction for these cusps, in the metrical form he employs: Let D, E, F be the respective mid points of the sides BC, CA, AB of the given triangle, and P, Q, R the points where these sides are, respectively, met by the perpendiculars from the opposite angular points; let N be the centre of the nine-point circle of the triangle. On the arc DP of the nine-point circle, take the point X, one-third of the way from D to P. Find points Y, Z similarly, on the arcs EQ, FR. Then X, Y, Z form an equilateral triangle, and, at each of these points, the envelope of the pedal line touches the nine-point circle. If, on the line XN, beyond N, the point X' be such that the length of NX' is three times the radius of the nine-point circle (so that the length of XNX' is twice the diameter of the circle), then X' is one of the cusps of the envelope, and the two other cusps are similarly found. In other, still more metrical, terms, referred to but not developed by Steiner, we may say: On the interior of the circle, with centre at N and radius three times the radius of the nine-point circle, let a circle, of radius equal to that of the nine-point circle, roll, without slipping. Then the locus of a proper point fixed in the circumference of the rolling circle is the curve enveloped by the pedal line; so that we speak of this curve as a three-cusped hypocycloid. The two points of view can be brought together, and Steiner's theorem proved, as follows: let H be the orthocentre of the fixed triangle, S a variable point on the circumcircle of this triangle, and M the mid point of H, S. Then M is on the nine-point circle, the line NM being parallel to that joining the circumcentre of the triangle to the point S; and the pedal line of S passes through M. As S describes the circumcircle, the line NM rotates about N, and the pedal line rotates about M. It can be shewn that the pedal line rotates in the opposite direction to NM, and at half the angular rate of NM. If the circumcentre be O, this can be proved by shewing that the pedal line is inclined, to any side, BC, of the triangle, at an angle equal to half the inclination of OS to OA. The angles must be measured in proper sense; precisely, if a denote the side BC, and k denote the pedal line, we have $\epsilon O\,(S, A) = [\epsilon\,(a, k)]^2$. The pedal line meets the rolling circle spoken of in M, where the rolling circle touches the nine-point circle, and in another point; this other

point is the point which is fixed in the rolling circle, and is the point of contact of the rolling circle with its envelope.

The following examples are a, by no means complete, selection of theorems relating to circles which may be considered in connexion with given lines of a plane. The study of them has probably been stimulated by Steiner's results. These occur in (7), (8), (11).

(2) Let the lines forming the sides BC, CA, AB, of a fixed triangle, be respectively denoted by a, b, c. Let d, e, f denote any lines, drawn respectively through A, B, C, subject to the conditions that $\epsilon\,(d,\,a) = \epsilon\,(e,\,b) = \epsilon\,(f,\,c)$, (that is, in metrical terms, so as to be equally inclined, in the same sense, to the respective sides). Then the triangle formed by the lines d, e, f has a fixed circumcentre, which is therefore at the orthocentre of the triangle ABC.

Of this elementary theorem we may give a proof founded on the ideas used in the foregoing chapter on length (Chap. xv), which are useful also for the proof of some of the following results.

Let O, I, J denote the symbols of the circumcentre of the given triangle, and of the Absolute points. Thus the symbols of the angular points A, B, C may (Chap. ix (17)) be taken to be

$$A = \alpha I + O + \alpha^{-1}J, \quad B = \beta I + O + \beta^{-1}J, \quad C = \gamma I + O + \gamma^{-1}J;$$

and the side BC contains the point of the Absolute line whose symbol is $\beta\gamma I - J$. Wherefore, the orthocentre H, of the triangle ABC, lies on the line joining A to the point $\beta\gamma I + J$; so that its symbol may be taken to be $H = (\alpha + \beta + \gamma)\,I + O + (\alpha^{-1} + \beta^{-1} + \gamma^{-1})\,J$; for this is $\alpha I + O + \alpha^{-1}J + (\beta^{-1} + \gamma^{-1})\,(\beta\gamma I + J)$. If m be the value of $\epsilon\,(d,\,a)$, the lines e, f, which, by definition, are those, respectively, which join $\beta I + O + \beta^{-1}J$ to $\gamma\alpha I - mJ$, and join $\gamma I + O + \gamma^{-1}J$ to $\alpha\beta I - mJ$, meet in the point P whose symbol is

$$P = (\beta + \gamma - \alpha m^{-1})\,I + O + (\beta^{-1} + \gamma^{-1} - m\alpha^{-1})\,J;$$

for, with this value, we have

$$P = \beta I + O + \beta^{-1}J + (\alpha^{-1} - m^{-1}\gamma^{-1})\,(\gamma\alpha I - mJ),$$

for e, and a similar equation for f. The three such intersections as P, for (e, f), (f, d), (d, e), have all, therefore, a symbol included in the form

$$[\alpha + \beta + \gamma + \theta\,(m^{-1} + 1)]\,I + O + [\alpha^{-1} + \beta^{-1} + \gamma^{-1} + \theta^{-1}\,(m + 1)]\,J,$$

namely, for $\theta = -\alpha$, $\theta = -\beta$, $\theta = -\gamma$, respectively. This last symbol, however, is $\theta\,(m^{-1} + 1)\,I + H + \theta^{-1}\,(m + 1)\,J$, and so represents, for varying θ, a variable point of a circle whose centre is H, and radius $m^{\frac{1}{2}} + m^{-\frac{1}{2}}$. In metrical terms, this radius is twice the radius of the circumcircle of ABC multiplied by the cosine of the angle which the lines d, e, f make with the sides of the triangle ABC.

This proves the theorem stated.

(3) Let C_1, C_2, C_3, C_4 be any four points of a circle, whose centre is T; and let $K_1, \ldots, K_4$ be the respective orthocentres of the triangles $C_2 C_3 C_4, \ldots, C_1 C_2 C_3$, formed by threes of the four points. We prove that

these orthocentres $K_1, \ldots, K_4$ lie on a circle, whose radius is equal to that of the circle (T) containing $C_1, \ldots, C_4$ (Steiner, *Ges. Werke*, I, p. 128). For this, remark that the mid points, of the four pairs of points $C_1, K_1; \ldots; C_4, K_4$, are the same point (in fact the centre of the rectangular hyperbola containing $C_1, \ldots, C_4$); denote this point by U. On the line TU, take the point O, such that U is the mid point of T, O; then O is a definite point, determined by $C_1, \ldots, C_4$. As TK_1 is parallel to C_1O, and OK_1 is parallel to C_1T, the points $K_1, \ldots, K_4$ lie on a circle, with centre at O, with radius equal to that of the original circle (T).

(4) Now suppose we are given four arbitrary lines, in general position. We may denote these lines by 1, 2, 3, 4, respectively; and we denote the respective intersections 23, 31, 12, 14, 24, 34 by P, Q, R, L, M, N. We know (Wallace, Steiner) that the four circumcircles, PQR, PMN, QNL, RLM meet in a point, say S; also that the centres of these circles, which we denote, respectively, by C_4, C_1, C_2, C_3, lie on a circle; and that this circle passes through S. Let T be the centre of this circle. Then, as C_4 is the centre of the circle QRS, the point C_4 is the common point of the lines, through the mid points of S, Q, and of S, R, respectively perpendicular to SQ, SR; and C_4 is on the line, through the mid point of S, P which is perpendicular to SP. In a similar way, C_2 is on the line through the mid point of S, Q which is perpendicular to SQ, and C_3 on the line through the mid point of S, R which is perpendicular to SR.

Thus, conversely, if we start from an arbitrary circle, of centre T, on which C_1, C_2, C_3, C_4 and S are arbitrary points, and then take the optical images of S in regard, respectively, to the lines C_4C_1, C_4C_2, C_4C_3, C_2C_3, C_3C_1, C_1C_2, calling these, respectively, P, Q, R, L, M, N, these points will be the common points of pairs of four lines, say 1, 2, 3, 4, which, respectively, contain LQR, MRP, NPQ and LMN. By what has been said, these are four quite general lines; and when they are given, the points $C_1, \ldots, C_4$ and S can be appropriately found.

It can be shewn that the triangles PQR and $C_1C_2C_3$ are, in metrical phraseology, similar to one another, and are in perspective, the lines PC_1, QC_2, RC_3 meeting at a point of the circle (T). If the symbols of C_1 and S, in terms of the symbols of T and the Absolute points, be, respectively, $\delta I + T + \delta^{-1}J$ and $\theta I + T + \theta^{-1}J$, the centre of perspective has the symbol $\theta^{-1}\delta^2 I + T + \theta\delta^{-2}J$. Similarly for other pairs of triangles. The remark that the triangles are in perspective is attributed to Hermes, *Nouv. Annales*, XVIII, 1859 (J. P. Gabbatt, *Proc. Camb. Phil. Soc.* XXI, 1923, p. 349).

(5) The point O, determined in (3) above from the points C_1, C_2, C_3, C_4, independently of the position of S on the circle, has particular relations with the quadrilateral of lines 1, 2, 3, 4. Let H_1, H_2, H_3, H_4 be the respective orthocentres of the triangles PMN, QNL, RLM, PQR, formed by the threes of the four lines; also denote the mid points, respectively, of C_1, H_1; C_2, H_2; C_3, H_3 and C_4, H_4 by N_1, N_2, N_3, N_4, these

being the centres of the nine-point circles of these triangles, respectively. Then we can prove that the lines through $N_1, ..., N_4$, respectively perpendicular to $C_1H_1, ..., C_4H_4$, all pass through O (Kantor, *Sitzungsber. Wiener Akad.* LXXVIII (1879), p. 209). Also, we know that the points $H_1, ..., H_4$ lie on a line, the directrix of the parabola touching the four lines 1, 2, 3, 4. Let the line through O, perpendicular to this directrix, meet this in X. It can be proved that the lines $XN_1, ..., XN_4$ are respectively perpendicular to the lines 1, 2, 3, 4 (Morley, *Inversive Geometry*, 1933, p. 247). Further, U being the mid point of T, O, the line from U, perpendicular to the directrix, is the median line of the quadrilateral 1234, that is, contains the mid points of the three complementary pairs, P, L; Q, M; R, N, of intersections of the lines. This is remarked by Kantor.

(6) The verifications, by the symbols, of the three results stated in (5), may be given in outline. For this, take the symbols of $C_1, ..., C_4$ and S, in terms of the symbol of T and of the Absolute points I, J, to be $-\alpha I + T - \alpha^{-1} J, ..., -\delta I + T - \delta^{-1} J$ and $\theta I + T + \theta^{-1} J$; and denote $\alpha\beta\gamma\delta/\theta$ by $-r$. Then, from the symbol of K_4, already remarked in (2), we find the symbol of U, and hence the symbol of O, in the form $O = -\sigma_1 I + T - \sigma_{-1} J$, where $\sigma_1 = \alpha + \beta + \gamma + \delta$ and $\sigma_{-1} = \alpha^{-1} + ... + \delta^{-1}$. Also, the symbol of P, the optical image of S in regard to the line C_4C_1, is $P = -(\alpha + \delta + \theta^{-1}\alpha\delta) I + T - (\alpha^{-1} + \delta^{-1} + \theta\alpha^{-1}\delta^{-1})J$; for, with this, it can be verified that

$$S - P = (\alpha^{-1} + \delta^{-1} + \theta^{-1} + \theta\alpha^{-1}\delta^{-1})(\alpha\delta I + J), \quad \tfrac{1}{2}(S+P) = \rho C_4 + \sigma C_1,$$

where the numbers ρ, σ are given by

$$\rho - \tfrac{1}{2} = -(\sigma - \tfrac{1}{2}) = \tfrac{1}{2}(\theta - \theta^{-1}\alpha\delta)/(\alpha - \delta);$$

the former equation shews that SP is perpendicular to C_4C_1, and the latter that the mid point of S, P lies on C_4C_1. In terms of O, instead of T, the symbol of P is then

$$P = (\beta + \gamma + r\beta^{-1}\gamma^{-1}) I + O + (\beta^{-1} + \gamma^{-1} + r^{-1}\beta\gamma) J;$$

and there are similar forms for the symbols of Q, R, L, M, N.

From these forms we find that the line QR, or 1, joins the point on the Absolute line whose symbol is $rI + \alpha J$, to the point, of the circle before considered with centre at O, whose symbol is $\alpha I + O + \alpha^{-1}J$; the other intersection of the line 1 with this circle is the point whose symbol is $\alpha_1 I + O + \alpha_1^{-1} J$, where $\alpha_1 = -r\alpha^{-2}$. To see this, remark that the symbols found for Q and R are, respectively, capable of the forms

$$Q = \alpha I + O + \alpha^{-1}J + \mu(rI + \alpha J), \quad R = \alpha I + O + \alpha^{-1}J + \nu(rI + \alpha J),$$

where μ, ν denote, respectively, $\gamma^{-1}\alpha^{-1} + r^{-1}\gamma$ and $\alpha^{-1}\beta^{-1} + r^{-1}\beta$. Of the two points thus determining the line 1, that on the circle (O), whose symbol in terms of T, instead of O, is $-pI + T - p'J$, in which p, p' momentarily stand respectively for $\beta + \gamma + \delta$, $\beta^{-1} + \gamma^{-1} + \delta^{-1}$, is the point K_1, the orthocentre of the triangle $C_2C_3C_4$ (see (2) above). It is in fact a well-known property of the pedal line of S in regard to the triangle $C_2C_3C_4$ that the line 1 contains K_1. And we have the relation among the symbols $K_1 - O = T - C_1$. We see then that the lines $1,2,3,4$ may be regarded as respectively determined by α, β, γ, δ; and each contains a point, on the circle (O), which does not depend on the position of S. We may remark, also, that in terms of O, instead of T, the symbol of C_1 is $pI + O + p'J$.

Now let s_1, s_2, s_3, respectively, denote $\alpha + \beta + \gamma$, $\beta\gamma + \gamma\alpha + \alpha\beta$, $\alpha\beta\gamma$. Then the orthocentre, H_4, of the triangle PQR has the symbol $H_4 = rs_1s_3^{-1}I + O + r^{-1}s_2J$. To see this we need only shew that, with this form of H_4, the line PH_4 contains the point $rI - \alpha J$, since the line 1, or QR, meets IJ in $rI + \alpha J$; and this is at once seen by computing $P - H_4$. The symbol of H_4 (as Kantor remarks) is unaltered by change of α, β, γ, respectively, to $-\alpha$, $-\beta$, $-\gamma$. With the notation suggested, the symbol of C_4 is $s_1I + O + s_2s_3^{-1}J$. Wherefore, we see that the line through the mid point of C_4, H_4, perpendicular to C_4H_4, contains O, the distance of this point from C_4 and H_4 being $(s_1s_2/s_3)^{\frac{1}{2}}$. We prove also easily that the directrix, containing $H_1, \ldots, H_4$, contains the point $rI + \theta J$, on the Absolute line.

Again, as K_4 is $\delta I + O + \delta^{-1}J$, the mid point, U, of C_4, K_4, is given by $U = \frac{1}{2}\sigma_1 I + O + \frac{1}{2}\sigma_{-1}J$, where, as before, $\sigma_1 = \alpha + \beta + \gamma + \delta$, etc. This shews that $\frac{1}{2}(P + L) - U = \frac{1}{2}(\beta^{-1}\gamma^{-1} + \alpha^{-1}\delta^{-1})(rI - \theta J)$, so that the line, from U, perpendicular to the directrix, contains the mid point of P, L. Similarly, it contains the mid points of Q, M and of R, N.

Introducing the further notations

$$\sigma_2 = \beta\gamma + \gamma\alpha + \alpha\beta + \alpha\delta + \beta\delta + \gamma\delta, \quad \sigma_4 = \alpha\beta\gamma\delta,$$

consider next the point X whose symbol is

$$X = \tfrac{1}{2}r\sigma_2\sigma_4^{-1}I + O + \tfrac{1}{2}r\sigma_2J.$$

This gives

$$X - H_4 = \sigma_4^{-1}\left(\tfrac{1}{2}\sigma_2 - s_1\delta\right)(rI + \theta J), \quad X - O = \tfrac{1}{2}\sigma_2\sigma_4^{-1}(rI - \theta J),$$

proving, respectively, that the point X is on the directrix, and that the line XO is perpendicular thereto. The point N_4, the mid point of C_4 and H_4, has the symbol

$$N_4 = \tfrac{1}{2}\left(s_1 + rs_1s_3^{-1}\right)I + O + \tfrac{1}{2}\left(s_2s_3^{-1} + r^{-1}s_2\right)J;$$

hence we find that $N_4 - X = k(rI - \delta J)$, where k stands for $\frac{1}{2}(r^{-1}s_1 - s_2\sigma_4^{-1})$. This shews that the line joining N_4 to X is perpendicular to the line 4, or LMN, which joins the points $\delta I + O + \delta^{-1}J$, $rI + \delta J$. Similarly for the joins of X to N_1, N_2, N_3.

(7) We consider now the line joining the two variable points, on the Absolute line and the circle (O), whose symbols are, respectively, $rI + \phi J$, $\phi I + O + \phi^{-1}J$, in which ϕ is a variable number. The lines 1, 2, 3, 4, considered, are such lines, for $\phi = \alpha$, β, γ, δ. Another such line, when ϕ has the value θ we have used to determine S, is that given by taking, on the circle (O), the point $K = \theta I + O + \theta^{-1}J$, obtained by drawing through O the line OK parallel to TS, and then the line through K parallel to the directrix. The points (ϕ), on the circle (O), through

which the lines to be considered are drawn, form a range, on the circle, which is related to the range of the points $rI + \phi J$, on the Absolute line, through which these lines are respectively drawn; with the condition that the points I, J of the circle correspond, respectively, to the points J, I on the Absolute line (for $\phi = \infty$, 0, respectively). Thus, if one of these lines is given, the others are determined by this relation.

These lines are, in fact, the tangents of a curve of class three and order four. To see this clearly, take the symbol of any point of the plane in the form $\xi I + O + \zeta J$, and write the equation of one of these lines (ϕ) in the form $u\xi + v\eta + w\zeta = 0$. From the two defining points of one of these lines we then find $u = -r^{-1}\phi$, $v = r^{-1}\phi^2 - \phi^{-1}$, $w = 1$. Hence $vwu = r^{-1}w^3 + ru^3$; as this is of order three in u, v, w, three of the lines pass through an arbitrary point of the plane, or the envelope is of class three. Two lines (ϕ), (ψ) meet in the point whose symbol is

$$(\phi + \psi + r\phi^{-1}\psi^{-1})\, I + O + (\phi^{-1} + \psi^{-1} + r^{-1}\phi\psi)\, J,$$

as is clear because this is $\phi I + O + \phi^{-1}J + k\,(rI + \phi J)$, where k stands for $r^{-1}\psi + \phi^{-1}\psi^{-1}$. Thus the line ($\phi$) touches the envelope at the point whose symbol is $(2\phi + r\phi^{-2})\, I + O + (2\phi^{-1} + r^{-1}\phi^2)\, J$; this is the harmonic conjugate, in regard to $\phi I + O + \phi^{-1}J$ and $rI + \phi J$, of the point (of parameter $-r\phi^{-2}$) which is the second point in which the line meets the circle (O). The third tangent to the envelope from the common point of the tangents (ϕ), (ψ) is the line (χ), where $\chi = r/\phi\psi$, the three tangents meeting in the point $(\phi + \psi + \chi)\, I + O + (\phi^{-1} + \psi^{-1} + \chi^{-1})\, J$. Corollaries are: ($a$) If d, e, f be three tangents of the envelope, forming the triangle D, E, F, and d', e', f' be the other tangents, respectively, from D, E, F, then $\epsilon\,(d, d') = \epsilon\,(e, e') = \epsilon\,(f, f')$ (cf. (2) above); (b) To a triangle formed by the tangents (θ), (ϕ), (ψ) corresponds another triangle formed by tangents $(-\theta)$, $(-\phi)$, $(-\psi)$, which are respectively perpendicular to the former tangents; the lines joining corresponding angular points of these two triangles are three other tangents of the curve; (c) The perpendicular tangents (θ), $(-\theta)$ meet on the fundamental circle (O). The line joining the points of contact of the two perpendicular tangents is itself a tangent (with the parameter $r\theta^{-2}$); and the points of contact are at constant distance, equal to twice the length of the diameter of the circle (O). The third tangent of the envelope, drawn from the common point of the two perpendicular tangents, is perpendicular to the line joining the points of contact of these.

(8) The envelope in (7) is the envelope of the pedal lines, not of one triangle only, but of an infinite series of triangles, each formed by three tangents of the curve, of which two may be taken arbitrarily, the pedal line being that of a point on the circumcircle of the triangle.

For take the tangents of parameters α, β, γ, subject to $\alpha\beta\gamma = -r$; the third tangent to the curve from the common point of any two of these tangents is then the perpendicular from this point to the other tangent. The point Z, with symbol

$$Z = (\alpha + \beta + \gamma - 2\theta)\, I + O + (\alpha^{-1} + \beta^{-1} + \gamma^{-1} - 2\theta^{-1})\, J,$$

as θ varies, describes a circle, of radius twice that of the circle (O), with centre at the point K whose symbol is

$$(\alpha + \beta + \gamma)\, I + O + (\alpha^{-1} + \beta^{-1} + \gamma^{-1})\, J,$$

as we see because $Z = -2\theta I + K - 2\theta^{-1} J$. This circle contains the angular points, D, E, F, of the triangle formed by the tangents (α), (β), (γ), namely, for $\theta = \alpha$, $\theta = \beta$, $\theta = \gamma$, respectively. If, from the point Z, the perpendicular be drawn to the side EF of the triangle DEF, it meets this side in the point L of symbol

$$L = (\alpha - \theta - r/\alpha\theta)\, I + O + (\alpha^{-1} - \theta^{-1} - \alpha\theta/r)\, J,$$

as we see by considering that, besides Z, this perpendicular must contain the point $rI - \alpha J$. This point L, however, lies on the line joining $-\theta I + O - \theta^{-1} J$ and $rI - \theta J$; and the corresponding points M, N, on FD, DE, derived from Z, equally lie on this line. This is then the pedal line of Z, in regard to the triangle DEF; and it is a tangent of the envelope, with the parameter $-\theta$. This proves the result enunciated.

It should be noticed that all the triangles DEF have the same nine-point circle, which is the original circle of centre O. For the mid point of E, F, whose symbols are

$$(\gamma + \alpha - \beta)\, I + O + (\gamma^{-1} + \alpha^{-1} - \beta^{-1})\, J$$

and
$$(\alpha + \beta - \gamma)\, I + O + (\alpha^{-1} + \beta^{-1} - \gamma^{-1})\, J,$$

is the point of symbol $\alpha I + O + \alpha^{-1} J$.

(9) Another result, due to Kantor (*loc. cit.* p. 215), may be noticed. In general, when we take five arbitrary lines in a plane, the foci of the five parabolas, which touch the fours of these lines, lie upon a circle. This has been already noticed; an incidental proof arises below. We have seen that the envelope here considered is determined as touching four arbitrary lines. When the fifth line is also a tangent of this envelope, the circle containing the five foci degenerates into a line (taken with the Absolute line). (Cf. also Misc. Ex. 254.)

For we have seen that the focus of the parabola touching the four tangents (α), (β), (γ), (δ) of the envelope, denoted in (6) by S, with the symbol $\theta I + T + \theta^{-1} J$, where $\alpha\beta\gamma\delta = -r\theta$, and $T = O + \sigma_1 I + \sigma_{-1} J$, has the symbol $(\sigma_1 - r^{-1}\sigma_4)\, I + O + (\sigma_{-1} - r\sigma_4^{-1})\, J$. Let ϵ be the symbol of a fifth tangent of the envelope, and denote the symmetric functions for α, β, γ, δ, ϵ by ρ_1, ρ_{-1}, ρ_5, where $\rho_1 = \sigma_1 + \epsilon$, $\rho_{-1} = \sigma_{-1} + \epsilon^{-1}$, $\rho_5 = \sigma_4\epsilon$. Then this focus S has the symbol

$$(\rho_1 - \psi - r^{-1}\psi^{-1}\rho_5)\, I + O + (\rho_{-1} - \psi^{-1} - r\psi\rho_5^{-1})\, J$$

for the particular value ϵ of ψ. This general form is, however,

$$\rho_1 I + O + \rho_{-1} J - k\,(I + r\rho_5^{-1}\, J),$$

where k stands for $\psi + r^{-1}\psi^{-1}\rho_5$; and this shews that S lies on the line joining the two points whose symbols are $\rho_1 I + O + \rho_{-1} J$ and $I + r\rho_5^{-1} J$. As these two points are symmetrical in regard to (α), (β), (γ), (δ), (ϵ), we see that the five foci lie on this line; as was stated.

(10) In connexion with the metrical summary given above in (1), we recall the two following results: (*a*) Let ABC, SIJ be two triangles, both inscribed in a conic σ, whose sides touch the conic ϖ. Let ϖ touch IJ at H, and BC meet IJ in U. Prove that the cross ratio $(U, H/I, J)$ is equal to that of the pencil of lines joining S, A, I, J to any point of the conic σ. This may be seen by regarding JH, HI, IS, SJ as a quadrilateral circumscribing the conic ϖ. (*b*) If p be the pedal line of a point P on the circumcircle of a triangle ABC, whose centre is O, then $\epsilon O(P, A) = [\epsilon (a, p)]^2$, where a denotes the line BC.

(11) We also state Steiner's construction for the cuspidal tangents of the envelope considered, which has been given in (1), in another form: Let K, H, O be, respectively, the circumcentre, the orthocentre, and the nine-point centre of a triangle ABC, and OQ_1 be one of the cuspidal tangents of the curve which is the envelope of the pedal lines of the triangle, for points on the circumcircle. Draw from A, besides the lines AK, AH, the line which is parallel to OQ_1; let these three lines meet the Absolute line, respectively, in K_1, H_1 and O_1. Prove that $[\epsilon (H_1, K_1)]^2 = [\epsilon (O_1, K_1)]^3$. The point Q_1, where the cuspidal tangent touches the envelope, is obtainable from the fact we have remarked (7), that three concurrent tangents of the envelope have parameters subject to $\theta\phi\psi = r$, so that a cuspidal tangent has a parameter θ subject to $\theta^3 = r$, and meets the Absolute line in $rI + \theta J$, namely $I + \epsilon r^{-\frac{2}{3}} J$, if $\epsilon^3 = 1$; this gives one of the three points Q_1; the points K_1, H_1, respectively, may easily be computed; taking the triangle to be as in (8), we have $K_1 = I + \alpha^{-2} J$, $H_1 = I - r^{-1}\alpha J$.

In metrical terms, therefore, the line from A parallel to one of the cuspidal tangents may be obtained by rotating the line AK about A, towards the line AH, through two-thirds of the angular distance of AH from AK. It can easily be seen, by elementary geometry, that this is equivalent with Steiner's rule quoted in (1); of which therefore we have an independent proof. (See also Misc. Ex. 265.)

(12) We pass now to a theorem for any number of lines in a plane, supposed to be in general positions. We have met the theorem that the circumcircles, of the triangles formed by the threes of four arbitrary lines, meet in a point, while the four centres of these circles lie on another circle. Denote the four circles momentarily by $C_1^{(3)}$, $C_2^{(3)}$, $C_3^{(3)}$, $C_4^{(3)}$; and the circle which contains the centres of these by $C^{(4)}$. It can be shewn that, if five arbitrary lines be taken, the circles, each containing four centres, which arise in this way from the fours of these lines, also meet in a point; while the five centres of these lie on another circle. We may say that the five circles $C_1^{(4)}$, $C_2^{(4)}$, ..., $C_5^{(4)}$ meet in a point, and the centres of these lie on a circle $C^{(5)}$.

For the case of five lines, this result is found in Kantor (*loc. cit.*). The extension to any number of lines is however given by de Longchamps, *Nouv. correspond. de Math.* (now continued as *Mathesis*), t. III (1877), pp. 306, 340. We sketch a proof of the general theorem, which is essentially due to J. H. Grace, *Proc. Lond. Math. Soc.* XXXIII (1901), pp. 193–197. Another proof has been given by F. Morley, *Trans. Amer. Math. Soc.* I (1900), p. 99. See also W. B. Carver, *Amer. Journ. of Math.* XLII (1920), p. 143. A proof by projection from a rational curve in higher space was given (*Proc. Camb. Phil. Soc.* XXIX (1933), p. 51) by H. Lob, whose recent death is deplored by all who have read his geometrical papers.

We consider lines in a plane, with equations $u_i = 0$, where

$$u_i = a_i x + b_i y + z,$$

the coordinates being such that the Absolute points I, J, lying in $y = 0$, have coordinates $(1, 0, 0)$, $(0, 0, 1)$, respectively. Then the equation $\Phi_3 = 0$, where

$$\Phi_3 = \begin{vmatrix} a_1 u_1, & a_2 u_2, & a_3 u_3 \\ u_1, & u_2, & u_3 \\ a_1, & a_2, & a_3 \end{vmatrix}$$

represents a conic, which, as we see by inserting the coordinates of I and J, is a circle; this is thus the circumcircle of the triangle formed by the three lines $u_1 = 0$, $u_2 = 0$, $u_3 = 0$. The centre of this circle, which is obtainable by differentiating the equation partially in regard to x and z, is then the point satisfying both the equations

$$\begin{vmatrix} a_1{}^2, & a_2{}^2, & a_3{}^2 \\ u_1, & u_2, & u_3 \\ a_1, & a_2, & a_3 \end{vmatrix} = 0, \qquad \begin{vmatrix} a_1 u_1, & a_2 u_2, & a_3 u_3 \\ 1, & 1, & 1 \\ a_1, & a_2, & a_3 \end{vmatrix} = 0.$$

Thus we may say that this centre is the point for which all determinants of three rows in the matrix, of four rows and three columns,

$$\begin{vmatrix} a_1{}^2 u_1, & a_2{}^2 u_2, & a_3{}^2 u_3 \\ u_1, & u_2, & u_3 \\ a_1{}^2, & a_2{}^2, & a_3{}^2 \\ a_1, & a_2, & a_3 \end{vmatrix}$$

vanish. The vanishing of all these determinants is secured by the vanishing of two of them, if we exclude the common zeros of these two which are not zeros of all the determinants. It is easy to see that these excluded zeros are irrelevant here. It follows thence that the circumcentre lies on the conic represented by $\Phi_4 = 0$, where

$$\Phi_4 = \begin{vmatrix} a_1{}^2 u_1, & a_2{}^2 u_2, & a_3{}^2 u_3, & a_4{}^2 u_4 \\ u_1, & u_2, & u_3, & u_4 \\ a_1{}^2, & a_2{}^2, & a_3{}^2, & a_4{}^2 \\ a_1, & a_2, & a_3, & a_4 \end{vmatrix} .$$

This conic is likewise a circle, and it is symmetrical in regard to the four lines $u_1 = 0$, ..., $u_4 = 0$. Thus we have Steiner's result that the four circumcentres lie on a circle (Steiner, *Ges. Werke*, I (1881), p. 223). Moreover, as Steiner also remarks, the four circumcircles such as $\Phi_3 = 0$ meet in a point. For they meet at the common zero of the determinants of three rows contained in the matrix, of three rows and four columns,

$$M_3 = \begin{vmatrix} a_1 u_1, & a_2 u_2, & a_3 u_3, & a_4 u_4 \\ u_1, & u_2, & u_3, & u_4 \\ a_1, & a_2, & a_3, & a_4 \end{vmatrix}.$$

Two of these determinants, say those formed from the columns 1, 2, 3 and 1, 2, 4, also both vanish at the common zero of the determinants of two rows which are contained in columns 1 and 2 of M_3, that is, at the common point of the lines $u_1 = 0$, $u_2 = 0$. This zero is irrelevant as not being common to all the determinants of three rows contained in M_3.

Differentiating Φ_4 partially in regard to x and z, we see that the centre of the circle $\Phi_4 = 0$ is the common zero of the determinants of four rows, contained in the first four columns of the determinant, of five rows and columns,

$$\Phi_5 = \begin{vmatrix} a_1{}^3 u_1, & a_2{}^3 u_2, & a_3{}^3 u_3, & a_4{}^3 u_4, & a_5{}^3 u_5 \\ u_1, & u_2, & u_3, & u_4, & u_5 \\ a_1{}^3, & a_2{}^3, & a_3{}^3, & a_4{}^3, & a_5{}^3 \\ a_1{}^2, & a_2{}^2, & a_3{}^2, & a_4{}^2, & a_5{}^2 \\ a_1, & a_2, & a_3, & a_4, & a_5 \end{vmatrix}.$$

As before, we shew that $\Phi_5 = 0$ is a circle; which thus contains the centres of all the five circles such as $\Phi_4 = 0$. Moreover, all these circles such as $\Phi_4 = 0$ have a common point, namely, the common zero of the determinants of four rows contained in the matrix

$$M_4 = \begin{vmatrix} a_1{}^2 u_1, & a_2{}^2 u_2, & a_3{}^2 u_3, & a_4{}^2 u_4, & a_5{}^2 u_5 \\ u_1, & u_2, & u_3, & u_4, & u_5 \\ a_1{}^2, & a_2{}^2, & a_3{}^2, & a_4{}^2, & a_5{}^2 \\ a_1, & a_2, & a_3, & a_4, & a_5 \end{vmatrix}.$$

Two of these determinants, say those formed by the columns 1234 and 1235, have also in common the common zero of the determinants, of three rows, contained in the first three columns of M_4. This point is the centre of the circle $\Phi_3 = 0$, formed for u_1, u_2, u_3, and is not a common zero of all the determinants in M_4; so that it is irrelevant.

Obviously the process continues indefinitely. The circle $\Phi_n = 0$, formed from n lines, contains the centres of the n circles $\Phi_{n-1} = 0$ formed for every $(n-1)$ of these lines; and these n circles meet in a point.

(13) It was noted by Steiner (*loc. cit.*) that the common point M_3 of the four circles $\Phi_3 = 0$, arising from four lines, lies on the circle $\Phi_4 = 0$ which contains the centres of the circumcircles of the four triangles formed by these lines. Also, we associate, here, with the name of Miquel, the further theorem that the five points of concurrence, each associated here with a matrix M_3 formed from four of five lines, of the four circles $\Phi_3 = 0$ arising from these four lines, lie on a circle; in other words, that the foci of the five parabolas, each touching four of five

lines, lie on a circle; for distinctness we call this the Miquel circle for five lines, though the theorem appears to be due to Wallace. Recently (*Proc. Camb. Phil. Soc.* xxxv (1939), p. 518), Dr Bath has proved that the common point of the five circles $\Phi_4=0$, arising from the fours of five lines, associated here with a matrix M_4, lies upon the Miquel circle of the five lines. We prove here these three results, employing the notation of (12); another more geometrical proof of Dr Bath's result is given in (14) below. See also Ex. 264 in the Misc. Examples.

Denote the product of $(n-1)$ differences $(a_i-a_1)(a_i-a_2) \dots (a_i-a_n)$ by $f_n'(a_i)$; recall that $\Sigma a_i^k/f_n'(a_i)=0$ for $k=0, 1, \dots, (n-2)$. Put F_2, F_1, respectively, for

$$F_2 = \sum_{i=1}^{4} \frac{a_i^2/f_4'(a_i)}{u_i}, \quad F_1 = \sum_{i=1}^{4} \frac{a_i/f_4'(a_i)}{u_i},$$

so that $u_1 u_2 u_3 u_4 F_2 = 0$ and $u_1 u_2 u_3 u_4 F_1 = 0$, which are integral polynomials of the third order in the coordinates, both represent cubic curves. For simplicity we denote these curves by $F_2=0$, $F_1=0$.

Both these curves, as we easily see, pass through the six common points of the four lines $u_1=0, \dots, u_4=0$, and also through the Absolute points I, J. As two cubic curves have nine common points, the curves $F_2=0$, $F_1=0$ have therefore another common point beside the eight specified points. We shew that this further common point is that in which the four circumcircles $\Phi_3=0$ meet. Consider in fact $F_2-a_4F_1$, which is $\sum_{i=1}^{3} a_i(a_i-a_4)/u_i f_4'(a_i)$ or $\sum_{i=1}^{3} a_i/u_i f_3'(a_i)$. If Φ_3 mean the form given in (12) for u_1, u_2, u_3, this difference is $-\Phi_3$ divided by $u_1 u_2 u_3 (a_2-a_3)(a_3-a_1)(a_1-a_2)$. We may similarly consider $F_2-a_1F_1$, $F_2-a_2F_1$, $F_2-a_3F_1$. The conclusion is that the remaining common point of the cubic curves $F_2=0$, $F_1=0$ is the common point of the four circumcircles arising from forms Φ_3, as was said. Now take the determinant Φ_4 of (12). Divide the columns in turn by the four respective numbers $u_i f_4'(a_i)$, and add then the first three columns to the fourth. The result is

$$\begin{vmatrix} a_1^2 e_1, & a_2^2 e_2, & a_3^2 e_3, & 0 \\ e_1, & e_2, & e_3, & 0 \\ a_1^2 e_1/u_1, & a_2^2 e_2/u_2, & a_3^2 e_3/u_3, & F_2 \\ a_1 e_1/u_1, & a_2 e_2/u_2, & a_3 e_3/u_3, & F_1 \end{vmatrix},$$

where e_i denotes $1/f_4'(a_i)$. This shews, neglecting common zeros of pairs of $u_1, \dots, u_4$, and the points I, J, that the circle $\Phi_4=0$ contains the ninth common point of $F_2=0$, $F_1=0$. This proves Steiner's theorem stated above.

To prove the result for what we have called the Miquel circle, consider the three functions

$$\psi_3 = \sum_{i=1}^{5} a_i^3 c_i/u_i, \quad \psi_2 = \sum_{i=1}^{5} a_i^2 c_i/u_i, \quad \psi_1 = \sum_{i=1}^{5} a_i c_i/u_i,$$

in which c_i denotes $1/f_5'(a_i)$. These, on multiplication by $u_1 u_2 u_3 u_4 u_5$, define three quartic curves, which we may denote by $\psi_3 = 0$, $\psi_2 = 0$, $\psi_1 = 0$. It is easy to see that these quartic curves have the twelve common points given by the ten common points of pairs of the five lines $u_1 = 0, \dots, u_5 = 0$, and the points I, J. Thus any two of these quartic curves have four other common points. From the obvious relations $\psi_3 - a_5 \psi_2 = F_2$, $\psi_2 - a_5 \psi_1 = F_1$, it appears that the two particular quartic curves denoted by $\psi_3 - a_5 \psi_2 = 0$, $\psi_2 - a_5 \psi_1 = 0$ break up into the line $u_5 = 0$ and the cubic curves $F_2 = 0$, $F_1 = 0$, respectively.

Consider now the curve which is the locus of the sets of four points which are given, as λ varies, by the common points of the two quartic curves $\psi_3 - \lambda \psi_2 = 0$, $\psi_2 - \lambda \psi_1 = 0$, the twelve points, consisting of I, J and the common points of pairs of $u_1 = 0, \dots, u_5 = 0$, not being regarded. This curve is of order eight, being given by the equation $\psi_3 \psi_1 - \psi_2^2 = 0$; but this leads to

$$\sum_{i=1}^{5} \sum_{j=1}^{5} a_i a_j (a_i - a_j)^2 c_i c_j / u_i u_j = 0,$$

the curve breaking up into the five lines $u_1 = 0, \dots, u_5 = 0$, together with the cubic curve represented by this latter equation. As $\psi_3 = 0$, $\psi_2 = 0$, $\psi_1 = 0$ all pass through I, J, this cubic curve has a double point at each of I and J, and so breaks up into the Absolute line taken with a circle. This circle contains, we have seen, the common point of $\psi_3 - \lambda \psi_2 = 0$, $\psi_2 - \lambda \psi_1 = 0$ for $\lambda = a_5$, namely, the common point (other than certain eight points) of $F_2 = 0$, $F_1 = 0$; and it equally contains the other four analogous points arising for $\lambda = a_1, \dots, a_4$. We infer then that this cubic curve, arising from $\psi_3 \psi_1 - \psi_2^2 = 0$, consists of the Absolute line, *and the Miquel circle*, as defined above. This circle then appears as what we may describe as the envelope, save for six lines, of the quartic curves given by $\psi_3 - 2\lambda \psi_2 + \lambda^2 \psi_1 = 0$, when λ varies.†

Now take the matrix M_4 of (12). Divide the columns in turn by the respective values of $u_i f_5'(a_i)$, and then consider the two determinants formed from the columns 1, 2, 3, 4 and 1, 2, 3, 5. From the vanishing of these determinants we thence infer

$$\begin{vmatrix} a_1^2 c_1, & a_2^2 c_2, & a_3^2 c_3, & 0 \\ c_1, & c_2, & c_3, & 0 \\ a_1^2 c_1/u_1, & a_2^2 c_2/u_2, & a_3^2 c_3/u_3, & \psi_2 \\ a_1 c_1/u_1, & a_2 c_2/u_2, & a_3 c_3/u_3, & \psi_1 \end{vmatrix} = 0;$$

if, however, we multiply the columns of M_4 by the respective values of $a_i c_i/u_i$, then proceed as before, and then divide the first three columns, respectively, by a_1, a_2, a_3, we obtain a vanishing determinant identical with that found, save that ψ_2, ψ_1 are, respectively, replaced by ψ_3 and ψ_2. In other words, the common zero of the determinants

---

† As Mr J. H. Grace remarks, these quartic curves may be described as those which pass through the ten intersections of the five lines, and through the points I, J, and, besides this, *touch* the line IJ.

1234 and 1235 in M_4 satisfies the two equations $\psi_2 - \lambda\psi_1 = 0$, $\psi_3 - \lambda\psi_2 = 0$, where λ is the quotient of the two determinants

$$\begin{vmatrix} a_1^2 u_1, & a_2^2 u_2, & a_3^2 u_3 \\ u_1, & u_2, & u_3 \\ a_1^2, & a_2^2, & a_3^2 \end{vmatrix}, \qquad \begin{vmatrix} a_1^2 u_1, & a_2^2 u_2, & a_3^2 u_3 \\ u_1, & u_2, & u_3 \\ a_1, & a_2, & a_3 \end{vmatrix}.$$

The four common zeros of these two determinants are the angular points of the triangle formed by the lines $u_1 = 0$, $u_2 = 0$, $u_3 = 0$, and the circumcentre of this triangle. These four points are not common zeros of all the determinants contained in M_4. Wherefore we infer that the point common to the five circles $\Phi_4 = 0$, associated with the matrix M_4, satisfies the equation $\psi_3\psi_1 - \psi_2^2 = 0$. The common zero of the determinants in M_4 does not lie on the Absolute line. Whence, *the five circles $\Phi_4 = 0$ meet on the Miquel circle*; and this is Dr Bath's result referred to.

(14) The reader familiar with the elements of geometry in higher space may welcome the following proof of the results in (13). Suppose that the plane, ϖ, which contains the lines $u_1 = 0$, ..., $u_5 = 0$, lies in a space of four dimensions. Let a space of three dimensions (a *prime*) be drawn, in general position, through each of these lines. The common points of the sets of four of these primes determine five points A_1, ..., A_5, in the space of four dimensions. It is a known fact that a rational quartic curve can be described, in this space, to contain the five points A_1, ..., A_5 and also the two Absolute points I, J of the plane ϖ; by taking five suitable points of reference, the coordinates of any point of this curve can be taken to be θ^4, θ^3, θ^2, θ, 1, where θ is a variable parameter. Let O, R be two points of this quartic curve. Planes drawn through the chord OR, and a variable point of the curve, will each meet the plane ϖ in a point; and, as the variable point describes the quartic curve, the point in the plane ϖ describes a conic, containing the points I, J of the curve. This conic we then describe as a circle, and speak of it as arising by projection of the quartic curve from the chord OR.

It can be shewn that to every point R, of the quartic curve, there corresponds another point R' of the curve, such that the chord RR' has a point in common with the plane ϖ; say the point R_0. Indeed, the space of three dimensions containing the plane ϖ, and the point R, meets the curve again in the point R', besides the three points R, I, J. Further, it can be shewn that the points R_0 so obtained in the plane ϖ, for all positions of R on the quartic curve, lie on a conic containing I and J. This conic we describe therefore as the circle, μ. The corresponding points R, R', of the quartic curve, belong to an involution thereon, being, in the range of points on the curve, harmonic conjugates in regard to the two points, existing on the curve, whereat the tangent of the curve meets the plane ϖ. Let us now project the quartic curve, in turn, from all the chords OR, which contain a definite point O and a varying point R of the curve, on to the plane ϖ. Thereby we obtain an infinite series of circles in the plane ϖ. And it is clear, if

O' be the point corresponding to O in the involution spoken of, that all these circles pass through the point (O_0) where OO' meets ϖ.

In particular, if O be taken at A_5, the circle, obtained by projection of the quartic curve from the chord $A_5 A_4$, is the circumcircle of the triangle obtained by projection of the points A_1, A_2, A_3, on the curve, which, we easily see, is the triangle formed by the lines $u_1 = 0$, $u_2 = 0$, $u_3 = 0$. It follows that the circumcircles of the four triangles, in the plane ϖ, formed by the lines $u_1 = 0$, ..., $u_4 = 0$, meet in the point, on the circle μ, in which the chord joining A_5, to its conjugate point $A_5{}'$, meets the plane ϖ. Likewise, the five points of concurrence, of the circumcircles of the four triangles formed from any four of the five lines $u_1 = 0$, ..., $u_5 = 0$, lie on the circle μ. This is the theorem we have agreed to associate with the name of Miquel. This proof of the theorem is found in *Principles of Geometry*, Vol. IV, p. 64 (1925, and 1940).

Consider again the infinite series of circles, in the plane ϖ, which arise by projection of the quartic curve from all the chords OR of the curve which contain the same point O. We can shew that the centres of these circles lie on a circle. For, the centre of the circle, arising from a particular chord OR, is the common point of the tangents of this circle at the points I, J. The tangent at I is the line, in the plane ϖ, which lies in the space of three dimensions defined by the chord OR, and the tangent, t_i, of the quartic curve at I. The two such threefold spaces, or primes, which give the two tangents at I, J of the circle (OR), pass, respectively, through the planes $[OI, t_i]$, $[OJ, t_j]$; as R varies on the quartic curve, these primes give two pencils of lines in the plane ϖ, with centres, respectively, at I and J, which are both projectively related to the range of points R on the quartic curve. Thus these centres lie on a circle, one of whose two intersections with the Miquel circle μ (other than I, J), is the point in which OO' meets the circle μ. Thus, in particular, we have Steiner's result, that the four centres of the circumcircles of the triangles, formed by the lines $u_1 = 0$, ..., $u_4 = 0$, lie on a circle containing the common point of these circumcircles.

It can now be shewn that the remaining common point, of the circle which is the locus of centres considered, with the Miquel circle μ, is independent of the point O. For, consider a second involution of pairs of points on the quartic curve, those namely which, in the range of points of the curve, are harmonic conjugates in regard to I and J. The lines joining the pairs of points of this involution (as of any other such involution) generate a ruled cubic surface in the space of four dimensions. The surface arising from the involution now considered we denote by ψ. The quartic curve lies on this surface. This surface is projected, from any point O of the curve, by means of the planes which join O to the generating lines of the surface. These planes form one system of planes of a quadric point-cone, with vertex at O, any point of any one of these planes being on this cone. Such a cone is obtained by projecting a quadric surface (lying in a space of three dimensions), not passing through O, with O as centre of projection. Such a cone, therefore, contains another system of planes, passing through O, each of which meets every plane of the first system in a line through O. We can specify this second system of planes. For, as before, let t_i, t_j be the tangent lines of the quartic curve at I, J, respectively, and R be any point of the curve. The two primes determined, respectively, by (OI, t_i, OR) and (OJ, t_j, OR), meet in a plane, passing through OR, which we may denote by ρ. The chords of the quartic curve which meet this plane ρ contain, as particular chords, the tangents t_i, t_j, because ρ is a plane lying in primes containing these tangents. The chords of the curve meeting any plane, such as ρ, which has two points on the curve, determine, as already remarked, an involution on the curve. Of this involution, we have just seen, I and J are the foci. Thus this involution, determined by the plane ρ, is the same as that previously considered; and the chords of the curve, which meet ρ, are the same as those by which we defined the surface ψ. Conversely, then, the plane ρ, which contains O, meets every plane joining O to a chord-generator of the surface ψ, namely, in a line through O. Which shews that the planes ρ, each determined by associating a point R of the curve with the point O, are the second system of planes of the quadric cone, whose vertex is O.

Now, the centre of the circle, in the plane ϖ, which is the projection of the quartic curve from the chord OR, is the point where the plane ρ meets ϖ. The circle, say ω, which is the locus of the centres of these circles, arising with O fixed, when R varies, is thus the section of the plane ϖ by the quadric cone spoken of. We have seen that, besides I and J, one common point of the circle ω, with the Miquel circle μ, arises by taking R to be on the chord of the quartic curve, drawn through O, which meets the plane ϖ. The ruled cubic surface ψ has three points on the plane ϖ; of these I, J are two. Through the third there passes a generator of ψ; the quadric cone spoken of, containing the surface ψ, therefore, also passes through this third point. This point is then the remaining common point of the circles ω, μ. Being determined by the surface ψ, this point is independent of O. It is the point, of the plane ϖ, where this plane is met by the single chord of the quartic curve, meeting ϖ, which joins points of the curve which are harmonic conjugates in regard to I and J. This proves Dr Bath's capital result, in a slightly more general form. In general there is one definite chord of the quartic curve from which the curve is projected into a circle, on the plane ϖ, whose centre is given. But when this centre is taken at the point found, there is an infinite series of such chords.

With use of equations the verification of the result is very brief. See Ex. 264, in the Misc. Examples above.

NOTE 2

A construction, given by Professor F. Morley, for an equilateral triangle, from any given triangle

(1) The theorem to be given here is, we shall see, intimately related to the theory of Steiner's hypocycloid given in Note 1. We begin with a lemma. Let A, B, C, T be any four general points of a plane; let the lines BC, CA, AB be denoted, respectively, by a, b, c. Through A let lines, x_1, x_2, be drawn, such that AT is one of the focal rays of the involution pencil determined by the two pairs (b, c) and (x_1, x_2). Similarly, let lines y_1, y_2 be drawn, through B, such that BT is one of the focal rays of the involution determined by (c, a) and (y_1, y_2); and lines z_1, z_2 be drawn, through C, for which CT is similarly related to (a, b) and (z_1, z_2). As yet we do not distinguish between the two ways in which the names x_1, x_2 are assigned; or between y_1 and y_2, or z_1 and z_2. Let the six points of intersection (y_1, z_2), (y_2, z_1), (z_1, x_2), (z_2, x_1), (x_1, y_2), (x_2, y_1) be, respectively, denoted by P, P', Q, Q', R, R'. It can then be shewn that the lines PP', QQ', RR' meet in a point, so that the triangles PQR, $P'Q'R'$ are in perspective. Also that the two triangles ABC, PQR are in perspective; and the two triangles ABC, $P'Q'R'$ are in perspective. And that the three centres of perspective are in a line.

The statement that PP', QQ', RR' meet in a point follows from a result already given (Chap. XIV (4)); for by the definitions of the six lines x_1, x_2, y_1, y_2, z_1, z_2, these lines touch a conic, and the concurrence of the three lines follows from Brianchon's theorem. Or we can use the symbols: let the symbol of T, in terms of the symbols of A, B, C, be $T = A + B + C$; let the points in which x_1, x_2 meet BC be of symbols $\xi B + C$, $B + \xi C$, respectively, which is in accordance with the definitions of these lines; similarly, let y_1, y_2, respectively, meet CA in points of symbols $\eta C + A$, $C + \eta A$; and z_1, z_2 meet AB, respectively, in points $\zeta A + B$, $A + \zeta B$. The symbols of P, Q, R, P', Q', R' will then, respectively, be

$$P = A + \zeta B + \eta C, \quad Q = \zeta A + B + \xi C, \quad R = \eta A + \xi B + C,$$
$$P' = A + \zeta^{-1}B + \eta^{-1}C, \quad Q' = \zeta^{-1}A + B + \xi^{-1}C, \quad R' = \eta^{-1}A + \xi^{-1}B + C;$$

and, as

$$\xi P + \eta \zeta P', \quad = (\xi + \eta \zeta) A + (\eta + \zeta \xi) B + (\zeta + \xi \eta) C,$$
$$= \eta Q + \zeta \xi Q' = \zeta R + \xi \eta R',$$

it follows that PP', QQ', RR' meet in the point whose coordinates are $(\xi + \eta \zeta, \eta + \zeta \xi, \zeta + \xi \eta)$. The lines AP, BQ, CR clearly meet in the point of symbol $\eta \zeta A + \zeta \xi B + \xi \eta C$, and the lines AP', BQ', CR' meet in the point of symbol $\xi A + \eta B + \zeta C$; and the three centres of perspective lie in a line.

(2) Next let I, J be the Absolute points of the plane. Lines x_1, x_2 can be drawn through A so that, using cross ratios,

$$(AB,\ x_1/AI,\ AJ) = (x_1,\ x_2/AI,\ AJ) = (x_2,\ AC/AI,\ AJ).$$

Let x_1, x_2, AI, AJ meet BC respectively in X_1, X_2, I_1, J_1. The equation $(AB,\ x_1/AI,\ AJ) = (x_2,\ AC/AI,\ AJ)$ gives $(B,\ X_1/I_1,\ J_1) = (X_2,\ C/I_1,\ J_1)$, which shews that the pairs B, C; X_1, X_2; I_1, J_1 are in involution. Thus the focal lines, say AT, AT', of the involution pencil containing the pairs AB, AC; x_1, x_2; AI, AJ, are the bisectors of the angles determined by the two lines AB, AC, while AT is one of the focal lines of the involution which is determined by the pairs AB, AC and x_1, x_2, as in (1). The equations

$$(AB,\ x_1/AI,\ AJ) = (x_1,\ x_2/AI,\ AJ) = (x_2,\ AC/AI,\ AJ)$$

do not determine x_1, x_2 uniquely; they determine three lines x_1, to each of which corresponds a definite x_2, as we may prove. In metrical terms, if x_1 be one of the positions of x_1, the other two positions of x_1 are inclined to this, and to one another, at the angle $2\pi/3$.

(3) Now assume the topological conventions of metrical geometry. Take the particular case of the figure in (1), (2), in which x_1, x_2 are the internal trisectors of the angle A of the triangle ABC, with AB, x_1, x_2, AC in order, and likewise y_1, y_2 the internal trisectors of the angle B, with BC, y_1, y_2, BA in order, and z_1, z_2 the internal trisectors of the angle C, with CA, z_1, z_2, CB in order. As before, let $P = (y_1,\ z_2)$ and $P' = (y_2,\ z_1)$, etc. Then, as we have shewn, the lines PP', QQ', RR' meet in a point. From this we can shew at once, by elementary metrical geometry, that *the triangle PQR is an equilateral triangle, with circumcentre at the common point of PP', QQ', RR'*. For this, let PP', QQ', RR' meet BC, CA, AB, respectively, in X, Y, Z, and meet one another in O. As BP, CP are the internal bisectors, respectively, of the angle $(BC,\ BP')$, and the angle $(CP',\ CB)$, it follows that $P'P$ is the internal bisector of the angle $(P'B,\ P'C)$, and P is the centre of the inscribed circle of the triangle $P'BC$. Thus the line $P'OPX$ is a diameter of the circle BPC. Whence it follows that the angles $(PB,\ PX)$, $(PX,\ PC)$ are, respectively, of Euclidean measures $\frac{1}{2}\pi - \frac{1}{3}C$, $\frac{1}{2}\pi - \frac{1}{3}B$, where C, B, as usual, measure the angles of the triangle at C and B, and also that the angle $(XC,\ XP)$ is of measure $\frac{1}{2}\pi - \frac{1}{3}(C - B)$. If H, K be the orthocentre and circumcentre, respectively, of the triangle ABC, the angle $(AK,\ AH)$ is easily found to be $C - B$; thus the line drawn through A, obtained by rotating AK about A through two-thirds of the angular distance of AH from AK, meets BC in a point X' such that the angle $(X'C,\ X'A)$ is $\frac{1}{2}\pi - \frac{1}{3}(C - B)$. In other words, recurring to a result obtained in Note 1 (in (1) and (11)), *the line P'OPX is parallel to one of the cuspidal tangents of Steiner's hypocycloid for the triangle*; thus it follows that the lines OP, OQ, OR are equally spaced about O, any two of them

being at angular distance $2\pi/3$. This, however, follows at once, without reference to Steiner's hypocycloid, from the fact that the angle (XC, XP) is of measure $\frac{1}{2}\pi - \frac{1}{3}(C-B)$, and the similar angles, made by OQ with CA, and by OR with AB, are, respectively, $\frac{1}{2}\pi - \frac{1}{3}(A-C)$, $\frac{1}{2}\pi - \frac{1}{3}(B-A)$. It can be proved, however, that the lines OP, OQ, OR are respectively perpendicular to QR, RP, PQ, and pass through the respective mid points of Q, R; of R, P; and of P, Q. For the angles (QP', QO), (RO, RP'), that is (QC, QY), (RZ, RB), are respectively equal to complements of the angles (AQ, AY), (AZ, AR), and are therefore equal to one another; while $P'O$ is the internal bisector of the angle $(P'R, P'Q)$. Whence the triangles OQP', ORP' are optical images of one another in the line OP'. Similarly for OQ' and RP; and for OR' and PQ. It follows then that the triangle PQR is equilateral, with its centre at O.

The Steiner hypocycloid of a triangle touches the nine-point circle at three points (the apses of the curve) whereat the tangents are perpendicular to the cuspidal tangents. It has been shewn, in Note 1, that, if the hypocycloid be given, there is an infinite series of circumscribed triangles of the curve, of which two sides are arbitrary tangents, all having the same nine-point circle, for all of which the hypocycloid is Steiner's envelope of the pedal lines. For every such triangle, then, the Morley equilateral triangle, just established, has its sides parallel to the apsidal tangents of the curve.†

(4) With use of the symbols we can compute the points P, Q, R, and shew that there are eighteen equilateral triangles arising from any given triangle, whose angular points lie on nine lines, consisting of three sets of three parallel lines, whose directions are those of the sides of the equilateral triangles.

After what is proved in Note 1, it will be quite general, and most interesting, to suppose the given triangle to be that formed by three tangents of Steiner's hypocycloid, with parameters α, β, γ, such that $\alpha\beta\gamma = -r$. Choose three definite numbers, l, m, n, such that $\alpha = l^3$, $\beta = m^3$, $\gamma = n^3$, so that $l^3 m^3 n^3 = -r$. Let u, v, w be numbers such that $u^3 = 1$, $v^3 = 1$, $w^3 = 1$. Draw from the angular point A, of the triangle ABC considered, two lines meeting BC in D_1, D_2, such that, in terms of the symbols, I, J, of the Absolute points, the lines AB, AD_1, AD_2, AC contain, respectively, the points of symbols $rI + n^3 J$, $rI + umn^2 J$, $rI + u^2 m^2 n J$, $rI + m^3 J$. Likewise, from the angular point B, draw two lines meeting CA in E_1, E_2, such that BC, BE_1, BE_2, BA meet the Absolute line, respectively, in points of symbols $rI + l^3 J$, $rI + vnl^2 J$, $rI + v^2 n^2 l J$, $rI + n^3 J$; and draw from C, similarly, lines CF_1, CF_2, such

---

† Mr J. H. Grace points out that an elementary proof of Morley's equilateral triangle, which begins by establishing (metrically) that the lines PP', QQ', RR' are concurrent, is found in *Proc. Edin. Math. Soc.* xxxii, 1914, p. 121, with a diagram, and developments. See also *Math. Gazette*, xxii, 1938, p. 56 (with a diagram), and *ibid*, xxiii, 1939, where many references are given. See further H. Lob, *Proc. Camb. Phil. Soc.* xxxvi, 1940, p. 401, who obtains the cardioids which Morley uses (*Inversive Geometry*, 1933, p. 244) by projection from higher space.

that CA, CF_1, CF_2, CB contain, respectively, the points $rI + m^3J$, $rI + wlm^2J$, $rI + w^2l^2mJ$, $rI + l^3J$. We have then

$$A\,(D_1,\,B/I,\,J) = A\,(D_2,\,D_1/I,\,J) = A\,(C,\,D_2/I,\,J) = um/n,$$

with corresponding cross ratios vn/l and wl/m in the case of the lines drawn from B, C, respectively. And the six intersections (BE_1, CF_2), (BE_2, CF_1), (CF_1, AD_2), (CF_2, AD_1), (AD_1, BE_2), (AD_2, BE_1) are respectively denoted by P, P', Q, Q', R, R'. Recalling, from Note 1, the symbols of the intersections B, C of two tangents of the hypocycloid, in terms of the symbols of I, J, and the symbol O, of the centre of the nine-point circle of the triangle, the symbol of P must be capable of both the forms

$$P = (\gamma + \alpha - \beta)\,I + O + (\gamma^{-1} + \alpha^{-1} - \beta^{-1})\,J + \mu\,(rI + vnl^2J),$$

$$P = (\alpha + \beta - \gamma)\,I + O + (\alpha^{-1} + \beta^{-1} - \gamma^{-1})\,J + \nu\,(rI + w^2l^2mJ);$$

where μ, ν are certain numbers obtainable by identifying these. Hence we find that the number multiplying I, in the symbol of P, is

$$\alpha + \frac{\beta - \gamma}{w^2m - vn}\,(w^2m + vn - 2l).$$

We similarly find that the number multiplying I, in the symbol of P', is

$$\alpha + \frac{\beta - \gamma}{wm^2 - v^2n^2}\,(wm^2 + v^2n^2 - 2l^2),$$

which, in fact, is obtainable from the former coefficient by replacing u, v, w, l, m, n respectively by u^2, v^2, w^2, l^2, m^2, n^2. From these, noticing that $wm^2 - v^2n^2$ is $(w^2m)^2 - (vn)^2$, we compute that the number multiplying I, in the symbol of $P - P'$, is

$$2\,\frac{\beta - \gamma}{wm^2 - v^2n^2}\,(l - w^2m)\,(l - vn).$$

Likewise, replacing u, l, ..., respectively by u^{-1}, l^{-1}, ..., the number multiplying J, in the symbol of $P - P'$, is

$$2\,\frac{\beta - \gamma}{wm^2 - v^2n^2}\,(l - w^2m)\,(l - vn)\,\frac{vw^2}{l^2m^2n^2}.$$

The line PP' thus meets IJ in the point whose symbol is $l^2m^2n^2I + vw^2J$. By what has been seen in Note 1, this point is on one of the cuspidal tangents of the hypocycloid. Similarly for QQ' and RR'. If we now suppose vw^2/wu^2, and wu^2/uv^2, to be both equal to ω, where ω is one of the two imaginary cube roots of unity, which is secured by taking $uvw = \omega$, then the three lines PP', QQ', RR' are, in metrical terms, at equal angular distances $2\pi/3$.

From the symbol of P we obtain the symbols of Q and R by cyclical change of l, m, n and u, v, w. Hence we can compute $Q - R$. Supposing $uvw = \omega$, we thence find

$$Q - R = -2wu^2\,(m - u^2 n)\,(n - v^2 l)\,(l - w^2 m)\left[I - \frac{vw^2}{l^2 m^2 n^2}\,J\right];$$

this shews that the line QR is perpendicular to PP'. From the form of $Q - R$, also, we can put down the length of the side QR. And we can carry the computation to further detail.

It is obvious that the equation $uvw = \omega$, in which u, v, w are cube roots of unity, and ω is an imaginary cube root, can only be satisfied when two of u, v, w are equal. For instance, if $u = v$, the three possible values of u, v, w are, respectively, $(1, 1, \omega)$, or $(\omega, \omega, \omega^2)$, or $(\omega^2, \omega^2, 1)$. Thus there are nine possible triangles PQR arising when $uvw = \omega$; and similarly there are nine triangles when $uvw = \omega^2$. By considering the symbols we have found for the angular points P, Q, R of such a triangle, we can easily see that the angular points of the eighteen triangles lie on nine lines, consisting of three sets of three parallel lines. For details of these lines, reference may be made to the papers referred to in the footnote to (3) above.

NOTE 3

On Pascal lines

If A, B', C, A', B, C' be six points lying on a conic, the three intersections of opposite pairs of sides of the hexagon formed by these six points, namely, the points $(BC', B'C)$, $(CA', C'A)$, $(AB', A'B)$, lie on a line, called a Pascal line. If we consider the two related ranges of points of the conic which are determined by the fact that the points A, B, C, of one range, correspond, respectively, to the points A', B', C', of the other range, there are two points of the conic of which each corresponds to itself when regarded as a point of both ranges; the join of these two points is the Pascal line in question. By considering all possible orders of the six points, sixty hexagons, and sixty Pascal lines, are obtained. The cointersections of these sixty lines have been the occasion of much study, from the time of Steiner onwards. It has finally been found that the best and simplest view of the figure of sixty lines is obtainable from a figure of fifteen lines in a space of four dimensions, which are such as to meet in threes in fifteen points, of which each line contains three of these points. For an exposition of this higher figure, we refer the reader to *Principles of Geometry*, Vol. ii, pp. 219–236.

But there are certain properties of the Pascal lines which may be stated here, with references only to the plane in which the conic lies:

(*a*) There are twenty points, called *Steiner points*, in each of which three of the sixty Pascal lines meet; one of these points exists on every one of the lines. The twenty Steiner points may be considered as ten pairs, those of a pair being conjugate to one another in regard to the fundamental conic.

(*b*) There are sixty points, called *Kirkman points*, in each of which, also, three Pascal lines meet; three of these points exist on every one of the lines. The distinction between these and the Steiner points will appear.

(*c*) There are forty-five points, in each of which four Pascal lines meet; of these points, three exist on every Pascal line.

Consider first the Steiner points. Denote the Pascal line, defined as above from the hexagon $AB'CA'BC'$, by $(A, B, C/A', B', C')$, this being the join of the common corresponding points of the two ranges $(A, B, C, ...)$, $(A', B', C', ...)$ on the conic. Taking A', B', C' in the three orders A', B', C'; B', C', A'; C', A', B', it can be shewn that the three Pascal lines

$$(A, B, C/A', B', C'), \quad (A, B, C/B', C', A'), \quad (A, B, C/C', A', B')$$

meet in a point. This point is one of the Steiner points. And another such point is found from the three Pascal lines

$$(A, B, C/A', C', B'), \quad (A, B, C/C', B', A'), \quad (A, B, C/B', A', C'),$$

which also meet in a point. These two points are conjugate to one another in regard to the fundamental conic. All the twenty Steiner points are found in this way, by variation of the order in which the six given points of the conic are considered.

Of Steiner's theorem we give here a brief analytical proof, suggested by a proof given by **Dr J. A.** Todd: The quadratic equation for the points of coincidence in a projective relation, in which the elements of parameters a, b, c correspond to the elements whose parameters are a', b', c', is

$$(x-a)(c-b)/(x-b)(c-a) = (x-a')(c'-b')/(x-b')(c'-a');$$

denote $xa+bc$, $xb+ca$, $xc+ab$, $xa'+b'c'$, $xb'+c'a'$, $xc'+a'b'$, respectively, by u, v, w, u', v', w'. Then the quadratic equation for the coincidence points can be written

$$u(v'-w')+v(w'-u')+w(u'-v')=0,$$

or, say $(u, v, w/u', v', w')=0$, as is easily verified. As two pairs of elements determine an involution, we can consider the involution which is determined by the two pairs which are the roots, respectively, of the two quadratic equations in x,

$$(u, v, w/v', w', u')=0, \quad (u, v, w/w', u', v')=0.$$

The two quadratics occurring on the left, in these equations, when added, lead to $-(u, v, w/u', v', w')$. If we regard a, b, c, a', b', c' as values of a parameter for points of a conic, we thus infer, since the joins of three pairs of an involution, on a conic, are concurrent lines, that the Pascal line $(A, B, C/A', B', C')$ passes through the common point of the Pascal lines $(A, B, C/B', C', A')$, $(A, B, C/C', A', B')$. By replacing u', v', w', respectively, by u', w', v', we obtain another point in which three Pascal lines meet. To prove Steiner's result that these two points are conjugate in regard to the conic, we consider the quadrangle I, J, I', J', of four points of the conic, in which I, J are the Hessian points of A, B, C, and I', J' are the Hessian points of A', B', C'. By what has been said in regard to the Hessian points (Misc. Ex. 108), the Hessian points of A, B, C are the common corresponding points of the two related ranges $(A, B, C, ...)$, $(B, C, A, ...)$ on the conic; thus the points I, J are given by the two parameters which are the roots of the quadratic polynomial $(u, v, w/v, w, u)$, and therefore by $(u + \omega v + \omega^2 w) (u + \omega^2 v + \omega w) = 0$, where ω is an imaginary cube root of unity. Whence, the identity

$$(\omega^2 - \omega) (u, v, w/u', v', w') = (u + \omega v + \omega^2 w) (u' + \omega^2 v' + \omega w')$$
$$- (u + \omega^2 v + \omega w) (u' + \omega v' + \omega^2 w')$$

shews that the first Steiner point found above is the vertex of the quadrangle I, J, I', J' which is the intersection of the complementary joins IJ', JI' (the notation for I', J' being properly chosen); and the change of u', v', w', respectively, into u', w', v', shews that the second Steiner point is that vertex of the quadrangle which is the common point of II' and JJ'. This vertex is conjugate to the former in regard to the conic. This proves Steiner's result. The third vertex of the quadrangle is the common point of the Hessian lines $IJ, I'J'$, respectively, of A, B, C and A', B', C'.

The twenty Steiner points do themselves lie in fours upon fifteen lines, forming a familiar configuration. This configuration consists of twenty points and fifteen lines, such that three of the lines pass through every one of the twenty points, and four of the points lie on each of the fifteen lines. The possibility of the existence of such a configuration, in general, follows from the following theorem: Let there be three lines in a plane, passing through a point O, which contain, respectively, the points A_1, A_2, A_3; B_1, B_2, B_3 and C_1, C_2, C_3; so that the three triangles $A_1B_1C_1, A_2B_2C_2, A_3B_3C_3$ are in perspective from O. The axis of perspective of the two latter triangles contains the three points $A_1' = (B_2C_2, B_3C_3)$, $B_1' = (C_2A_2, C_3A_3)$, $C_1' = (A_2B_2, A_3B_3)$. The theorem is that the three such axes of perspective meet in a point, which we may call O'. A proof may be constructed in the plane; or by projection of the points and lines common to two triads of planes in space, the point common to the planes of one triad projecting into O, and the other such point into O'. The existence of this configuration being understood, let the six fundamental points on the conic be denoted by 1, 2, 3, 4, 5, 6, instead of A, B, C, A', B', C'; then let the Steiner point common to the Pascal lines $(1, 2, 3/4, 5, 6)$,

(1, 2, 3/5, 6, 4), (1, 2, 3/6, 4, 5) be denoted by [1, 2, 3/4, 5, 6], and the conjugate Steiner point, obtained by putting 4, 6, 5, respectively, for 4, 5, 6, be denoted by [1, 2, 3/4, 6, 5]. Alternative symbols may be used for these points, because any one of the sixty hexagons can be described in order-starting from any one of its six points, and, then, in either of two reverse directions. With this notation, the points A_1, A_2, etc. of this configuration, for the twenty Steiner points, may (according to Schröter's edition of Steiner's *Lectures on Conics*, already referred to) be taken to be

$$O=[1, 3, 5/2, 4, 6], \quad A_1=[1, 3, 6/2, 4, 5], \quad A_2=[1, 4, 5/2, 3, 6],$$
$$A_3=[1, 4, 6/2, 3, 5],$$
$$O'=[1, 3, 5/2, 6, 4], \quad A_1'=[1, 3, 6/2, 5, 4], \quad A_2'=[1, 4, 5/2, 6, 3],$$
$$A_3'=[1, 4, 6/2, 5, 3],$$
$$B_1=[1, 2, 6/3, 4, 5], \quad B_2=[1, 2, 3/4, 5, 6], \quad B_3=[1, 5, 6/2, 3, 4],$$
$$C_1=[1, 2, 5/3, 4, 6], \quad C_2=[1, 2, 4/3, 5, 6], \quad C_3=[1, 3, 4/2, 5, 6],$$
$$B_1'=[1, 2, 6/3, 5, 4], \quad B_2'=[1, 2, 3/4, 6, 5], \quad B_3'=[1, 5, 6/2, 4, 3],$$
$$C_1'=[1, 2, 5/3, 6, 4], \quad C_2'=[1, 2, 4/3, 6, 5], \quad C_3'=[1, 3, 4/2, 6, 5].$$

We can look at the Steiner theorem, for the intersection of three Pascal lines, from another point of view, namely, in connexion with the theorem already stated, that there are forty-five points through each of which there pass four Pascal lines.

Name the six points of the conic, as before, A, B', C, A', B, C'; through the common point of the joins of two of these six points which do not meet on the conic, say the point (BB', CC'), there clearly pass the four Pascal lines which are obtained by taking the six points in the respective orders

$$(B, B', A, C, C', A'); \quad (B', B, A, C, C', A');$$
$$(B, B', A, C', C, A'); \quad (B', B, A, C', C, A'),$$

or, what is the same, the orders

$$(B', B, A', C', C, A); \quad (B, B', A', C', C, A);$$
$$(B', B, A', C, C', A); \quad (B, B', A', C, C', A).$$

There are forty-five such possibilities; for, the pair A, A', not lying on either of the joins BB', CC', can be selected in fifteen ways, and there are three pairs of complementary joins of the other four points B, B', C, C'. Further, any one of the four Pascal lines specified, passes through the common point of two joins, of the original six points, which do not meet on the conic, and in three ways. For instance, the Pascal line arising from the order (B, B', A, C, C', A') contains the three points $(BB', CC'), (B'A, C'A'), (BA', CA)$. Wherefore, the forty-five points obtained enumerate $4 . 15$, that is, all the Pascal lines. Thus we have proved that there are forty-five points, each lying on four Pascal lines, three such points lying on any Pascal line, which was our statement (c).

Now consider the triangle formed by the three lines AA', BB', CC'. Then, to any one of the four Pascal lines we have seen to pass through one angular point of this triangle, say (BB', CC'), there corresponds a definite Pascal line through each of the other two angular points; and the three so associated Pascal lines meet in a point, which is one of the twenty Steiner points. So that the triangle formed by the three lines AA', BB', CC' gives rise to four Steiner points. To see this, put down in detail the four lines through the angular points of this triangle, naming them by the order of the points,

$$(BB'ACC'A'), \quad (CC'BAA'B'), \quad (AA'CBB'C'),$$
$$(B'BACC'A'), \quad (C'CBAA'B'), \quad (A'ACBB'C'),$$
$$(BB'AC'CA'), \quad (CC'BA'AB'), \quad (AA'CB'BC'),$$
$$(B'BAC'CA'), \quad (C'CBA'AB'), \quad (A'ACB'BC'),$$

those written in a column belonging to one angular point. Let these columns be named p, q, r, and the lines in a column have the suffixes 1, 2, 3, 4, in the order written. It can then be seen that each of the sets of three Pascal lines (p_1, q_4, r_2), (p_2, q_1, r_4), (p_3, q_3, r_3), (p_4, q_2, r_1) consists of three lines which meet in a Steiner point. For instance, p_1, q_4, r_2 are the lines given by the hexagons $(BB'ACC'A')$, $(C'CBA'AB')$, $(A'ACBB'C')$, that is, the hexagons $(AB'BA'C'C)$, $(AA'BCC'B')$, $(ACBB'C'A')$; they are therefore the lines denoted, in a previous notation, respectively by $(A, C', B/A', B', C)$, $(A, C', B/C, A', B')$, $(A, C', B/B', C, A')$; so that they meet in a Steiner point. By interchanging B, B', or interchanging C, C', and, by interchanging B, B' and C, C' simultaneously, the other three Steiner points are obtained. Now, there are fifteen triangles such as that formed by the lines AA', BB', CC'; through each angular point of such a triangle there pass four Pascal lines. This would give 15 . 3 . 4 Pascal lines. Thus we infer that any one of the Pascal lines is obtainable in a similar way from three of the fifteen triangles; so that, of the forty-five points, in which four Pascal lines meet, there are three on any one of the Pascal lines. Each of the fifteen triangles gives four Steiner points, which would make sixty Steiner points, were it not that each Pascal line arises from three of the triangles.

We have seen that, on any one of the four Pascal lines through the point (BB', CC'), besides this point there is one point lying also on a Pascal line through each of the points (CC', AA'), (AA', BB'), the point being a Steiner point. There is, however, on this same Pascal line through (BB', CC') another point, *not* a Steiner point, lying also on a Pascal line from each of (CC', AA'), (AA', BB'). If we assume this, for a moment, and recall that the Pascal line we have considered, through (BB', CC'), passes through angular points of three of the fifteen triangles such as (AA', BB', CC'), we can infer that there are three points on this Pascal line, at each of which three Pascal lines meet, which do not form a Steiner triad; of such so-called *Kirkman points* there are then sixty in all. We consider now the theorem of the existence of the Kirkman points: For the triangle (AA', BB', CC') we remarked, in the notation employed, the Steiner triad of three lines (p_1, q_4, r_2). But it is also true, in this notation, that the lines (p_1, q_3, r_4) meet in a point. These are the Pascal lines of the hexagons

$$(BB'ACC'A'),\quad (CC'BA'AB'),\quad (A'ACB'BC'),$$

and are *not* a Steiner triad. To prove that these three lines meet in a point we name another triangle which is in perspective with the triangle (AA', BB', CC'), the three lines in question being the joins of corresponding angular points of these two triangles. This other triangle is that formed by the Pascal lines l, m, n, which are respectively given by the hexagons $(AA'BC'B'C)$, $(BB'ACA'C')$, $(CC'A'BAB')$. The three common points of corresponding sides of these two triangles, namely (AA', l), (BB', m), (CC', n), are obviously the three points $(AA', B'C')$, (BB', CA'), (CC', AB), which lie on the Pascal line $(A'ABB'C'C)$, so that the triangles are in perspective; this proves the statement that the Pascal lines (p_1, q_3, r_4) meet, provided we shew that these lines are those joining the corresponding angular points of the two triangles. For instance, we must shew that the line p_1, which is the Pascal line $(BB'ACC'A')$, passes through the point (BB', CC'), and also through the common point of the two lines m, n, namely, the two Pascal lines $(BB'ACA'C')$, $(CC'A'BAB')$; this point (m, n) is $(AB', A'C')$. So that this fact is clear. The point (n, l) is $(A'B, B'C)$; the line q_3, or $(CC'BA'AB')$, contains this, and also the point (CC', AA'). Finally, the point (l, m) is (AC, BC'); the line r_4, or $(A'ACB'BC')$, contains this, and also the point (AA', BB'). The proof, which is given by Salmon, *Conics*, 1879, p. 380, is thus completed.

We have now considered the three results named (a), (b), (c) at starting.

Ex. 1. If the fundamental conic has the equation $xz - y^2 = 0$, and the coordinates of the points $A, B', \ldots$ thereon be $(a^2, a, 1)$, $(b'^2, b', 1)$, $\ldots$, prove that the Pascal line of the hexagon $(AB'CA'BC')$ has the equation $\Delta = 0$, where Δ is the determinant of four rows in which the elements in the first row are x, y, y, z, and the elements in the other three rows are, respectively, $(aa', a, a', 1)$, $(bb', b, b', 1)$, $(cc', c, c', 1)$.

Ex. 2. Denote the linear function $x - y(a + a') + zaa'$ by (AA'), or $(A'A)$, etc., and the difference $b - c$ by (bc), etc. Then prove that a Steiner triad of concurrent Pascal lines through the respective angular points of the triangle (AA', BB', CC') is given by $v - w = 0$, $w - u = 0$, $u - v = 0$, where u, v, w are, respectively, $(bc)\,(b'c')\,(AA')$, $(ca)\,(c'a')\,(BB')$, $(ab)\,(a'b')\,(CC')$. But concurrent Pascal lines through the angular points meeting in a Kirkman point are given by $v' - w' = 0$, $w' - u' = 0$, $u' - v' = 0$, where u', v', w' are, respectively, $(bc')\,(b'c)\,(AA')$, $(ca)\,(c'a')\,(BB')$, $(ab)\,(a'b')\,(CC')$, of which v', w' are the same as v, w. (Salmon, *Conics*, 1879, p. 383.)

Ex. 3. A general statement of the incidences of the lines may be added. The figure involves

60 k-lines, the Pascal lines, and 60 K-points, the Kirkman points,

20 g-lines, the Cayley-Salmon lines, and 20 G-points, the Steiner points,

and 15 i-lines, the Steiner-Plücker lines, and 15 I-points, the Salmon points.

Each k-line contains one G-point, and also three K-points; each K-point lies on one g-line, and on three k-lines. Each g-line contains one G-point, three K-points and three I-points; each G-point lies on one g-line, and on three k-lines, and also on three i-lines. Each i-line contains four G-points; each I-point lies on four g-lines.

There are also forty-five points through each of which pass four k-lines, every one of the lines containing three of these points. Further, the sixty Pascal lines can be divided into six sets of ten, meeting in threes in ten Kirkman points, of which three lie on any one of the Pascal lines; the configuration in each of the partial figures is that of two triangles in perspective, with their centre and axis of perspective.

NOTE 4

On relations between the angles of intersection of circles

(1) In this Note we give an incomplete account of a wide literature dealing with what are commonly called the angles of intersection of circles in a plane. We approach the matter in what seems the most descriptive way, by using some elementary theorems in space, which we explain.

First we make a remark in regard to two lines, say l and m, which meet in a point C (not lying on the Absolute line). Let the lines meet the Absolute line in L, M, and I, J be the Absolute points. We have denoted the cross ratio $(L, M/I, J)$ by $\epsilon(l, m)$; this is changed into the inverse of itself either by the interchange of l and m, or by the interchange of I and J. Denoting this cross ratio by ϵ, consider the expression $(\epsilon \pm 1)^2/4\epsilon$; being unchanged when ϵ^{-1} is put for ϵ, this is independent of the order in which l, m are taken, and also independent of the interchange of I and J. It depends, therefore, only on the two lines l, m. Whence, the function $(\epsilon + 1)/2\epsilon^{\frac{1}{2}}$ also depends only on the two lines, but is ambiguous in sign. The two lines therefore give rise to two functions, which we denote by γ and $-\gamma$, where γ is one of the two functions $\frac{1}{2}(\epsilon^{\frac{1}{2}} + \epsilon^{-\frac{1}{2}})$. We can interpret these functions in terms of length: on the lines l, m, respectively, take two points A, B, with symbols, in terms of the symbols of C, I, J, respectively,

$$A = C + \xi_1 I + \eta_1 J, \quad B = C + \xi_2 I + \eta_2 J.$$

Supposing I, J to be conjugate imaginaries, as explained previously, ξ_1 and η_1 will be conjugate imaginaries, as will ξ_2, η_2. The cross ratio ϵ is then $\eta_1\xi_1^{-1}/\eta_2\xi_2^{-1}$, or $\eta_1\xi_2/\eta_2\xi_1$; let a, b be the lengths of CA and CB, which are respectively the positive square roots $(\xi_1\eta_1)^{\frac{1}{2}}$ and $(\xi_2\eta_2)^{\frac{1}{2}}$; then the functions γ, $-\gamma$, for the two lines, will be $(\xi_1\eta_2 + \xi_2\eta_1)/2ab$, and the negative of this. Let c be the length of AB, which is the positive square root of $(\xi_1 - \xi_2)(\eta_1 - \eta_2)$. Then we have $a^2 + b^2 - c^2 = \xi_1\eta_2 + \xi_2\eta_1$, which is $\pm 2ab\gamma$. Thus γ is $\pm(a^2+b^2-c^2)/2ab$. Now we can take another point A' on the line l, so that C is the mid point of A', A, and then the length of $A'C$, as of CA, will be a. If the length of BA' be c', we can similarly prove that γ is $\pm(a^2+b^2-c'^2)/2ab$. And we can prove that $a^2+b^2-c'^2 = -(a^2+b^2-c^2)$. Thus the geometrical distinction between γ and $-\gamma$ is expressed by saying that if $c^2 = a^2+b^2-2ab\gamma$, with a proper sign for γ, then $c'^2 = a^2+b^2+2ab\gamma$.

For a circle, with centre O, the real positive distance from O to any point P of the circle, which is the same for every P, is called the radius of the circle. If two circles be given, with centres O, O', and radii r, r', which meet in two real points, we may consider the function γ which arises from the tangents of the circles at either of these points. This is then $\pm(d^2-r^2-r'^2)/2rr'$, where d is the distance of O and O'. We shall define the function γ, *for the two circles*, whether they intersect in real points or not, by $\gamma = -(d^2-r^2-r'^2)/2rr'$. This leads to

$$\tfrac{1}{2}(1-\gamma) = [d^2-(r-r')^2]/4rr', \quad \tfrac{1}{2}(1+\gamma) = -[d^2-(r+r')^2]/4rr'.$$

We shall define t, t', save for sign, by $t^2 = d^2-(r-r')^2$ and $t'^2 = d^2-(r+r')^2$. When the circles do not intersect in real points, the positive values of t, t' are the lengths, on the common tangents of the two circles, between the points of contact of such a tangent (as follows from the so-called Pythagoras theorem proved earlier, Chap. xv (4)). Thence we have, save for sign,

$$t/2\,(rr')^{\frac{1}{2}} = [\tfrac{1}{2}(1-\gamma)]^{\frac{1}{2}}, \quad t'/2\,(rr')^{\frac{1}{2}} = i\,[\tfrac{1}{2}(1+\gamma)]^{\frac{1}{2}}.$$

Ex. 1.   Shew that the function γ^2, for two circles, has the same value as for the circles obtained from these by inversion in regard to a third circle.

Ex. 2.   A line drawn from a given point O meets a given circle in P, P'; and lines from P, P', both perpendicular to OPP', meet another line l, drawn through O, in L and L', respectively. Another line from O meets the circle in Q, Q'; and lines from Q, Q', both perpendicular to OQQ', meet another line, m, drawn through O, in M and M', respectively. Let the distances PL, QM, LM be, respectively, p, q, c; and the distances $P'L'$, $Q'M'$, $L'M'$ be p', q', c', respectively. Prove that

$$(p^2+q^2-c^2)/pq = (p'^2+q'^2-c'^2)/p'q'.$$

(2) Next we enunciate certain elementary properties for geometry in space. We suppose an arbitrary plane in space to be taken for *Absolute plane*, and an arbitrary non-degenerate conic lying in this plane to be taken for *Absolute conic*. The plane is supposed to be real, and the conic is supposed to be such that any real line in the Absolute plane meets the Absolute conic in two conjugate imaginary points of that plane. Two lines in space, not lying in the Absolute plane, are said to be perpendicular when they meet the Absolute plane in two points which are conjugate in regard to the Absolute conic; also, a line and a plane are said to be perpendicular when they meet the Absolute plane, respectively, in a point and a line, of which the former is the pole of the latter in regard to the Absolute conic; and, two planes are said to be perpendicular when they meet the Absolute plane in two lines

which are conjugate to one another in regard to the Absolute conic. Two lines, or two planes, are said to be parallel when their respective intersections with the Absolute plane are identical. For the Absolute points of any plane we take the two points in which the plane meets the Absolute conic.

A quadric surface, in space, is a locus which is met by an arbitrary line in two points, and is met by an arbitrary plane in a conic. If this conic degenerate into two lines, any line of the plane meets the quadric surface in two points lying on these two lines, respectively; in particular, any line of the plane which contains the common point of the two lines, into which the conic section degenerates, meets the quadric surface in two points which coincide at this common point, and does not meet the surface again; in this case, the plane is called a *tangent plane* of the surface, the common point of the two lines into which the section degenerates being the *point of contact* of this tangent plane. A particular case of a quadric surface is a *cone*, which contains an infinite series of lines all meeting in a point, the *vertex* of the cone; we suppose the quadric surface considered not to be a cone. Then, conversely, through any point of the surface, there passes a plane which meets the surface in two lines meeting at this point; this is the tangent plane of the surface at this point. The lines so found, lying on the surface, are called the *generators* of the surface; there are then two generators through any point of the surface. Let these generators, through a point O of the surface, be called l and m; let an arbitrary line, lying in the tangent plane $[l, m]$, meet these respectively in L and M, which are therefore points of the surface. Thus through L, besides l, will pass another generator, say m'; and the plane $[l, m']$, which is the tangent plane at L, will contain O; its complete intersection with the surface will consist of the lines l, m'. Similarly, there is another generator, besides m, through the point M, say l'; and the plane $[m, l']$, the tangent plane at M, also passes through O. The two planes $[l, m']$, $[m, l']$ meet in a line, containing O; and this line meets the surface in another point, say O'. This point O', lying on both planes $[l, m']$, $[m, l']$, and lying on the surface, must, on the one hand, lie either on l or m', and, on the other hand, lie either on m or l'. It must, therefore, be one of the four points, (l, m), which is O, or (l, l'), or (m', m), or (m', l'), if these exist. But, m' meets the plane $[l, m]$ on l, and, therefore, does not meet m, unless it lies in the plane $[l, m]$, whose only intersection with the surface consists of the lines l, m. Thus we infer that m', m do not meet; and, similarly, that l, l' do not meet. Wherefore O' is the point (l', m'), the lines l', m' having a common point; and the point O' is the point of contact of a second tangent plane $[l', m']$ of the surface. This, and the plane $[l, m]$, are then two tangent planes which pass through the line LM. Conversely, then, we infer that, through an arbitrary line, meeting the quadric surface in two points L, M, there can be drawn two tangent planes of the surface, $[l, m]$ and $[l', m']$, with points of contact at O and O', say; and, also, through the line OO', there can be drawn two tangent planes, $[l, m']$ and $[m, l']$, with points of contact at L and M. Further, that any generator l is met by an infinite series of generators m', no two of which meet. There are thus two systems of generators on the quadric surface, with the property that no two generators of the same system meet one another, but any generator of either system meets every generator of the other system.

The lines LM and OO' are said to be *polars* of one another in regard to the quadric surface. Let H be any point not lying on the surface, and an arbitrary line be drawn through H to meet the surface in P and P', and thereon be taken the point $K=(P, P')/H$. It can then be proved that the locus of K, when the line HPP' varies about H, is a plane; this is called the *polar plane* of H in regard to the surface. In particular, it can be proved, further, that if the point H lie on the line LM spoken of, the polar plane of H contains the line OO'; and, also, that when H is at L, its polar plane is the tangent plane at L, containing the generators l, m', which pass, respectively, through O and O'. The relation of the lines LM, OO' is reciprocal, the polar plane of any point of either containing the other line. Conversely, if an arbitrary plane be given, there is a definite point, called the *pole* of the plane, for which it is the polar plane; this pole is common to the tangent planes of the surface at all the points of the section of the surface by the plane. And the pole of any plane, which passes through either of the lines OO', LM, is a point of the other line. Moreover, the range of points on the line LM, taken for positions of H, is related to the pencil of corresponding polar planes which pass through the line OO', and is therefore related to the range of points which these polar planes determine on an arbitrary line.

(3) A *sphere* is a quadric surface which contains the Absolute conic; the pole of the Absolute plane, in regard to the sphere, is called the *centre* of the sphere; the tangent planes of the sphere, at the points thereon which lie on the Absolute conic, all pass through its centre. Any plane section of the sphere, as containing the two points where the plane meets the Absolute conic, which are the Absolute points of the plane, is a circle; so that, in particular, the generators, at any point of the sphere, pass through the Absolute points of the tangent plane at this point. Let T' be any point of the sphere, and T the point of the sphere for which the mid point of T', T is the centre of the sphere, so that the tangent planes at T', T are parallel, meeting on the Absolute plane in the polar line of the line $T'T$; it is a frequent operation to project the points of the sphere, from the point T', into points of the tangent plane at T; this tangent plane we may call τ. In particular, consider any plane section of the sphere, not containing T'; the generators, say i and j, of the sphere, at T', meet the plane section in two points which project from T' into the Absolute points I, J of the plane τ. Thus any plane section of the sphere projects, from T', into a circle in the plane τ. Conversely, any circle of the plane τ,

determined say by three points A_1, B_1, C_1 of this plane (besides the points I, J), arises, by projection from T', from the plane section of the sphere which contains the points A, B, C in which $T'A_1$, $T'B_1$, $T'C_1$ respectively, meet the sphere (besides T'). Moreover, if P be the pole of any plane section of the sphere, the point P_1, of the plane τ, into which P projects from T', is the centre of the circle into which the plane section projects. For let the generators i, j, at T', meet the section respectively in I_0, J_0. The tangent plane at I_0 contains the generator i, and contains also the pole P of the section; this plane meets the plane τ in a line which touches the circle at the point I, arising from I_0, and which contains the point P_1, arising from P. The point P_1 equally lies on the tangent which touches the circle at J. So that P_1 is the centre of the circle.

Now consider two plane sections of the sphere, meeting one another at the points O, O' in which the sphere is met by the line common to the planes. These sections, projected from T', give two circles in the plane τ, meeting, say, in O_1 and O_1'. The tangents of the sections at O project into the tangents of the circles at O_1; the generators of the sphere at O each meet one of the generators i, j at T'; these generators then project into the lines joining O_1 to the points I, J of the plane τ. Consider now the cross ratio of the four lines, all lying in the tangent plane at O, which consist of the tangent lines of the two sections at O, and the generators there, which respectively meet i, j. This is equal to the cross ratio of the pencil of lines into which these lines project, namely, the tangents of the two circles at O_1, and the lines O_1I, O_1J. This cross ratio, of the tangents of the two circles, taken with respect to the lines O_1I, O_1J, we denote by ϵ. Further, if the tangent planes of the sphere at O and O' meet in the line LM, as above, where L is the common point of one generator, l, at O, and one generator, m', at O', and M is a similar common point (m, l'), then the four lines spoken of, in the tangent plane at O, are the sections, by this tangent plane, of the pencil of four planes, passing through the axis OO', which consist of the two tangent planes OLO', OMO', which can be drawn from OO' to touch the sphere, with the planes of the two sections of the sphere which are under consideration. Thus the cross ratio spoken of is the cross ratio of the range which these four planes determine on an arbitrary line. Moreover, the pole of the plane of either of these sections, these planes meeting in OO', is on the line LM, which is the polar line of OO'; and L, M are themselves the poles of the tangent planes which can be drawn to the sphere through OO'. Wherefore, the cross ratio spoken of may be defined, without reference to the generators of the sphere, by taking the points which are the poles, in regard to the sphere, of the planes of the two given sections, and then, on the line joining these poles, the cross ratio of these poles in regard to the two points where the line meets the sphere. This is a convenient way in which to express the cross ratio, ϵ, for the two circles, in the plane τ, which arise from the sections of the sphere. Two particular cases may be mentioned: (*a*) When the circles are perpendicular; then $\epsilon = -1$; and the poles, of the planes of the corresponding sections of the sphere, are such that either pole lies on the plane of the other section, since their join is divided harmonically by the sphere. Two such plane sections are said to be *conjugate*, in regard to the sphere; (*b*) When the sections of the sphere touch one another; then the points O, O' coincide, say in O, and the two tangent planes OLO', OMO' coincide in the tangent plane of the sphere at O. The join of the poles of the planes of the sections is then a line which touches the sphere at O, and $\epsilon = 1$. In general, the cross ratio ϵ is changed into ϵ^{-1}, either by interchange of the two sections, or by interchange of the two points where the line, joining the poles of the sections, meets the sphere. We consider then, instead of ϵ, the function $\frac{1}{2}(\epsilon^{\frac{1}{2}} + \epsilon^{-\frac{1}{2}})$; though this is ambiguous in sign, it is unaltered, save for this, by the interchanges spoken of, and depends only on the two sections taken. In the particular cases (*a*), (*b*), the values of γ are respectively 0 and ± 1.

(4) Suppose now three circles to be given in a plane. We consider the problem of finding a circle, in this plane, which, in the ordinary metrical phraseology, cuts these circles at three given angles respectively. In the phraseology we use, the circle to be found is to be such that the functions γ, for it and the given circles respectively, shall have three given values. To avoid the discussion of signs, we seek the circles for which the function γ^2, taken for any one such, respectively with the given circles, shall have three given values; and we suppose that no one of these values is zero. The reason for this condition will appear.

Instead of considering three circles in a plane we may, after what has been said, consider three given plane sections of the sphere; and seek a section of the sphere for which the functions γ_1^2, γ_2^2, γ_3^2 of this, taken in turn with the given sections, shall have given values. Let A_1, A_2, A_3 be the respective poles of the planes of the given sections; by the definition of the polar plane of a point in regard to a sphere, the plane $A_1A_2A_3$ is the polar plane of the point in which the planes of the three given sections meet. Let this point be O; we suppose the sections to be such that this common point O is not on the Absolute plane. The plane $A_1A_2A_3$ is conjugate to the plane of every one of the given sections; thus this plane meets the sphere in a section which projects (from T' on to the plane τ) into the circle which is perpendicular to all the circles in the plane τ corresponding to the given sections of the sphere; the centre O_1 of this perpendicular circle is the projection of O, the pole of the plane $A_1A_2A_3$; and O_1 is the radical centre of the three circles in the plane τ. Let P be the pole of a section of the sphere, which we seek, with given values of γ_1^2, γ_2^2, γ_3^2. The condition then, for P, is that the lines PA_1, PA_2, PA_3 meet the sphere in points, say, respectively, L_1, M_1; L_2, M_2; L_3, M_3, such that the cross ratios, $(P, A_1/L_1, M_1)$, $(P, A_2/L_2, M_2)$, $(P, A_3/L_3, M_3)$, are all given; and this is the only condition for P.

We assume now that, if a point P be such, in regard to a given point A_1, that the cross ratio $(P, A_1/L_1, M_1)$ is given, then the locus of P is a quadric surface, which touches the sphere at all points of a plane section, the plane which determines this section being the polar plane of A_1. An algebraic proof of this result is indicated below; but the result follows from our knowledge of conics. For consider the section of the sphere by an arbitrary plane through A_1, and the points of the locus sought which lie in this plane; these are the points P such that P, A_1 have a given cross ratio in regard to the points where the line PA_1 meets this section; thus the locus of such points P is a conic which touches the section at two points lying on the polar of A_1. The aggregate of such conics, lying in the various planes through A_1, defines the quadric surface spoken of. We shall see that this surface depends on the assigned cross ratio only through the function γ_1^2. Thus, in the problem under consideration, with three given sections of the sphere, the point P will be a common point of three quadric surfaces, each having *ring-contact* with the sphere, respectively on the planes of the given sections. Now, three quadric surfaces have eight points in common; there are therefore eight possible positions for P; and the section we are in search of, by the polar plane of P, has eight possible positions. It can, however, be proved, and the simple algebraic proof is given below, that, if three quadric surfaces have each ring contact with a sphere, then their eight common points consist of four pairs, such that the line joining the points of a pair passes through the common point, O, of the planes of contact. Thus, to any one, P, of the eight possible points, there corresponds another, say P', such that PP' passes through O. It follows from this that the polar planes of P and P', in which are two of the eight sections of the sphere which we are seeking, meet in a line which lies on the polar plane of O; the two common points of these sections then lie on the section of the sphere by the polar plane of O. The result for three quadric surfaces which have ring contact with a sphere, is analogous to a well-known result in the theory of conics: Let two conics σ_1, σ_2, in a plane, both have double contact with another conic ω; then the point O common to the two chords of contact has the same polar line in regard to all three conics; hence there is a complementary pair of common chords of σ_1 and σ_2 passing through O.

From what we have said we infer that, the problem under discussion, of finding a circle, in the plane τ, to cut three given circles (A_1), (A_2), (A_3) so that the functions γ_1^2, γ_2^2, γ_3^2, belonging to the circle to be found and the given circles, in turn, shall have given values, is capable of eight solutions. The numbers γ_1^2, γ_2^2, γ_3^2, together, determine four pairs of points lying on the common perpendicular circle (O) of (A_1), (A_2), (A_3); and through each such pair there pass two circles, say (P) and (P'), which are solutions of the problem. We can shew that such a pair of solutions, (P), (P'), are circles which are inverse to one another in regard to the circle (O). For, any one of the circles (A_1),

(A_2), (A_3), say (A_i), is inverse to itself in regard to the perpendicular circle (O); thus, as remarked in Ex. 1 of (1), the function $\gamma_i{}^2$, determined for the pair (P), (A_i), is equal to the function $\gamma_i{}^2$, determined for the pair $(\overline{P})$, (A_i), if $(\overline{P})$ denote the circle inverse to (P) in regard to (O). We may also remark, in regard to the three circles (P), (P'), (A_i), that the radical axis of (P) and (P'), which is one of the four chords of the circle (O) which we have obtained, contains the common point of the radical axes of the two pairs (P), (A_i) and (P'), (A_i). The existence of the four chords of the circle (O) depends on the fact that, when $\gamma_1{}^2$, $\gamma_2{}^2$, $\gamma_3{}^2$ are given, there are four sets of values for the signs of γ_1, γ_2, γ_3; as we shall see algebraically below.

(5) A particular case of (4) is when $\gamma_1{}^2 = \gamma_2{}^2 = \gamma_3{}^2$. Then we are to find circles (P) which, in the usual phraseology, cut three given circles all at the same (or supplementary) angles. In this case the four chords of the circle (O), arising in general, are the same whatever be the common value of $\gamma_1{}^2$, $\gamma_2{}^2$, $\gamma_3{}^2$, as we shall prove. And, from a property of coaxal circles, every circle through the common points, on (O), of two circles (P), (P') which have $\gamma_1{}^2 = \gamma_2{}^2 = \gamma_3{}^2$, is also a circle satisfying these conditions; so that the number of solutions of the problem is infinite, if the common value of $\gamma_1{}^2$, $\gamma_2{}^2$, $\gamma_3{}^2$ is not given. In particular, the radical axis of the pair (P), (P') is part of a degenerate circle belonging to the series of solutions. The four chords of the circle (O) thus arising can be geometrically specified; for it can be proved that any one of the four lines which exist, each containing a centre of similitude of three pairs from three given circles (A_1), (A_2), (A_3), has the property that the functions $\gamma_1{}^2$, $\gamma_2{}^2$, $\gamma_3{}^2$, formed for this line and the circles (A_1), (A_2), (A_3) in turn, have equal values. From this it follows that the four series of circles satisfying the condition are those which are coaxal with the circle (O), taken in turn with these four axes of similitude. A modified case is when the common value of $\gamma_1{}^2$, $\gamma_2{}^2$, $\gamma_3{}^2$ is given and equal to 1. Then the problem is to find the circles which touch three given circles. These then are eight in number, and consist of four pairs (P), (P'); and a remark made above, in (4), gives a construction for the points of contact, with (A_1), (A_2), (A_3), of the two circles (P), (P') of a pair. For, as (P), (P') are inverse to one another in regard to (O), which is perpendicular to (A_1), (A_2), (A_3), and (A_i) is inverse to itself, it follows that the points of contact with (A_i) of (P) and (P') lie on a line through the centre O; the pole of this line, in regard to (A_i), being the common point of the tangents at the two points of contact, is the radical centre of the three circles (P), (P'), (A_i); by the remark in question, this pole then lies on the chord of (O) which is the radical axis of (P) and (P'); and, we have seen, this chord is one of the four axes of similitude of (A_1), (A_2), (A_3). Wherefore, the pole of this axis, taken in regard to (A_i), when joined to the centre of (O), gives the two points of contact of (A_i) with (P) and (P'). By suitable selection of sets of three points of contact, thus found respectively on (A_1), (A_2), (A_3), the eight circles (P) may be constructed.

(6) We give now a brief algebraic version of the theory. Let D be the centre of the sphere, and A, B, C be points on the Absolute plane which form a self-polar triangle in regard to the Absolute conic. We easily prove, then, that the symbol of any point of space can be expressed in terms of the symbols of A, B, C, D, in the form $xA + yB + zC + tD$, in such a way that the equations of the Absolute conic are $t = 0$, $x^2 + y^2 + z^2 = 0$, and the equation of the sphere is $x^2 + y^2 + z^2 + t^2 = 0$. If it is preferred to use what are called rectangular cartesian coordinates, we can suppose the equation of the sphere to be $\xi^2 + \eta^2 + \zeta^2 - 1 = 0$. In general, the three plane sections of the sphere considered will be given by equations

$$a_i x + b_i y + c_i z + d_i t = 0 \quad (i = 1, 2, 3),$$

the poles P_1, P_2, P_3 of the planes of these sections being of coordinates (a_i, b_i, c_i, d_i). The two points, of symbol $\lambda P_1 + P_2$, in which the line $P_1 P_2$ meets the sphere, arise by taking λ to be a root of the quadratic equation $\lambda^2 \sigma_1 + 2\lambda \sigma_{12} + \sigma_2 = 0$, where $\sigma_1 = a_1^2 + b_1^2 + c_1^2 + d_1^2$, and $\sigma_{12} = a_1 a_2 + b_1 b_2 + c_1 c_2 + d_1 d_2$, etc. If these two points be L, M, the cross ratio ϵ_{12}, $= (P_1, P_2/L, M)$, or $(L, M/P_1, P_2)$, is the ratio λ_1/λ_2 of the two roots of this equation, and the function γ_{12}, or $(\epsilon_{12} + 1)/2\epsilon_{12}^{\frac{1}{2}}$, is

$$(\lambda_1 + \lambda_2)/2 \,(\lambda_1 \lambda_2)^{\frac{1}{2}}, \text{ or } -\sigma_{12}/\sigma_1^{\frac{1}{2}} \sigma_2^{\frac{1}{2}}.$$

Consider now a point P, of coordinates (x, y, z, t), not lying on the sphere, which is such that the functions γ_1^2, γ_2^2, γ_3^2, formed for P and the given points P_1, P_2, P_3 in turn, have given values. This condition is equivalent to the three conditions for the coordinates of P expressed by the equations, such as

$$\gamma_i^2 \,(x^2 + y^2 + z^2 + t^2) = (a_i x + b_i y + c_i z + d_i t)^2/\sigma_i, \quad \ldots\ldots(1)$$

for $i = 1, 2, 3$. This equation represents a quadric surface having ring-contact with the given sphere at the points of the section given by $a_i x + b_i y + c_i z + d_i t = 0$, or, say, $u_i = 0$. The points common to the three quadric surfaces satisfy the set of equations

$$u_1/\gamma_1 \sigma_1^{\frac{1}{2}} = u_2/\gamma_2 \sigma_2^{\frac{1}{2}} = u_3/\gamma_3 \sigma_3^{\frac{1}{2}},$$

where there are four possibilities in regard to the signs of the ratios of γ_1, γ_2, γ_3. For one choice of these signs, these two equations represent planes, meeting in a definite line. This line contains two of the eight common points of the three quadric surfaces (1); these two points may be found, if we denote the equal functions, such as $u_1/\gamma_1 \sigma_1^{\frac{1}{2}}$, by θt, by solving for the ratios of x, y, z, t, and substituting the values in the equation (1); this leads to two values of θ. The four lines clearly meet in the point, O, common to the planes $u_1 = 0$, $u_2 = 0$, $u_3 = 0$; each of the lines is determined by one set of ratios of γ_1, γ_2, γ_3, with assigned signs for the ratios of $\gamma_1 \sigma_1^{\frac{1}{2}}$, $\gamma_2 \sigma_2^{\frac{1}{2}}$, $\gamma_3 \sigma_3^{\frac{1}{2}}$. When the ratios of γ_1, γ_2, γ_3 are those of 1, ± 1, ± 1, we still obtain four lines; but then each line is determined solely by the signs, and is the same whatever be the common value of γ_1^2, γ_2^2, γ_3^2; in this case then there is an infinite series of pairs of points on each of the four lines, a pair being determined by the common value of $\gamma_1{}^2, \gamma_2{}^2, \gamma_3{}^2$. These remarks provide proof of the assumptions made in the theory given.

Ex. 1. Let D be any point, and δ any plane not containing D. By the harmonic inverse of an arbitrary point P, in regard to D and δ, is meant the point P', on the line DP, which is the harmonic conjugate of P in regard to the two points consisting of D and the point where DP meets δ. When δ is the polar plane of D in regard to the sphere, and P is on the sphere, P' is the second point common to DP and the sphere. In this case, projecting points of the sphere, as before, from a point T' of this, on to the tangent plane τ at the opposite point T, prove that the points P, P' project into points which are inverse to one another in regard to the circle (δ) in the plane τ, where (δ) is obtained by projection of the section of the sphere by the plane δ, and the centre of (δ) is the projection of D.

Let α_1, α_2 be any two plane sections of the sphere. Prove that the harmonic inverses of these sections, in regard to any point D and its polar plane δ, are also plane sections of the sphere; and that the function γ^2 for the sections α_1, α_2, respectively, is equal to that for these inverse sections. It may be seen that the coordinates can be chosen, without loss of generality, so that two inverse plane sections are given by planes with equations

$$ax + by + cz \pm dt = 0.$$

Ex. 2. Consider any two circles in a plane. Two other circles can then be defined, coaxal with the given circles, having their centres at the centres of similitude of the two given circles; these are commonly called the *circles of antisimilitude* of the two given circles. Consider a circle for which the functions $\gamma_1{}^2, \gamma_2{}^2$, of this, taken in turn with the given circles, have equal values. Prove that such a circle belongs to one of the two families of circles, those, namely, which are perpendicular to one, or the other, of the two circles of antisimilitude. These two families have in common the coaxal series of circles which are perpendicular to both the given circles; the functions $\gamma_1{}^2, \gamma_2{}^2$, for any circle of this common series, with the two given circles in turn, are zero.

Ex. 3. Through two given points, two circles pass which touch a given circle; and one circle passes which is perpendicular to the given circle.

Ex. 4. We consider now the condition to be satisfied by four circles when one of the eight circles which touch three of these also touches the fourth; or, more briefly, the condition that four circles should all be touched by another circle.

As before, denote $a_i a_j + b_i b_j + c_i c_j + d_i d_j$ by σ_{ij}, and σ_{ii} simply by σ_i. By multiplying the vanishing determinants, of five rows, in which the general row of the first consists of the elements $a_i, b_i, c_i, d_i, 0$, and the general column of the second consists of the same elements, for $i = 1, 2, 3, 4, 5$, we obtain the identity, homogeneous in each of the sets a_i, b_i, c_i, d_i,

$$\begin{vmatrix} \sigma_1, & \sigma_{12}, & \cdots, & \sigma_{15} \\ \cdots\cdots\cdots\cdots\cdots\cdots\cdots \\ \sigma_{51}, & \sigma_{52}, & \cdots, & \sigma_5 \end{vmatrix} = 0.$$

Adopt now definite signs for $\sigma_1{}^{\frac{1}{2}}, \ldots, \sigma_5{}^{\frac{1}{2}}$, and denote $\sigma_{ij}/\sigma_i{}^{\frac{1}{2}}\sigma_j{}^{\frac{1}{2}}$ by γ_{ij}. Then this identity is the same as

$$\begin{vmatrix} 1, & \gamma_{12}, & \cdots, & \gamma_{15} \\ \cdots\cdots\cdots\cdots\cdots\cdots\cdots \\ \gamma_{51}, & \gamma_{52}, & \cdots, & 1 \end{vmatrix} = 0,$$

and this relation connects any five sections (A_i) of the sphere, given by planes $u_i = 0$. Herein, suppose that the section (A_5) touches the other four sections, for which the conditions are $\gamma_{15}{}^2 = \gamma_{25}{}^2 = \gamma_{35}{}^2 = \gamma_{45}{}^2 = 1$. We denote

γ_{15}, γ_{25}, γ_{35}, γ_{45}, respectively, by ζ_1, ζ_2, ζ_3, ζ_4, each of these being ± 1. Then, by first dividing the i-th row and the i-th column of the determinant by ζ_i, for $i = 1, 2, 3, 4$, and then subtracting the last column from all others, we see that the identity is obtained by equating to zero the product of the four factors $\lambda_{23}^{\frac{1}{2}}\lambda_{14}^{\frac{1}{2}} \pm \lambda_{31}^{\frac{1}{2}}\lambda_{24}^{\frac{1}{2}} \pm \lambda_{12}^{\frac{1}{2}}\lambda_{34}^{\frac{1}{2}}$, where λ_{ij} is $1 - \zeta_i\zeta_j\gamma_{ij}$.

Conversely, the vanishing of one of these four factors is sufficient to ensure that the sections (A_1), (A_2), (A_3), (A_4) are all touched by another properly chosen section (A_5). That, as here, there are essentially eight ways in which this can happen, may be illustrated, if we assume the topological relations of Euclidean geometry, by considering the ways in which four circles in a plane can touch a single line, according to the side of the line on which any one of the circles may lie. Indeed, if we use the equations, found in (1), $t/2(rr')^{\frac{1}{2}} = [\frac{1}{2}(1-\gamma)]^{\frac{1}{2}}$, $t'/(2rr')^{\frac{1}{2}} = i[\frac{1}{2}(1+\gamma)]^{\frac{1}{2}}$, together with the fact that the function γ^2 of two circles is unaltered by inversion, we can obtain the irrational equation above by inversion of the figure of four circles touching a line, recalling that for four points A, B, C, D, in order upon a line (these being the points of contact of four circles with the line), the distances are connected by $\overline{AB}.\overline{CD} - \overline{AC}.\overline{BD} + \overline{AD}.\overline{BC} = 0$.

If we change the signs of $\sigma_1^{\frac{1}{2}}$, ..., $\sigma_4^{\frac{1}{2}}$ appropriately, and then denote $\sigma_{ij} - \sigma_i^{\frac{1}{2}}\sigma_j^{\frac{1}{2}}$ by μ_{ij}, the irrational form of the condition found is

$$\mu_{23}^{\frac{1}{2}}\mu_{14}^{\frac{1}{2}} \pm \mu_{31}^{\frac{1}{2}}\mu_{24}^{\frac{1}{2}} \pm \mu_{12}^{\frac{1}{2}}\mu_{34}^{\frac{1}{2}} = 0.$$

Ex. 5. For a particular case of Ex. 4, suppose that the planes of the four sections $u_i = 0$ touch the sphere, so that the poles of these sections lie on the sphere, and (the real parts of) the sections reduce to the points (a_i, b_i, c_i, d_i). The condition, for the four sections to be touched by another section, reduces then to the fact that these four points lie on a plane. Then $\sigma_1 = \sigma_2 = \sigma_3 = \sigma_4 = 0$, and the condition found reduces to $\sigma_{23}^{\frac{1}{2}}\sigma_{14}^{\frac{1}{2}} \pm \sigma_{31}^{\frac{1}{2}}\sigma_{24}^{\frac{1}{2}} \pm \sigma_{12}^{\frac{1}{2}}\sigma_{34}^{\frac{1}{2}} = 0$. Because the four points lie in a plane, the determinant of four rows, of which the general row consists of the elements a_i, b_i, c_i, d_i, vanishes. If we multiply this determinant into the determinant obtained from this by transposition of rows and columns, we obtain the rationalised form of the identity found.

As the common tangent of two circles in a plane, which both reduce to points, is the segment joining these points, and one case of the relation connecting the lengths of common tangents of four circles in the plane, which all touch another circle, has been seen to be $-t_{23}t_{14} + t_{31}t_{24} + t_{12}t_{34} = 0$, we infer that the particular relation, in this example, leads to that connecting the distances of four points on a circle. This is generally associated with the name of Ptolemy, and may be written $-ap + bq + cr = 0$.

It is, however, the case that the distances of *any* four points in a plane are connected by a relation; and, combining this with that just remarked, the distances of four points on a circle are connected by two relations. The relation connecting the distances a, b, c, p, q, r of any four points in the plane (see Chap. xv (9), Ex. 17) is that expressed by the vanishing of the determinant of five rows whose respective rows consist of the elements $(0, c^2, b^2, p^2, 1)$, $(c^2, 0, a^2, q^2, 1)$, $(b^2, a^2, 0, r^2, 1)$, $(p^2, q^2, r^2, 0, 1)$, $(1, 1, 1, 1, 0)$. This relation can be shewn to be

$$\text{where} \qquad (ap - bq - cr)\,U + [a\,(bc + qr) - p\,(br + cq)]^2 = 0;$$

$$U = (a^2 + p^2)(ap + bq + cr) - (b^2 + q^2)(ap + bq - cr) - (c^2 + r^2)(ap - bq + cr).$$

If we suppose the four points to be on a circle, and $ap - bq - cr = 0$, this leads to $a\,(bc + qr) - p\,(br + cq) = 0$. Hence we can deduce (cf. Misc. Ex. 174)

$$a^2 = (bq + cr)(br + cq)/(bc + qr), \quad p^2 = (bq + cr)(bc + qr)/(br + cq),$$

by which two of the six distances are expressed in terms of the other four.

The reader may also shew that the relation connecting the distances of any four points in a plane can be written in a form in which there enter four quite arbitrary distances, r_1, r_2, r_3, r_4, namely as the vanishing of the symmetrical determinant of five rows in which the elements of the respective rows are $(1, \gamma_{12}, \gamma_{13}, \gamma_{14}, r_1^{-1})$, $(\gamma_{21}, 1, \gamma_{23}, \gamma_{24}, r_2^{-1})$, $(\gamma_{31}, \gamma_{32}, 1, \gamma_{34}, r_3^{-1})$, $(\gamma_{41}, \gamma_{42}, \gamma_{43}, 1, r_4^{-1})$ and $(r_1^{-1}, r_2^{-1}, r_3^{-1}, r_4^{-1}, 0)$. Here γ_{12}, γ_{13} stand, respectively, for $(r_1^2 + r_2^2 - c^2)/2r_1r_2$, and $(r_1^2 + r_3^2 - b^2)/2r_1r_3$, etc., while γ_{14} stands for $(r_1^2 + r_4^2 - p^2)/2r_1r_4$, etc. This may be interpreted as expressing a relation for any four circles in the plane.

Also, taking any four points 1, 2, 3, 4 in a plane, and denoting by t_4 the length of the tangent drawn from the point 4 to the circle containing the points 1, 2, 3, with t_4^2 taken negatively when the point 4 is interior to this circle, etc., the reader may prove that $t_1^{-2} + t_2^{-2} + t_3^{-2} + t_4^{-2} = 0$.

Further, taking any conic, and any four points P_1, P_2, P_3, P_4, in a plane, if the line P_1P_2 meet the conic in U, U', and ϵ_{12} denote the cross ratio $(P_1, P_2/U, U')$, and δ_{12} denote $(\epsilon_{12} + 1)/2\epsilon_{12}^{\frac{1}{2}}$, and so for any pair of the four points, it may be proved that the determinant of four rows whose respective rows consist of the elements $(1, \delta_{12}, \delta_{13}, \delta_{14})$, $(\delta_{21}, 1, \delta_{23}, \delta_{24})$, $(\delta_{31}, \delta_{32}, 1, \delta_{34})$ $(\delta_{41}, \delta_{42}, \delta_{43}, 1)$ has the value zero. Shew, using metrical notions, that this leads to the relation above which connects the distances of any four points in a plane; it leads, in fact, directly to a relation connecting the distances of any four points which lie on a sphere, and this sphere can be supposed to become a plane.

Ex. 6. Another particular case of Ex. 4 arises when we consider three points, and a circle, such that this circle is touched by the circle containing the three points. In metrical terms, let d, e, f be the lengths of the sides EF, FD, DE of a triangle DEF, and t_1, t_2, t_3 be the lengths of the tangents drawn respectively to another circle from D, E, F; then, if this other circle touches the circle DEF, there exists a relation of one of the forms $t_1 d \pm t_2 e \pm t_3 f = 0$. In particular, when D, E, F are the respective mid points of the sides BC, CA, AB of a triangle ABC, and these sides be of respective lengths a, b, c, if we take for t_1, t_2, t_3 the lengths of the tangents drawn from D, E, F to a circle which touches BC, CA, AB, we obtain a proof of Feuerbach's theorem that the nine-point circle of ABC touches the inscribed and escribed circles of the triangle.

Ex. 7. Let a plane be drawn through the centre of a sphere, so as to touch two plane sections of the sphere, respectively at P_1 and P_2. Let T_{12} denote the cross ratio of P_1 and P_2 taken in regard to the points where the line P_1P_2 meets the asymptotic cone of the sphere, and denote $T_{12}^{\frac{1}{4}} - T_{12}^{-\frac{1}{4}}$ by t_{12}. Now suppose four sections of the sphere to be taken, with a similar notation for each pair; and suppose these four sections to be all touched by a single section of the sphere. Then (cf. Ex. 15 below), the relation of Ex. 4 is equivalent to $t_{23}t_{14} \pm t_{31}t_{24} \pm t_{12}t_{34} = 0$. In metrical terms, consider any four sections of the sphere; for any pair take a *great* circle common tangent, whose length is measured, as in spherical trigonometry, by the Euclidean angle which the arc of this great circle, between the points of contact, subtends at the centre of the sphere. The lengths of the common tangents are then connected by a relation, which is that just put down, t_{12} denoting the sine of half the length of the great circle tangent to the sections 1, 2, etc.

Ex. 8. We have shewn that a circle can be described, in a plane, to cut three given circles so as to have given values for the three functions γ^2, of the given circles taken in turn with the described circle. Now consider the two identities.

$$\sin \tfrac{1}{2}\alpha \sin \tfrac{1}{2}(\beta - \gamma) - \sin \tfrac{1}{2}\beta \sin \tfrac{1}{2}(\alpha - \gamma) + \sin \tfrac{1}{2}\gamma \sin \tfrac{1}{2}(\alpha - \beta) = 0,$$

$$\sin \tfrac{1}{2}\alpha \sin \tfrac{1}{2}(\beta - \gamma) + \cos \tfrac{1}{2}\beta \cos \tfrac{1}{2}(\alpha - \gamma) - \cos \tfrac{1}{2}\gamma \cos \tfrac{1}{2}(\alpha - \beta) = 0,$$

and suppose we have three circles in a plane for which the functions γ^2, of the pairs of these circles, are, respectively, $\chi_{23}^2 = \cos^2 \alpha$, $\gamma_{31}^2 = \cos^2 \beta$, $\gamma_{12}^2 = \cos^2 \gamma$.

Then the two identities are of the forms

$$(1-\gamma_{23})^{\frac{1}{2}}\,(1-\gamma_{14})^{\frac{1}{2}}-(1-\gamma_{31})^{\frac{1}{2}}\,(1-\gamma_{24})^{\frac{1}{2}}+(1-\gamma_{12})^{\frac{1}{2}}\,(1-\gamma_{34})^{\frac{1}{2}}=0,$$

$$(1-\gamma_{23})^{\frac{1}{2}}\,(1-\gamma_{14})^{\frac{1}{2}}+(1+\gamma_{31})^{\frac{1}{2}}\,(1+\gamma_{24})^{\frac{1}{2}}-(1+\gamma_{12})^{\frac{1}{2}}\,(1+\gamma_{34})^{\frac{1}{2}}=0,$$

for appropriate values of γ_{14}, γ_{24}, γ_{34}; and we infer that if, in addition to the three given circles, there be another circle whose functions γ^2, when this circle is taken in turn with the three given circles, are $\gamma_{14}{}^2=\cos^2(\beta-\gamma)$, $\gamma_{24}{}^2=\cos^2(\gamma-\alpha)$, $\gamma_{34}{}^2=\cos^2(\alpha-\beta)$, then the *four* circles are all touched by *four* other circles. For, there are three identities of the form of the second of those written down.

Ex. 9. The necessary and sufficient condition that four plane sections of a sphere should all be touched by four other planes, which, by what we have shewn, leads to the condition that four given circles in a plane should all be touched by four other circles, can be expressed as follows: Let the given sections be called α, β, γ, δ; and let the planes of these sections form a tetrahedron X, Y, Z, T, of which T is the angular point opposite to the section δ, etc. The sphere on which the sections lie will meet each of the six edges of the tetrahedron in two points; denote these points, on the edges TX, TY, TZ, YZ, ZX, XY, respectively, by A, A'; B, B'; C, C'; D, D'; E, E'; F, F'. Thus A, A' are the two common points of the sections β, γ; and D, D' are the two common points of the sections α, δ, etc. On the edge TX take the two points $A_1=(T,X)/A$ and $A_1'=(T,X)/A'$; on the edge YZ take the two points $D_1=(Y,Z)/D$ and $D_1'=(Y,Z)/D'$; and similarly for each edge. Then the necessary and sufficient condition in question is that the twelve points A_1, A_1', D_1, D_1', etc., should lie in two planes, each meeting the six edges in six of these points. This can be expressed differently: For, it can be shewn that the necessary and sufficient condition that the points A_1, B_1, C_1, D_1, E_1, F_1, on the respective edges, should lie in a plane, is that the three lines AD, BE, CF should meet in a point. If this is so, and the points A', ..., F' lie on the sphere as described, then also the three lines $A'D'$, $B'E'$, $C'F'$ meet in a point, and the condition is satisfied. A line such as AD joins a point of the sphere lying on one edge of the tetrahedron to a point of the sphere lying on the opposite edge; and any change of notation, for the various pairs of points of the sphere which lie on the edges, furnishes therefore another sufficient condition. Also, the condition that AD, BE, CF meet in a point is better expressed, for our purpose, by saying that each of the sets of four points B, C, E, F; C, A, F, D; A, B, D, E lies in a plane. A proof of the necessary and sufficient condition is given in *Principles of Geometry*, vol. IV, pp. 82 ff.

When, as in the preceding theory, the sections of the sphere are projected from a point K', of the sphere, on to the tangent plane κ at the opposite point K, the four sections become four circles in the plane κ; and the pairs A, A'; D, D'; etc., become intersections of pairs of these circles. When the lines AD, BE, CF meet in a point, the planes, joining K' to these lines, meet in a line through K', so that the lines, in the plane κ, obtained by projection of the lines AD, BE, CF, meet in a point. Moreover, as the points B, C, E, F then lie in the plane of the intersecting lines BC, EF, the four points obtained by projection of B, C, E, F, on to the plane κ, lie on a circle; and so for C, A, F, D and A, B, D, E. Conversely, suppose that, in a plane, we have three circles (α), (β), (γ), of which the pair (β), (γ) meet in the points (A) and (A'), the pair (γ), (α) meet in (B), (B'), and the pair (α), (β) meet in (C), (C'); and, let points (D), (E), (F), lying respectively on the circles (α), (β), (γ), be such that the four points (B), (C), (E), (F) lie on a circle, say (ξ), while (C), (A), (F), (D) also lie on a circle, say (η), and (A), (B), (D), (E) lie on a circle, say (ζ); then the three lines joining respectively (A), (D); (B), (E); (C), (F), being the radical axes of the pairs of circles (ξ), (η), (ζ), will meet in a point. It follows then that the circle (δ), containing the points

(D), (E), (F), will form, with (α), (β), (γ), a set of four circles with the property of being all touched by four other circles. For, on projecting back to the sphere from the point K', the points A, B, C, D, E, F will be such that each pair of lines from the three, AD, BE, CF, lie in a plane; in general, however, these three lines do not lie in a plane; so that they meet in a point. Thus, also, if the circle containing the points (D), (E), (F) meet the circles (α), (β), (γ) further in the respective points (D'), (E'), (F'), there will be other circles, (ξ'), (η'), (ζ'), containing, respectively, the four points (B'), (C'), (E'), (F'), etc., and the three lines, joining (A'), to (D') etc., being radical axes of the pairs of (ξ'), (η'), (ζ'), will meet in a point.

It can be shewn that, essentially, the circle, here described to contain (D), (E), (F), is related to the circles (α), (β), (γ), as is the fourth circle in Ex. 8, to the three first circles there taken (cf. Ex. 13 below).

Ex. 10. The condition for four circles α, β, γ, δ, in a plane, to be touched by another circle, is one condition. We have, in Ex. 9, found *three* conditions for four circles α, β, γ, δ to be touched by four circles α', β', γ', δ'. It is to be expected then that, if four circles α, β, γ, δ be all touched by three circles α', β', γ', then they are equally touched by another circle. That is, if α, β, γ, δ be four of the eight circles which, by the theory given earlier, touch three circles α', β', γ', then these circles α, β, γ, δ are touched by another circle. In the particular case when α', β', γ' are three lines, this is true; in this case the circles α, β, γ, δ are the inscribed and escribed circles of the triangle formed by the lines α', β', γ', and the fourth circle, δ', touching α, β, γ, δ, is the nine-point circle of the triangle.

We have seen, in Ex. 4, that the condition for four circles to be touched by another circle, is $\lambda u_{23}^{\frac{1}{2}} u_{14}^{\frac{1}{2}} + \mu u_{31}^{\frac{1}{2}} u_{24}^{\frac{1}{2}} + \nu u_{12}^{\frac{1}{2}} u_{34}^{\frac{1}{2}} = 0$, where $u_{ij} = 1 - \zeta_i \zeta_j \gamma_{ij}$, and each of λ, μ, ν, ζ_1, ζ_2, ζ_3, ζ_4 is ± 1. Let this condition be denoted by $(\zeta_1, \zeta_2, \zeta_3, \zeta_4 \chi \lambda, \mu, \nu) = 0$; it is unaltered by simultaneous change of the signs of ζ_1, ζ_2, ζ_3, ζ_4. It can then be proved, taking only one case, that

$$(-1, 1, 1, 1\chi - \lambda, \mu, \nu) + (1, -1, 1, 1\chi - \lambda, -\mu, -\nu)$$
$$+ (1, 1, -1, 1\chi\lambda, -\mu, -\nu) + (1, 1, 1, -1\chi\lambda, \mu, \nu) = 0;$$

thus, if the three first terms of this equation vanish, so does the fourth term, and the suggested theorem is proved in this case. We do not enter into a complete investigation.

Ex. 11. Prove that, if three plane sections of the sphere whose equation is $x^2 + y^2 + z^2 - 1 = 0$ be given by the equations $v_i = 0$ or $a_i x + b_i y + c_i z - 1 = 0$, for $i = 1, 2, 3$, then, any point $(x, y, z, 1)$, of the sphere, which lies on a plane touching these three circles, satisfies an equation of the form

$$\mu_{23}^{\frac{1}{2}} v_1^{\frac{1}{2}} + \mu_{31}^{\frac{1}{2}} v_2^{\frac{1}{2}} + \mu_{12}^{\frac{1}{2}} v_3^{\frac{1}{2}} = 0,$$

where $\mu_{23} = \sigma_{23} - \sigma_2^{\frac{1}{2}} \sigma_3^{\frac{1}{2}}$, $\sigma_{23} = a_2 a_3 + b_2 b_3 + c_2 c_3 - 1$ (cf. Ex. 4), etc., for the pairs of the given sections, are as before. When rationalised, this equation leads to a cone with vertex at the centre of the sphere.

Ex. 12. A simple application of Ex. 9 is the following: Let α, β, γ be three circles in a plane, of which the pairs (β, γ), (γ, α), (α, β) meet, respectively, in A, A'; B, B'; C, C'. Let the circles $AB'C$, $BC'A'$, $CA'B'$, ABC be, respectively, α', β', γ', δ'. Then, on the circular arcs $B'C'$, $C'A'$, $A'B'$ there lie, respectively, the points A, B, C. Also, the four points B', C', B, C lie on the circle α, as do C', A', C, A on β, and A', B', A, B on γ. And the circle ABC is δ'. It follows then that α', β', γ', δ' are four circles which are all touched by four circles. And, if we invert the figure, in regard to the circle which is perpendicular to the circles α, β, γ, whose centre is the common point of the lines AA', BB', CC', we see that the four circles $A'BC$, $B'CA$, $C'AB$, $A'B'C'$ are also a set of four circles which are all touched by four circles. By Ex. 9, it follows that any set of four circles, with the property of being touched by four circles, can be obtained in this way.

As a particular application, if H be the orthocentre of a plane triangle ABC, prove that the circles having AH, BH, CH as diameters form, with the circumcircle ABC, a system of four circles which are all touched by four other circles. And find these four other circles.

Ex. 13. Let two lines a, b, in a plane, meet a circle, respectively, in A, A' and in B, B'. If $\epsilon(A, B')$, $\epsilon(B, A')$ be cross ratios computed on the circle, prove that $\epsilon(a, b) = \epsilon(A, B')/\epsilon(B, A')$. Or, in Euclidean phrase, which is ambiguous without the topological conventions exhibited in a diagram, any one of the four sides, taken in order, of an inscribed quadrangle of a circle, makes angles θ, ϕ, ϕ' with the opposite side, and with the two contiguous sides, respectively, which are connected by an equation $\pm\theta + \phi - \phi' = 0$. Hence, if three circles α, β, γ, in a plane, be such that β, γ meet in A, A', etc., as in Ex. 12, and the angles of intersection of the pairs β, γ; γ, α; α, β be, respectively, ϕ_1, ϕ_2, ϕ_3, prove that the angle of intersection of the circle $AB'C'$ with the circle ABC is $\pm(\phi_2 - \phi_3)$, etc. In accord with Ex. 8 above, we can thus prove, in another way, that the circles $AB'C'$, $BC'A'$, $CA'B'$, ABC are all touched by four other circles.

Ex. 14. Using Euclidean phraseology, if, of two circles, one become a line, and r be the radius of the other, while p is the length of the perpendicular drawn to the line from the centre of the circle, prove that the cosine of the angle of intersection of the line and circle is p/r.

Hence, using the relation we have found connecting the cosines of the angles of intersection of pairs of five circles in a plane (Ex. 4, above), prove the following result, which is recorded in Pappus' *Collection*: On the diameter of a semicircle, and interior to it, let two semicircles (of less radii) be described, each touching the given semicircle at one end of its diameter, and touching one another. Now let a circle be described touching the original semicircle internally, and touching the other two semicircles externally. Prove that the length of the radius of this circle is half the distance of its centre from the common diameter of the three circles.

Several results of this kind are found in Pappus (*Collectio Math.*), and extensions of Pappus' theorems were given by Steiner (*Ges. Werke*, i). The reader may compare Heath, *History of Greek Mathematics*, ii, 1921; and for further references, *Principles of Geometry*, vol. ii (1930), p. 249. The artifice, of obtaining the relation connecting five circles, by multiplication of two vanishing determinants, is the subject of the first paper (1841) in Cayley's *Collected Papers*; the writer was born in 1821.

Ex. 15. To any cross ratio ϵ, $=(A, B/C, D)$, we have associated a cosine $\frac{1}{2}(\epsilon + 1)/\epsilon^{\frac{1}{2}}$, which is independent of the order in which A, B are taken, and of the order in which C, D are taken, but is ambiguous in sign.

Now consider two conics, ω_1, ω_2, both having double contact with a conic σ; let P_1, P_2 be the poles of the respective chords of contact. If P_1 be joined to any point H of ω_1, the cross ratio of P_1, H, in regard to the points where the line $P_1 H$ meets σ, is independent of H; denote the cosine associated with this cross ratio by $\cos r_1$; for ω_2 we similarly define $\cos r_2$. To the cross ratio of P_1, P_2, in regard to the two points where $P_1 P_2$ meets σ, the associated cosine is denoted by $\cos d$. At a point common to ω_1, ω_2, we consider the cross ratio of the tangents of these conics, taken in regard to the two tangents drawn from this point to σ; the associated cosine is the same for two of these common points of ω_1, ω_2, and is denoted by $\cos \phi$; for the complementary pair of common points, we likewise have $\cos \psi$. For a common tangent of ω_1 and ω_2, we consider the cross ratio of the points of contact, taken in regard to the two points where the tangent meets σ. The associated cosine we denote by $\cos t$; this is the same for two of the common tangents; for the complementary pair of common tangents of ω_1 and ω_2, we similarly have $\cos u$. With a proper correspondence of t and ϕ, and of u and ψ, prove the eight formulae

$$\cos d = \cos r_1 \cos r_2 + \sin r_1 \sin r_2 \cos \phi = \sin r_1 \sin r_2 + \cos r_1 \cos r_2 \cos t,$$
$$-\cos d = \cos r_1 \cos r_2 - \sin r_1 \sin r_2 \cos \psi = \sin r_1 \sin r_2 - \cos r_1 \cos r_2 \cos u,$$
$$\sin^2 \tfrac{1}{2} t = \tan r_1 \tan r_2 \sin^2 \tfrac{1}{2}\phi, \quad \cos^2 \tfrac{1}{2} t = -\tan r_1 \tan r_2 \sin^2 \tfrac{1}{2}\psi,$$
$$\sin^2 \tfrac{1}{2} u = -\tan r_1 \tan r_2 \cos^2 \tfrac{1}{2}\phi, \quad \cos^2 \tfrac{1}{2} u = \tan r_1 \tan r_2 \cos^2 \tfrac{1}{2}\psi.$$

Any plane section of a sphere, when projected *from the centre* of the sphere, on to the Absolute plane, becomes a conic with double contact with the Absolute conic, σ, along which the sphere and the asymptotic cone, whose vertex is at the centre of the sphere, have ring contact. Hence the formulae put down give relations for any two plane circular sections on a sphere.

Ex. 16. Prove that the necessary and sufficient condition that the two sections, of the quadric surface whose equation is

$$x^2 + y^2 + z^2 + 2fyz + 2gzx + 2hxy = 1,$$

made by the planes whose equations are $x + y + z = 1$ and $lx + my + nz = 1$, should touch one another, is the vanishing of one of the four functions $\lambda^{\frac{1}{2}} \pm \mu^{\frac{1}{2}} \pm \nu^{\frac{1}{2}}$, where $\lambda = (1-f)(1-l)$, $\mu = (1-g)(1-m)$, $\nu = (1-h)(1-n)$. Hence it follows that the sections of the quadric surface whose equation is

$$x^2 + y^2 + z^2 + 2yz \cos A + 2zx \cos B + 2xy \cos C = 1,$$

made by the planes whose equations are

$$x + y + z = 1 \text{ and } x \cos (B-C) + y \cos (C-A) + z \cos (A-B) = 1,$$

touch one another.

If we change, respectively, B, C, y, z into $B - \pi$, $C - \pi$, $-y$, $-z$, the quadric surface, and the second plane, are both unaltered. Thus the section of the quadric surface by the second plane touches the four sections given by the planes $x + y + z = 1$, $x - y - z = 1$, $-x + y - z = 1$, $-x - y + z = 1$. Again, the quadric surface is unaltered by the change of A into $-A$. Thus the four sections given by the equations just written are all touched by the four sections which are given by

$$x \cos (B-C) + y \cos (C-A) + z \cos (A-B) = 1,$$
$$x \cos (B-C) + y \cos (C+A) + z \cos (A+B) = 1,$$

and by the two obtained from the first, respectively, by changing B into $-B$ and C into $-C$.

Hence, if AOA', BOB', COC' be three diameters of a sphere whose centre is O, the four planes $AB'C'$, $BC'A'$, $CA'B'$, ABC give sections of the sphere which are all touched by four other sections of the sphere.

It follows, in a plane, that the four conics, which can be drawn through three given points to have double contact with a given conic, are all touched by four other conics which have double contact with the given conic. (Salmon, *Conics*, 1879, p. 359. Cf. also, *Principles of Geometry*, vol. IV, p. 82.)

NOTE 5

Proof, with the use of elliptic functions, of a theorem given by Steiner, for two quadrilaterals whose sides touch a conic

Of the theorem in question, in dual form, a geometrical proof has been given above (Misc. Ex. 125). The theorem is found in Schröter's edition of Steiner's lectures on Conics (1867, p. 546), and obtained as a consequence of properties of a cubic curve there investigated. The necessary properties of this cubic curve are summarised very briefly if we employ the elliptic functions developed by Legendre, and Abel, and others. Some readers may be glad to make acquaintance with this point of view. Or, still using the cubic curve, the necessary steps may be expressed in terms of the theory of residuation thereon; many of the consequences of this theory were known to Maclaurin.

The cubic curve in question arises as the locus of a point from which the pairs of tangents to three given conics belong to a pencil in involution. Or, dually, if a line moves so that the pairs of its common points with three given conics belong to an involution, then the line envelops a curve, having the property that three tangents pass through an arbitrary point.

We may give a brief algebraic proof of this theorem of the envelope: The condition that three pairs of points on a line, given respectively by the roots of the quadratic equations $a_0x^2 + 2a_1x + a_2 = 0$, $b_0x^2 + 2b_1x + b_2 = 0$, $c_0x^2 + 2c_1x + c_2 = 0$, should belong to an involution, is that the determinant, of three rows, whose rows consist respectively of the elements (a_0, a_1, a_2), (b_0, b_1, b_2), (c_0, c_1, c_2) should vanish. This condition can be expressed in a symbolical form which facilitates the proof of the theorem of the envelope. Namely, it can be expressed by $(\beta\gamma)(\gamma\alpha)(\alpha\beta) = 0$, where $(\beta\gamma)$ stands for $\beta_1\gamma_2 - \beta_2\gamma_1$, also $(\gamma\alpha)$ stands for $\gamma_1\alpha_2 - \gamma_2\alpha_1$, and $(\alpha\beta)$ stands for $\alpha_1\beta_2 - \alpha_2\beta_1$; and, in the expanded form of the product $(\beta\gamma)(\gamma\alpha)(\alpha\beta)$, we are to replace α_1^2, $\alpha_1\alpha_2$, α_2^2, respectively, by a_0, a_1, a_2, with similar replacements for β_1^2, $\beta_1\beta_2$, β_2^2 and γ_1^2, $\gamma_1\gamma_2$, γ_2^2. Now, with a similar convention, let the equations of three conics be respectively represented by $(p_1x_1 + p_2x_2 + p_3x_3)^2 = 0$, or $p_x^2 = 0$, and $q_x^2 = 0$, $r_x^2 = 0$, where x_1, x_2, x_3 are the coordinates, and, after expansion of the squared linear forms, p_1^2, p_1p_2, ... are to be replaced by the coefficients in the equations of the conics. Let then (y_1, y_2, y_3) and (z_1, z_2, z_3) be the coordinates of any two points. Any point on the line joining these has coordinates of the forms $(\lambda y_1 + z_1, \lambda y_2 + z_2, \lambda y_3 + z_3)$, where λ varies from point to point of the line. Thus, the points where the joining line meets the conic $p_x^2 = 0$ are given by the two values of λ which satisfy the quadratic equation in λ, $\lambda^2 p_y^2 + 2\lambda p_y p_z + p_z^2 = 0$. Hence, by what was said, the condition that the line joining the points (y_1, y_2, y_3), (z_1, z_2, z_3) should meet the three conics in pairs belonging to an involution, is

$$(q_y r_z - q_z r_y)(r_y p_z - r_z p_y)(p_y q_z - p_z q_y) = 0.$$

But, if the equation of the line be written $u_1x_1 + u_2x_2 + u_3x_3 = 0$, and this contains two points (y_1, y_2, y_3), (z_1, z_2, z_3), we may suppose

$$u_1 = y_2z_3 - y_3z_2, \quad u_2 = y_3z_1 - y_1z_3, \quad u_3 = y_1z_2 - y_2z_1.$$

Then the determinant of three rows, whose rows consist respectively of the elements (q_1, q_2, q_3), (r_1, r_2, r_3), (u_1, u_2, u_3), is easily seen to be equal to $q_y r_z - q_z r_y$. Denoting this determinant by (qru), we thence conclude that the tangential equation of the envelope is

$$(qru)(rpu)(pqu) = 0,$$

wherein, after expansion, p_1^2, p_1p_2, ... are to be replaced as before by coefficients of the equation of the first conic, with similar replacements for the second and third conics. The tangential equation is thus of the third order in u_1, u_2, u_3 and represents a curve of class three; and it is of the first degree in the coefficients of every one of the conics. In what follows we use the dual of the theorem we have proved, namely, that the locus of a point, from which the pairs of tangents to three given conics form a pencil in involution, is a curve of order three. It is obvious geometrically that the six common points of the pairs of the common tangents, of any two of the conics, lie on the locus.

A particular case is that in which two of the three conics are degenerate, one of these consisting of the two pencils of lines whose centres are the points E, F, say, the other consisting of the two pencils of lines whose centres are the points E', F'. Let the undegenerate conic be U. Then, the locus of a point P, which is such that the tangents drawn from it, to a conic U, belong to the pencil in involution determined by the two pairs of lines PE, PF and PE', PF', is such a cubic curve as we have found. The four points, common to the two tangents of U drawn from E and the two tangents of U drawn from F, lie on the locus, as do the points E, F; and there is a similar statement for E', F'. The locus also contains the two points (EE', FF') and (EF', FE'). Thus we infer that, *if we have two quadrilaterals, each constituted by four tangents of a conic, U, then the twelve points consisting of the six common points of pairs of lines of one quadrilateral, and the similar six points from the other, lie on a cubic curve*; in general a cubic curve is determined by the assignment of nine points of it (for the general equation of such a curve contains ten terms). From the first quadrilateral, any two complementary intersections of the lines may be taken for the two points we have named E and F; likewise any two complementary intersections from the second for E' and F'; any such choice leads also to two other points, (EE', FF') and (EF', FE'), lying on the cubic curve.

We now introduce the elliptic functions. We assume that the coordinates of a general point of a plane cubic curve, which is without a double point, can be expressed as elliptic functions of a parametric argument. This has already been remarked in the Appendix to Chapter XVI, relating to Poncelet's theorem. It is analogous to the fact that the coordinates of a point, of a conic, can be expressed as trigonometric functions of an argument; whereas, however, the trigonometric functions have only one period (2π), the elliptic functions have two distinct periods, whose ratio is not real; and, just as the statement that a point of a conic is associated with an argument θ is the same as that it is associated with an argument $\theta + 2k\pi$, where k is an arbitrary integer, so the argument associated with a point of the cubic curve is ambiguous; it may be increased by the sum of two arbitrary integral multiples of the two periods of the elliptic functions. This being understood, we assume, further, that the elliptic argument can be so chosen that the necessary and sufficient condition, for three points of the curve to lie on a line, is that the sum of the arguments belonging to these points should be zero (or a sum of integral multiples of the two periods); likewise, that the necessary and sufficient condition, for six points of the curve to lie on a conic, is that the sum of the arguments of these points should be zero.

Consider then a quadrilateral whose three pairs of complementary intersections lie on the cubic curve, the four lines of the quadrilateral being tangents of a conic U. Denote one of the pairs of complementary intersections by E, F, as before, the two other pairs being A, C and B, D, and the four lines of the quadrilateral being DCE, ABE, ADF, BCF. Let the elliptic arguments of A, D, E be, respectively, u, v, w. Then, by what has been said, the arguments for C, B, F, on the lines DCE, ABE, ADF, are respectively $-v-w$, $-u-w$, $-u-v$. As C, B, F are in a line, the sum of these arguments is zero, or a sum of integral multiples of the periods. This sum is $-2(u+v+w)$; and this is not zero because A, D, E are not in a line. Thus $u+v+w$ is half a sum of integral multiples of the periods, that is, is half a period; we denote it by ω. It follows that, for each of the four triangles formed by the four lines, the sum of the arguments of the three angular points is ω (or $-\omega$, which, by adding the period 2ω, is the same thing). Further, the tangent line of the cubic curve, at a point at which the argument is u, meets the curve again in the point for which the argument is $-2u$. Thus, the tangents of the curve, at the three angular points of any one of the four triangles, meet the curve again in points which lie on a line (the sum of the arguments of these being -2ω). And also, the tangents of the curve, at any two complementary intersections of the four lines of the quadrilateral, meet in a point which lies on the curve, and the three such points of meeting lie in a line. These points of meeting, for the respective pairs E, F; A, C; D, B, have the respectively equal pairs of arguments $-2w, 2u+2v$; $-2u, 2v+2w$; $-2v, 2w+2u$. Again, a conic can be drawn to touch the cubic curve at the angular points of any one of the four triangles formed by the four lines; for the doubles of the arguments of these angular points have the sum 2ω or zero.

Next, take another quadrilateral formed by tangents of the same conic U; and suppose the cubic curve under consideration to be that which contains the six intersections, A', D', E', C', B', F', of the pairs of lines of the second quadrilateral, as well as the six points A, D, E, C, B, F of the first quadrilateral. Then, as the two triangles ADE, $A'D'E'$ are circumscribed to the same conic, their six angular points lie on a conic. Whence, the respective arguments, say u', v', w', of the points A', D', E', having the sum $-u-v-w$, are such that $u'+v'+w'$ is (save for a period, possibly) the same half period, ω, as $u+v+w$.

From this it follows that the lines EE', FF', joining corresponding intersections for the two quadrilaterals, meet on the curve. For EE' meets the curve in the point whose argument is $-w-w'$; and FF' meets the curve in the point whose argument is $(u+v)+(u'+v')$, which is $\omega-w+\omega-w'$, or $2\omega-(w+w')$. Denote the point (EE', FF') by L. Similarly, the point (AA', CC'), say M, is on the curve, with the argument $-u-u'$; and the point (BB', DD'), say N, is on the curve, with the argument $-v-v'$. Thus the points L, M, N, whose arguments have the sum $-w-w'-u-u'-v-v'$, or -2ω, lie on a line. We have thus proved: *If E, F; A, C; B, D be the pairs of complementary intersections of the lines of a quadrilateral, which are tangents of a conic; and E', F'; A', C'; B', D' be likewise the complementary pairs of intersections of the lines of another quadrilateral, which are tangents of the same conic, then the three points $L=(EE', FF')$, $M=(AA', CC')$, $N=(BB', DD')$ lie in a line.*

Dually, now, let P, Q, R, T and P', Q', R', T' be two quadrangles inscribed in a conic. Let e, or QR, and f, or PT, be one pair of complementary joins of the points of the first quadrangle, the two other such pairs being a, or RP, with c, or QT; and b, or PQ, with d, or RT. Then, with a similar notation for the second quadrangle, the lines $l=(ee', ff')$, $m=(aa', cc')$, and $n=(bb', dd')$, meet in a point.

The points (ee'), (aa'), (bb') are the intersections $(QR, Q'R')$, $(RP, R'P')$, $(PQ, P'Q')$, respectively, and form a triangle. The points (ff'), (cc'), (dd') are the intersections $(PT, P'T')$, $(QT, Q'T')$, $(RT, R'T')$, respectively, and form another triangle. Essentially the theorem is that these two triangles are in perspective.

NOTE 6

The reduced form of a linear substitution, for three homogeneous variables; and of the equations of two conics

It has been said that, for two conics in general position, points of reference for the coordinates can be so chosen that the equations of the conics become $x^2+y^2+z^2=0$, $ax^2+by^2+cz^2=0$ (Chap. XVI (3), Ex. 15). Two conics may, however, be such that this is not possible. As we have been led (chapter on length, XV (11)) to refer to linear transformation of coordinates in a plane, it may be desirable to collect the complete results; the formulation is capable of extension to the case of four variables, for quadric surfaces, or to any number. We suppose that both the conics considered are undegenerate.

We denote by ϑ_p a square matrix, of p rows and columns, which has the number ϑ in every place of the diagonal, and has 1 in every one of the $(p-1)$ places of the matrix whose positions are given by $(1, 2)$, $(2, 3)$, ..., $(p-1, p)$, where (r, s) means the element in the rth row, and sth column, this matrix having zero in every other place. Also, we denote by $u(\vartheta, p)$ a matrix, of p rows and columns, which has 1 in every one of the places $(1, p)$, $(2, p-1)$, ..., $(p, 1)$, every other element being zero. And we denote by $v(\vartheta, p)$ a matrix of p rows and columns

which has ϑ in every one of the places $(1, p)$, $(2, p-1)$, ..., $(p, 1)$, and
1 in every one of the $(p-1)$ places $(2, p)$, $(3, p-1)$, ..., $(p, 2)$, all other
elements being zero. Examples of this notation are (for $p=3$)

$$\vartheta_3 = \begin{pmatrix} \vartheta, & 1, & 0 \\ 0, & \vartheta, & 1 \\ 0, & 0, & \vartheta \end{pmatrix}, \quad u(\vartheta, 3) = \begin{pmatrix} 0, & 0, & 1 \\ 0, & 1, & 0 \\ 1, & 0, & 0 \end{pmatrix}, \quad v(\vartheta, 3) = \begin{pmatrix} 0, & 0, & \vartheta \\ 0, & \vartheta, & 1 \\ \vartheta, & 1, & 0 \end{pmatrix}.$$

In general, we find, on multiplication, that $\vartheta_p = u(\vartheta, p)\, v(\vartheta, p)$.

We have considered a linear transformation, say that by which the
new variables x', y', ... are each expressed linearly as homogeneous
functions of the original variables, with coefficients of such generality
that, conversely, each of x, y, ... can be linearly expressed in terms of
x', y', ... Such a transformation can be simplified by suitably expressing x, y, ... homogeneously in terms of variables ξ, η, ..., and
expressing, *by the same transformation*, x', y', ... in terms of variables
ξ', η', ... The original transformation then becomes a linear transformation from the variables ξ, η, ... to the variables ξ', η', ... The
process is that which arises when, in place of two points of coordinates
$(x, y, \ldots)$, $(x', y', \ldots)$, we consider the same points with coordinates
referred to a new set of reference points, the same for both points. The
first question we are concerned with in this Note is, What is the simplest
form in which the linear transformation from ξ, η, ... to ξ', η', ... can
be taken, by a proper transformation applied to both $(x, y, \ldots)$ and
$(x', y', \ldots)$? Any such linear transformation, from $(\xi, \eta, \ldots)$ to
$(\xi', \eta', \ldots)$, is specified by the matrix formed by the coefficients of the
equations by which ξ', η', ... are, in turn, expressed by ξ, η, ... The
answer to the question put down is then that, for the linear transformation of n variables, the matrix of the simplified transformation
is of such a form that it can be represented by a symbol

$$\vartheta_p \vartheta_{p'} \ldots \phi_q \phi_{q'} \ldots \psi_r \psi_{r'} \ldots,$$

in which ϑ_p is as described, $\vartheta_{p'}$ is also formed from the number ϑ, but
with p' instead of p, while ϕ_q is like ϑ_p, but with the number ϕ put for
ϑ, and q put for p, and so on. In this symbol, there may be equalities
among the integers p, p', ..., as also among the integers q, q', ..., and
among the integers r, r', ..., and so on. But in general p, p', ... are
supposed to be arranged in descending order of magnitude, as are
q, q', ..., and r, r', ..., etc. The integers p, p', etc., are such that
$p + p' + \ldots + q + q' + \ldots + r + r' + \ldots + \text{etc.} = n$.

The matrix for the simplified form of the linear transformation,
which is of n rows and columns, may be denoted by A. It can be
described by the symbol we have associated with it. It is formed
by placing the matrix ϑ_p to have its diagonal along the diagonal of A,
followed by the matrix $\vartheta_{p'}$, also with its diagonal along the diagonal
of A, ..., and so on, with all the matrices of the symbol in turn. All
elements of A not occurring in one of these placed matrices ϑ_p, ..., ϕ_q ...
are to be zero.

With the matrix A we consider also two other matrices, U and V, also of n rows and columns. Of these U is formed by placing the matrix $u\,(\vartheta,\,p)$ to have its diagonal along the diagonal of U, followed by $u\,(\vartheta,\,p')$, placed to have its diagonal along the diagonal of U, and so on, for all the matrices indicated by the symbol above. The matrix V is similarly formed with the matrices $v\,(\vartheta,\,p),\,v\,(\vartheta,\,p'),\,\dots,\,v\,(\phi,\,q),\,\dots$. It is easy to see that the inverse matrix U^{-1} is equal to U. An example of U, V, corresponding to the symbol $\vartheta_2\phi_1$, is

$$U=\begin{pmatrix}0,\,1,\,0\\1,\,0,\,0\\0,\,0,\,1\end{pmatrix},\qquad V=\begin{pmatrix}0,\,\vartheta,\,0\\\vartheta,\,1,\,0\\0,\,0,\,\phi\end{pmatrix}.$$

If now we have two quadratic polynomials, homogeneous in n variables $x,\,y,\,\dots$, it is possible to simplify the forms of these by transforming the variables $x,\,y,\,\dots$, so that the polynomials are expressed in terms of the variables $\xi,\,\eta,\,\dots$, already considered, which are homogeneous linear functions of $x,\,y,\,\dots$. A quadratic polynomial,

$$\Sigma\,(a_{ii}x_i{}^2+2a_{ij}x_ix_j+\dots),$$

is competently represented by the matrix in which the element in the $(r,\,s)$-th place is a_{rs}. For both the quadratic polynomials we consider here, we suppose that the determinant of the associated matrix does not vanish; it will therefore not vanish for the simplified form. We then state the theorem that, with proper choice of the transformed variables $\xi,\,\eta,\,\dots$, for both the given quadratic polynomials, these become those for which the associated matrices are, respectively, U and V. And, from the identity $\vartheta_p=u\,(\vartheta,\,p)\,v\,(\vartheta,\,p)$, remarked above, it is easy to see that the matrices U, V, A are such that $A=U^{-1}V$. This gives the capital result: *The most general linear transformation, in the n variables, is obtainable by first taking the linear polar locus of the original point, in regard to the quadric locus whose equation is given by the matrix V, and then the point, its pole, of which this polar is the polar locus in regard to the quadric locus whose equation is given by the matrix U.* This has been stated in the text (Chap. xv), for the case of three variables. For instance, in the example just given, for a linear transformation of three variables associated with the symbol $\vartheta_2\phi_1$, the conics associated with the matrices U, V are, respectively, $2xy+z^2=0$, and $2\vartheta xy+y^2+\phi z^2$, and the linear transformation $U^{-1}V$ is $x'=\vartheta x+y$, $y'=\vartheta y$, $z'=\phi z$.

We now put down the six possible cases which can arise for the case of three variables. From the equation

$$p+p'+\dots+q+q'+\dots+\dots=n,$$

we can easily see that the possible symbols are

(1) $\vartheta_1\phi_1\psi_1$; (2) $\vartheta_2\phi_1$; (3) ϑ_3; (4) $\vartheta_2\vartheta_1$; (5) $\vartheta_1\vartheta_1\phi_1$; (6) $\vartheta_1\vartheta_1\vartheta_1$.

For these six cases the matrices U, V are, respectively,

$$(1)\ U=\begin{pmatrix}1,0,0\\0,1,0\\0,0,1\end{pmatrix},\ V=\begin{pmatrix}\vartheta,0,0\\0,\phi,0\\0,0,\psi\end{pmatrix};\quad (2)\ U=\begin{pmatrix}0,1,0\\1,0,0\\0,0,1\end{pmatrix},\ V=\begin{pmatrix}0,\vartheta,0\\\vartheta,1,0\\0,0,\phi\end{pmatrix};$$

$$(3)\ U=\begin{pmatrix}0,0,1\\0,1,0\\1,0,0\end{pmatrix},\ V=\begin{pmatrix}0,0,\vartheta\\0,\vartheta,1\\\vartheta,1,0\end{pmatrix};\quad (4)\ U=\begin{pmatrix}0,1,0\\1,0,0\\0,0,1\end{pmatrix},\ V=\begin{pmatrix}0,\vartheta,0\\\vartheta,1,0\\0,0,\vartheta\end{pmatrix};$$

$$(5)\ U=\begin{pmatrix}1,0,0\\0,1,0\\0,0,1\end{pmatrix},\ V=\begin{pmatrix}\vartheta,0,0\\0,\vartheta,0\\0,0,\phi\end{pmatrix};\quad (6)\ U=\begin{pmatrix}1,0,0\\0,1,0\\0,0,1\end{pmatrix},\ V=\begin{pmatrix}\vartheta,0,0\\0,\vartheta,0\\0,0,\vartheta\end{pmatrix}.$$

The corresponding forms of the reduced equations of a pair of conics are, in these cases, using x, y, z, not ξ, η, ζ, for the final variables,

$$(1)\ x^2+y^2+z^2=0,\quad \vartheta x^2+\phi y^2+\psi z^2=0;$$
$$(2)\ 2xy+z^2=0,\quad 2\vartheta xy+y^2+\phi z^2=0;$$
$$(3)\ 2xz+y^2=0,\quad 2\vartheta xz+\vartheta y^2+2yz=0;$$
$$(4)\ 2xy+z^2=0,\quad 2\vartheta xy+y^2+\vartheta z^2=0;$$
$$(5)\ x^2+y^2+z^2=0,\quad \vartheta(x^2+y^2)+\phi z^2=0;$$
$$(6)\ x^2+y^2+z^2=0,\quad \vartheta(x^2+y^2+z^2)=0.$$

The geometrical relations of the two conics are respectively: (1) The conics have four distinct common points; (2) The conics touch at one point, $y=z=0$, the common tangent being $y=0$; (3) The conics have three of their common points coincident, at $y=z=0$, the common tangent being $z=0$; (4) The conics have their four common points coincident, at $y=z=0$, the common tangent being $y=0$; (5) The conics touch at two points, on $z=0$; (6) The conics are wholly coincident.

The corresponding reduced forms of a linear transformation in the six cases are:

(1) x', y', z' are respectively equal to ϑx, ϕy, ψz. There are three points, and three lines (the joins of these) which are unaltered by the transformation.

(2) x', y', z' are respectively equal to $\vartheta x+y$, ϑy, ϕz. In this case there are two points unaltered, namely, $x=y=0$ and $y=z=0$; and two unaltered lines, $y=0$ and $z=0$. The former line contains the two unaltered points; one of the unaltered points is the common point of the two unaltered lines.

(3) x', y', z' are respectively equal to $\vartheta x+y$, $\vartheta y+z$, ϑz. There is a single unaltered point $y=z=0$, and single unaltered line $z=0$, passing through this.

(4) x', y', z' are respectively equal to $\vartheta x+y$, ϑy, ϑz. In this case there is a line, $y=0$, of which every point is unaltered; say, a range of unaltered points; and there is a pencil of each unaltered lines, through the point $y=z=0$. This point is one of the range of unaltered points, and the line containing the range is one of the unaltered lines. The transformation is that particular perspectivity in which the centre is on the axis.

(5) x', y', z' are respectively equal to ϑx, ϑy, ϕz. Here there is a range of unaltered points, on $z=0$; and a pencil of unaltered lines through the point $x=y=0$. This is the case of a general perspectivity.

(6) x', y', z' are respectively equal to ϑx, ϑy, ϑz. This is the identical transformation, leaving every point unaltered. Every case is self-dual; as may be seen to be necessarily true. Also, it may be noticed, another notation, for the six cases in turn, is [1, 1, 1], [2, 1], [3], [(2, 1)], [(1, 1), 1], [(1, 1, 1)], of which the rule for the change from one notation to the other is easily recognised. Further, the identical equations satisfied by the matrix of transformation A, in the six respective cases, are

$$(A-\theta)(A-\phi)(A-\psi)=0, \ (A-\theta)^2(A-\phi)=0, \ (A-\theta)^3=0,$$
$$(A-\theta)^2=0, \ (A-\theta)(A-\phi)=0, \ (A-\theta)=0.$$

For two quadratic polynomials in four homogeneous variables there are fourteen cases, if we suppose the determinantal discriminant of both forms to be other than zero. The symbols for these are, respectively,

$$\vartheta_1\phi_1\psi_1\sigma_1; \ \vartheta_2\phi_1\psi_1; \ \vartheta_3\phi_1; \ \vartheta_2\phi_2; \ \vartheta_4; \ \vartheta_1\vartheta_1\phi_1\psi_1; \ \vartheta_1\vartheta_1\phi_2; \ \vartheta_2\vartheta_1\phi_1;$$
$$\vartheta_3\vartheta_1; \ \vartheta_1\vartheta_1\phi_1\phi_1; \ \vartheta_2\vartheta_2; \ \vartheta_1\vartheta_1\vartheta_1\phi_1; \ \vartheta_2\vartheta_1\vartheta_1; \ \vartheta_1\vartheta_1\vartheta_1\vartheta_1.$$

The reader who wishes will find the geometrical relations of two quadric surfaces corresponding to these various cases specified in *Principles of Geometry*, iii (1934), pp. 123, 124; and the corresponding forms of the linear transformations are at once found by taking the poles of a plane in regard to the two surfaces, as explained.

INDEX